Water
Quality
Assessments

Water Quality Assessments

A guide to the use of biota, sediments and
water in environmental monitoring

Second edition

Edited by
Deborah Chapman

Published on behalf of

 UNESCO United Nations Educational, Scientific
and Cultural Organization

 WHO World Health Organization

 UNEP United Nations Environment Programme

 Taylor & Francis
Taylor & Francis Group
LONDON AND NEW YORK

First published 1992 by Taylor & Francis
Second edition 1996
Reprinted 1997

Reprinted 1998 by Taylor & Francis, an imprint of Routledge
2 Park Square, Milton Park, Abingdon, Oxon, OX14 4RN
270 Madison Ave, New York NY 10016

Transferred to Digital Printing 2007

British Library Cataloguing in Publication Data
A catalogue record for this book is available from the British Library

ISBN 0–419–21590–5 (hbk)
ISBN 0–419–21600–6 (pbk)

Publisher's Note
The publisher has gone to great lengths to ensure the quality of this reprint
but points out that some imperfections in the original may be apparent

TABLE OF CONTENTS

FOREWORD TO THE FIRST EDITION

Hydrological problems related to artificial and natural changes in the quality of inland water bodies were discussed by the Co-ordinating Council of the International Hydrological Decade (IHD) in the late 1960s. As a result, the Secretariats of UNESCO (United Nations Educational, Scientific and Cultural Organization) and WHO (World Health Organization), with the assistance of FAO (Food and Agriculture Organization of the United Nations) and IAHS (International Association for Hydrological Sciences), established an international working group, primarily to:

- identify and define the hydrological processes and phenomena directly concerned with the means of entry, distribution and self-purification of pollutants in surface and groundwater;
- review the known effects of such pollutants on any aspect of these processes and phenomena.

The outcome of the IHD working group and their collaborators was not meant to constitute a treatise on water chemistry or water pollution problems, but was a document attempting to link water quality considerations to aspects of the quantitative hydrology of surface and groundwater bodies. Advice was also included on the organisation of hydrological services, methods of conducting water quality surveys, and interpretation and evaluation of water quality data for hydrological purposes. An attempt was also made to meet the needs of developing regions by describing methods likely to be applied in these regions, both from the point of view of practicability and economy. On the other hand the report also aimed to be attractive to industrialised countries by including references to sophisticated methods.

It appeared that many hydrologists found difficulty in coping with water quality problems, and that hydrological surveys and water quality studies were not often adequately linked. The joint UNESCO/WHO publication *Water Quality Surveys* (1978) was, therefore, intended to harmonise these aspects and to synthesise the assessment of the hydrological regime and quality changes brought about by nature and man. The publication became a success world-wide and soon ran out of stock. The two Secretariats of UNESCO and WHO considered a re-print of the 1978 version, but decided to compile a completely new edition in view of the following:

(a) The progress in water quality research had been enormous over the past years and this needed to be taken into account.

(b) Water quality had become a regional, if not a global, concern encompassing more pollutants than in the past; an ecological approach could combine the physical, chemical, biological and microbiological aspects;

heavy metals and synthetic organic compounds have called for a change in the strategies for water quality surveys and monitoring.

(c) There is no need to describe the operational aspects of water quality monitoring and the laboratory procedures since they are mostly contained in the *GEMS/WATER Operational Guide*, a revised third edition of which appeared in 1991.

(d) Basic guidance on methodology is given in the *GEMS/WATER Handbook for Water Quality Monitoring in Developing Countries* which will be available by the end of 1991.

In October 1987, the two Secretariats compiled an annotated outline for the revised *Water Quality Surveys* on the understanding that the new book would describe, in a much broader way, the application and interpretation of water quality information in water resource management. The methodological and technical aspects could be largely omitted since the reader could be referred to the above-mentioned GEMS/WATER literature.

Authors were designated in 1988 and a first meeting of authors and contributors, supported by the United Nations Environment Programme (UNEP) and the USSR Centre for International Projects, took place in Sochi (former USSR) from 14 to 20 November 1988, followed by a second editorial meeting at Baikalsk (former USSR) from 3 to 10 August 1990. A final editorial panel meeting was then convened in Geneva, 22–23 November 1990. The result of these meetings is this guidebook, now renamed *Water Quality Assessments*.

FOREWORD TO THE SECOND EDITION

Much has happened in the water sector at national and international level since the preparation of the first edition of this guidebook. One major event was the International Conference on Water and the Environment which was held in January 1992 in Dublin, Ireland. In dealing with the protection of water resources, water quality and aquatic ecosystems, the conference made rather specific requests regarding the need for more and better water quality assessments, including:

- Purpose-orientated water assessments and predictions, taking into account the specificity of both surface and groundwaters, water quality and water quantity, and addressing all pollution types.
- Assessments harmonised for natural basins or catchments (including station networks, field and laboratory techniques, methodologies and procedures, and data handling) and leading to basin-wide data systems.
- New appropriate assessment and prediction techniques and methodologies, such as low-cost field measurements, continuous and automatic monitoring, use of biota and sediment for micro-pollution monitoring, remote sensing, and geographic information systems.

In June 1992 in Rio de Janeiro, Brazil, the United Nations Conference on Environment and Development resulted in an agreement on the action plan known as Agenda 21 which, in its chapter on freshwater, largely endorsed the recommendations from the Dublin conference. The stated objectives of Agenda 21 include issues which this guidebook aims to address, specifically:

- to make available to all countries water resources assessment technology that is appropriate to their needs, irrespective of their level of development, and
- to have all countries establish the institutional arrangements needed to ensure the efficient collection, processing, storage, retrieval and dissemination to users of information about the quality and quantity of available water resources, at the level of catchments and groundwater aquifers, in an integrated manner.

The concerns expressed at these conferences, together with the feedback from readers and users of the first edition of this guidebook, have guided the editor and authors in preparing the second edition. Latest developments in strategies, as well as on technologies and methods, have been taken into account to make the book useful for water resources managers charged with the monitoring, assessment and control of water quality for a variety of purposes. Thus this guidebook should contribute to the capacity building initiatives launched in a number of countries in the aftermath of the Rio de

Janeiro conference by supporting the scientifically-sound assessment of water resources which are tending to become more sparse and polluted.

One major change from the first edition, which is in addition to the general review and updating, is the introduction of a new chapter on reservoirs. The construction of dams along many rivers has increased rapidly over recent years, including some more controversial large dam projects. The multiple use of the resulting reservoirs requires a sound water quality assessment component to their management strategies. A separate chapter has been devoted to reservoirs because many have complex hydrodynamic features and all are subject to potential or actual human intervention in their natural chemical and physical processes. The original rivers and lakes chapters of the first edition have been modified accordingly.

The other major development in preparing the second edition concerns the production of the companion handbook *Water Quality Monitoring: a practical guide to the design and implementation of freshwater quality studies and monitoring programmes*. The manuscript for the *Water Quality Monitoring* handbook emerged and was finalised in parallel to the second edition of this guidebook. *Water Quality Monitoring* provides the practical and methodological details whereas *Water Quality Assessments* gives the overall strategy for assessments of the quality of the main types of water body. Together the two books cover all the major aspects of water quality, its measurement and its evaluation.

SUMMARY AND SCOPE

This guidebook concentrates on the process of setting up monitoring programmes for the purpose of providing a valid data base for water quality assessments. The choice of variables to be measured in the water, the sediment and in biota are described in Chapters 3, 4 and 5 and the common procedures for data handling and presentation in Chapter 10. Interpretation of these data for the purpose of assessing water quality in rivers, lakes, reservoirs and groundwaters is presented in Chapters 6, 7, 8 and 9 respectively. These chapters, specific to the type of water body, focus on monitoring strategies, requirements for water quality and quantity data and interpretative techniques. The choice of the appropriate methods is illustrated by case studies for typical water pollution situations. In view of the varying levels of resources which countries can put at the disposal of this activity, the strategies for water quality assessment are developed according to three different levels of monitoring operations: simple, intermediate and advanced.

For the purpose of this presentation of water quality assessment techniques the following types of water resources have been taken into consideration:

- Rivers and streams of all sizes from source to tidal limit (i.e. the influence of salt water intrusion). Canals and inter-connecting river systems are also included.
- Lakes of all sizes and types, including marshes and bogs.
- Reservoirs of various types, especially river impoundments.
- Groundwaters of various types, shallow or deep, and phreatic or confined.

These types of water bodies include all major freshwater resources subject to anthropogenic influences or intentionally used for municipal or industrial supply, irrigation, recreation, cooling or other purposes. However, certain types of waters are outside the scope of this book, including: estuaries, coastal lagoons, salt marshes and other saline waters, wastewaters of different origins, thermal and mineral springs, saline aquifers, brines and atmospheric precipitation such as rain and snow.

Within the range of water quality issues addressed in this guidebook efforts have been concentrated on major areas of vital importance. Several complementary publications are readily available which cover other specific aspects in great detail and, when referred to, their use in conjunction with the present guide is strongly recommended.

There are no geographical limits imposed on the applicability of the guidance provided in this book since an effort was made by the authors to address all kinds of environmental conditions occurring in aquatic ecosystems world-wide. Thus, the specific situations of humid and dry tropics, as well as of mountainous and lowland waters, in water abundant or semi-arid and arid climatic zones, are all covered by means of examples. Attempts have been made to find examples from all world regions but, inevitably, there is more literature available from the developed world than the developing world and this is reflected in the reference lists attached to each chapter. Water quality, and its monitoring and assessment, is also greatly influenced by the size of the water body and, therefore, relevant guidance is provided for different levels of magnitude.

Within this guidebook various water quality problems (organic pollution, eutrophication, acidification, toxic contamination etc.) and their related descriptors are discussed at various levels of complexity. Chemical constituents and contaminants, as well as the biological characteristics of water bodies, are covered extensively. However, the consequences of temperature changes due to thermal discharges are only addressed in relation to their effects on aquatic life.

Human health is affected, in many world regions, by vector-transmitted diseases associated with vector organisms which breed in the aquatic environment. This problem is enormous since there are 200 million people suffering from one such disease alone, i.e. schistosomiasis. However, since the occurrence of such diseases, and their containment, is closely linked to water resource development projects, rather than to pollution sources and effects, this issue is not dealt with in this book. Further information on these topics can be found in the internationally recognised literature on the subject (WHO, 1980,1982,1983).

Pathogenic agents causing water-borne diseases include bacteria and viruses as well as protozoa and helminths. Although they interfere only marginally with aquatic life in general, they cause severe public health problems and are considered responsible for most of the infant mortality in developing countries. Monitoring is usually done indirectly by identifying and quantifying indicators of faecal pollution such as the coliform groups. This guidebook follows the same concept and interested readers are referred to further background information in the relevant literature published by the World Health Organization (WHO, 1976,1985,1993).

Radioactive isotopes, natural or man-made, are not included in this publication because the monitoring of radiation is covered by the work of the International Atomic Energy Agency (IAEA).

The basic methods, procedures, techniques, field equipment and analytical instruments required to monitor water quality have been developed and field-tested in a wide range of situations over the last two decades. A wealth of experience has been accumulated and communicated through guidebooks and reports on water quality. As a consequence, the authors of this book felt that the monitoring methods and procedures already published adequately cover the necessary techniques. Therefore, they have concentrated more on the principles, approaches and design for water quality assessment and on the interpretation of the resulting data.

With respect to the field operations for monitoring, a comprehensive and practical booklet has been produced by the World Meteorological Organization (WMO, 1988). It describes essential factors to consider in monitoring such as the location of sampling sites, the collection of surface water samples, field measurements, sampling for biological analysis, shipment of samples, field safety and training programmes related to all of the above. Similar publications are not widely available for groundwaters but a description of techniques is given in Barcelona *et al.* (1985). The international project on global freshwater quality monitoring, GEMS/WATER (WHO, 1991), has based its monitoring operations on a practical guidebook, the *GEMS/WATER Operational Guide* (WHO, 1992) which gives in detail information on site selection, sampling, analysis, quality control and data processing. Most chemical analyses required for water quality monitoring are adequately covered by such reference books, whereas non-standardised methods for biological monitoring have to be developed for local or regional situations. However, although analytical reference methods are given in several general publications, it is also necessary to consult the International Standards Organization (ISO) "Standard methods" series of publications and to refer to recognised national publications, such as the standard methods produced by the American Public Health Association (APHA, 1989), the German standard methods (Deutsche Einheitsverfahren zur Wasser-, Abwasser und Schlammuntersuchung (DIN)) and those of the USSR State Committee for Hydrometeorology and Environmental Control (1987, 1989) which are now used in Russia and other CIS countries.

Hydrological measurements are an indispensable accompaniment to any surface water quality monitoring operation. Groundwater quality data also require adequate hydrological information for any meaningful interpretation. The World Meteorological Organization has developed practical guidelines as part of its Operational Hydrology Programme (WMO, 1994) and the United Nations Educational, Scientific and Cultural Organization (UNESCO) has also issued groundwater hydrology guidebooks. These publications provide methodology for water quality data collection, interpretation and presentation (UNESCO, 1983).

References

APHA 1989 *Standard Methods for the Examination of Water and Wastewater*. 17th edition, American Public Health Association, Washington D.C., 1,268 pp.

Barcelona, M.J., Gibb, J.P., Helfrich, J.A. and Garske, E.E. 1985 *Practical Guide for Groundwater Sampling*. ISWS Contract Report 374, Illinois State Water Survey, Champaign, Illinois, 94 pp.

UNESCO 1983 *International Legend for Hydrogeological Maps*. Technical Documents in Hydrology, SC.841/S7, United Nations Educational Scientific and Cultural Organization, Paris, 51 pp.

USSR State Committee for Hydrometeorology and Environmental Control 1987 *Methods for Bioindication and Biotesting in Natural Waters*. Volume 1. Hydrochemical Institute, Leningrad, 152 pp.

USSR State Committee for Hydrometeorology and Environmental Control 1989 *Methods for Bioindication and Biotesting in Natural Waters*. Volume 2. Hydrochemical Institute, Leningrad, 275 pp, [In Russian].

WHO 1976 *Surveillance of Drinking-Water Quality*. WHO Monograph Series No. 63, World Health Organization, Geneva, 128 pp.

WHO 1980 *Environmental Management for Vector Control*. Fourth report of the WHO Expert Committee on Vector Biology and Control, Technical Report Series No. 649, World Health Organization, Geneva, 67 pp.

WHO 1982 *Manual for Environmental Management for Mosquito Control, with Special Emphasis on Malaria Vectors*. WHO Offset Publication No. 66, World Health Organization, Geneva, 281 pp.

WHO 1983 *Integrated Vector Control*. Seventh report of the WHO Expert Committee on Vector Biology and Control, Technical Report Series No. 688, World Health Organization, Geneva, 72 pp.

WHO 1985 *Guidelines for Drinking-Water Quality, Volume 3, Drinking-Water Quality Control in Small-Community Supplies*. World Health Organization, Geneva, 120 pp.

WHO 1991 *GEMS/WATER 1990-2000. The Challenge Ahead*. WHO/PEP/91.2, World Health Organization, Geneva.

WHO 1992 *GEMS/WATER Operational Guide*. Third edition. World Health Organization, Geneva.

WHO 1993 *Guidelines for Drinking-Water Quality, Volume 1, Recommendations*. Second edition. World Health Organization, Geneva, 130 pp.

WMO 1988 *Manual on Water Quality Monitoring*. WMO Operational Hydrology Report, No. 27, WMO Publication No. 680, World Meteorological Organization, Geneva, 197 pp.

WMO 1994 *Guide to Hydrological Practices*. Fifth edition, WMO Publication No. 168, World Meteorological Organization, Geneva, 735 pp.

ACKNOWLEDGEMENTS

The co-sponsoring organisations would like to thank and acknowledge the time and effort contributed by many people, all of whom have helped to ensure the smooth and efficient production of this book. A number of authors provided material and, in many cases, several authors and their collaborators have worked on the same sections. As it is difficult to identify adequately the contribution of each individual author in the chapter headings, the names of the principal contributors, to whom we are greatly indebted, are listed below:

Albert Beim, Institute of Ecological Toxicology, Ministry of Environmental Protection and Natural Resources of Russia, Baikalsk, Russia (Chapters 5 and 7)

Deborah Chapman, Environment Consultant, Kinsale, Ireland (Chapters 3, 5 and 6)

John Chilton, British Geological Survey, Wallingford, UK (Chapter 8)

Adrian Demayo, Environment Canada, Ottawa, Canada (Chapter 9)

Günther Friedrich, Landesamt für Wasser und Abfall N.W., Dusseldorf, Germany (Chapters 5 and 6)

Richard Helmer, World Health Organization, Geneva, Switzerland (Chapters 1 and 2)

Vitaly Kimstach, Arctic Monitoring and Assessment Programme, Oslo, Norway (Chapters 2 and 3)

Michel Meybeck, Université de Pierre et Marie Curie, Paris, France (Chapters 1, 2, 4, 6 and 7)

Walter Rast, United Nations Environment Programme, Nairobi, Kenya (Chapter 8)

Alan Steel, Water Supply Consultant, Kinsale, Ireland (Chapters 8 and 10)

Richard Thomas, Waterloo Centre for Groundwater Research, Waterloo, Canada (Chapters 4, 6 and 7)

Jeff Thornton, International Environmental Management Services Limited, Waukesha, USA (Chapter 8)

Messrs R. Helmer (WHO) and W.H. Gilbrich (UNESCO) provided the Secretariat services for the meetings and were responsible for the arrangements for the initial compilation of manuscripts and all meetings. Ms J. Kenny (WHO) provided secretarial and administrative assistance throughout. The co-sponsoring organisations are also greatly indebted to Michel Meybeck for his advice and efforts in the preparation of the revision of the outline for this book and to Richard Thomas for chairing the authors' meetings.

Thanks are also due to Dave Rickert (U.S. Geological Survey, Reston, VA) for his thorough review of the final manuscript, to Walter Rast (UNEP) for his advice on the second edition, and to Deborah Chapman who undertook the laborious tasks of editing the whole manuscript and management of the production of camera-ready copy and publication of the book.

The additional services of Imogen Bertin (typesetting, design and layout) and Alan Steel (graphics, editorial assistance), as well as those who reviewed and supplied material for several chapters, are gratefully acknowledged.

The printing of colour Figures 6.34 and 6.35 was financially supported by the IHP/OHP National Committee of the Federal Republic of Germany, and of colour Figures 7.10 and 7.11 by the National Water Research Institute, Environment Canada.

The book is a contribution to UNESCO's International Hydrological Programme and to the UNEP/WHO/WMO/UNESCO co-sponsored GEMS/WATER programme. The UN agencies are greatly indebted to the authors, and to the former USSR institutes which hosted their meetings.

ABBREVIATIONS USED IN TEXT

AAS	Atomic absorption spectrophotometry
AES	Atomic emission spectrophotometry
ANC	Acid neutralising capacity
ANOVA	Analysis of variance
AOX	Adsorbable organic halides
APHA	American Public Health Association
AQC	Analytical quality control
ASPT	Average Score Per Taxon
BMWP	Biological Monitoring Working Party-score
BOD	Biochemical oxygen demand
CEC	Commission of the European Communities
CIPEL	International Surveillance Commission of Lake Geneva
COD	Chemical oxygen demand
DDT	Dichlorodiphenyltrichloroethane
DIN	Deutsche Einheitsverfahren zur Wasser-, Abwasser und Schlammuntersuchung
DO	Dissolved oxygen
DOC	Dissolved organic carbon
DON	Dissolved organic nitrogen
EDTA	Ethylenediaminetetraacetic acid
EIFAC	European Inland Fisheries Advisory Commission
EQI	Ecological Quality Index
EU	European Union (formerly European Community)
FAO	Food and Agriculture Organization of the United Nations
GC	Gas chromatograph(y)
GC/MS	Gas chromatography/mass spectrometry
GEMS	Global Environment Monitoring System
GIS	Geographic information systems
GLOWDAT	GLObal Water DATa Management System
IAEA	International Atomic Energy Agency
IAHS	International Association for Hydrological Science
IC	Ion chromatography
ICP/AES	Inductively coupled plasma atomic emission spectrometry
ICPS	Inductively coupled plasma spectroscopy
IHD	International Hydrological Decade
IR	Infra red

IRPTC	International Register of Potentially Toxic Chemicals
ISO	International Standards Organization
JTU	Jackson turbidity units
LC	Liquid chromatography
LOD	Limit-of-detection
LOWESS	Locally weighted scatter plot smoothing
LT	Less than values
MAC	Maximum allowable concentration
MATC	Maximum allowable toxic concentration
MCNC	Most common natural concentrations
MLE	Maximum likelihood estimator
MPN	Most probable number
MS	Mass spectrometer (spectrometry)
NAQUADAT	The NAtional Water QUALity Accounting DATa Bank (Canada)
NCPB	National Contaminant Biomonitoring Program (USA)
ND	Not detected
NOEC	No observed effect concentration
NTU	Nephelometric turbidity units
OECD	Organisation for Economic Co-operation and Development
OII	Odour intensity index
OPP	Oxygen Production Potential
PA	Apatitic phosphorus
PAH	Polychlorinated aromatic hydrocarbons
PCA	Principal components analysis
PCBs	Polychlorinated biphenyls
PCs	Personal computers
PFU	Plaque forming units
PINA	Non-apatitic inorganic phosphorus
PM	Particulate matter
PO	Organic phosphorus
POC	Particulate organic carbon
PON	Particulate organic nitrogen
PTFE	Polytetrafluoroethylene
RAISON	Regional Analysis by Intelligent Systems on a Microcomputer
RCRA	Resource Conservation and Recovery Act, USA
RIVPACS	River InVertebrate Prediction and Classification System

SA	Sediment accretion
SAR	Sodium adsorption ratio
SEF	Sediment enrichment factor
SM	Suspended matter
SOE	State of the Environment
SR	Settling rate
SRP	Soluble reactive phosphorus
TAC	Total absorbance colour
TCDD	Tetra chlorinated dibenzo dioxin
TDS	Total dissolved solids
TOC	Total organic carbon
TP	Total phosphorus
TSS	Total suspended solids
UNEP	United Nations Environment Programme
UNESCO	United Nations Educational, Scientific and Cultural Organization
US EPA	United States Environmental Protection Agency
UV	Ultra violet
VA	Voltammetry
WASP	Water Analysis Simulation Programme
WHO	World Health Organization
WMO	World Meteorological Organization

Chapter 1*

AN INTRODUCTION TO WATER QUALITY

1.1 Characterisation of water bodies

Water bodies can be fully characterised by the three major components: hydrology, physico-chemistry, and biology. A complete assessment of water quality is based on appropriate monitoring of these components.

1.1.1 Hydrodynamic features

All freshwater bodies are inter-connected, from the atmosphere to the sea, via the hydrological cycle. Thus water constitutes a continuum, with different stages ranging from rainwater to marine salt waters. The parts of the hydrological cycle which are considered in this book are the inland freshwaters which appear in the form of rivers, lakes or groundwaters. These are closely inter-connected and may influence each other directly, or through intermediate stages, as shown in Table 1.1 and Figure 1.1. Each of the three principal types of water body has distinctly different hydrodynamic properties as described below.

Rivers are characterised by uni-directional current with a relatively high, average flow velocity ranging from 0.1 to 1 m s^{-1}. The river flow is highly variable in time, depending on the climatic situation and the drainage pattern. In general, thorough and continuous vertical mixing is achieved in rivers due to the prevailing currents and turbulence. Lateral mixing may take place only over considerable distances downstream of major confluences.

Lakes are characterised by a low, average current velocity of 0.001 to 0.01 m s^{-1} (surface values). Therefore, water or element residence times, ranging from one month to several hundreds of years, are often used to quantify mass movements of material. Currents within lakes are multi-directional. Many lakes have alternating periods of stratification and vertical mixing; the periodicity of which is regulated by climatic conditions and lake depth.

Groundwaters are characterised by a rather steady flow pattern in terms of direction and velocity. The average flow velocities commonly found in aquifers range from 10^{-10} to 10^{-3} m s^{-1} and are largely governed by the porosity and permeability of the geological material. As a consequence mixing is rather poor and, depending on local hydrogeological features, the groundwater dynamics can be highly diverse.

There are several transitional forms of water bodies which demonstrate features of more than one of the three basic types described above and are

*This chapter was prepared by M. Meybeck and R. Helmer

Table 1.1   The hydrological cycle: water volumes, residence times and fluxes

	Total cycle volume		Freshwater volume only (%)	Freshwater volume without icecaps and glaciers (%)	Residence times
	(10^6 km^3)	(%)			
Oceans and seas	1,370	94			~4,000 years
Lakes and reservoirs	0.13	< 0.01	0.14	0.21	~10 years
Swamps and marshes	< 0.01	< 0.01	< 0.01	< 0.01	1–10 years
River channels	< 0.01	< 0.01	< 0.01	< 0.01	~2 weeks
Soil moisture	0.07	< 0.01	0.07	0.11	2 weeks–1 year
Groundwater	60	4	66.5	99.65	2 weeks–50,000 years
Icecaps and glaciers	30	2	33.3		10–1,000 years
Atmospheric water	0.01	<0.01	0.01	0.02	~10 days
Biospheric water	< 0.01	< 0.01	< 0.01	< 0.01	~1 week

Fluxes	
Evaporation from oceans	425
Evaporation from land	71
Precipitation from oceans	385
Precipitation from land	111
Run-off to oceans	37.4
Glacial ice	2.5

Source: Modified from Nace, 1971 and various sources

characterised by a particular combination of hydrodynamic features. The most important transitional water bodies are illustrated in Figure 1.1 and are described below.

Reservoirs are characterised by features which are intermediate between rivers and lakes. They can range from large-scale impoundments, such as Lake Nasser, to small dammed rivers with a seasonal pattern of operation and water level fluctuations closely related to the river discharge, to entirely constructed water bodies with pumped in-flows and out-flows. The cascade of dams along the course of the River Dnjepr is an example of the interdependence between rivers and reservoirs. The hydrodynamics of reservoirs are greatly influenced by their operational management regime.

Flood plains constitute an intermediate state between rivers and lakes with a distinct seasonal variability pattern. Their hydrodynamics are, however, determined by the river flow regime.

Marshes are characterised by the dual features of lakes and phreatic aquifers. Their hydrodynamics are relatively complex.

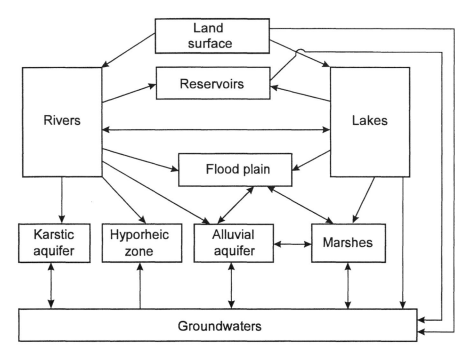

Figure 1.1 Inter-connections between inland freshwater bodies (intermediate water bodies have mixed characteristics belonging to two or three of the major water bodies)

Alluvial and karstic aquifers are intermediate between rivers and groundwaters. They differ, generally, in their flow regime which is rather slow for alluvial and very rapid for karstic aquifers. The latter are often referred to as underground rivers.

As a consequence of the range of flow regimes noted above, large variations in water residence times occur in the different types of inland water bodies (Figure 1.2). The hydrodynamic characteristics of each type of water body are highly dependent on the size of the water body and on the climatic conditions in the drainage basin. The governing factor for rivers is their hydrological regime, i.e. their discharge variability. Lakes are classified by their water residence time and their thermal regime resulting in varying stratification patterns. Although some reservoirs share many features in common with lakes, others have characteristics which are specific to the origin of the reservoir. One feature common to most reservoirs is the deliberate management of the inputs and/or outputs of water for specific purposes. Groundwaters greatly depend upon their recharge regime, i.e. infiltration through the unsaturated aquifer zone, which allows for the renewal of the

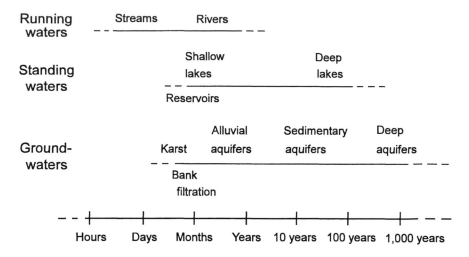

Figure 1.2   Water residence time in inland freshwater bodies (After Meybeck *et al.*, 1989)

groundwater body. Further details for each of these water bodies are available in Chapters 6, 7, 8 and 9.

It cannot be over-emphasised that thorough knowledge of the hydrodynamic properties of a water body must be acquired before an effective water quality monitoring system can be established. Interpretation of water quality data cannot provide meaningful conclusions unless based on the temporal and spatial variability of the hydrological regime.

1.1.2 Physical and chemical properties

Each freshwater body has an individual pattern of physical and chemical characteristics which are determined largely by the climatic, geomorphological and geochemical conditions prevailing in the drainage basin and the underlying aquifer. Summary characteristics, such as total dissolved solids, conductivity and redox potential, provide a general classification of water bodies of a similar nature. Mineral content, determined by the total dissolved solids present, is an essential feature of the quality of any water body resulting from the balance between dissolution and precipitation. Oxygen content is another vital feature of any water body because it greatly influences the solubility of metals and is essential for all forms of biological life. For a complete description of chemical water quality variables see Chapter 3.

The chemical quality of the aquatic environment varies according to local geology, the climate, the distance from the ocean and the amount of soil cover, etc. If surface waters were totally unaffected by human activities, up to 90–99 per cent of global freshwaters, depending on the variable of interest, would have natural chemical concentrations suitable for aquatic life and most human uses. Rare (between 1 and 10 per cent and between 90 and 99 per cent of the global distribution) and very rare (< 1 per cent and > 99 per cent of the global distribution — see section 1.3, Figure 1.4) chemical conditions in freshwaters, such as occur in salt lakes, hydrothermal waters, acid volcanic lakes, peat bogs, etc., usually make the water unsuitable for human use (see section 1.3). Nonetheless, a range of aquatic organisms have adapted to these extreme environments. In many regions groundwater concentrations of total dissolved salts, fluoride, arsenic, etc., may also naturally exceed maximum allowable concentrations (MAC) (see section 9.2.6).

Particulate matter (PM) is a key factor in water quality, regulating adsorption–desorption processes. These processes depend on: (i) the amount of PM in contact with a unit water volume, (ii) the type and character of the PM (e.g. whether organic or inorganic), and (iii) the contact time between the water and the PM. The time variability of dissolved and particulate matter content in water bodies results mainly from the interactions between hydrodynamic variability, mineral solubility, PM characteristics and the nature and intensity of biological activity.

1.1.3 Biological characteristics

The development of biota (flora and fauna) in surface waters is governed by a variety of environmental conditions which determine the selection of species as well as the physiological performance of individual organisms. A complete description of biological aspects of water quality is presented in Chapter 5. The primary production of organic matter, in the form of phytoplankton and macrophytes, is most intensive in lakes and reservoirs and usually more limited in rivers. The degradation of organic substances and the associated bacterial production can be a long-term process which can be important in groundwaters and deep lake waters which are not directly exposed to sunlight.

In contrast to the chemical quality of water bodies, which can be measured by suitable analytical methods, the description of the biological quality of a water body is a combination of qualitative and quantitative characterisation. Biological monitoring can generally be carried out at two different levels:
- the response of individual species to changes in their environment or,
- the response of biological communities to changes in their environment.

Water quality classification systems based upon biological characteristics have been developed for various water bodies. The chemical analysis of selected species (e.g. mussels and aquatic mosses) and/or selected body tissues (e.g. muscle or liver) for contaminants can be considered as a combination of chemical and biological monitoring. Biological quality, including the chemical analysis of biota, has a much longer time dimension than the chemical quality of the water since biota can be affected by chemical, and/or hydrological, events that may have lasted only a few days, some months or even years before the monitoring was carried out.

1.2   Definitions related to water quality

In view of the complexity of factors determining water quality, and the large choice of variables used to describe the status of water bodies in quantitative terms, it is difficult to provide a simple definition of water quality. Furthermore, our understanding of water quality has evolved over the past century with the expansion of water use requirements and the ability to measure and interpret water characteristics. Figure 1.3 demonstrates the evolutionary nature of chemical water quality issues in industrialised countries. For the purposes of this guidebook the following definitions have been accepted:

Term	Definition
QUALITY of the aquatic environment	■ Set of concentrations, speciations, and physical partitions of inorganic or organic substances. ■ Composition and state of aquatic biota in the water body. ■ Description of temporal and spatial variations due to factors internal and external to the water body.
POLLUTION of the aquatic environment	Introduction by man, directly or indirectly, of substances or energy which result in such deleterious effects as: ■ harm to living resources, ■ hazards to human health, ■ hindrance to aquatic activities including fishing, ■ impairment of water quality with respect to its use in agricultural, industrial and often economic activities, and ■ reduction of amenities[1]

[1]   as defined by GESAMP (1988)

The physical and chemical quality of pristine waters would normally be as occurred in pre-human times, i.e. with no signs of anthropogenic impacts. The natural concentrations (governed by factors described in section 1.1.2) could, nevertheless, vary by one or more orders of magnitude between different drainage basins. In practice, pristine waters are very difficult to find as a result of atmospheric transport of contaminants and their subsequent deposition in locations far distant from their origin. Before pristine waters reach the polluted condition, two phases of water quality degradation occur.

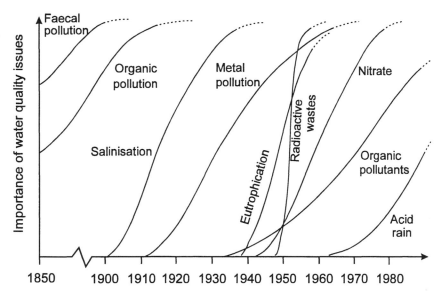

Figure 1.3 The sequence of water quality issues arising in industrialised countries (After Meybeck and Helmer, 1989)

The first phase shows an alteration in water quality with evidence of human impact but without any harm to the biota or restriction of water use. Such changes may only be detectable by repeated chemical measurements over long time spans. Typical examples are when Cl⁻ concentrations change from a few mg l⁻¹ to 10 mg l⁻¹ (as in Lake Geneva where average concentrations went from 2 mg l⁻¹ in 1960 to 6 mg l⁻¹ at present) or when N-NO_3^- concentrations change from 0.1 mg l⁻¹ to 0.2 mg l⁻¹. The next phase consists of some degradation of water quality and possible restriction of specific water uses because recommended water quality guidelines (local, regional or global) may be exceeded. Once maximum acceptable concentrations for selected variables in relation to water use have been exceeded, or the aquatic habitat and biota have been markedly modified, the water quality is usually defined as polluted (see example in section 6.7.3).

Description of the quality of the aquatic environment can be carried out in a variety of ways. It can be achieved either through quantitative measurements, such as physico-chemical determinations (in the water, particulate material, or biological tissues) and biochemical/biological tests (BOD measurement, toxicity tests, etc.), or through semi-quantitative and qualitative descriptions such as biotic indices, visual aspects, species inventories, odour, etc. (see Chapters 3, 4 and 5). These determinations are carried out in the field and in the laboratory and produce various types of data which lend themselves to different interpretative techniques (see section 10.3.1). For the purpose of simplicity the term "water quality" is used throughout this book, although it refers to the overall quality of the aquatic environment.

The terms monitoring and assessment are frequently confused and used synonymously. For the purpose of this guidebook and its companion hand-book (Bartram and Ballance, 1996) the following definitions are used:

Term	Definition
Water quality ASSESSMENT	The overall process of evaluation of the physical, chemical and biological nature of water in relation to natural quality, human effects and intended uses, particularly uses which may affect human health and the health of the aquatic system itself.
Water quality MONITORING	The actual collection of information at set locations and at regular intervals in order to provide the data which may be used to define current conditions, establish trends, etc.

Water quality assessment includes the use of monitoring to define the condition of the water, to provide the basis for detecting trends and to provide the information enabling the establishment of cause–effect relationships. Important aspects of an assessment are the interpretation and reporting of the results of monitoring and the making of recommendations for future actions (see Chapter 2). Thus there is a logical sequence consisting of three components: monitoring, followed by assessment, followed by management. In addition, there is also a feedback loop because management inevitably requires compliance monitoring to enforce regulations, as well as assessments at periodic intervals to verify the effectiveness of management decisions. The principal objective of the global freshwater quality monitoring project, GEMS/WATER, provides an illustrative example of the complexity of the assessment task and its relation to management (WHO, 1991):

- To provide water quality assessments to governments, the scientific community and the public, on the quality of the world's freshwater relative to human and aquatic ecosystem health, and global environmental concerns, specifically:
  - to define the status of water quality;
  - to identify and quantify trends in water quality;
  - to define the cause of observed conditions and trends;
  - to identify the types of water quality problems that occur in specific geographical areas; and

- to provide the accumulated information and assessments in a form that resource management and regulatory agencies can use to evaluate alternatives and make necessary decisions.

1.3   Anthropogenic impacts on water quality

With the advent of industrialisation and increasing populations, the range of requirements for water have increased together with greater demands for higher quality water. Over time, water requirements have emerged for drinking and personal hygiene, fisheries, agriculture (irrigation and livestock supply), navigation for transport of goods, industrial production, cooling in fossil fuel (and later also in nuclear) power plants, hydropower generation, and recreational activities such as bathing or fishing. Fortunately, the largest demands for water quantity, such as for agricultural irrigation and industrial cooling, require the least in terms of water quality (i.e. critical concentrations may only be set for one or two variables). Drinking water supplies and specialised industrial manufacturers exert the most sophisticated demands on water quality but their quantitative needs are relatively moderate. In parallel with these uses, water has been considered, since ancient times, the most suitable medium to clean, disperse, transport and dispose of wastes (domestic and industrial wastes, mine drainage waters, irrigation returns, etc.).

Each water use, including abstraction of water and discharge of wastes, leads to specific, and generally rather predictable, impacts on the quality of the aquatic environment (see Chapter 3). In addition to these intentional water uses, there are several human activities which have indirect and undesirable, if not devastating, effects on the aquatic environment. Examples are uncontrolled land use for urbanisation or deforestation, accidental (or unauthorised) release of chemical substances, discharge of untreated wastes or leaching of noxious liquids from solid waste deposits. Similarly, the uncontrolled and excessive use of fertilisers and pesticides has long-term effects on ground and surface water resources.

Structural interventions in the natural hydrological cycle through canalisation or damming of rivers, diversion of water within or among drainage basins, and the over-pumping of aquifers are usually undertaken with a beneficial objective in mind. Experience has shown, however, that the resulting long-term environmental degradation often outweighs these benefits. The most important anthropogenic impacts on water quality, on a global scale, are summarised in Table 1.2, which also distinguishes between the severity of the impairment of use in different types of water bodies.

Table 1.2   Major freshwater quality issues at the global scale[1]

Issue	Rivers	Lakes	Reservoirs	Groundwaters
			Water body	
Pathogens	xxx	x^2	x^2	x
Suspended solids	xx	na	x	na
Decomposable organic matter[3]	xxx	x	xx	x
Eutrophication[4]	x	xx	xxx	na
Nitrate as a pollutant	x	0	0	xxx
Salinisation	x	0	x	xxx
Trace elements	xx	xx	xx	xx^5
Organic micropollutants	xxx	xx	xx	xxx^5
Acidification	x	xx	xx	0
Modification of hydrological regimes[6]	xx	x		x

A full discussion of the sources and effects of each of these pollution issues is available in the relevant chapters of Meybeck et al., 1989

xxx   Severe or global deterioration found
xx   Important deterioration
x   Occasional or regional deterioration
0   Rare deterioration
na   Not applicable

[1]   This is an estimate for the global scale. At a regional scale these ranks may vary greatly according to the stage of economic development and land-use. Radioactive and thermal wastes are not considered here.
[2]   Mostly in small and shallow water bodies
[3]   Other than resulting from aquatic primary production
[4]   Algae and macrophytes
[5]   From landfill, mine tailings
[6]   Water diversion, damming, overpumping, etc.

Pollution and water quality degradation interfere with vital and legitimate water uses at any scale, i.e. local, regional or international (Meybeck et al., 1989). As shown in Table 1.3, some types of uses are more prone to be affected than others. Water quality criteria, standards and the related legislation are used as the main administrative means to manage water quality in order to achieve user requirements. The most common national requirement is for drinking water of suitable quality, and many countries base their own standards on the World Health Organization (WHO) guidelines for drinking water quality (WHO, 1984, 1993). In some instances, natural water quality (particularly conditions which occur very rarely; see section 1.1.2) is inadequate for certain purposes as defined by recommended or guideline concentrations (Figure 1.4B). However, other water bodies may still be perfectly usable for some activities even after their natural conditions have been altered by pollution. A very comprehensive collection and evaluation of water quality criteria for a variety of uses has been made, and is being regularly updated, by Canadian scientists (Environment Canada, 1987).

Table 1.3 Limits of water uses due to water quality degradation

				Use			
Pollutant	Drinking water	Aquatic wildlife, fisheries	Recreation	Irrigation	Industrial uses	Power and cooling	Transport
Pathogens	xx	0	xx	x	xx^1	na	na
Suspended solids	xx	xx	xx	x	x	x^2	xx^3
Organic matter	xx	x	xx	+	xx^4	x^5	na
Algae	$x^{5,6}$	x^7	xx	+	xx^4	x^5	x^8
Nitrate	xx	x	na	+	xx^1	na	na
Salts^9	xx	xx	na	xx	xx^{10}	na	na
Trace elements	xx	xx	x	x	x	na	na
Organic micropollutants	xx	xx	x	x	?	na	na
Acidification	x	xx	x	?	x	x	na

xx	Marked impairment causing major treatment or excluding the desired use	3	Sediment settling in channels
		4	Electronic industries
x	Minor impairment	5	Filter clogging
0	No impairment	6	Odour, taste
na	Not applicable	7	In fish ponds higher algal biomass can be accepted
+	Degraded water quality may be beneficial for this specific use	8	Development of water hyacinth (*Eichhornia crassipes*)
?	Effects not yet fully realised		
		9	Also includes boron, fluoride, etc.
1	Food industries	10	Ca, Fe, Mn in textile industries, etc.
2	Abrasion		

Due to the complexity of factors determining water quality, large variations are found between rivers or lakes on different continents or in different hydroclimatic zones. Similarly, the response to anthropogenic impacts is also highly variable. As a consequence, there is no universally applicable standard which can define the baseline chemical or biological quality of waters. At best, a general description of some types of rivers, lakes or aquifers can be given.

Although the major proportion of all water quality degradation world-wide is due to anthropogenic influences, there are natural events and environmental catastrophes which can lead, locally, to severe deterioration of the aquatic environment. Hurricanes, mud flows, torrential rainfalls, glacial outbursts and unseasonal lake overturns are just a few examples. Some natural events are, however, aggravated by human activities, such as soil erosion associated with heavy rainfall in deforested regions. Restoration of the natural water quality often takes many years, depending on the geographical scale

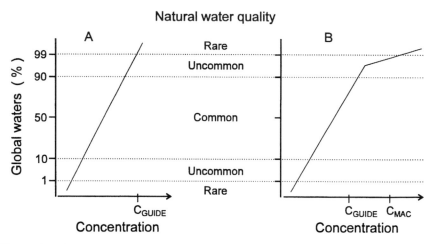

Figure 1.4 A schematic representation of the statistical distribution of natural waters on a global scale and their suitability for different uses as defined by guideline and maximum allowable concentrations (MAC). **A**. An element of single natural origin (e.g. K^+) and with concentrations always within guideline values. **B**. An element of more than one natural origin (e.g. Na^+) which can occur in concentrations which restrict its use or are too high for most purposes

and intensity of the event. The eruption of Mount Saint Helens, USA in 1980, and the subsequent mud flows, are still having a profound effect on down-stream water quality (D. Rickert, US Geological Survey, pers. comm.).

1.4   Pollutant sources and pathways

In general, pollutants can be released into the environment as gases, dis-solved substances or in the particulate form. Ultimately pollutants reach the aquatic environment through a variety of pathways, including the atmosphere and the soil. Figure 1.5 illustrates, in schematic form, the principal pathways of pollutants that influence freshwater quality.

Pollution may result from point sources or diffuse sources (non-point sources). There is no clear-cut distinction between the two, because a diffuse source on a regional or even local scale may result from a large number of individual point sources, such as automobile exhausts. An important differ-ence between a point and a diffuse source is that a point source may be collected, treated or controlled (diffuse sources consisting of many point sources may also be controlled provided all point sources can be identified). The major point sources of pollution to freshwaters originate from the col-lection and discharge of domestic wastewaters, industrial wastes or certain agricultural activities, such as animal husbandry. Most other agricultural ac-tivities, such as pesticide spraying or fertiliser application, are considered as

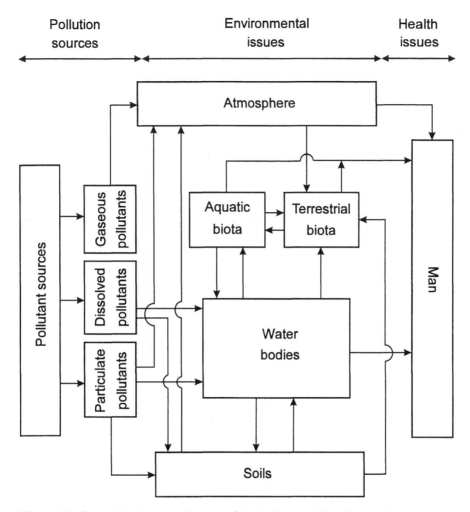

Figure 1.5 Potential pollutant pathways related to the aquatic environment

diffuse sources. The atmospheric fall-out of pollutants also leads to diffuse pollution of the aquatic environment. The various sources of major pollutant categories are summarised in Table 1.4 and examples of pollution sources for groundwater are presented in Table 9.7.

Atmospheric sources
The atmosphere is proving to be one of the most pervasive sources of pollutants to the global environment. Significant concentrations of certain contaminants are even being observed in Arctic and Antarctic snow and ice, with high levels of bioaccumulation magnified through the food chain to

Table 1.4   Anthropogenic sources of pollutants in the aquatic environment

Source	Bacteria	Nutrients	Trace elements	Pesticides/ herbicides	Industrial organic micropollutants	Oils and greases
Atmosphere		x	xxxG	xxxG	xxxG	
Point sources						
Sewage	xxx	xxx	xxx	x	xxx	
Industrial effluents		x	xxxG		xxxG	xx
Diffuse sources						
Agriculture	xx	xxx	x	xxxG		
Dredging		x	xxx	xx	xxx	x
Navigation and harbours	x	x	xx		x	xxx
Mixed sources						
Urban run-off and waste disposal	xx	xx	xxx	xx	xx	xx
Industrial waste disposal sites		x	xxx	x	xxx	x

x   Low local significance   xxx High local/regional significance
xx Moderate local/regional significance   G   Globally significant

mammals and native human populations (see Chapter 5).

Sources of anthropogenic materials to the atmosphere include:

- combustion of fossil fuels for energy generation,
- combustion of fossil fuels in automobiles, other forms of transport, heating in cold climates and industrial needs (e.g. steel making),
- ore smelting, mainly sulphides,
- wind blown soils from arid and agricultural regions, and
- volatilisation from agriculture, from waste disposal and from previously polluted regions.

These sources, together, provide an array of inorganic and organic pollutants to the atmosphere which are then widely dispersed by weather systems and deposited on a global scale. For example, toxaphene and PCBs (poly-chlorinated biphenyls) have been described in remote lake sediments from Isle Royale, Lake Superior (Swaine, 1978) and in high Arctic ice (Gregor and Gummer, 1989). In the former case, the source was postulated as the southern USA and Central America, whereas in the latter case, the source was believed to be Eastern Europe and the former USSR. Deposition of pollutants from the atmosphere, either as solutes in rain or in particulate form, occurs evenly over a wide area; covering soils, forests and water surfaces, where they become entrained in both the hydrological and sedimentary (erosion, transport and deposition) cycles. This may be termed secondary cycling, as distinct from the primary cycle of emission into the atmosphere, transport and deposition.

Point sources
By definition a point source is a pollution input that can be related to a single outlet. Untreated, or inadequately treated, sewage disposal is probably still the major point source of pollution to the world's waters. Other important point sources include mines and industrial effluents.

As point sources are localised, spatial profiles of the quality of the aquatic environment may be used to locate them. Some point sources are characterised by a relatively constant discharge of the polluting substances over time, such as domestic sewers, whereas others are occasional or fluctuating discharges, such as leaks and accidental spillages. A sewage treatment plant serving a fixed population delivers a continuous load of nutrients to a receiving water body. Therefore, an increase in river discharge causes greater dilution and a characteristic decrease in river concentration. This contrasts with atmospheric deposition and other diffuse sources where increased land run-off often causes increased pollutant concentrations in the receiving water system.

Non-atmospheric diffuse sources
Diffuse sources cannot be ascribed to a single point or a single human activity although, as pointed out above, they may be due to many individual point sources to a water body over a large area. Typical examples are:
- Agricultural run-off, including soil erosion from surface and sub-soil drainage. These processes transfer organic and inorganic soil particles, nutrients, pesticides and herbicides to adjacent water bodies.
- Urban run-off from city streets and surrounding areas (which is not channelled into a main drain or sewer). Likely contaminants include derivatives of fossil fuel combustion, bacteria, metals (particularly lead) and industrial organic pollutants, particularly PCBs. Pesticides and herbicides may also be derived from urban gardening, landscaping, horticulture and their regular use on railways, airfields and roadsides. In the worst circumstances pollutants from a variety of diffuse sources may be diverted into combined storm/sewer systems during storm-induced, high drainage flow conditions, where they then contribute to major point sources.

- Waste disposal sites which include municipal and industrial solid waste disposal facilities; liquid waste disposal (particularly if groundwater is impacted); dredged sediment disposal sites (both confined and open lake). Depending on the relative sizes of the disposal sites and receiving water bodies, these sources of pollution can be considered as either diffuse or point sources, as in the case of groundwater pollution (see Table 9.7).
- Other diffuse sources including waste from navigation, harbour and marina sediment pollution, and pollution from open lake resource exploitation, in particular oil and gas (e.g. Lakes Erie and Maracaibo).

The time variability of pollutant release into the aquatic environment falls into four main categories. Sources can be considered as permanent or continuous (e.g. domestic wastes from a major city and many industrial wastes), periodic (e.g. seasonal variation associated with the influx of tourist populations, or food processing wastes), occasional (e.g. certain industrial waste releases), or accidental (e.g. tank failure, truck or train accidents, fires, etc.). The effects of these various types of pollutants on receiving water bodies are rather different. The continuous discharge of municipal sewage, for example, may be quite acceptable to a river during high discharge periods when dilution is high and biodegradation is sufficient to cope with the pollution load. During low discharges, however, pollution levels and effects may exceed acceptable levels in downstream river stretches. Figure 1.6A shows these two seasonal situations for rivers. The example of the effects of an episodic pollution event on a lake is given in Figure 1.6B which shows the influence of residence time on the elimination of the pollutant from the lake, as measured at its natural outlet. Lake volume and initial dilution are also factors co-determining the prevalence of the pollutant in the lake.

1.5   Spatial and temporal variations

Spatial variation in water quality is one of the main features of different types of water bodies, and is largely determined by the hydrodynamic characteristics of the water body. Water quality varies in all three dimensions (see section 2.2.1) which are further modified by flow direction, discharge and time. Consequently, water quality cannot usually be measured in only one location within a water body but may require a grid or network of sampling sites.

For practical purposes, i.e. to limit the number of sampling sites and to facilitate the presentation of data, some simplifications to the ideal sampling grid are used. Examples include longitudinal or vertical profiles as shown in Figure 1.7. Two-dimensional profiles are most suitable for observing plumes

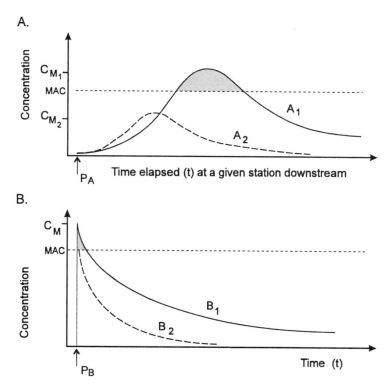

Figure 1.6 The influence of hydrodynamic characteristics on the environmental fate of pollutants (C_M maximum concentration reached, MAC maximum allowable concentration) **A**. Schematic response observed at a given river station downstream of a chronic point source of pollution (P_A) (non-reactive dissolved substances). High (A_2) and low (A_1) river discharge. **B**. Schematic response observed at lake outlets following a single episode of pollution (P_B) (non-reactive dissolved substances) for long (B_1) and short (B_2) residence times in lakes of equal volumes

of pollution from a source, presenting the information either with depth or horizontally in the form of maps. These are particularly applicable to lakes, reservoirs and groundwater aquifers.

The temporal variation of the chemical quality of water bodies can be described by studying concentrations (also loads in the case of rivers) or by determining rates such as settling rates, biodegradation rates or transport rates. It is particularly important to define temporal variability. Five major types are considered here:

- Minute-to-minute to day-to-day variability resulting from water mixing, fluctuations in inputs, etc., mostly linked to meteorological conditions and water body size (e.g. variations during river floods).
- Diel variability (24 hour variations) limited to biological cycles, light/dark

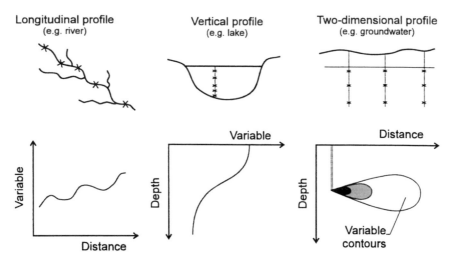

Figure 1.7   Examples of the description of spatial variations in water quality

cycles etc. (e.g. O$_2$, nutrients, pH), and to cycles in pollution inputs (e.g. domestic wastes).

- Days-to-months variability mostly in connection with climatic factors (river regime, lake overturn, etc.) and to pollution sources (e.g. industrial wastewaters, run-off from agricultural land).
- The seasonal hydrological and biological cycles (mostly in connection with climatic factors).
- Year-to-year trends, mostly due to human influences.

Once the cause of water quality degradation has been removed or reduced (such as the treatment of point sources or the regulation of diffuse sources), the restoration or recovery period of the aquatic environment may take weeks, or even millennia (e.g. some rivers in Wales are still influenced by mine tailings from the Roman period) (Figure 1.8). The temporal and spatial scales of many water quality issues are associated with water residence time (see Figure 1.2). Other issues, however, are hardly linked to water residence time or to water body size; for example, the changes in aquatic habitat downstream of river dams typically last more than 100 years. From the human perspective, a recovery period between 10 and 100 years can be considered as a limited form of reversibility, whereas recovery taking over 100 years can be considered as irreversible degradation of the aquatic environment.

1.6   Economic development and water quality
The continuing increase in socio-economic activities world-wide has been accompanied by an even faster growth in pollution stress on the aquatic

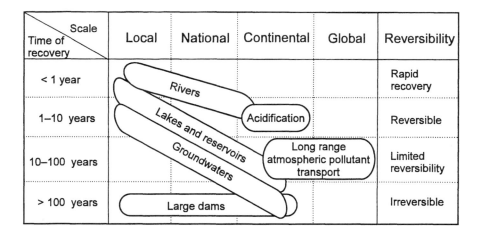

Figure 1.8 Schematic representation of the relationship between the spatial scale of water quality issues in different water bodies and the period taken for recovery of aquatic systems after remedial measures have been taken

environment. Only after a considerable time lapse, allowing for the public perception of water quality deterioration, have the necessary remedial measures been taken. This historic evolution of water pollution control is reflected schematically in Figure 1.9 which illustrates the general deterioration of the quality of the aquatic environment without any control (AB) and which generally accelerates during industrialisation (BC). If public concern starts early (point B) it takes some time (B–C) for the relevant authorities to initiate control measures. If these measures are insufficient, the rate of increase in pollution is lowered (C–D2), but if the economic activity is still growing, or if the assimilation capacity of the environment (storage, dilution, self purification) is limited, the pollution rapidly reaches the threshold concentration (C–D1) where severe or irreversible damage occurs. If proper action is taken, the pollution reaches a maximum (E) after a time-lag (C–E) which depends on the effectiveness of the control, on the water residence time (evaluation of pollution) and on the pollutant interaction with other "sinks and reservoirs" (including storage). Finally a tolerable environmental level (F) may eventually be reached, although this is not generally equivalent to the pristine level (O).

Four phases of environmental problem development (I to IV) can be identified in relation to progress in socio-economic development (Figure 1.9). In general, this sequence of phases is applicable not only to different types of pollution problems but also to countries at different levels of socio-economic

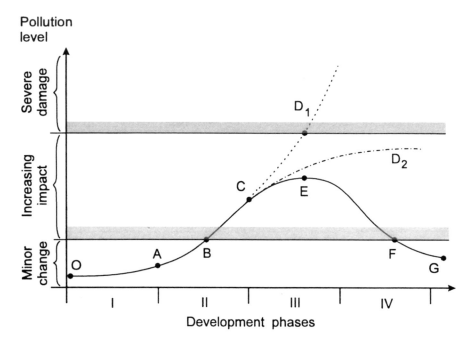

Figure 1.9 Long-term impact and control of pollution of the aquatic environment (Modified from Meybeck *et al.*, 1989)
Development phases:
Phase I — a linear increase in low-level pollution with population number (typical pattern for traditional agricultural society)
Phase II — exponential pollution increase with industrial production, energy consumption and agricultural intensification (typical pattern for newly industrialising countries)
Phase III — containment of pollution problems due to the implementation of control strategies (typical pattern for highly industrialised countries)
Phase IV — reduction of pollution problems, principally at the source, to a level which is ecologically tolerable and does not interfere with water uses (desired ultimate situation)

development. A simplified global scheme with three categories of countries can be used for this purpose as follows:

- Highly industrialised countries, which encountered the four phases over a long period of time, starting at about 1850 for some of the issues listed in Figure 1.3.
- Newly industrialising countries, which faced most of the problems in the 1950s or even more recently.
- Low-development countries (with predominantly traditional agricultural economies) which have not yet faced most water quality problems except faecal and organic pollution.

For each of these three categories the occurrence and control of the domestic sewage pollution problem has followed a different time schedule, as indicated in Figure 1.9. The extent to which environmental management services have been installed, and how far they are commensurate with the pollution problems, largely determines the resulting state of the quality of a country's water resources. Furthermore, the different types of pollution problems occur in developing countries in much more rapid succession than in Europe, due to the modern international trade of chemicals, ubiquitous dispersion of persistent contaminants and changing hydrological cycles, etc. Thus developing countries are, and will be, faced more and more with situations where second and third generation pollution issues appear before much control over "traditional" pollution sources has been achieved.

1.7   References

Bartram, J. and Ballance, R. [Eds] 1996 *Water Quality Monitoring; A Practical Guide to the Design and Implementation of Fresh Water Quality Studies and Monitoring Programmes*. Chapman & Hall, London.

Environment Canada 1987 *Canadian Water Quality Guidelines*. Prepared by the Task Force on Water Quality Guidelines of the Canadian Council of Resource Ministers, Environment Canada, Ottawa.

GESAMP 1988 *Report of the Eighteenth Session, Paris 11-15 April 1988*. GESAMP Reports and Studies No. 33, United Nations Educational, Scientific and Cultural Organization, Paris.

Gregor, D.M. and Gummer, W.D. 1989 Evidence of atmospheric transport and deposition of organochlorine pesticides and polychorinated biphenyls in Canadian Arctic Snow. *Environ. Sci. Technol.*, **23**, 561-565.

Meybeck, M., Chapman, D. and Helmer, R. [Eds] 1989 *Global Freshwater Quality: A First Assessment*. Blackwell Reference, Oxford, 306 pp.

Meybeck, M. and Helmer, R. 1989 The quality of rivers: from pristine state to global pollution. *Paleogeog. Paleoclimat. Paleoecol.* (Global Planet. Change Sect.), **75**, 283-309.

Nace, R.L. [Ed.] 1971 *Scientific Framework of World Water Balance*. Technical Papers in Hydrology 7, United Nations Educational, Scientific and Cultural Organization, Paris, 27 pp.

Swaine, W.R. 1978 Chlorinated organic residues in fish, water and precipitation from the vicinity of Isle Royale, Lake Superior. *J. Great Lakes Res.*, **4**, 398-407.

WHO 1984 *Guidelines for Drinking-Water Quality, Volume 2, Health Criteria and other Supporting Information*. World Health Organization, Geneva, 335 pp.

WHO   1991   *GEMS/WATER   1990-2000.   The   Challenge   Ahead.* WHO/PEP/91.2, World Health Organization, Geneva.

WHO 1993 *Guidelines for Drinking-Water Quality, Volume 1, Recommendations.* Second edition, World Health Organization, Geneva, 188 pp.

Chapter 2*

STRATEGIES FOR WATER QUALITY ASSESSMENT

2.1   Introduction

The operations involved in water quality assessment are many and complex. They can be compared to a chain of about a dozen links and the failure of any one of them can weaken the whole assessment. It is imperative that the design of these operations must take into account the precise objectives of the water quality assessment. During the 1950s, in the early days of modern water quality monitoring, activities were rarely focused on particular issues. However, the water quality assessment process has now evolved into a set of sophisticated monitoring activities including the use of water chemistry, particulate material and aquatic biota (e.g. Hirsch *et al.*, 1988). Many manuals on water quality monitoring methods already exist (e.g. Alabaster, 1977; UNESCO/WHO, 1978; Krenkel and Novotny, 1980; Sanders *et al.*, 1983; Barcelona *et al.*, 1985; WMO, 1988; Yasuno and Whitton, 1988; WHO, 1992) although most of these consider only one type of water body (i.e. rivers, lakes or groundwaters) or one approach to monitoring (e.g. chemical or biological methods). Few manuals exist which consider all water bodies (e.g. Hem, 1989). This guidebook presents the combined use of water, particulate matter and biological monitoring in order to produce comprehensive water quality assessments for the principal types of freshwater bodies. However, economic constraints frequently mean that the variables to be monitored, and the methods to be used, must be chosen carefully to ensure water quality assessment objectives are met as efficiently as possible.

2.2   The water quality assessment process

As defined in Chapter 1, water quality assessment is the overall process of evaluation of the physical, chemical and biological nature of the water, whereas water quality monitoring is the collection of the relevant information. This guidebook concentrates on the whole assessment process, in different types of water bodies. The details of monitoring methods and approaches which can be applied in the field are given in the practical handbook by Bartram and Ballance (1996).

2.2.1 Monitoring, survey and surveillance

The main reason for the assessment of the quality of the aquatic environment has been, traditionally, the need to verify whether the observed water quality

*This chapter was prepared by M. Meybeck, V. Kimstach and R. Helmer

is suitable for intended uses. The use of monitoring has also evolved to help determine trends in the quality of the aquatic environment and how that quality is affected by the release of contaminants, other anthropogenic activities, and/or by waste treatment operations (impact monitoring). More recently monitoring has been carried out to estimate nutrient or pollutant fluxes discharged by rivers or groundwaters to lakes and oceans, or across international boundaries. Monitoring to determine the background quality of the aquatic environment is also now widely carried out, as it provides a means of comparison with impact monitoring. It is also used simply to check whether any unexpected change is occurring in otherwise pristine conditions, for example, through the long range transport of atmospheric pollutants (note, however, that natural water quality is very variable depending on local conditions).

General definitions for various types of environmental observation programmes have been proposed which may also be modified and interpreted for the aquatic environment as follows:

Term	Definition
MONITORING	Long-term, standardised measurement and observation of the aquatic environment in order to define status and trends.
SURVEY	A finite duration, intensive programme to measure and observe the quality of the aquatic environment for a specific purpose.
SURVEILLANCE	Continuous, specific measurement and observation for the purpose of water quality management and operational activities.

These different definitions are often not distinguished from one another and all three may be referred to as monitoring, since they all involve collection of information at set locations and intervals (see definition of monitoring in section 1.2). They do, nevertheless, differ in relation to their principal use in the water quality assessment process.

Monitoring, survey and surveillance are all based on data collection. Data are principally collected at given geographical locations in the water body (see section 1.5). Water quality variables are often described by the longitude and latitude of the sampling or measurement site (x and y co-ordinates) and further characterised by the depth at which the sample is taken (vertical co-ordinate z). Monitoring data must also be characterised and recorded with regard to the time t at which the sample is taken or the *in situ* measurement made. Thus any physical, chemical or biological variable will be measured as a concentration c, or number, which is a function of the above parameters: $c = f(x,y,z,t)$. In rivers, the flux determination and the data interpretation also require the knowledge of water discharge Q, thus: $c = f(x,y,z,t,Q)$. Monitoring data must, therefore, provide a clear determination of these parameters in order to be used for data interpretation and water quality assessments.

2.2.2 Objectives of water quality assessment

No assessment programme should be started without scrutinising critically the real need for water quality information (i.e. the "need to know" as opposed to "it would be nice to know"). Since water resources are usually put to several competing beneficial uses, monitoring which is used to acquire necessary information should reflect the data needs of the various users involved (Helmer, 1994). Consequently, there are two different types of monitoring programmes, depending on how many assessment objectives have to be met:

- *Single-objective monitoring* which may be set up to address one problem area only. This involves a simple set of variables, such as: pH, alkalinity and some cations for acid rain; nutrients and chlorophyll pigments for eutrophication; various nitrogenous compounds for nitrate pollution; or sodium, calcium, chloride and a few other elements for irrigation.
- *Multi-objective monitoring* which may cover various water uses and provide data for more than one assessment programme, such as drinking water supply, industrial manufacturing, fisheries or aquatic life, thereby involving a large set of variables. The Commission of the European Communities has a list in excess of 100 micropollutants to be considered in drinking water alone.

The implementation of the assessment programme objectives may focus on the spatial distribution of quality (high station number), on trends (high sampling frequency), or on pollutants (in-depth inventories). Full coverage of all three requirements is virtually impossible, or very costly. Consequently preliminary surveys are necessary in order to determine the necessary focus of an operational programme. Table 2.1 summarises the existing types of water quality operations in relation to their main objectives.

The process of determining objectives should start with an in-depth investigation of all factors and activities which exert an influence, directly or indirectly, on water quality. Inventories have to be prepared on:

- the geographical features of the area, including: topography, relief, lithology, pedology, climate, land-use, hydrogeology, hydrology etc.,
- water uses, including: dams, canals, water withdrawal for cities and

Table 2.1   Typical objectives of water quality assessment operations

	Type of operation	Major focus of water quality assessment
Common operations		
1	Multipurpose monitoring	Space and time distribution of water quality in general
2	Trend monitoring	Long-term evolution of pollution (concentrations and loads)
3	Basic survey	Identification and location of major survey problems and their spatial distribution
4	Operational surveillance	Water quality for specific uses and related water quality descriptors (variables)
Specific operations		
5	Background monitoring	Background levels for studying natural processes; used as reference point for pollution and impact assessments
6	Preliminary surveys	Inventory of pollutants and their space and time variability prior to monitoring programme design
7	Emergency surveys	Rapid inventory and analysis of pollutants, rapid situation assessment following a catastrophic event
8	Impact surveys	Sampling limited in time and space, generally focusing on few variables, near pollution sources
9	Modelling surveys	Intensive water quality assessment limited in time and space and choice of variables, for example, eutrophication models or oxygen balance models
10	Early warning surveillance	At critical water use locations such as major drinking water intakes or fisheries; continuous and sensitive measurements

industries, agricultural activities, navigation, recreation, fisheries, etc., and

- pollution sources (present and expected), including: domestic, industrial and agricultural, as well as their stage of pollution control and waste treatment facilities.

An example of a pollutant source inventory is presented in Figure 2.1 for the watershed of Lake Vättern in Sweden. The emphasis in these inventories should be put on water uses and their specific water quality requirements, particularly in the future. Economic trends should be predicted for at least five years ahead since monitoring design, implementation and data interpretation take a long time, although specific surveys can give "snapshots" of the quality of the aquatic environment which can also be used in environmental planning.

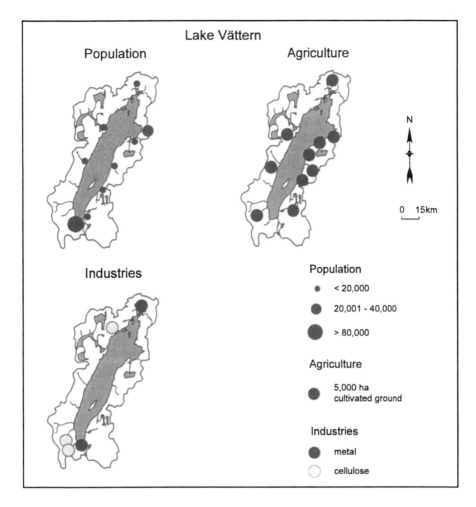

Figure 2.1 A pollutant source inventory for the Lake Vättern basin, Sweden (After Hakanson, 1977)

In addition to the above investigations, preliminary water quality surveys may be carried out for the following specific purposes:
- to determine the time and space variability of the quality of the aquatic environment in order to select sampling stations and frequencies,
- to determine the key descriptors to be considered, and
- to assess the feasibility and cost of a monitoring programme.

It cannot be over-emphasised that the benefits for an optimal monitoring operation, drawn from careful preliminary planning and investigation, by far outweigh the efforts spent during this initial phase. Mistakes and over-sights during this part of the programme may lead to costly deficiencies, or over-spending, during many years of routine monitoring!

2.2.3 Elements of water quality assessment

The possible types of water quality assessment programmes are numerous. These should be designed or adopted according to objectives set on the basis of environmental conditions, water uses (actual and future), water legislation, etc. Once the objectives have been set, a review of existing water quality data, sometimes supported by preliminary surveys, determines the monitoring design. Following the implementation of the various assessment activities an important step which is often underestimated, if not omitted, is data interpretation. This should be followed by recommendations to relevant water authorities for water management, water pollution control, and eventually the adjustment or modification of monitoring activities.

There are certain standard elements which are common to all water quality assessment programmes. They are more, or less, extensively developed depending on the type of assessment required. The generalised structure of a water quality assessment programme is given in Figure 2.2 and described below:

The key elements of an assessment programme

1 Objectives	These should take into account the hydrological factors, the water uses, the economic development, the legislative policies etc. Necessary decisions involve whether the emphasis should be put on concentrations or loads, or spatial or time distributions, and the most appropriate monitoring media.
2 Preliminary surveys	These are short-term, limited activities to determine the water quality variability, the type of monitoring media and pollutants to be considered, and the technical and financial feasibility of a complete monitoring programme.
3 Monitoring design	This includes the selection of types of pollutants, station location, sampling frequency, sampling apparatus, etc.
4 Field monitoring operations	These include *in situ* measurements, sampling of appropriate media (water, biota, particulate matter), sample pretreatment and conservation, identification and shipment.
5 Hydrological monitoring	This includes water discharge measurements, water levels, thermal profiles, etc., and should always be related to the water quality assessment activities.
6 Laboratory activities	These include concentration measurements, biological determinations, etc.
7 Data quality control	This must be undertaken by using analytical quality

	assurance within each laboratory, and amongst all laboratories participating in the same programme, and by checking field operations and hydrological data.
8 Data storage treatment and reporting	This is now widely computerised and involves the use of databases, statistical analysis, trend determinations, multifactorial correlation, etc., and presentation and dissemination of results in appropriate forms (graphs, tabulated data, data diskettes, etc.).
9 Data interpretation	This involves comparison of water quality data between stations (water quality descriptors, fluxes), analysis of water quality trends, development of cause–effect relationships between water quality data and environmental data (geology, hydrology, land use, pollutant sources inventory), and judgement of the adequacy of water quality for various uses etc. For specific problems, and the evaluation of the environmental significance of observed changes, external expertise may be needed. Publication and dissemination of data and reports to relevant authorities, the public, and the scientific community is the necessary final stage of assessment activities.
10 Water management recommendations	These decisions should be taken at various levels from local government to international bodies, by water authorities as well as by other environmental authorities. An important decision is the re-design of assessment operations, to improve the monitoring programme and to make it more cost-effective.

The duration of each segment in the assessment chain is highly variable. The initial step of preliminary surveys and design may take months to years. A field sampling mission may last some days, shipment can take days, and chemical analysis of major ions, nutrients and a few other key variables in a set of samples typically takes one week. Data treatment may take some months, while interpretation and publication may take from a few months to several years. For surveillance purposes, however, field work, data assessments and reporting (including preliminary reports if necessary) need to be accomplished within a time-scale governed by the operational requirements.

In the case of an emergency water quality assessment operation, the whole chain, from design to interpretation, should be shortened to a few days, or even less. Assessment of trends often cannot be provided before ten years or so, depending on the variability of the quality and the nature of the programmes. Impact surveys (Table 2.1) are usually accomplished in less than one year and multi-purpose monitoring is a steady annual exercise which may be assessed and published within one to two years.

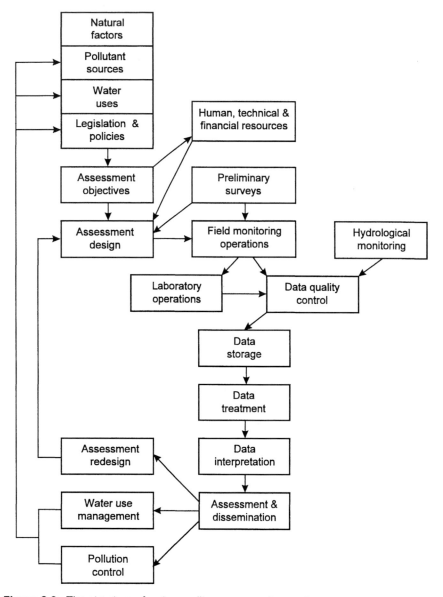

Figure 2.2   The structure of water quality assessment operations

2.2.4 Levels of water quality assessment

Experience shows that it is not easy to choose the optimum level of assessment needed to generate precisely the data required for specific purposes and
objectives. Most of the time, assessment is achieved principally through

monitoring activities (i.e. long-term standardised measurement and observation). The monitoring phase must generate the data which are essential for meaningful interpretation and management decisions, but must not lead to a vast collection of unnecessary data which are costly to obtain, but do not contribute to the required understanding of water quality. Furthermore, monitoring activities, by necessity, must be commensurate with the socio-economic and technical and scientific development of the country. For example, some water quality variables (e.g. organic micropollutants and trace elements) require highly trained technicians and costly laboratory facilities. In a simplified manner, it is possible to distinguish three levels:

- *Simple monitoring* based on a limited number of samples, simple analyses or observations, and data treatment which can be performed by pocket calculator.
- *Intermediate-level monitoring* requiring some specific laboratory facilities and more financial support to increase the number of stations, samples, analytical variables, etc. Personal computers (PCs) are recommended for processing data.
- *Advanced-level monitoring* involving sophisticated techniques and highly trained technicians and engineers. The analytical facilities can perform any pollutant determination required, with an increasing number of variables per sample, and of samples taken. Large computer data storage and handling facilities, such as a mainframe computer, are required for data treatment.

All elements of assessment, from objective setting to data interpretation, will be related to these three levels. The ultimate aim is to promote monitoring operations from the basic level to the more elaborate ones, in accordance with increasing complexity of water quality problems and the capabilities to perform scientific programmes. The above three monitoring levels will be referred to throughout this guidebook.

2.3   Typical water quality monitoring programmes

In principle, there could be as many types of monitoring programmes as there are objectives, water bodies, pollutants and water uses, as well as any combination thereof. In practice, assessments are limited to about ten different types of operations which are summarised in Tables 2.1 and 2.2, giving details of the different types of monitoring operations. It should be noted that, in the past, many countries or water authorities have installed multi-purpose or multi-objective monitoring programmes without conducting the necessary preliminary surveys which are considered necessary. Critical scrutiny of results after several years of operation has led to a second generation of

Table 2.2   Categories and principal characteristics of water quality assessment operations

Type of operation	Station density and location	Sampling or observation frequencies	Number of variables considered	Duration	Interpretation lag
Multi-purpose monitoring	medium	medium (12 per year)	medium	medium (> 5 years)	medium (1 year)
Other common water quality operations					
Trend monitoring	low: major uses and international stations	**very high**	low for single objective; high for multiple objective	**> 10 years**	> 1 year
Basic survey	**high**	depending on media considered	medium to high	once per year to once every 4 years	1 year
Operational surveillance	low: at specific uses	medium	**specific**	variable	short (month/week)
Specific water quality operations					
Background monitoring	**low**	low	low to high	variable	medium
Preliminary surveys	high	usually low	low to medium (depending on objectives)	**short** **< 1 year**	short (months)
Emergency surveys	medium to high	high	**pollutant inventory**	very short (days–weeks)	**very short (days)**
Impact surveys	limited down-stream pollut-ion sources	medium	**specific**	variable	short to medium
Modelling surveys	**specific** (e.g. profiles)	**specific** (e.g. diel cycles)	**specific** (e.g. O_2, BOD)	short to, medium two periods: calibration and validation	short
Early warning surveillance	very limited	**continuous**	very limited	unlimited	instantaneous

The levels (high, medium, low) of all operation characteristics (frequency, density, number of variables, duration, interpretation lag) are given in relation to multi-purpose monitoring, which has been taken as a reference.
Important monitoring characteristics are emphasised in **bold**.

programmes with more differentiated objectives such as impact assessment, trend analysis or operational management decisions.

Background monitoring (principally in unpolluted areas) has usually been developed to help the interpretation of trend monitoring (time variations over a long period) and the definition of natural, spatial variations. Models and their related surveys have usually been set up to predict the water quality for management purposes prior to pollution treatment, or to test the impact of a new water pollution source, and are thus closely connected to operational surveillance and impact surveys. Early warning surveillance is undertaken for specific uses in the event of any sudden and unpredictable change in water quality, whereas emergency surveys of a catastrophic event should be followed in the medium and long-term by impact surveys. For practical reasons, several types of regular monitoring are often combined and some sampling stations will belong to several programmes.

Multi-dimensional approaches to water quality assessment have become an inevitable necessity. About a hundred years ago the quality of the aquatic environment was defined by a few analyses of water, but this definition has now reached a level of complexity which requires simultaneous consideration of multiple aspects. One dimension concerns physico-chemical variables, such as nitrates, chromium and polyaromatic hydrocarbons, another concerns guidelines and criteria for given water uses and another concerns different monitoring media (water, suspended or deposited sediments, colloids, whole organisms, selected biological tissues). Different aquatic environments (streams, rivers, wetlands, lakes, reservoirs, all types of groundwaters) react differently to pollution with different temporal (from minutes to years) and spatial (vertical, transverse, longitudinal, upstream/downstream) scales of variability. In addition, aquatic organisms have to be considered, from bacteria to mammals, including various types of biological features, such as community structure or bioassays.

In reality, water quality is never considered in all its dimensions. In most cases the assessment approach is determined by the perceived importance of the aquatic environment, on the objectives of the operation and on the human and financial resources available. For example, lake eutrophication is mostly considered through nutrient analysis, biomass estimates by means of chlorophyll measurements, dissolved oxygen profiles, and a few indicator species of phytoplankton (see Chapters 7 and 8).

Monitoring activities are also greatly dependent on the level of deterioration of the aquatic environment. At each stage there is a specific type of monitoring as described in Table 2.1. Background monitoring determines the natural concentrations and verifies that xenobiotic organic substances were not found at that stage. Specific monitoring of early change is required to assess the subtle changes that characterise anthropogenic impacts. The fundamental question at this stage is what forthcoming issues should be considered within basic monitoring which is routinely performed on all water bodies. Often, water quality issues are studied by explanatory and/or predictive models, which need intensive, specific monitoring and which are necessary for defining the basis for mitigating measures and regulations. Once the issue is more fully understood, environmental impact assessments can be focused on the issue until control measures are successful. If they are not successful, water quality degradation becomes a major concern and early warning monitoring might have to be installed. This type of monitoring is often used at drinking water intakes to warn of upstream, accidental pollution events.

2.4   Design of assessment programmes

Once the objectives have been clearly identified, four steps are essential in the good design of an assessment programme: (i) the selection of the appropriate media to sample, (ii) the determination of water quality variability through preliminary surveys, (iii) the integration of hydrological and water quality monitoring, and (iv) the periodic review and modification of the design of the programme.

2.4.1   Selection of media to sample

Three principal media can be used for aquatic monitoring: water, particulate matter and living organisms. The quality of water and particulate matter is estimated through physical and chemical analysis. Biological quality can be determined through: (i) specific ecological surveys (e.g. invertebrate species or bacterial counts) which can lead to the elaboration of biotic indices, (ii) specific bioassays using one or several species (e.g. bacteria, crustacea, algae) such as toxicity tests, algal growth tests, respiration rates, etc., (iii) histological and enzymatic studies in selected organisms, and (iv) chemical analysis of body tissues from selected organisms (see Chapter 5).

Each aquatic medium has its own set of characteristics for monitoring purposes, e.g. applicability to water bodies, inter-comparability, specificity to given pollutants, possibility of quantification such as fluxes and rates, sensitivity to pollution with the possibility of amplifying the pollution signal by several orders of magnitude, sensitivity to sample contamination, time-integration of the information received from instantaneous (for point sampling of the water) to integrated (for biotic indices), required level of field personnel, storage facility of samples, and length of the water quality determination process from field operation to result. The most significant characteristics of each medium are given tentatively in Table 2.3, i.e. water with dissolved components, particles in the deposited or suspended state, living organisms used for chemical analyses, ecological surveys, biotests, and physiological analyses. Water, itself, is by far the most common monitoring medium used to date, and the only one directly relevant to groundwaters. Particulate matter is widely used in lake studies and in trend monitoring, whereas biological indices based on ecological methods are used more and more for river and lake assessments. An example of selection of appropriate media for monitoring selected organic micropollutants in rivers and lakes is given in Table 2.4.

The information obtained from samples of water, particulate matter and aquatic organisms reflects different environmental time spans. The analysis of water from a grab sample represents a very short time period in the history of the aquatic environment, sometimes less than one hour for many chemical measurements, such as ammonia. Nevertheless, when river discharge is stable, a grab sample is also supposed to be stable for at least a day, if not more. Diurnal fluctuations, which are now commonly studied in eutrophic lakes and rivers, are responsible for marked variations in concentrations of nutrients, dissolved oxygen, pH, conductivity, calcium and bicarbonates. Careful interpretation of grab sample analysis is, therefore, necessary even for bioassays performed with grab sample material.

Dynamic toxicity tests, such as fish tests (see Figure 5.5) are representative of a slightly longer period of time (possibly several hours) depending on the resistance of the test organisms to the pollutants present. Algal production in a water body is commonly assessed by pigment concentration and is an integrated measure of two to three days prior to sampling, since the day to day variation in chlorophyll concentrations is rarely greater than 20 per cent.

During periods of steady water discharge, the analysis of suspended matter integrates environmental conditions for at least one or two weeks, but it can take months, or a even a few years, to observe changes in the quality of deposited sediments.

The representative time span of many types of biological methods is often unclear to non-biologists, although it is implied by their principal suitability for indicating changes over long time periods. Biotic indices based on benthic communities usually integrate conditions for a few weeks to several months (i.e. the time required for the organisms' life cycles or development). In temperate climates, the seasonal loss of some groups of organisms makes the use and interpretation of some biotic indices much more difficult or less reliable. One method providing a very long-term integration of environmental conditions, particularly the presence of toxic residues, is the analysis

Table 2.3   Principal characteristics of media used for water quality assessments

Characteristics	Water	Particulate matter		Living organisms			
		Suspended	Deposited	Tissue analyses	Biotests	Ecological surveys[1]	Physiol. determin.[2]
Type of analysis or observation	<———physical———> <———————chemical————————[3] > <——————————biological——————————>						
Applicability to water bodies	rivers, lakes, groundwater	mostly rivers	lakes, rivers	rivers, lakes	rivers, lakes	rivers, lakes	rivers, lakes
Intercomparability[4]	<——————global——————>			depends on species occurrence	global	<-local to regional->	
Specificity to given pollutant	<————————specific————————>				<————integrative————>		
Quantification	<——complete——> quantification of concs & loads		concentrations only	quantitative	semi-quantitative	<—— relative——>	
Sensitivity to low levels of pollution	low	<———————high———————>			variable	medium	variable
Sample contamination risk	high[5]	medium	<————low————>		medium	<———low———>	
Temporal span of information obtained	instant	short	long to very long (continuous record)	medium (1 month) to long (>1 year)	instant. to continuous[6]	<—medium to long—>	
Levels of field operators	untrained to highly trained[7]	trained	untrained to trained	trained	<——medium to highly——> trained		
Permissible sample storage duration[7]	low	high	high	high	very low	high	na
Minimum duration of determination	instant. (*in situ* determ.) to days	days	days to weeks	days	days to months	weeks to months	days to weeks

na Not applicable

[1] Including algal biomass estimates
[2] Histologic, enzymatic, etc.
[3] Including BOD determination
[4] Most biological determinations depend on the natural occurrence of given species and are, therefore, specific to a given geographic region. Chemical and physical descriptors are globally representative
[5] For dissolved micropollutants
[6] e.g. organisms continuously exposed to water
[7] Depending on water quality descriptors

Table 2.4   Appropriate media for monitoring some organic micropollutants in rivers and lakes

	Monitoring media				
	Filtered water	Unfiltered water	Particulate matter	Filter[1] feeders	Fish[2]
Appropriate characteristics of compound to be monitored	Highly soluble in water	Volatile	Low solubility in water	Lipid soluble	Lipid soluble
Solvents[3]		xx			
Phenols	x	x			
Hydrocarbons	x	x[4]	x	xx	x
DDT and PCBs[5]		x	xx	x	x
PAH		x	xx	x	x
Soluble pesticides[6]	x	x			
Lindane	x	x	x	x	x
Other pesticides[7]	x	x			

x to xx Appropriate to highly appropriate monitoring media

[1] Direct ingestion of particulate associated pollutants

[2] Indirect ingestion and biomagnification effects

[3] Benzene, tri- and tetrachloroethylene, toluene etc.

[4] For low molecular weights (< C16), mostly volatile

[5] Also aldrin, heptachlor, chlordane

[6] TCA, paraquat, metham-sodium, carbamates

[7] Such as triazine compounds, isoproturon, chlortoluron, 2,4-D, etc.

of biological tissues which can concentrate such residues over life spans of many years (especially fish and mammals).

Chemical approaches should not automatically be preferred to biological approaches because both have advantages and shortcomings (Table 2.5). Instead, the two approaches should be regarded as complimentary. Each category of water quality assessment operations (see Table 2.2) has its own requirements in relation to the characteristics given in Table 2.3. Some practically relevant examples are:

- All monitoring activities should take into account such characteristics as continuity of the monitoring chain and the required levels of field operators.
- International monitoring programmes should consider, among others, the inter-comparability, the quantification, and the sample storage requirements.
- Trend monitoring should consider the signal amplification, the duration of information obtained, the sample storage capacity, etc.
- Basic surveys within a given region should consider biological monitoring (chemical analyses of tissues, biotic indices, physiological determinations) and/or particulate matter monitoring.

Table 2.5   Advantages and shortcomings of biological and chemical water quality monitoring

Biological monitoring	Chemical monitoring
Advantages	
Good spatial and temporal integration	Possibility of very fine temporal variations
Good response to chronic, minor pollution events	Possibility of precise pollutant determination
Signal amplification (bioaccumulation, biomagnification)	Determination of pollutant fluxes
Real time studies (in-line bioassays)	Valid for all water bodies, including groundwaters
Measures the physical degradation of the aquatic habitat	Standardisation possible
Shortcomings	
General lack of temporal sensitivity	High detection limits for many routine analyses (micropollutants)
Many semi-quantitative or quantitative responses possible	No time-integration for water grab samples
Standardisation difficult	Possible sample contamination for some micropollutants (e.g. metals)
Not valid for pollutant flux studies	High costs involved in surveys
Not yet adapted to groundwaters	Limited use for continuous surveillance

- Operational surveillance for a specific use is usually focused on water analysis and the duration of the determinations should be short.
- Impact surveys may consider field biotic indices.
- Emergency surveys imply the sampling of each medium for rapid or delayed chemical analysis and physiological effects.
- Early warning surveillance mostly relies on the continuous exposure of sensitive organisms to water and/or on the continuous measurement of some chemicals (e.g. ammonia).
- Background monitoring and surveys undertaken for modelling purposes may involve all three monitoring media.

2.4.2   Water quality variability and sampling frequency

Water quality is a highly variable aspect of any water body, although it is more variable in rivers than in lakes, but much less so in aquifers. Variabilities occur not only with regard to their spatial distribution but also over time. As indicated in section 2.2.1 any variable is a function of space and time: $c = f(x,y,z,t)$. Not all of these four parameters have equal influence in the different types of water bodies as demonstrated in Table 2.6 for surface and groundwaters. Rivers generally have only one spatial dimension (i.e.

Table 2.6   Characteristics of spatial and temporal variations in water quality

Rivers	Lakes and reservoirs	Groundwaters
Characteristics of spatial variations[1]		
In fully mixed rivers variability only in x	No variability at overturn[2]	Usually high variability in x and y
In locations downstream of confluences or effluent discharges, variability in x and y	High variability in z for most systems[3]	In some multi-layer aquifers and in the unsaturated zone, high variability in x, y and z
	High variability in x, y and z in some irregularly shaped lakes	
Characteristics of time variations		
Depends on river discharge regime	Some predictable variability (hydrodynamic and biological variations)	Low variability[4]
	Medium to low variability in large systems	

[1] x – longitudinal dimension;
y – transverse dimension;
z – vertical dimension;
[2] One sample can describe the whole water body

[3] Two dimensions (x and z) if poor lateral mixing
[4] Except for some alluvial aquifers and for karstic aquifers

longitudinal) and a pronounced time variability whereas groundwaters are characterised by low, to very low, time variability and two to three spatial dimensions. Therefore, it is essential to have sufficient knowledge of the hydrodynamics of each water body.

Field monitoring activities, including the choice of sampling location and sampling frequency, are highly dependent on the type of aquatic environment (for further details see Bartram and Ballance, 1996). Oxygen concentrations in a temperate lake are a good illustration of this (see Chapter 7). During the summer period of planktonic growth, oxygen concentrations show day and night variations (diel cycles) in the layer which receives light (euphotic zone). The vertical gradient is large from top to bottom, sometimes with a marked depletion at the thermocline (zone of rapid temperature change) and in shallow waters and sheltered bays, the oxygen cycle may be different from that at the lake centre. The oxygen variability may, therefore, have four dimensions: longitudinal, transverse, vertical and time. During mixing periods (generally once or twice per year depending on the thermal regime), lakes are

Table 2.7 Trend monitoring of water quality: frequency optimisation

	Streams and rivers	Large rivers[1]	Lakes	Groundwaters
Water	< 24 per year	< 12 per year	1 per year at overturn[2] or at each overturn[3]	1 to 4 per year[4]
Particulate matter	1 per year[5]	1 per year[5]	1 per year[6]	not relevant
Biological monitoring	1 per year[7]	1 per year biotic indices[8]	8–12 per year[9] 0.2 per year[10]	

[1] Basin area >100,000 km^2. The required frequency actually depends on water discharge variability
[2] If water residence time >1 year
[3] If water residence time <1 year
[4] Depending on water residence time
[5] A composite of 12 total suspended solids (TSS) samples weighted according to TSS discharge or, if not possible, to water discharge
[6] A composite sample of 2 to 4 sediment trap samples
[7] Organisms introduced for monitoring bioaccumulation of elements or compounds
[8] At low flow
[9] For summer chlorophyll samples
[10] Macrophytes inventory; once every 5 years

usually fully mixed during the day and sometimes this may last for weeks (e.g. during winter mixing). Therefore, one single sample may be representative of the whole lake for a long time period (see Chapters 7 and 8).

Operational surveillance should focus on the worst possible conditions, such as summer dry periods and on extreme values of concentration. Impact surveys which must also determine pollutant fluxes should, in the case of rivers, include surveys during high flow periods. The same is true for trend monitoring. A tentative frequency optimisation with respect to trend monitoring on the basis of chemical analysis is presented in Table 2.7 for streams and small rivers, large rivers, lakes and groundwater, and for all three media. It also takes into account the general variability of chemical components. Attention can be focused on: (i) the possibility of mixing and storing composite samples of particulate matter which then enables the analysis of only one sample per year, (ii) sampling lakes during overturn when the water body is well mixed, (iii) determining biotic indices during the low water stage when quality is usually at its worst or (iv) taking composite water samples through automatic samplers. Other examples of sampling strategies are detailed more fully in Chapters 6, 7, 8 and 9 and statistical sampling design is discussed in Appendix 10.1.

Table 2.8   Hydrological information required for water quality assessment

Level[1]	Rivers	Lakes/reservoirs	Groundwaters
Basic information			
A	Watershed map	Thermal regime	Major aquifer type
B	River seasonal regime	Bathymetric map	Aquifer map
C	Flow duration statistic	Water balance and current patterns	Hydrodynamic characteristics
Hydrological monitoring			
A	River level at sampling	Lake level at sampling	Piezometric level
B	River discharge at sampling	Lake level between sampling	Piezometric level between sampling
C	Continuous river discharge	Tributary discharge and lake water budget	Full knowledge of ground-water hydrodynamics

[1]   Levels A,B,C are increasing orders of assessment programme complexity

2.4.3   Integrating hydrological and water quality monitoring

No meaningful interpretation of analytical results for the assessment of water quality is possible without the corresponding hydrometric data base. Consequently, all aquatic environment monitoring should take into account the hydrological characteristics of the water bodies, which should be determined by preliminary inventories and surveys. All field observations and samples should be associated with the relevant hydrological measurements such as the ones given in Table 2.8 for different types of water bodies.

The combined evaluation of water quantity and quality data sets should also take into account spatial and temporal variabilities. The hydrological features of water bodies follow their own variability patterns which may be quite separate from natural and/or man-made water quality fluctuations. In practical terms, however, they tend to be rather closely inter-linked.

2.4.4   Review of water quality assessment design

The design of any water quality assessment programme should be examined periodically (see section 2.2.3 and Figure 2.2). When new approaches become available and when new issues are found, the objectives and procedures of water quality assessments must be reviewed. It may be necessary, for example, to drop some objectives or procedures and add others. Most water quality assessments which were started in the late 1960s and early 1970s were multi-purpose programmes, frequently based on analysis of water only. Since then many changes have occurred, including a reduced reliance on metal analysis in water samples alone, which has been replaced by a greater use of suspended or deposited particles and of mosses and fish tissues, as well as the development of specific monitoring approaches for issues such as lake eutrophication, acidification and groundwater pollution.

This complexity can be illustrated by the example of river monitoring in France over the past years. Monitoring is performed at three levels (Meybeck, 1994):

- Research and development of new methods by specialised institutions; current research is carried out on diatom indices, cytochrome P-450 in fish and on-line bioassays, etc.
- Testing of some new approaches by river basin authorities.
- Standardisation and general application at the country level.

The use of biotic indices, based on benthic communities, started with the beginning of the regular, national monitoring programme in 1971 (now the Reseau National de Bassin). This approach was gradually improved and is now standardised. After an initial period of basic multi-purpose monitoring, the French national network evolved to handle eutrophication and micropollutants through the monitoring of aquatic mosses, deposited sediments and, more recently, suspended matter, as well as specific monitoring for modelling in a few lakes. Specific early warning monitoring systems are being developed by the water supply industries.

2.5   Implementation of water quality assessment programmes

When installing national programmes for the assessment of water quality, it should be recognised that the quality aspects of water resources cannot be seen in isolation. There are also other media which have an influence on water quality or are, themselves, affected by water pollution. A general presentation of the influence of water pollution problems on other media is given in Table 2.9. As a consequence, assessment of these compartments has to take place concurrently with water quality assessment in order to obtain a complete picture of the sources, pathways and links between environmental contaminants. To achieve maximum benefit from the various assessment efforts, data transfers among the different sectors are necessary. Some sectors have to provide information to water quality assessment activities, and some derive data from them. The most important transfers are shown in Figure 2.3.

2.5.1   Administrative co-ordination of water quality assessment

Development of a universally applicable system to solve all problems of water quality assessment and prediction is not realistic. A system designed for the assessment of water quality at the national level would not be ideal for local assessment. Similarly a system designed for evaluation of long-term

Table 2.9   Links between inland freshwater pollution and other assessment media

	Pathways/links				
Issues	Air to Water	Water to Soil	Water to Tap water	Water to Food	Water to Coastal marine waters
Pathogens	x	x	xxx	xxx	xxx
Organic matter	0	x	x	na	xxx
Eutrophication[1]	x	x	0	0	xxx
Nitrate pollution	0	0	xx	xxx	xxx
Salinisation	0	xxx	x	0	na
Trace elements	xx	xx	xx	xxx	xxx
Organic micropollutants	xx	xx	xxx	xxx	xxx
Acidification	xxx	xx	xx	0	na
Suspended solids	0	na	x	na	xx

xxx Severe effects or impairments
xx   Important effects or impairments
x   Some effects or impairments

0   No effects or impairments
na   Not applicable
[1] Excluding nitrate

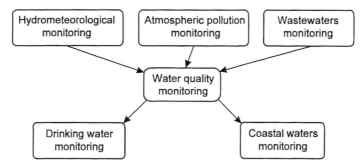

Figure 2.3   The interrelationship between major environmental monitoring systems

trends does not enable description of short-term variations in water quality. Some industrialised countries have several independent sub-systems. There are, for example, eight sub-programmes functioning in Sweden (Ahl, 1985). In Israel, assessment is carried out by 9 agencies (Collin and Sheffer, 1985) and in the USA by 12 federal agencies and more than 100 non-federal groups (Quindy-Hunt *et al.*, 1986). Therefore, when designing a national water quality assessment system, it is first necessary to set the goals and estimate the problems which need to be solved, and then to compare them with the possible information available from the existing systems.

As a rule, national assessment systems are designed, and operate, under the supervision of government agencies responsible for environmental protection and the rational use of water resources, often with the active participation of research institutions belonging to these agencies. However, in a number of countries there are several government agencies which are responsible for different aspects of water quality protection and management. In this case, the government appoints the particular agency responsible for the organisation of the national assessment system. As it is important in the process of water quality monitoring to correlate the data obtained with hydrological information, and to compute fluxes, this work is often carried out by hydrometeorological services as in Chile (Merino *et al.*, 1985), Panama (Muschett, 1985) and the former USSR (Izrael, 1984). The responsibilities of the head agency include, not only the organisation of the basic national monitoring programmes, but also the co-ordination of the work of other programmes supplying information to the general data bank. With the increasing number of independent agencies providing different functions within the monitoring system the management gets more difficult. Therefore, the involvement of all agencies should be well defined.

An assessment system should be designed in such a way that it can provide information, at all levels, i.e. international, national (federal), regional and local. The requirement for additional information from other assessment programmes or agencies also depends on the level of assessment undertaken. The national (or federal) level of assessment, which is based on the programmes of the principal agency, can include data taken from other agencies as well as summarised information taken at regional and local levels.

From a geographical point of view it is ideal if water quality information can be obtained in relation to major river basins. In some countries, e.g. New Zealand, the UK and France (McBride *et al.*, 1985), national assessment programmes are undertaken in this way. However, in a number of countries where the local authorities are organised within regional administrations, the assessments are carried out according to the administrative and territorial divisions of the country, as in the former USSR (Izrael *et al.*, 1978) and Finland (Laaksonen, 1985). In this situation, the information can be transformed into river basin data sets. Ideally this should be done by one of the institutions responsible for processing the data at the national level.

2.5.2 Field operations

The personnel in charge of sample collection, field handling and field measurements must be specially trained for these activities. The choice of personnel for sampling depends on a number of factors which include the geographic features of the region and the system of transportation. For example, in a small country such as Panama (Muschett, 1985), sampling is carried out by laboratory personnel going to the field for nearly a week to take samples, conduct field analyses, and transport samples back to the laboratory by public transport. In countries of larger size which possess a more developed monitoring system, specially trained field personnel often conduct the sampling, as well as train and organise the work of additional personnel. This approach is practised in Canada (Whitlow, 1985). In large countries which have a poor transportation system, relatively more personnel are required. In this situation specialists from hydrometeorological and hydrological stations, for example, may be used, although these personnel often do not possess the necessary training in water sampling for water quality monitoring.

When sampling is carried out by a specialised team, it is desirable that the means of transportation can also be used as a mobile "laboratory" for filtration, field measurements and sample storage. When working in lakes and reservoirs different types of vessels, including small boats, may be used. Special ships with on-board laboratories are only justified for large water bodies for which the travel time exceeds 48 hours and/or if systematic investigations of high intensity are required.

Recently, helicopters have come into use for sampling in remote areas deprived of roads. In addition, special multi-parametric, portable instruments which enable measurements of pH, redox potential, oxygen, conductivity, turbidity, etc., now exist. Automatic water quality stations are also available commercially. However, they still require regular, and sometimes sophisticated, maintenance. Therefore, they are more commonly used for operational surveillance or early warning monitoring.

Everyone participating in sampling should be thoroughly instructed on safety regulations appropriate to their work and assisted in the case of an emergency. Such instruction sessions should be organised routinely and systematically. Special attention should be paid to the careful use of boats. Specific safety requirements should also be followed when sampling during winter ice cover.

2.5.3 Laboratory organisation

Laboratories must be selected or set up to meet the objectives of each assessment programme. Attention should be paid mainly to the choice of analytical methods. The range of concentrations measured by the chosen methods must correspond to the concentrations of the variable in a water body and to the concentrations set by any applicable water quality standards. For example, water quality standards often require the total absence of certain toxic pollutants. In such cases the detection limit of the analytical method must be very low to determine whether the water quality meets the standard. The less sensitive the method, the greater the permitted deviation of the pollutant concentration from the predetermined standard.

During the initial stages of development of a national monitoring system, it is reasonable to focus on the basic variables of water quality which, as a rule, do not require expensive, sophisticated equipment. Gradually, the number of variables measured can be increased in relation to the financial resources of the monitoring agency. The elaborate equipment and technical skills necessary for the measurement of complex variables are not needed in every laboratory.

In many countries, monitoring laboratories are organised on two tiers: regional laboratories (lower level) to conduct basic determinations not requiring very complex equipment, and central laboratories (higher level) to conduct more complex analyses requiring elaborate equipment and well-trained personnel. In addition, the central laboratories often provide the regional laboratories with methodologies and analytical data quality control. The two tier approach is used for laboratories in industrialised countries such as Canada (Whitlow, 1985) and Finland (Laaksonen, 1985) as well as in some developing countries such as Tanzania (Gumbo, 1985). However, in large countries three levels may be necessary, allowing local laboratories to make routine analyses nearer to sampling sites. For example, in Russia there are about 80 local laboratories, headed by 24 regional laboratories. Management of the entire network for national monitoring in Russia is undertaken by the Hydrochemical Institute. This Institute acts as a central reference laboratory whenever it is necessary to conduct either complex or arbitration analyses.

Comparability of water quality data from different laboratories can only be ensured if identical or, at least, similar methods are used. There are many comprehensive standard manuals and guidebooks describing laboratory methods in detail, such as the *GEMS/WATER Operational Guide* (WHO, 1992) and the practical guide to the methods discussed in this volume (Bartram and Ballance, 1996). The use of such guidebooks helps ensure the compatibility of the data supplied to national and global monitoring systems. Manuals developed in countries which have great experience in monitoring design, for example the USA (APHA, 1989), are in common use in a number of countries such as Uruguay (Balmelli and Alciaturi, 1985), Egypt (El-Gamal and Shafik, 1985) and the Fiji Islands (Chandra, 1985).

Progress in analytical chemistry has stimulated the appearance of new, more elaborate and more efficient methods. However, replacement of existing methods often results in data comparability problems with older methods which can cause problems in the statistical analysis of data-time series. Studies of the compatibility of new with old analytical methods must be undertaken. If a new method is accepted into a monitoring programme an overlap period is required where samples are analysed by both the new and the old methods.

2.6   Data processing

Analytical data collected by laboratories, together with the information on sampling sites and hydrological parameters, are usually sent (mostly in coded form) to a data treatment centre. The function, allocation and number of these centres are determined by the volume of information processed, the density of the monitoring network and the geographical peculiarities of the country. Nevertheless, it is reasonable to have a common information centre which summarises the data from the entire country. As some countries participate in the GEMS/WATER project, their data are stored in the GLOWDAT (GLObal Water DATa Management System) format (WHO, 1992) or a similar system (see Chapter 10). The GLOWDAT format was developed on the basis of NAQUADAT (The NAtional Water QUALity Accounting DATa Bank) used in Canada (Whitlow, 1985).

For phase two of GEMS/WATER, which started in the early 1990s, a new data processing software was introduced called RAISON/GEMS (Lam and Swayne, 1991). The RAISON (Regional Analysis by Intelligent Systems on a Microcomputer) system is a fully integrated database, spreadsheet and graphic interpretation package with geographic information (GIS) and expert system capabilities for personal computers. This software is specifically designed for GEMS/WATER applications and is used widely by national centres participating in this project (UNEP/WHO, 1992).

The main objective of a modern data processing centre is the development, replenishment and management of the data bank and the functioning of the automated information system on which it is based. The objectives of such a centre with a data bank already functioning are:

- information coding (for cases where it is submitted in a decoded form),
- logical control of acquired information,
- exchange of coded information with the data processing centre of higher priority,
- entering the data into an information fund or data bank,
- preparation of regular, standard, information materials on the state of water bodies (reviews, annual reports, etc.), and
- satisfaction of consumers' special demands.

During the development of a data bank, the centre must conduct intensive work on the preparation and management of the information and software systems. These tasks determine both the quality of the data handling and of the various forms of output e.g. statistics, trends, charts, bar graphs, etc. However, at the initial stage of the data bank development, attention should be focused principally on the accuracy of the stored information because the forms of output may be improved in the process of data bank exploitation. Chapter 10 describes more fully these data processing activities.

2.7   Data quality control

Data quality control is a complex and time-consuming activity which must be undertaken continuously to ensure meaningful water quality assessments. This is particularly crucial for some of the chemical analyses carried out on water samples, such as dissolved trace elements, pesticides or even ammonia and phosphates. Errors can occur in many operations of the assessment process as indicated in Table 2.10.

2.7.1   Quality control of field work

Detailed descriptions of methods and the appropriate recommendations for field work, field sampling and sample storage are given in Bartram and Ballance (1996). Some basic principles are highlighted in the following sub-sections.

Sampling and sample representativeness
When sampling, it is always necessary to follow recommended procedures to avoid collection of unrepresentative samples. Each method, or piece of sampling apparatus, has appropriate procedures which should be followed, accurately and at every sampling occasion. In addition, simple, basic rules such as avoiding any unnecessary disturbance of the site prior to sampling (e.g. by standing downstream and collecting the sample upstream) must be followed. A full discussion of sampling procedures and precautions is beyond the scope of this chapter as it depends largely on the nature of the monitoring programme and the media to be sampled. Further information is given in the appropriate chapters and in Bartram and Ballance (1996).

Strict observance of the sampling requirements developed for a given site (type of sampler, sampling depth, cross-sectional samples, etc.) usually enables collection of representative samples. Nevertheless, to assure representativeness, it is recommended that replicate samples be taken occasionally to determine temporal (at one point in a certain time interval) and spatial (simultaneously at different points of the given water body, e.g. river

Table 2.10   Some possible sources of errors in the water quality assessment process with special reference to chemical methods

Assessment step	Operation	Possible source of error	Appropriate actions
Monitoring design	Site selection	Station not representative (e.g. poor mixing in rivers)	Preliminary surveys
	Frequency determination	Sample not representative (e.g. unexpected cycles or variations between samples)	
Field operations	Sampling	Sample contamination (micropollutant monitoring)	Decontamination of sampling equipment, containers, preservatives
	Filtration	Contamination or loss	Running field blanks
	Field measurement	Uncalibrated operations (pH, conduct., temperature) Inadequate understanding of hydrological regime	Field calibrations Replicate sampling Hydrological survey
Sample shipments to laboratory	Sample conservation and identification	Error in chemical conservation Lack of cooling Error in biological conservation Error and loss of label Break of container	Field spiking Appropriate field pretreatment Field operator training
Laboratory	Preconcentration	Contamination or loss	Decontamination of laboratory equipment and facilities
	Analysis	Contamination	Quality control of laboratory air, equipment and distilled water
		Lack of sensitivity Lack of calibration	Quality assurance tests (analysis of control sample; analysis of standards)
		Error in data report	Check internal consistency of data (e.g. with adjacent sample, ionic balance etc.)
Computer facility	Data entry and retrieval	Error in data handling	Checks by data interpretation team
Interpretation	Data interpretation	Lack of basic knowledge Ignorance of appropriate statistical methods Omission in data report	Appropriate training of scientists
Publication	Data publication	Lack of communication and dissemination of results to authorities, the public, scientists, etc.	Setting of goals and training to meet the need of decision makers

cross-section) variability. Temporal variability is usually determined in pre-liminary surveys to check diel variations, seasonal variations and the influence of river floods, etc. Groundwater systems are usually more complex and less accessible than rivers or lakes, and obtaining representative samples is often very difficult, as described in Chapter 9.

Sample treatment and storage
All water quality variables should be grouped according to the specific operations preceding analysis (filtration, preservation, types of bottles for storage and transportation of the sample, conditions and permissible time of storage). Some advanced monitoring may require as many as 20 different storage vessels. For each analytical category, accurate observance of the pre-determined requirements of the sample handling is necessary; deviation from these requirements could result in serious errors. The recording of each field operation step is important for quality control, especially if the operator might deviate from predetermined procedures.

Collected samples can be contaminated by inadequately or inappropriately cleaned glassware, filters, filter apparatus, chemicals used for preservation, etc. Thus great care must be taken in the cleaning of equipment and in the checking for purity of chemicals used. Water quality variables which should be determined in the field immediately after sampling need individual aliquots which cannot be used for further analytical work. In addition, field analytical operations should follow a defined sequence in order to avoid contamination. For example, conductivity must not be determined after measurements of pH in the same water sample because concentrated electrolytes from the reference electrode used in the pH determination may enter the sample and affect the conductivity measurement.

During field operations, periodic blank samples (one blank for every ten water samples) are required to determine errors arising from contamination. Usually, for this purpose a distilled water sample is subjected to all the operations carried out for the environmental sample such as filtration, storage and preservation. The blank is then shipped, with the other samples, to the laboratory for analysis. When blank tests show evidence of contamination, additional investigations must be made during the next round of sampling; one blank should be taken for each separate step in the field and laboratory operations to determine the source of contamination. Contamination during sampling is often difficult to detect. Cross checks with highly cleaned, simple samplers (such as PTFE (polytetrafluoroethylene) bottles) should be carried out regularly. To determine the reproducibility of field operations, periodic analysis of duplicate aliquots collected from one sample (i.e. split samples) is recommended.

2.7.2 Analytical quality control

Experts agree that 10 to 20 per cent of resources, including manpower, should be directed towards ensuring the quality of analytical determinations for common water quality variables (WHO, 1992). When trace pollutants (e.g. pesticides and trace elements) are measured, the resources required for quality control may reach 50 per cent. Unfortunately, in many countries this problem is not given adequate attention. This results in unreliable data and hence, unsatisfactory solutions to the water quality problems addressed by the monitoring programme.

To provide high quality analyses, it is necessary to fulfil a number of basic requirements:

(i)   The analytical methods should have characteristics (range of measured concentrations, sensitivity, selectivity) which are adequate for the water body being monitored and must pass an inter-laboratory calibration test.

(ii) The instrumental equipment of the laboratory and the available technical accessories must correspond to the set of analytical methods chosen.

(iii) The laboratory must have adequate conditions for the maintenance of analytical instruments.

(iv) A reliable and steady supply of laboratory reagents, solvents, gases of special grade, as well as standard samples and mixtures must be provided.

(v)   The laboratory personnel should be sufficiently trained and qualified to carry out the necessary analytical operations properly.

(vi) A programme of systematic quality control must be organised.

Each of the above is obligatory for the proper functioning of an analytical laboratory. Requirements (i) to (v) would enable a laboratory to undertake water analyses, but requirement (vi) (analytical quality control (AQC), also known as analytical quality assurance) is necessary to ensure the quality of the data. There are two parts to AQC: intra-laboratory control carried out systematically within a laboratory, which has been well developed in many water quality manuals, and external (inter-laboratory) control carried out periodically (once or twice a year). External (inter-laboratory) control is checked by the laboratory or research institute responsible for the functioning of the monitoring system. For this purpose, several "unknown" control samples are sent to the participating laboratories for analysis and subsequent comparison of the data. If differences in the analytical results are found, the laboratories must then identify and correct the problem. Although these samples may not actually differ from those used in an intra-laboratory control exercise, the concentrations of each component of the samples should be known only to the organisers of the quality control exercise. For further details see Bartram and Ballance (1996).

2.7.3 Control of data storage and treatment

Every stage of data handling increases the risk of introduced errors. Most risk is associated with human error during written transcription or during "keying-in" via a computer keyboard. Such errors can be reduced by using direct electronic recording and data transfer processes, but such options are expensive. Whenever suspect data are encountered (at whatever stage of the data handling process), they must be checked against the original records of the sample analysis. In cases where the data are handled by a central facility, checking involves going back to the original laboratory.

Errors arising during the transcription of data from laboratory notebooks to record books or computer databases can only be reduced by careful checking of the original and copied data for mistakes, and correcting them immediately. This time is well spent, particularly when the data have been entered onto a computer system. Where entry of the data occurs via keyboards, careful and integrated design of raw data recording forms and a computer entry template can substantially reduce input errors. The likelihood of new errors is reduced when data are also handled by electronic means for interpretation and reporting procedures. Some modern analytical equipment eliminates errors by storing its output directly onto computer diskettes.

In many organisations, individuals responsible for data entry have no knowledge of water quality. Therefore, it is important to have the database checked periodically by an expert who is capable of spotting obvious errors. Some computer database systems, either commercially available or custom designed, have the ability to automatically check and flag values which are outside the normally expected ranges (or those set by the user) for each variable. The manipulation of data and the production of routine outputs such as graphs, may also be done by individuals with no special knowledge of water quality. It is, therefore, important that appropriate experts, or specially trained individuals, are used for the final analysis and interpretation of data. This enables a further check to be made for unusual or unlikely values during the interpretation phase.

Another common problem is loss of data due to accidental erasure of computer files. Fortunately, there are several precautions which can be taken to avoid this problem. The raw, unprocessed data should be held on a master file with limited access, possibly controlled by the use of a password system, or protected from manipulation other than copying. In addition, there should always be at least a second, and possibly even a third, copy of the master file (e.g. on diskette) in a location away from the computer, and where it is safe from fire, theft, etc. The back-up file needs to be updated frequently to minimise loss if the main file is erased. Updating can occur daily (for databases receiving daily additions) or weekly (for databases receiving occasional inputs).

2.8   Interpretation and dissemination of data

Interpretation of data and communication of results are the final two steps in an assessment programme. Correctly interpreted data will not be of much use if they are not disseminated to all relevant authorities, scientists, and the public in a form which is readily understandable by, and acceptable to, the target audience. The form and level of data presentation is, therefore, crucial. Often it is advisable to produce two types of data publications: (i) a comprehensive, detailed report containing all relevant data and interpretations thereof and (ii) an executive summary (in an illustrated and simple form) which highlights the major findings. Usually, the interpretation of data is undertaken by specialised professionals such as: (i) the relevant scientists, e.g. hydrologists, hydrobiologists, chemists or geologists, (ii) the data treatment team and (iii) professionals from other organisations such as environmental protection agencies, health authorities, national resource agencies (e.g. energy, fisheries), and in some countries, transportation and agriculture departments.

A tentative list of the major means of data analysis pertinent to each type of water body, (with increasing complexity of monitoring operations) is presented in Table 2.11, specifically for the assessment of chemical water quality. Interpretations should always refer to the objectives, and should also propose improvements, including simplifications, in the monitoring activities, as well as needs for further research and guidelines for environmental planning and economic development. Subsequently, these findings should be discussed with the appropriate local, regional or national authorities and, as required, others such as the industrial development and/or national planning boards. Besides these authorities, results should be communicated to water resource managers, the public, associations for environmental protection, educational institutions, other countries (in the case of transboundary water bodies, e.g. Danube Commission, Great Lakes International Joint Commission, International Rhine Commission) and to international organisations such as the World Health Organization (WHO) and the United Nations Environment Programme (UNEP) for the purpose of international activities such as the GEMS/WATER programme.

During the designing and functioning of a national monitoring system, scientific problems constantly arise which need solutions which cannot be provided by the network personnel (such events occur even when the design has been based on the general experience of the international community). Research teams should be associated with the monitoring system and provide

Table 2.11   Principal means of analysing and reporting chemical data

Level[1]	Rivers	Lakes	Reservoirs	Groundwaters
A	Basic statistics on $C(t)$ at station	<———Vertical $C(z)$ profiles at ———> max. stratification		Time variations at station $C(t)$
		<———Trends ($\overline{C}$ at over-turn) ———>		
B	Concentration vs water discharge relationships $C = f(Q)$	<———Sediment mapping ———> $C(x,y)$		Mapping $C(x,y)$ trend $C(t)$
	Loads $(C.Q)$ Statistics on loads Trends $C(t)$	Seasonal variations $C(z,t)$	Seasonal variations $C(x,z,t)$	
C	<——————————— Trends (flow corrected and/or seasonally adjusted)————————————>			
	Profiles $C(x)$ Maps $C(x,y)$	<——— Water quality mapping ———> $C(x,y,z,t)$		Mapping evolution $C(x,y,t)$
	Models	<——— Models (eutrophication) ———>		Models (dispersion)

$C(t)$ Temporal variation in concentrations
$\overline{C}$ Average concentration
$C(x)$ Longitudinal variation in concentrations
$C(z)$ Vertical variation in concentrations
Q Water discharge

[1]   Levels A,B,C are increasing orders of assessment programme complexity

an important function in problem solving. They can be grouped into an institute specially founded for this purpose or dispersed in different research units. The basic functions of such teams are: (i) to give the necessary methodological and advisory assistance in order to optimise the monitoring system and take into account major scientific advances, (ii) to develop and refine methods, and (iii) to train personnel.

2.9   Recommendations

Water quality assessment should always be seen in the wider context of the management of water resources, encompassing both the quality and quantity aspects. The usefulness of the information obtained from monitoring is severely limited unless an administrative and legal framework (together with an institutional and financial commitment to appropriate follow-up action) exists at local, regional, or even international, level. Four main reasons for obtaining inadequate information from assessment programmes have been defined, for groundwaters, by Wilkinson and Edworthy (1981). These reasons are equally applicable to surface waters and are as follows:

- The objectives of the assessment were not properly defined.
- The monitoring system was installed with insufficient knowledge of the water body.
- There was inadequate planning of sample collection, handling, storage and analysis.
- Data were poorly archived.

A further reason could be added:

- Data were improperly interpreted and reported.

To ensure that these mistakes are avoided the basic rules for a successful assessment programme are proposed below.

The ten basic rules for a successful assessment progamme

1. The objectives must be defined first and the programme adapted to them and not *vice versa* (as was often the case for multi-purpose monitoring in the past). Adequate financial support must then be obtained.

2. The type and nature of the water body must be fully understood (most frequently through preliminary surveys), particularly the spatial and temporal variability within the whole water body.

3. The appropriate media (water, particulate matter, biota) must be chosen.

4. The variables, type of samples, sampling frequency and station location must be chosen carefully with respect to the objectives.

5. The field, analytical equipment and laboratory facilities must be selected in relation to the objectives and not *vice versa*.

6. A complete, and operational, data treatment scheme must be established.

7. The monitoring of the quality of the aquatic environment must be coupled with the appropriate hydrological monitoring.

8. The analytical quality of data must be regularly checked through internal and external control.

9. The data should be given to decision makers, not merely as a list of variables and their concentrations, but interpreted and assessed by experts with relevant recommendations for management action.

10. The programme must be evaluated periodically, especially if the general situation or any particular influence on the environment is changed, either naturally or by measures taken in the catchment area.

2.10   References

Ahl, K.T. 1985 Water quality monitoring in Sweden. *Wat. Qual. Bull.*, **10**(2), 82-90.

Alabaster, J.S. [Ed.] 1977 *Biological Monitoring of Inland Fisheries*. Applied Science Publishers Ltd., London, 226 pp.

APHA 1989 *Standard Methods for the Examination of Water and Wastewater*. 17th edition, American Public Health Association, Washington DC., 1268 pp.

Balmelli, L.J. and Alciaturi, F.A. 1985 Water quality monitoring in Uruguay.

Wat. Qual. Bull., **10**(2), 96-97.

Barcelona, M.J., Gibb, J.P., Helfrich, J.A. and Garske, E.E. 1985 *Practical Guide for Groundwater Sampling*. ISWS Contract Report 374, Illinois State Water Survey, Champaign, Illinois, 94 pp.

Bartram, J. and Ballance, R. [Eds] 1996 *Water Quality Monitoring: A Practical Guide to the Design and Implementation of Freshwater Quality Studies and Monitoring Programmes*. Chapman and Hall, London.

Chandra, A. 1985 Water quality monitoring in the Fiji Islands. *Wat. Qual. Bull.*, **10**(4), 197-203, 217.

Collin, M. and Sheffer, S.M. 1985 Water quality monitoring in Israel. *Wat. Qual. Bull.*, **10**(4), 186-188.

El-Gamal, A. and Shafik, Y. 1985 Monitoring of pollutants discharging to the River Nile and their effect on river water quality. *Wat. Qual. Bull.*, **10**(3), 111-115, 161.

Gumbo, F.J. 1985 Water quality monitoring in Tanzania. *Wat. Qual. Bull.*, **10**(4), 174-180, 215-216.

Hakanson, L. 1977 *Sediments as Indicators of Contamination. Investigation in the Four Largest Swedish Lakes*. Naturvarsverkets Kimnologiska Undersökning Report 92, Uppsala, 159 pp.

Helmer, R. 1994 Water quality monitoring: national and international approaches. In: *Hydrological, Chemical and Biological Processes of Transfer, Motion and Transport of Contaminants in Aquatic Environments* (Proceedings of the Rostov-on-Don Symposium, May 1993). IAHS Publication No. **219**, International Association of Hydrological Sciences, Wallingford, UK.

Hem, J.D. 1989 *Study and Interpretation of the Chemical Characteristics of Natural Water*. Water Supply Paper 2254, 3rd edition, US Geological Survey, Washington, D.C., 263 pp.

Hirsch, R.M., Alley, W.M. and Wilber, W.G. 1988 *Concepts for a National Water-Quality Assessment Program*. U.S. Geological Survey Circular 1021. United States Geological Survey, Denver, CO, 42 pp.

Izrael, Yu.A. 1984 *Ecology and Control of the Environment State.* Gidrometeoizdat, Moscow, 560 pp, [In Russian].

Izrael, Yu.A., Gasilina, N.K., Rovinskiy, F.Ya. and Philippova, L.M. 1978 *Realization of the Environmental Pollution Monitoring System in the USSR*. Gidrometeoizdat, Leningrad, 115 pp, [In Russian].

Krenkel, P.A. and Novotny, V. 1980 *Water Quality Management*. Academic Press, New York.

Laaksonen, R. 1985 Water quality monitoring in Finland. *Wat. Qual. Bull.*, **10**(3), 140-145, 161.

Lam, D.C.L. and Swayne, D.A. 1991 Integrating database, spreadsheet, graphics, GIS, statistics, simulation models and expert systems: experiences with the RAISON system on microcomputers. In: *Decision Support Systems*, NATO ASI Series, Vol. G26, Springer-Verlag, Berlin, Heidelberg.

McBride, G.B., Smith, D.G. and Pridmore, R.D. 1985 Water quality monitoring in New Zealand. *Wat. Qual. Bull.*, **10**(2), 91-95, 105.

Merino, R., Sandoval, R. and Grilli, A. 1985 Monitoring of water quality in Chile. *Wat. Qual. Bull.*, **10**(3), 116-119.

Meybeck, M. 1994 De la qualité des eaux à l'état de santé des écosystèmes aquatiques: pourquoi, comment, où? In: *Les variables biologiques: des indicateurs de l'état de santé des écosystèmes aquatiques*, Actes du colloque 2-3 Nov. 1994, Min. de l'Environnement, Paris, 187-199.

Muschett, D. 1985 Water quality monitoring in Panama. *Wat. Qual. Bull.*, **10**(2), 80-81, 105.

Quindy-Hunt, M.S., McLaughlin, R.D. and Quintanilha, A.T. 1986 *Instrumentation for Environmental Monitoring. Volume 2, Water.* 2nd edition, Lawrence Bercley Lab. Environmental Instrumentation Survey, John Wiley and Sons, New York, 982 pp.

Sanders, T.G., Ward, R.C., Loftis, J.C., Steele, T.D., Adrian, D.D. and Yevjevich, V. 1983 *Design of Networks for Monitoring Water Quality.* Water Resources Publications, Littleton, Colorado, 323 pp.

UNEP/WHO 1992 *Report of the RAISON/GEMS Software Expert Review Meeting, May11-15, 1992, Burlington, Ontario, Canada.* GEMS Report Series No. 14. United Nations Environment Programme, Nairobi.

UNESCO/WHO 1978 *Water Quality Surveys. A Guide for the Collection and Interpretation of Water Quality Data.* Studies and Reports in Hydrology 23, United Nations Educational, Scientific and Cultural Organization, Paris, 350 pp.

Whitlow, S.H. 1985 Water quality assessment in Canada. *Wat. Qual. Bull.*, **10**(2), 75-79.

WHO 1992 *GEMS/WATER Operational Guide.* Third edition. World Health Organization, Geneva.

Wilkinson, W.B. and Edworthy, K.J. 1981 Groundwater quality systems — money wasted? In: W. van Duijvenbooden, P. Glasbergen and H. van Lelyveld [Eds] *Quality of Groundwater.* Proceedings of an International Symposium, Noordwijkerhout. Studies in Environmental Science No. 17, Elsevier, Amsterdam, 629-642.

WMO 1988 *Manual on Water Quality Monitoring.* WMO Operational Hydrology Report, No. 27, WMO Publication No. 680, World Meteorological Organization, Geneva, 197 pp.

Yasuno, M. and Whitton, B.A. 1988 [Eds] *Biological Monitoring of Environmental Pollution.* Tokai University Press, Tokyo.

Chapter 3*

SELECTION OF WATER QUALITY VARIABLES

3.1 Introduction

The selection of variables for any water quality assessment programme depends upon the objectives of the programme (see Chapters 1 and 2). Appropriate selection of variables will help the objectives to be met, efficiently and in the most cost effective way. The purpose of this chapter is to provide information which helps the appropriate selection of variables. Each variable is discussed with respect to its origins, sources, behaviour and transformations in the aquatic system, the observed ranges in natural and polluted freshwaters, the role of the variable in assessment programmes, and any special handling or treatment of samples that is required. The final section of this chapter suggests some combinations of variables which might be used for different water quality assessment purposes. These can be used as a basis for developing individual programmes.

The methods employed to measure the selected variables depend on access to equipment and reagents, availability of technical staff and their degree of expertise, and the level of accuracy required by the objectives of the programme (see Chapter 2). A summary of the principal analytical methods for major variables is given in Table 3.1 and a summary of pre-treatment and storage of samples for different analyses is given in Table 3.2. Detailed descriptions of sampling and analytical methods are available in the companion volume to this guidebook by Bartram and Ballance (1996) and in a number of standard reference guides published by various international organisations and programmes, or national agencies (e.g. Semenov, 1977; WHO, 1992; NIH, 1987–88; Keith, 1988; APHA, 1989; AOAC, 1990). In addition a world-wide federation of national standards bodies and international organisations, the International Standards Organization (ISO), publishes a series of approved "International Standards" which includes methods for determining water quality. Further detailed information on the study and interpretation of chemical characteristics in freshwaters is available in Hem (1989), Environment Canada (1979) and many other specialist texts.

3.2 Hydrological variables

Determining the hydrological regime of a water body is an important aspect of a water quality assessment. Discharge measurements, for example, are

This chapter was prepared by D. Chapman and V. Kimstach

Table 3.1 Analytical methods for determination of major chemical variables

Variable	Simple							Advanced						Sophisticated			
	Gravimetric	Titrimetric	Visual	Photometric	Electrochem. probe	Flame photometry	UV-VIS and IR	Fluorimetry	AES	AAS	GC	Flow injection	Stripping VA	ICP-AES	IC	LC	GC/MS
Residue	L																
Suspended matter			F	FL													
Conductivity					FL												
pH			F		FL												
Acidity, alkalinity		L			FL												
Eh					F												
Dissolved oxygen		L			F												
CO$_2$		L			F												
Hardness		L		L													
Chlorophyll a				L				FL									
Nutrients			F	L	FL							L			L		
Organic matter (TOC, COD,BOD)		L					L										
Major cations			F	L	FL	L	L			L				L			
Major anions			F	L	FL										L		
Sulphide		L		L	FL										L		
Silica		L		L													
Fluoride		L		L	FL										L		
Boron		L		L										L			
Cyanide		L		L								L					
Trace elements			F	L					L	L		L	L	L		L	

Continued

Table 3.1 Continued

Variable	Simple						Advanced						Sophisticated				
	Gravimetric	Titrimetric	Visual	Photometric	Electrochem. probe	Flame photometry	UV-VIS and IR	Fluorimetry	AES	AAS	GC	Flow injection	Stripping VA	ICP-AES	IC	LC	GC/MS
Mineral oil	L						L	L									
Phenols				L							L					L	L
Pesticides				L							L					L	L
Surfactants			F	L									L				
Other organic micropollutants								L			L					L	L

F   Field methods
L   Laboratory methods

UV-VIS   Ultraviolet and visual spectrophotometry
IR   Infra-red spectrography
AES   Atomic emission spectrography
AAS   Atomic absorption spectrophotometry
GC   Gas chromatography
VA   Voltammetry

ICP-AES   Inductively coupled plasma atomic emission spectrometry
IC   Ion chromatography
LC   Liquid chromatography
GC/MS   Gas chromatography/mass spectrometry

TOC   Total organic carbon
COD   Chemical oxygen demand
BOD   Biochemical oxygen demand

Table 3.2 Pretreatment and storage requirements of samples for laboratory determination of chemical variables (see text for further details)

Variable	Pretreatment						Type of bottles		Conditions of storage			Max. time of storage prior to analysis				
	None	Filtration	Chemical stabilisation	Acidi-fication	Alkalini-sation	Solvent extraction	Glass	Polyethylene	Dark	Cold (approx. 4 °C)	Frozen (max. −15 °C)	Minimum possible	24 hours	3 days	1 week	3 weeks
Residue	x						x					x				
Suspended matter	x						x					x				
Conductivity		x					x	x					x			
pH	x						x	x					x			
Acidity, alkalinity	x						x	x				x				
DO (Winkler method)			x				x	x	x			x				
CO_2	x								x			x				
Hardness (general)		x					x									
Chlorophyll a		x	x				x	x	x	x		x				
Chlorophyll a and POC		x						x	x	x	x					x[1]
Nutrients[2]			x				x		x	x						
TOC							x		x	x			x			
COD				x			x		x	x					x	
BOD	x						x		x	x			x			
Na^+, K^+		x						x				x				
Ca^{2+}, Mg^{2+}		x		x				x								
Major anions	x						x	x	x							
Sulphide			x				x	x	x							
Silica		x						x	x					x		
Fluoride	x							x		x					x	
Boron	x							x							x	
Cyanide					x			x								
Trace elements (dissolved)		x		x				x		x		x				

Continued

Table 3.2 Continued

Variable	Pretreatment						Type of bottles		Conditions of storage			Max. time of storage prior to analysis				
	None Filtration	Chemical stabilisation	Acidi-fication	Alkalini-sation	Solvent extraction		Glass	Polyethylene	Dark	Cold (approx. 4 °C)	Frozen (max. −15 °C)	Minimum possible	24 hours	3 days	1 week	3 weeks
Mineral oil					x		x		x	x						
Phenols		x					x		x	x				x		
Pesticides			x				x				x					x
Other organic micropollutants					x		x			x				x		

Where no indication is given under column headings, no special conditions of pretreatment or storage are necessary.

Sample bottles for many variables require special cleaning, particularly those for trace metals and organic micropollutants. Requirements for special cleaning are described in operational manuals for analytical methods (e.g. WHO, 1992).

COD   Chemical oxygen demand
BOD   Biochemical oxygen demand
DO   Dissolved oxygen
POC   Particulate organic carbon
TOC   Total organic carbon
1   When frozen
2   NO_3^-, NH_4^+, PO_4^{3-}, total P

necessary for mass flow or mass balance calculations and as inputs for water quality models.

3.2.1 Velocity

The velocity (sometimes referred to as the flow rate) of a water body can significantly affect its ability to assimilate and transport pollutants. Thus measurement of velocity is extremely important in any assessment programme. It enables the prediction of movement of compounds (particularly pollutants) within water bodies, including groundwaters. For example, knowledge of water velocity enables the prediction of the time of arrival downstream, of a contaminant accidentally discharged upstream.

Water velocity can vary within a day, as well as from day to day and season to season, depending on hydrometeorological influences and the nature of the catchment area. It is important, therefore, to record the time when measurements are taken and every attempt should be made to measure velocity at the same sites as other water quality samples are collected. Velocity is determined (in m s^{-1}) with current meters or tracers, such as dyes. Measurements are usually averaged over a period of 1–2 minutes.

3.2.2 Discharge

The discharge is the volume flowing for a given period of time. For rivers, it is usually expressed as m^3 s^{-1} or m^3 a^{-1}. The amount of suspended and dissolved matter in a water body depends on the discharge and is a product of the concentration and the discharge. Natural substances arising from erosion (suspended matter) increase in concentration exponentially with increased discharge (see Figure 6.11A and section 6.3.3). Substances introduced artificially into a water body, such as trace elements and organic matter, tend to occur at decreasing concentrations with increasing river discharge. If a pollutant is introduced into a river at a constant rate, the concentration in the receiving water can be estimated from the quantity input divided by the river discharge (see the example in Figure 6.13). Sedimentation and resuspension (see Chapter 4) can, however, affect this simple relationship.

Discharge can be estimated from the product of the velocity and the cross-sectional area of the river. It should be measured at the time of sampling and preferably at the same position as water samples are taken. As cross-sectional area varies with different discharges, a series of measurements are needed in relation to the different discharges. Measurements of depth across a transect of the water body can be used to obtain an approximate cross-sectional area. Specific methods for calculating discharge are available in WMO (1974, 1980).

3.2.3   Water level

Measurement of water level is important to determine the hydrological regime of lakes, reservoirs and groundwaters and the interaction between groundwaters and surface waters. Measurement of water level is necessary for mass flow calculations in lakes and groundwaters and must be measured at the time and place of water sampling.

Water can flow to or from an aquifer which is in continuity with a river, depending on the relative water levels in the river and aquifer. Low water levels in the river can induce groundwater flow to the river, and high water levels can reverse the flow and produce losses from the river to the aquifer. Similarly, when groundwater levels are low (or deep) surface water infiltrates downwards to the water table (see Chapter 9). Depending on the relative water levels in the aquifer and river, stretches which gain or lose may occur in the same river. Also a particular stretch may be gaining at one time of year and losing at another, as river levels change with the seasons. As the river water and groundwater may be of very different qualities, significant variations in water quality may be experienced in wells close to rivers, and in the river itself. Measurement of groundwater levels is particularly important in relation to saline intrusion.

3.2.4   Suspended matter dynamics

Suspended particulate matter consists of material originating from the surface of the catchment area, eroded from river banks or lake shores and resuspended from the bed of the water body. Measurement of suspended matter transport is particularly important where it is responsible for pollutant transport and in such cases its measurements should be undertaken frequently (see Chapter 4). Usually sediment concentration and load increase exponentially with discharge (see Figure 6.11A). Particles may also settle, or be resuspended, under different discharge conditions.

Suspended matter concentrations should be measured along with the other hydrological variables. In rivers of uniform cross-section, a single sample point may be adequate, whereas for other rivers, multiple point or multiple depth, integrated sampling is necessary. Such samples should be taken at the same points as water velocity measurements and other water quality samples. In addition to analysing suspended matter as described in sections 3.3.4 and 3.3.5, grain size should be determined. Whenever possible, samples from bottom sediments should also be examined.

3.3   General variables

3.3.1   Temperature

Water bodies undergo temperature variations along with normal climatic fluctuations. These variations occur seasonally and, in some water bodies, over periods of 24 hours. Lakes and reservoirs may also exhibit vertical stratification of temperature within the water column (see Chapters 7 and 8).

The temperature of surface waters is influenced by latitude, altitude, season, time of day, air circulation, cloud cover and the flow and depth of the water body. In turn, temperature affects physical, chemical and biological processes in water bodies and, therefore, the concentration of many variables. As water temperature increases, the rate of chemical reactions generally increases together with the evaporation and volatilisation of substances from the water. Increased temperature also decreases the solubility of gases in water, such as O_2, CO_2, N_2, CH_4 and others. The metabolic rate of aquatic organisms is also related to temperature, and in warm waters, respiration rates increase leading to increased oxygen consumption and increased decomposition of organic matter. Growth rates also increase (this is most noticeable for bacteria and phytoplankton which double their populations in very short time periods) leading to increased water turbidity, macrophyte growth and algal blooms, when nutrient conditions are suitable.

Surface waters are usually within the temperature range 0 °C to 30°C, although "hot springs" may reach 40 °C or more. These temperatures fluctuate seasonally with minima occurring during winter or wet periods, and maxima in the summer or dry seasons, particularly in shallow waters. Abnormally high temperatures in surface water can arise from thermal discharges, usually from power plants, metal foundries and sewage treatment plants. Groundwater usually maintains a fairly constant temperature which, for surficial aquifers, is normally close to the mean annual air temperature. However, deep aquifers have higher temperatures due to the earth's thermal gradient.

Temperature should be measured *in situ*, using a thermometer or thermistor. Some meters designed to measure oxygen or conductivity can also measure temperature. As temperature has an influence on so many other aquatic variables and processes, it is important always to include it in a sampling regime, and to take and record it at the time of collecting water samples. For a detailed understanding of biological and chemical processes in water bodies it is often necessary to take a series of temperature measurements throughout the depth of the water, particularly during periods of temperature stratification in lakes and reservoirs (see Chapters 7 and 8). This can be done with a recording thermistor linked to a pressure transducer, directly reading temperature with depth, or by reversing thermometers built into a string of sampling bottles, or by direct, rapid measurements of water samples taken at discrete depths.

3.3.2   Colour

The colour and the turbidity (see section 3.3.5) of water determine the depth to which light is transmitted. This, in turn, controls the amount of primary productivity that is possible by controlling the rate of photosynthesis of the algae present. The visible colour of water is the result of the different wave-lengths not absorbed by the water itself or the result of dissolved and particulate substances present. It is possible to measure both true and appar-ent colour in water. Natural minerals such as ferric hydroxide and organic substances such as humic acids give true colour to water. True colour can only be measured in a sample after filtration or centrifugation. Apparent col-our is caused by coloured particulates and the refraction and reflection of light on suspended particulates. Polluted water may, therefore, have quite a strong apparent colour.

Different species of phyto- and zooplankton can also give water an appar-ent colour. A dark or blue-green colour can be caused by blue-green algae, a yellow-brown colour by diatoms or dinoflagellates and reds and purples by the presence of zooplankton such as *Daphnia* sp. or copepods.

Colour can be measured by the comparison of water samples with a series of dilutions of potassium chloroplatinate and crystalline cobaltous chloride. The units are called platinum-cobalt units based on 1 mg l^{-1}Pt. Natural waters can range from < 5 in very clear waters to 300 units in dark peaty waters. The total absorbance colour (TAC) method measures integrated absorbance of the filtered sample (pH 7.6) between 400 and 700 nm and the true colour (TUC) is determined by measuring the absorbance at 465 nm. One TAC unit is equivalent to the colour of 2 mg l^{-1} Pt. The TAC units range from 1 to 250. As the compounds determining the colour of the water are not very stable, measurements should be made within two hours of collection.

3.3.3   Odour

Water odour is usually the result of labile, volatile organic compounds and may be produced by phytoplankton and aquatic plants or decaying organic matter. Industrial and human wastes can also create odours, either directly or as a result of stimulating biological activity. Organic compounds, inorganic chemicals, oil and gas can all impart odour to water although an odour does not automatically indicate the presence of harmful substances.

Usually, the presence of an odour suggests higher than normal biological activity and is a simple test for the suitability of drinking water, since the human sense of smell is far more sensitive to low concentrations of substances than human taste. Warm temperatures increase the rate and production of odour-causing metabolic and decay products. Different levels of pH may also affect the rate of chemical reactions leading to the production of odour.

Odour can be measured in terms of the greatest dilution of a sample, or the number of times a sample has to be halved with odour-free water, that yields the least definitely perceptible odour. The former method is known as the Threshold Odour Number (TON) and the latter method as the Odour Intensity Index (OII). Both methods suffer from the subjective variability of different human judges.

3.3.4   Residue and total suspended solids

The term "residue" applies to the substances remaining after evaporation of a water sample and its subsequent drying in an oven at a given temperature. It is approximately equivalent to the total content of dissolved and suspended matter in the water since half of the bicarbonate (the dominant anion in most waters) is transformed into CO_2 during this process. The term "solids" is widely used for the majority of compounds which are present in natural waters and remain in a solid state after evaporation (some organic compounds will remain in a liquid state after the water has evaporated). Total suspended solids (TSS) and total dissolved solids (TDS) correspond to non-filterable and filterable residue, respectively. "Fixed solids" and "volatile solids" correspond to the remainder after oven-drying, and to the loss after oven-drying at a given temperature, respectively. The latter two determinations are now less frequently carried out.

Residue determination is based on gravimetric measurement after following the appropriate procedures, i.e. filtration, evaporation, drying and ignition. The results of residue determination depend on the precise details of these procedures. Total suspended solids are the solids retained on a standard filter (usually a glass fibre "GF/C" grade) and dried to a constant weight at 105 °C (Bartram and Ballance, 1996).

To achieve reproducibility and comparability, care must be taken in following the appropriate methods. For further details see WHO (1992) and Bartram and Ballance (1996). Samples should preferably be kept in hard-glass bottles until analysis can be performed, although polythene bottles can be used if the suspended material does not stick to the walls of the bottle. To help prevent precipitation occurring in the sample bottles they should be completely filled and then analysed as soon as possible after collection.

3.3.5 Suspended matter, turbidity and transparency

The type and concentration of suspended matter controls the turbidity and transparency of the water. Suspended matter consists of silt, clay, fine particles of organic and inorganic matter, soluble organic compounds, plankton and other microscopic organisms. Such particles vary in size from approximately 10 nm in diameter to 0.1 mm in diameter, although it is usually accepted that suspended matter is the fraction that will not pass through a 0.45 μm pore diameter filter (see Chapter 4). Turbidity results from the scattering and absorption of incident light by the particles, and the transparency is the limit of visibility in the water. Both can vary seasonally according to biological activity in the water column and surface run-off carrying soil particles. Heavy rainfall can also result in hourly variations in turbidity. At a given river station turbidity can often be related to TSS, especially where there are large fluctuations in suspended matter. Therefore, following an appropriate calibration, turbidity is sometimes used as a continuous, indirect measurement for TSS.

Transparency can be measured easily in the field and is, therefore, included in many regular sampling programmes, particularly in lakes and reservoirs, to indicate the level of biological activity. It is determined by lowering a circular disc, called a Secchi disc, on a calibrated cable into the water until it just disappears. The depth at which it disappears, and just reappears, is recorded as the depth of transparency. A Secchi disc is usually 20–30 cm in diameter (although the result is not affected by the disc diameter), and coloured white or with black and white sectors.

Turbidity should be measured in the field but, if necessary, samples can be stored in the dark for not more than 24 hours. Settling during storage, and changes in pH leading to precipitation, can affect the results during storage. The most reliable method of determination uses nephelometry (light scattering by suspended particles) by means of a turbidity meter which gives values in Nephelometric Turbidity Units (NTU). Normal values range from 1 to 1,000 NTU and levels can be increased by the presence of organic matter pollution, other effluents, or run-off with a high suspended matter content. A visual method of determination is also available in Jackson Turbidity Units (JTU), which compares the length of the light path through the sample against a standard suspension mixture.

3.3.6 Conductivity

Conductivity, or specific conductance, is a measure of the ability of water to conduct an electric current. It is sensitive to variations in dissolved solids (see section 3.3.4), mostly mineral salts. The degree to which these dissociate into ions, the amount of electrical charge on each ion, ion mobility and the temperature of the solution all have an influence on conductivity. Conductivity is expressed as microsiemens per centimetre (μS cm^{-1}) and, for a given water body, is related to the concentrations of total dissolved solids and major ions (see Figure 10.14). Total dissolved solids (in mg l^{-1}) may be obtained by multiplying the conductance by a factor which is commonly between 0.55 and 0.75. This factor must be determined for each water body, but remains approximately constant provided the ionic proportions of the water body remain stable. The multiplication factor is close to 0.67 for waters in which sodium and chloride dominate, and higher for waters containing high concentrations of sulphate.

The conductivity of most freshwaters ranges from 10 to 1,000 μS cm^{-1} but may exceed 1,000 μS cm^{-1}, especially in polluted waters, or those receiving large quantities of land run-off. In addition to being a rough indicator of mineral content when other methods cannot easily be used, conductivity can be measured to establish a pollution zone, e.g. around an effluent discharge, or the extent of influence of run-off waters. It is usually measured *in situ* with a conductivity meter, and may be continuously measured and recorded. Such continuous measurements are particularly useful in rivers for the management of temporal variations in TDS and major ions.

3.3.7   pH, acidity and alkalinity

The pH is an important variable in water quality assessment as it influences many biological and chemical processes within a water body and all processes associated with water supply and treatment. When measuring the effects of an effluent discharge, it can be used to help determine the extent of the effluent plume in the water body.

The pH is a measure of the acid balance of a solution and is defined as the negative of the logarithm to the base 10 of the hydrogen ion concentration. The pH scale runs from 0 to 14 (i.e. very acidic to very alkaline), with pH 7 representing a neutral condition. At a given temperature, pH (or the hydrogen ion activity) indicates the intensity of the acidic or basic character of a solution and is controlled by the dissolved chemical compounds and biochemical processes in the solution. In unpolluted waters, pH is principally controlled by the balance between the carbon dioxide, carbonate and bicarbonate ions (see Figure 3.1) as well as other natural compounds such as humic and fulvic acids. The natural acid–base balance of a water body can be affected by industrial effluents and atmospheric deposition of acid-forming substances. Changes in pH can indicate the presence of certain effluents, particularly when continuously measured and recorded, together with the conductivity of

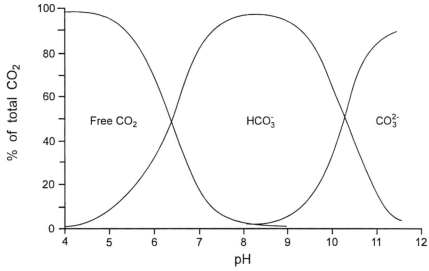

Figure 3.1   The relative proportions of different forms of inorganic carbon in relation to the pH of water under normal conditions

a water body. Diel variations in pH can be caused by the photosynthesis and respiration cycles of algae in eutrophic waters (see Figure 6.19). The pH of most natural waters is between 6.0 and 8.5, although lower values can occur in dilute waters high in organic content, and higher values in eutrophic waters, groundwater brines and salt lakes.

Acidity and alkalinity are the base- and acid-neutralising capacities (ANC) of water and are usually expressed as mmol l^{-1}. When the water has no buffering capacity they are inter-related with pH. However, as most natural waters contain weak acids and bases, acidity and alkalinity are usually determined as well as pH. The acidity of water is controlled by strong mineral acids, weak acids such as carbonic, humic and fulvic, and hydrolising salts of metals (e.g. iron, aluminium), as well as by strong acids. It is determined by titration with a strong base, up to pH 4 (free acidity) or to pH 8.3 (total acidity). The alkalinity of water is controlled by the sum of the titratable bases. It is mostly taken as an indication of the concentration of carbonate, bicarbonate and hydroxide, but may include contributions from borate, phosphates, silicates and other basic compounds. Waters of low alkalinity (< 24 ml l^{-1} as $CaCO_3$) have a low buffering capacity and can, therefore, be susceptible to alterations in pH, for example from atmospheric, acidic deposition. Alkalinity is determined by titration. The amount of strong acid needed to lower the pH of a sample to 8.3 gives the free alkalinity, and to pH 4 gives the total alkalinity (see also sections 3.3.10 and 3.6.5).

Ideally, pH should be determined *in situ*, or immediately after the sample is taken, as so many natural factors can influence it. Accurate measurement of pH is usually undertaken electrometrically with a glass electrode, many of which are suitable for field use and for continuous measurement and recording. A rough indication of pH can be obtained colorimetrically with indicator dyes. As pH is temperature dependent, the water temperature must also be measured in order to determine accurately the pH. If field measurement is not possible, samples must be transported to the laboratory in completely full, tightly stoppered bottles with no preservatives added.

3.3.8   Redox potential

The redox potential (Eh) characterises the oxidation–reduction state of natural waters. Ions of the same element but different oxidation states form the redox-system which is characterised by a certain value. Organic compounds can also form redox-systems. The co-existence of a number of such systems leads to an equilibrium which determines the redox-state of the water and is, in turn, characterised by the Eh value. Oxygen, iron and sulphur, as well as some organic systems are the most influential in determining Eh. For example, Eh values increase and may reach + 700 mV when dissolved oxygen concentrations increase. The presence of hydrogen sulphide is usually associated with a sharp decrease in Eh (down to – 100 mV or more) and is evidence of reducing conditions.

The Eh may vary in natural waters from – 500 mV to + 700 mV. Surface waters and groundwaters containing dissolved oxygen are usually characterised by a range of Eh values between + 100 mV and + 500 mV. The Eh of mineral waters connected with oil deposits is significantly lower than zero and may even reach the limit value of – 500 mV.

Redox potential is determined potentiometrically and may be measured *in situ* in the field. Considerable difficulty has been experienced by many workers in obtaining reliable Eh measurements. Therefore, the results and interpretation of any Eh measurements should be treated with caution. As Eh depends on the gas content of the water it can be very variable when the water is in contact with air. Therefore, determination of Eh should be made immediately after sampling whenever *in situ* determination is not possible, and for groundwater it is recommended that Eh is measured "in-line" in the flowing discharge of a pump.

3.3.9   Dissolved oxygen

Oxygen is essential to all forms of aquatic life, including those organisms responsible for the self-purification processes in natural waters. The oxygen content of natural waters varies with temperature, salinity, turbulence, the photosynthetic activity of algae and plants, and atmospheric pressure. The solubility of oxygen decreases as temperature and salinity increase. In freshwaters dissolved oxygen (DO) at sea level ranges from 15 mg l^{-1} at 0 °C to 8 mg l^{-1} at 25 °C. Concentrations in unpolluted waters are usually close to, but less than, 10 mg l^{-1}. Dissolved oxygen can also be expressed in terms of percentage saturation, and levels less than 80 per cent saturation in drinking water can usually be detected by consumers as a result of poor odour and taste.

Variations in DO can occur seasonally, or even over 24 hour periods, in relation to temperature and biological activity (i.e. photosynthesis and respiration) (see Figures 6.19 and 6.20). Biological respiration, including that related to decomposition processes, reduces DO concentrations. In still waters, pockets of high and low concentrations of dissolved oxygen can occur depending on the rates of biological processes (see Figure 7.8). Waste discharges high in organic matter and nutrients can lead to decreases in DO concentrations as a result of the increased microbial activity (respiration) occurring during the degradation of the organic matter (see Figures 6.17 and 6.20B). In severe cases of reduced oxygen concentrations (whether natural or man-made), anaerobic conditions can occur (i.e. 0 mg l^{-1} of oxygen), particularly close to the sediment–water interface as a result of decaying, sedimenting material.

Determination of DO concentrations is a fundamental part of a water quality assessment since oxygen is involved in, or influences, nearly all chemical and biological processes within water bodies. Concentrations below 5 mg l^{-1} may adversely affect the functioning and survival of biological communities and below 2 mg l^{-1} may lead to the death of most fish. The measurement of DO can be used to indicate the degree of pollution by organic matter, the destruction of organic substances and the level of self-purification of the water. Its determination is also used in the measurement of biochemical oxygen demand (BOD) (see section 3.5.3).

Dissolved oxygen is of much more limited use as an indicator of pollution in groundwater, and is not useful for evaluating the use of groundwater for normal purposes. In addition, the determination of DO in groundwater requires special equipment and it has not, therefore, been widely carried out. Nevertheless, measurement of DO is critical to the scientific understanding of the potential for chemical and biochemical processes in groundwater. Water that enters groundwater systems as recharge can be expected to contain oxygen at concentrations similar to those of surface water in contact with the atmosphere. Organic matter or oxidisable minerals present in some aquifers rapidly deplete the dissolved oxygen. Therefore, in aquifers where organic materials are less plentiful, groundwater containing measurable concentrations of DO (2–5 mg l^{-1}) can be found.

There are two principal methods for determination of dissolved oxygen. The older, titration method (often called the Winkler method) involves the chemical fixation of the oxygen in a water sample collected in an air-tight bottle. Fixation is carried out in the field and the analysis, by titration, is carried out in the laboratory. The method is time-consuming but can give a high degree of precision and accuracy. It is suitable for most kinds of water and enables samples to be taken and stored. The alternative membrane-electrode, or oxygen probe, method is quick and can be used *in situ* or for continuous monitoring, although a high degree of accuracy may be difficult to maintain.

Samples taken for analysis by titration must be taken with great care to ensure no air bubbles are trapped in the bottle, which must be filled to overflowing and stoppered. The necessary reagents must be added for oxygen fixation immediately the sample is taken and the bottles must be protected from sunlight until the determination is carried out, which should be as soon as possible. Regardless of the analytical method, the water temperature must be measured at the time of sampling.

3.3.10 Carbon dioxide

Carbon dioxide (CO_2) is highly soluble in water and atmospheric CO_2 is absorbed at the air–water interface. In addition, CO_2 is produced within water bodies by the respiration of aquatic biota, during aerobic and anaerobic heterotrophic decomposition of suspended and sedimented organic matter. Carbon dioxide dissolved in natural water is part of an equilibrium involving bicarbonate and carbonate ions (see section 3.6.5). The concentrations of these forms are dependent to some extent on the pH, as indicated in Figure 3.1.

Free CO_2 is that component in gaseous equilibrium with the atmosphere, whereas total CO_2 is the sum of all inorganic forms of carbon dioxide, i.e. CO_2, H_2CO_3, HCO_3^- and CO_3^{2-}. Both CO_2 and HCO_3^- can be incorporated into organic carbon by autotrophic organisms. Free CO_2 comprises the concentrations of CO_2 plus H_2CO_3, although the latter carbonate form is minimal in most surface waters as they rarely exceed pH 9. At high concentrations of free carbonic acid (pH 4.5 or lower), water becomes corrosive to metals and concrete as a result of the formation of soluble bicarbonates. The ability to affect the calcium carbonate component of concrete has led to the term aggressive carbonic acid or aggressive CO_2, which is also termed free CO_2.

Determination of free CO_2 is usually by titration methods and total CO_2 by calculation from pH and alkalinity estimates. The latter method is subject to some interferences and can be rather inaccurate.

Table 3.3 Conversion factors for various national grades of water hardness

		mmol l^{-1}	Germany °DH	UK °Clark	France degree F	USA ppm
	mmol l^{-1}	1	5.61	7.02	10	100
Germany	°DH	0.178	1	1.25	1.78	17.8
UK	°Clark	0.143	0.80	1	1.43	14.3
France	degree F	0.1	0.56	0.70	1	10
USA	ppm	0.01	0.056	0.07	0.1	1

Source: ISO, 1984

3.3.11 Hardness

The hardness of natural waters depends mainly on the presence of dissolved calcium and magnesium salts. The total content of these salts is known as general hardness, which can be further divided into carbonate hardness (determined by concentrations of calcium and magnesium hydrocarbonates), and non-carbonate hardness (determined by calcium and magnesium salts of strong acids). Hydrocarbonates are transformed during the boiling of water into carbonates, which usually precipitate. Therefore, carbonate hardness is also known as temporary or removed, whereas the hardness remaining in the water after boiling is called constant. Different countries have different hardness units as indicated in Table 3.3.

Hardness may vary over a wide range. Calcium hardness is usually prevalent (up to 70 per cent), although in some cases magnesium hardness can reach 50–60 per cent. Seasonal variations of river water hardness often occur, reaching the highest values during low flow conditions and the lowest values during floods. Groundwater hardness is, however, less variable. Where there are specific requirements for water hardness in relation to water use it is usually with respect to the properties of the cations forming the hardness.

Samples for hardness determination must be filtered but not preserved. If during storage a calcium carbonate sediment appears, it must be dissolved with a small volume of hydrochloric acid (1:1) after decanting the clear liquid above the sediment. General hardness is usually determined by EDTA complexometric titration. Depending on the indicator used, either general hardness (using eriochrome black T) or calcium hardness (using murexide) can be determined. Magnesium hardness is calculated from the difference between the two determinations. Carbonate hardness is determined by acid–base titration. Hardness may also be determined from the sum of the divalent ions analysed individually (e.g. by atomic absorption spectrophotometry).

3.3.12 Chlorophyll

The green pigment chlorophyll (which exists in three forms: chlorophyll *a*, *b* and *c*) is present in most photosynthetic organisms and provides an indirect measure of algal biomass and an indication of the trophic status of a water body. It is usually included in assessment programmes for lakes and reservoirs and is important for the management of water abstracted for drinking water supply, since excessive algal growth makes water unpalatable or more difficult to treat.

In waters with little input of sediment from the catchment, or with little resuspension, chlorophyll can give an approximate indication of the quantity of material suspended in the water column. The growth of planktonic algae in a water body is related to the presence of nutrients (principally nitrates and phosphates), temperature and light. Therefore, concentrations of chlorophyll fluctuate seasonally and even daily, or with water depth, depending on environmental conditions. Water bodies with low levels of nutrients (e.g. oligotrophic lakes) have low levels of chlorophyll (< 2.5 μg l^{-1}) whereas waters with high nutrient contents (especially those classed as eutrophic) have high levels of chlorophyll (5–140 μg l^{-1}), although levels in excess of 300 μg l^{-1} also occur.

Chlorophyll fluoresces red when excited by blue light and this property can be used to measure chlorophyll levels and indicate algal biomass. Direct, and continuous, measurement of chlorophyll fluorescence can be made with a fluorimeter which can be used *in situ* by pumping water through it or, for some specially designed instruments, by lowering it into the water. Samples taken for chlorophyll analysis in the laboratory should be collected in polythene bottles and 0.1 to 0.2 ml of magnesium carbonate suspension added immediately as a preservative. Samples should also be filtered immediately although they can be stored in a cool dark place for up to 8 hours. However, once filtered through a glass fibre (GF/C grade) filter, the filter can be stored frozen for a short period prior to analysis. The chlorophyll pigments are solvent-extracted and measured spectrophotometrically using one of the methods described by Strickland and Parsons (1972). The most common determination is for chlorophyll *a*, although some methods allow for the combined measurements of chlorophylls *a*, *b* and *c*. The presence of chlorophyll degradation products, such as phaeophytin, can interfere with the estimate of chlorophyll concentrations in the solvent extract. This can be overcome by reading the optical density before and after acidification of the extract, using the method based on Lorenzen (1967). A rough estimate of phytoplankton organic carbon can be obtained from a measurement of total pigments (i.e. chlorophyll *a* + phaeopigments). The minimum organic carbon present (in mg l^{-1}) is approximately equal to 30 times the total pigments (in mg l^{-1}), although this relationship has only been tested on western European rivers (Dessery *et al.*, 1984).

3.4   Nutrients

3.4.1   Nitrogen compounds

Nitrogen is essential for living organisms as an important constituent of proteins, including genetic material. Plants and micro-organisms convert inorganic nitrogen to organic forms. In the environment, inorganic nitrogen occurs in a range of oxidation states as nitrate (NO_3^-) and nitrite (NO_2^-), the ammonium ion (NH_4^+) and molecular nitrogen (N_2). It undergoes biological and non-biological transformations in the environment as part of the nitrogen cycle. The major non-biological processes involve phase transformations such as volatilisation, sorption and sedimentation. The biological transformations consist of: a) assimilation of inorganic forms (ammonia and nitrate) by plants and micro-organisms to form organic nitrogen e.g. amino acids, b) reduction of nitrogen gas to ammonia and organic nitrogen by micro-organisms, c) complex heterotrophic conversions from one organism to another, d) oxidation of ammonia to nitrate and nitrite (nitrification), e) ammonification of organic nitrogen to produce ammonia during the decomposition of organic matter, and f) bacterial reduction of nitrate to nitrous oxide (N_2O) and molecular nitrogen (N_2) under anoxic conditions (denitrification). For a better understanding of the nitrogen cycle it is strongly recommended that all nitrogen species are reported in moles per litre or as mg l^{-1} of nitrogen (e.g. NO_3-N, NH_4-N), rather than as mg l^{-1} of NO_3^- or NH_4^+.

Ammonia
Ammonia occurs naturally in water bodies arising from the breakdown of nitrogenous organic and inorganic matter in soil and water, excretion by biota, reduction of the nitrogen gas in water by micro-organisms and from gas exchange with the atmosphere. It is also discharged into water bodies by some industrial processes (e.g. ammonia-based pulp and paper production) and also as a component of municipal or community waste. At certain pH levels, high concentrations of ammonia (NH_3) are toxic to aquatic life and, therefore, detrimental to the ecological balance of water bodies.

In aqueous solution, un-ionised ammonia exists in equilibrium with the ammonium ion. Total ammonia is the sum of these two forms. Ammonia also forms complexes with several metal ions and may be adsorbed onto colloidal particles, suspended sediments and bed sediments. It may also be exchanged

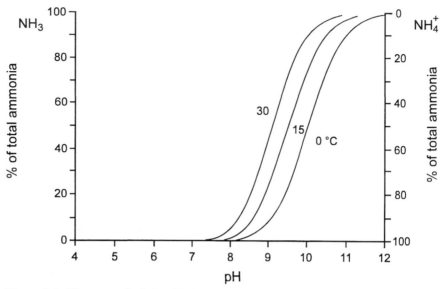

Figure 3.2 The general relationship between the percentage of un-ionised and free ammonia and varying pH in pure freshwaters

between sediments and the overlying water. The concentration of un-ionised ammonia is dependent on the temperature, pH and total ammonia concentration. The change in percentage of the two forms at different pH values is shown in Figure 3.2. Substantial losses of ammonia can occur via volatilisation with increasing pH.

Unpolluted waters contain small amounts of ammonia and ammonia compounds, usually < 0.1 mg l^{-1} as nitrogen. Total ammonia concentrations measured in surface waters are typically less than 0.2 mg l^{-1} N but may reach 2–3 mg l^{-1} N. Higher concentrations could be an indication of organic pollution such as from domestic sewage, industrial waste and fertiliser run-off. Ammonia is, therefore, a useful indicator of organic pollution. Natural seasonal fluctuations also occur as a result of the death and decay of aquatic organisms, particularly phytoplankton and bacteria in nutritionally rich waters. High ammonia concentrations may also be found in the bottom waters of lakes which have become anoxic.

Samples intended for the detection of ammonia should be analysed within 24 hours. If this is not possible the sample can be deep frozen or preserved with 0.8 ml of sulphuric acid (H_2SO_4) for each litre of sample and then stored at 4 °C. Prior to analysis any acid used as a preservative should be neutralised. There are many methods available for measuring ammonia ions. The simplest, which are suitable for waters with little or no pollution, are colorimetric methods using Nessler's reagent or the phenate method. For high concentrations of ammonia, such as occur in wastewaters, a distillation and titration method is more appropriate. Total ammonia nitrogen is also determined as part of the Kjeldahl method (see below).

Nitrate and nitrite

The nitrate ion (NO_3^-) is the common form of combined nitrogen found in natural waters. It may be biochemically reduced to nitrite (NO_2^-) by denitrification processes, usually under anaerobic conditions. The nitrite ion is rapidly oxidised to nitrate. Natural sources of nitrate to surface waters include igneous rocks, land drainage and plant and animal debris. Nitrate is an essential nutrient for aquatic plants and seasonal fluctuations can be caused by plant growth and decay. Natural concentrations, which seldom exceed 0.1 mg l^{-1} NO_3-N, may be enhanced by municipal and industrial wastewaters, including leachates from waste disposal sites and sanitary landfills. In rural and suburban areas, the use of inorganic nitrate fertilisers can be a significant source.

When influenced by human activities, surface waters can have nitrate concentrations up to 5 mg l^{-1} NO_3-N, but often less than 1 mg l^{-1} NO_3-N. Concentrations in excess of 5 mg l^{-1} NO_3-N usually indicate pollution by human or animal waste, or fertiliser run-off. In cases of extreme pollution, concentrations may reach 200 mg l^{-1} NO_3-N. The World Health Organization (WHO) recommended maximum limit for NO_3^- in drinking water is 50 mg l^{-1} (or 11.3 mg l^{-1} as NO_3-N) (Table 3.4), and waters with higher concentrations can represent a significant health risk. In lakes, concentrations of nitrate in excess of 0.2 mg l^{-1} NO_3-N tend to stimulate algal growth and indicate possible eutrophic conditions.

Nitrate occurs naturally in groundwaters as a result of soil leaching but in areas of high nitrogen fertiliser application it may reach very high concentrations (~500 mg l^{-1} NO_3-N). In some areas, sharp increases in nitrate concentrations in groundwaters over the last 20 or 30 years have been related to increased fertiliser applications, especially in many of the traditional agricultural regions of Europe (Hagebro et al., 1983; Roberts and Marsh, 1987). Increased fertiliser application is not, however, the only source of nitrate leaching to groundwater. Nitrate leaching from unfertilised grassland or natural vegetation is normally minimal, although soils in such areas contain sufficient organic matter to be a large potential source of nitrate (due to the activity of nitrifying bacteria in the soil). On clearing and ploughing for cultivation, the increased soil aeration that occurs enhances the action of nitrifying bacteria, and the production of soil nitrate.

Table 3.4 Examples of maximum allowable concentrations of selected water quality variables for different uses

Use	Drinking water					Fisheries and aquatic life		
Variable	WHO[1]	EU	Canada	USA	Russia[2]	EU	Canada[1]	Russia
Colour (TCU)	15	20 mg l⁻¹ Pt-Co	15	15	20			
Total dissolved solids (mg l⁻¹)	1,000		500	500	1,000			
Total suspended solids (mg l⁻¹)						25	inc. of 10 or 10%[3]	
Turbidity (NTU)	5	4 JTU	5	0.5–1.0				
pH	<8.0[4]	6.5[1]–8.5	6.5–8.5	6.5–8.5	6.0–9.0	6.0–9.0	6.5–9.0	
Dissolved oxygen (mg l⁻¹)					4.0	5.0–9.0	5.0–9.5	4.0[5]–6.0
Ammoniacal nitrogen (mg l⁻¹)					2.0	0.005–0.025	1.37–2.2[6,7]	0.05
Ammonium (mg l⁻¹)	0.5	0.5			2.0	0.04–1.0		0.5
Nitrate as N (mg l⁻¹)			10.0	10.0				
Nitrate (mg l⁻¹)	50	50			45			40
Nitrite as N (mg l⁻¹)			1.0	1.0				
Nitrite (mg l⁻¹)	3(P)	0.1			3.0	0.01–0.03	0.06	0.08
Phosphorus (mg l⁻¹)		5.0						
BOD (mg l⁻¹ O2)					3.0	3.0–6.0		3
Sodium (mg l⁻¹)	200	150						120
Chloride (mg l⁻¹)	250	25[1]	250	250	350			300
Chlorine (mg l⁻¹)	5						0.002	
Sulphate (mg l⁻¹)	250	250	500	250	500			100
Sulphide (mg l⁻¹)			0.05					
Fluoride (mg l⁻¹)	1.5	1.5	1.5	2.0	< 1.5			0.75
Boron (mg l⁻¹)	0.3	1.0[1]	5.0		0.3			
Cyanide (mg l⁻¹)	0.07	0.05	0.2	0.2(PP)	0.07		0.005	0.05
Trace elements								
Aluminium (mg l⁻¹)	0.2	0.2			0.5		0.005–0.1[7]	
Arsenic (mg l⁻¹)	0.01(P)	0.05	0.05	0.05	0.01		0.05	
Barium (mg l⁻¹)	0.7	0.1[1]	1.0	2.0	0.7			

Continued

Table 3.4 Continued

Use	Drinking water					Fisheries and aquatic life		
Variable	WHO[1]	EU	Canada	USA	Russia[2]	EU	Canada[1]	Russia
Cadmium (mg l^{-1})	0.003	0.005	0.005	0.005	0.003		0.0002–0.0018[8]	0.005
Chromium (mg l^{-1})	0.05(P)	0.05	0.05	0.1	0.05		0.02–0.002	0.02–0.005
Cobalt (mg l^{-1})					0.1			0.01
Copper (mg l^{-1})	2(P)	0.1–3.0[1]	1.0	1	2.0	0.005–0.112[8,9]	0.002–0.004[8]	0.001
Iron (mg l^{-1})	0.3	0.2	0.3	0.3	0.3		0.3	0.1
Lead (mg l^{-1})	0.01	0.05	0.05	0.015	0.01		0.001–0.007[8]	0.1
Manganese (mg l^{-1})	0.5(P)	0.05	0.05	0.05	0.5			0.01
Mercury (mg l^{-1})	0.001	0.001	0.001	0.002	0.001		0.0001	0.00001
Nickel (mg l^{-1})	0.02	0.05			0.02		0.025–0.15[8]	0.01
Selenium (mg l^{-1})	0.01	0.01	0.01	0.05	0.01		0.001	0.0016
Zinc (mg l^{-1})	3	0.1–5.0[1]	5.0	5	5.0	0.03–2.0[8,10]	0.03	0.01
Organic contaminants[11]								
Oil and petroleum products (mg l^{-1})		0.01						0.05
Total pesticides (µg l^{-1})		0.5	100		0.1			0.05
Aldrin & dieldrin (µg l^{-1})	0.03		0.7				4 ng l^{-1} dieldrin	
DDT (µg l^{-1})	2		30.0		2.0		1 ng l^{-1}	
Lindane (µg l^{-1})	2		4.0	0.2	2.0			
Methoxychlor (µg l^{-1})	20		100	40				
Benzene (µg l^{-1})	10			5				
Pentachlorophenol (µg l^{-1})	9(P)			10	10		300	
Phenols (µg l^{-1})		0.5	2		1.0		1.0	1.0
Detergents (mg l^{-1})		0.2		0.5[12]	0.5			0.1

Continued

Table 3.4 Continued

Use	Drinking water					Fisheries and aquatic life		
Variable	WHO[1]	EU	Canada	USA	Russia[2]	EU	Canada[1]	Russia
Microbiological variables								
Faecal coliforms (*E. coli*) (No. per 100 ml)	0	0	0		0			
Total coliforms (No. per 100 ml)	0	0	10^{13}	1	0.3			

WHO World Health Organization
EU   European Union
BOD Biochemical oxygen demand
TCU True colour units
NTU Nephelometric turbidity units
(P)   Provisional value
(PP) Proposed value
1   Guideline value
2   Some values not yet adopted but already

3   applied
    i.e. above background concentrations of
4   ≤ 100.0 mg l^{-1} or > 100 mg l^{-1} respectively
5   For effective disinfection with chlorine
6   Lower level acceptable under ice cover
7   Total ammonia
8   Depending on pH
9   Depending on hardness
10 Dissolved only
    Total zinc

11 For some groups values are also set for individual compounds
12 Foaming agents
13 For a single sample

Sources: Environment Canada, 1987
CEC, 1978, 1980
Committee for Fisheries, 1993
Gray, 1994
WHO, 1993

Nitrite concentrations in freshwaters are usually very low, 0.001 mg l^{-1} NO_2-N, and rarely higher than 1 mg l^{-1} NO_2-N. High nitrite concentrations are generally indicative of industrial effluents and are often associated with unsatisfactory microbiological quality of water.

Determination of nitrate plus nitrite in surface waters gives a general indication of the nutrient status and level of organic pollution. Consequently, these species are included in most basic water quality surveys and multi-purpose or background monitoring programmes, and are specifically included in programmes monitoring the impact of organic or relevant industrial inputs. As a result of the potential health risk of high levels of nitrate, it is also measured in drinking water sources. However, as little nitrate is removed during the normal processes for drinking water treatment, the treated drinking water should also be analysed when nitrate concentrations are high in the source water.

Samples taken for the determination of nitrate and/or nitrite should be collected in glass or polyethylene bottles and filtered and analysed immediately. If this is not possible, 2–4 ml of chloroform per litre can be added to the sample to retard bacterial decomposition. The sample can be cooled and then stored at 3–4 °C. As determination of nitrate is difficult, due to interferences from other substances present in the water, the precise choice of method may vary according to the expected concentration of nitrate as N. Alternatively, one portion of the sample can be chemically analysed for total inorganic nitrogen and the other for nitrite, and the nitrate concentration obtained from the difference between the two values. Nitrite concentrations can be determined using spectrophotometric methods. Some simple field determinations, of limited accuracy, can be made using colorimetric comparator methods available as kits.

Organic nitrogen
Organic nitrogen consists mainly of protein substances (e.g. amino acids, nucleic acids and urine) and the product of their biochemical transformations (e.g. humic acids and fulvic acids). Organic nitrogen is naturally subject to the seasonal fluctuations of the biological community because it is mainly formed in water by phytoplankton and bacteria, and cycled within the food chain. Increased concentrations of organic nitrogen could indicate pollution of a water body.

Organic nitrogen is usually determined using the Kjeldahl method which gives total ammonia nitrogen plus total organic nitrogen (Kjeldahl N). The difference between the total nitrogen and the inorganic forms gives the total organic nitrogen content. Samples must be unfiltered and analysed within

24 hours, since organic nitrogen is rapidly converted to ammonia. This process can be retarded if necessary by the addition of 2–4 ml of chloroform or approximately 0.8 ml of concentrated H_2SO_4 per litre of sample. Storage should be at 2–4 °C, and when this is necessary, the condition and duration of preservation should be stated with the results. Photochemical methods can also be used in place of the Kjeldahl method. These methods oxidise all organic nitrogen (as well as ammonia) to nitrates and nitrites and, therefore, the measurements of these must already have been carried out on the sample beforehand. If samples are filtered total dissolved nitrogen is determined instead of total organic nitrogen.

3.4.2   Phosphorus compounds

Phosphorus is an essential nutrient for living organisms and exists in water bodies as both dissolved and particulate species. It is generally the limiting nutrient for algal growth and, therefore, controls the primary productivity of a water body. Artificial increases in concentrations due to human activities are the principal cause of eutrophication (see Chapter 7).

In natural waters and in wastewaters, phosphorus occurs mostly as dissolved orthophosphates and polyphosphates, and organically bound phosphates. Changes between these forms occur continuously due to decomposition and synthesis of organically bound forms and oxidised inorganic forms. The equilibrium of the different forms of phosphate that occur at different pH values in pure water is shown in Figure 3.3. It is recommended that phosphate concentrations are expressed as phosphorus, i.e. mg l^{-1} PO_4-P (and not as mg l^{-1} PO_4^{3-}).

Natural sources of phosphorus are mainly the weathering of phosphorus-bearing rocks and the decomposition of organic matter. Domestic wastewaters (particularly those containing detergents), industrial effluents and fertiliser run-off contribute to elevated levels in surface waters. Phosphorus associated with organic and mineral constituents of sediments in water bodies can also be mobilised by bacteria and released to the water column.

Phosphorus is rarely found in high concentrations in freshwaters as it is actively taken up by plants. As a result there can be considerable seasonal fluctuations in concentrations in surface waters. In most natural surface waters, phosphorus ranges from 0.005 to 0.020 mg l^{-1} PO_4-P. Concentrations as low as 0.001 mg l^{-1} PO_4-P may be found in some pristine waters and as high as 200 mg l^{-1} PO_4-P in some enclosed saline waters. Average groundwater levels are about 0.02 mg l^{-1} PO_4-P.

As phosphorus is an essential component of the biological cycle in water bodies, it is often included in basic water quality surveys or background

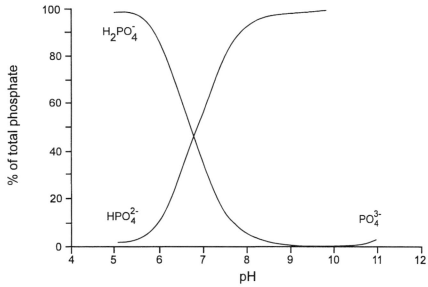

Figure 3.3 The equilibrium of different forms of phosphate in relation to the pH of pure freshwaters

monitoring programmes. High concentrations of phosphates can indicate the presence of pollution and are largely responsible for eutrophic conditions. The management of a lake or reservoir, particularly for drinking water supply, requires a knowledge of the levels of phosphate in order to help interpret the rates of algal growth.

Phosphorus concentrations are usually determined as orthophosphates, total inorganic phosphate or total phosphorus (organically combined phosphorus and all phosphates). The dissolved forms of phosphorus are measured after filtering the sample through a pre-washed 0.45 μm pore diameter membrane filter. Particulate concentrations can be deduced by the difference between total and dissolved concentrations. Phosphorus is readily adsorbed onto the surface of sample containers and, therefore, containers should be rinsed thoroughly with the sample before use. Samples for phosphate analysis can be preserved with chloroform and stored at 2–4 °C for up to 24 hours. Samples for total phosphorus determinations can be stored in a glass flask with a tightly fitting glass stopper, provided 1 ml of 30 per cent sulphuric acid is added per 100 ml of sample. For dissolved phosphorus, it is important that samples are filtered as soon as possible after collection. Determination of phosphate involves conversion to orthophosphate which is then measured colorimetrically.

3.5   Organic matter

Most freshwaters contain organic matter which can be measured as total organic carbon (TOC). For comparative purposes an indication of the amount of organic matter present can be obtained by measuring related properties, principally the biochemical oxygen demand (BOD) or the chemical oxygen demand (COD). The COD usually includes all, or most, of the BOD as well as some other chemical demands. In most samples, COD > BOD > TOC. However, in some situations this relationship may not be true, such as when the sample contains toxic substances (see section 3.5.3).

3.5.1   Total organic carbon

Organic carbon in freshwaters arises from living material (directly from plant photosynthesis or indirectly from terrestrial organic matter) and also as a constituent of many waste materials and effluents. Consequently, the total organic matter in the water can be a useful indication of the degree of pollution, particularly when concentrations can be compared upstream and downstream of potential sources of pollution, such as sewage or industrial discharges or urban areas. In surface waters, TOC concentrations are generally less than 10 mg l^{-1}, and in groundwater less than 2 mg l^{-1}, unless the water receives municipal or industrial wastes, or is highly coloured due to natural organic material, as in swamps. In such situations, TOC concentrations may exceed 100 mg l^{-1} (TOC concentrations in municipal wastewaters range from 10 to > 100 mg l^{-1}, depending on the level of wastewater treatment). Total organic carbon consists of dissolved and particulate material and is, therefore, affected by fluctuations in suspended solids, which can be quite pronounced in rivers. The dissolved and particulate organic carbon (DOC and POC respectively) can be determined separately after filtering the sample through a glass fibre filter (approximately 0.7 μm pore diameter), and this is recommended for river studies. In most surface waters, DOC levels exceed POC levels and are in the range 1–20 mg l^{-1}. During river floods, and throughout the year in many turbid rivers, POC is the most abundant form (see Table 6.3).

   Total organic carbon is determined without filtration of the sample. Samples for TOC determination should be stored in dark glass bottles, with minimum exposure to light or air, at 3–4 °C for no more than seven days prior to analysis. Alternatively, samples can be acidified with sulphuric acid to pH 2 or less.

   There are various methods available for determining organic carbon depending on the type of sample to be analysed. Methods are based on the principle of oxidation of the carbon in the sample to carbon dioxide (e.g. by combustion, chemical reaction or ultra violet irradiation) which is then determined by one of several methods (e.g. volumetric determination, thermal conductivity or specific CO_2 electrode).

3.5.2   Chemical oxygen demand

The chemical oxygen demand (COD) is a measure of the oxygen equivalent of the organic matter in a water sample that is susceptible to oxidation by a strong chemical oxidant, such as dichromate. The COD is widely used as a measure of the susceptibility to oxidation of the organic and inorganic materials present in water bodies and in the effluents from sewage and industrial plants. The test for COD is non-specific, in that it does not identify the oxidisable material or differentiate between the organic and inorganic material present. Similarly, it does not indicate the total organic carbon present since some organic compounds are not oxidised by the dichromate method whereas some inorganic compounds are oxidised. Nevertheless, COD is a useful, rapidly measured, variable for many industrial wastes and has been in use for several decades.

The concentrations of COD observed in surface waters range from 20 mg l^{-1} O_2 or less in unpolluted waters to greater than 200 mg l^{-1} O_2 in waters receiving effluents. Industrial wastewaters may have COD values ranging from 100 mg l^{-1} O_2 to 60,000 mg l^{-1} O_2.

Samples for COD analysis should be collected in bottles which do not release organic substances into the water, such as glass-stoppered glass bottles. Ideally samples should be analysed immediately, or if unpolluted, within 24 hours provided they are stored cold. If analysis cannot be carried out immediately, the samples should be preserved with sulphuric acid. For prolonged storage samples should be deep frozen. If appropriate, samples can be filtered prior to analysis using glass fibre filters. Unfiltered samples containing settleable solids should be homogenised prior to sub-sampling. The standard method for measurement of COD is oxidation of the sample with potassium dichromate in a sulphuric acid solution (although other oxidants can be used which may have different oxidation characteristics) followed by a titration. It is extremely important that the same method is followed each time during a series of measurements so that the results are comparable.

3.5.3   Biochemical oxygen demand

The biochemical oxygen demand (BOD) is an approximate measure of the amount of biochemically degradable organic matter present in a water sample. It is defined by the amount of oxygen required for the aerobic micro-organisms present in the sample to oxidise the organic matter to a stable inorganic form. The method is subject to various complicating factors such as the oxygen demand resulting from the respiration of algae in the sample and the possible oxidation of ammonia (if nitrifying bacteria are also present). The presence of toxic substances in a sample may affect microbial activity leading to a reduction in the measured BOD. The conditions in a BOD bottle usually differ from those in a river or lake. Therefore, interpretation of BOD results and their implications must be done with great care and by experienced personnel. Further discussion of the BOD test, together with case history results, is given in Velz (1984).

Standardised laboratory procedures are used to determine BOD by measuring the amount of oxygen consumed after incubating the sample in the dark at a specified temperature, which is usually 20 °C, for a specific period of time, usually five days. This gives rise to the commonly used term "BOD_5". The oxygen consumption is determined from the difference between the dissolved oxygen concentrations in the sample before and after the incubation period. If the concentration of organic material in the samples is very high, samples may require dilution with distilled water prior to incubation so that the oxygen is not totally depleted.

As noted above, BOD measurements are usually lower than COD measurements. Unpolluted waters typically have BOD values of 2 mg l^{-1} O_2 or less, whereas those receiving wastewaters may have values up to 10 mg l^{-1} O_2 or more, particularly near to the point of wastewater discharge. Raw sewage has a BOD of about 600 mg l^{-1} O_2, whereas treated sewage effluents have BOD values ranging from 20 to 100 mg l^{-1} O_2 depending on the level of treatment applied. Industrial wastes may have BOD values up to 25,000 mg l^{-1} O_2.

Water samples collected for BOD measurement must not contain any added preservatives and must be stored in glass bottles. Ideally the sample should be tested immediately since any form of storage at room temperature can cause changes in the BOD (increase or decrease depending on the character of the sample) by as much as 40 per cent. Storage should be at 5 °C and only when absolutely necessary.

3.5.4   Humic and fulvic acids

Organic matter arising from living organisms makes an important contribution to the natural quality of surface waters. The composition of this organic matter is extremely diverse. Natural organic compounds are not usually toxic, but exert major controlling effects on the hydrochemical and biochemical processes in a water body. Some natural organic compounds significantly affect the quality of water for certain uses, especially those which depend on organoleptic properties (taste and smell). During chlorination for drinking water disinfection, humic and fulvic acids act as precursor substances in the formation of trihalomethanes such as chloroform. In addition, substances included in aquatic humus determine the speciation of heavy metals and some other pollutants because of their high complexing ability. As a result, humic substances affect the toxicity and mobility of metal complexes. Therefore, measurement of the concentrations of these substances can be important for determining anthropogenic impacts on water bodies.

Humus is formed by the chemical and biochemical decomposition of vegetative residues and from the synthetic activity of micro-organisms. Humus enters water bodies from the soil and from peat bogs, or it can be formed directly within water bodies as a result of biochemical transformations. It is operationally separated into fulvic and humic acid fractions, each being an aggregate of many organic compounds of different masses. Fulvic acid has molecular masses mostly in the range 300–5,000 whereas the dominant masses in humic acid exceed 5,000. The relative content of fulvic acid in the dissolved humic substances present in freshwaters is between 60 and 90 per cent. Humic and fulvic acids are fairly stable (i.e. their BOD is low). However, these substances are chemically oxidisable and, therefore, can readily affect the results of COD determinations.

Fulvic and humic acid concentrations in river and lake waters are highly dependent on the physico-geographical conditions and are usually in the range of tens and hundreds of micrograms of carbon per litre. However, concentrations can reach milligrams of carbon per litre in waters of marshy and woodland areas. In natural conditions fulvic and humic acids can comprise up to 80 per cent of the DOC, which can be used as an approximate estimate of their concentrations.

Samples for fulvic and humic acid determination are not usually filtered or preserved. They can be stored for some months in a refrigerator (3–4 °C). Total fulvic and humic acid content can be determined photometrically and their separate determination can be made with spectrophotometric methods.

3.6 Major ions
Major ions (Ca^{2+}, Mg^{2+}, Na^+, K^+, Cl^-, SO_4^{2-}, HCO_3^-) are naturally very variable in surface and groundwaters due to local geological, climatic and geographical conditions (see Tables 6.2, 6.3 and 9.4).

3.6.1 Sodium
All natural waters contain some sodium since sodium salts are highly water soluble and it is one of the most abundant elements on earth. It is found in the ionic form (Na^+), and in plant and animal matter (it is an essential element for living organisms). Increased concentrations in surface waters may arise

from sewage and industrial effluents and from the use of salts on roads to control snow and ice. The latter source can also contribute to increased sodium in groundwaters. In coastal areas, sea water intrusion can also result in higher concentrations.

Concentrations of sodium in natural surface waters vary considerably depending on local geological conditions, wastewater discharges and seasonal use of road salt. Values can range from 1 mg l^{-1} or less to 10^5 mg l^{-1} or more in natural brines. The WHO guideline limit for sodium in drinking water is 200 mg l^{-1} (Table 3.4). Many surface waters, including those receiving wastewaters, have concentrations well below 50 mg l^{-1}. However, groundwater concentrations frequently exceed 50 mg l^{-1}.

Sodium is commonly measured where the water is to be used for drinking or agricultural purposes, particularly irrigation. Elevated sodium in certain soil types can degrade soil structure thereby restricting water movement and affecting plant growth. The sodium adsorption ratio (SAR) is used to evaluate the suitability of water for irrigation. The ratio estimates the degree to which sodium will be adsorbed by the soil. High values of SAR imply that the sodium in the irrigation water may replace the calcium and magnesium ions in the soil, potentially causing damage to the soil structure. The SAR for irrigation waters is defined as follows:

$$SAR = \frac{Na^+}{\sqrt{(Ca^{2+} + Mg^{2+})/2}}$$

where the concentrations of sodium, magnesium and calcium are expressed in milliequivalents per litre (meq l^{-1}).

Samples for sodium analysis should be stored in polyethylene bottles to avoid potential leaching from glass containers. Samples should be analysed as soon as possible because prolonged storage in polyethylene containers can lead to evaporation losses through the container walls or lid. Filtration may be necessary if the sample contains solid material. Analysis is best performed using flame atomic emission and absorption.

3.6.2 Potassium

Potassium (as K$^+$) is found in low concentrations in natural waters since rocks which contain potassium are relatively resistant to weathering. However, potassium salts are widely used in industry and in fertilisers for agriculture and enter freshwaters with industrial discharges and run-off from agricultural land.

Potassium is usually found in the ionic form and the salts are highly soluble. It is readily incorporated into mineral structures and accumulated by aquatic biota as it is an essential nutritional element. Concentrations in natural waters are usually less than 10 mg l^{-1}, whereas concentrations as high as 100 and 25,000 mg l^{-1} can occur in hot springs and brines, respectively.

Samples for potassium analysis should be stored in polyethylene containers to avoid potential contamination as a result of leaching from glass bottles. However, samples should be analysed as soon as possible as prolonged storage in polyethylene containers can lead to evaporation losses through the container walls or lid. Samples containing solids may require filtration prior to storage. Analysis is best carried out using atomic absorption spectrophotometry as for sodium.

3.6.3   Calcium

Calcium is present in all waters as Ca^{2+} and is readily dissolved from rocks rich in calcium minerals, particularly as carbonates and sulphates, especially limestone and gypsum. The cation is abundant in surface and groundwaters. The salts of calcium, together with those of magnesium, are responsible for the hardness of water (see section 3.3.11). Industrial, as well as water and wastewater treatment, processes also contribute calcium to surface waters. Acidic rainwater can increase the leaching of calcium from soils.

Calcium compounds are stable in water when carbon dioxide is present, but calcium concentrations can fall when calcium carbonate precipitates due to increased water temperature, photosynthetic activity or loss of carbon dioxide due to increases in pressure. Calcium is an essential element for all organisms and is incorporated into the shells of many aquatic invertebrates, as well as the bones of vertebrates. Calcium concentrations in natural waters are typically < 15 mg l^{-1}. For waters associated with carbonate-rich rocks, concentrations may reach 30–100 mg l^{-1}. Salt waters have concentrations of several hundred milligrams per litre or more.

Samples for calcium analysis should be collected in plastic or borosilicate glass bottles without a preservative. They should be analysed immediately, or as soon as possible, after collection and filtration. If any calcium carbonate precipitate forms after filtration and during storage, it must be re-dissolved with hydrochloric or nitric acid and then neutralised before analysis. Acidification of unfiltered waters prior to analysis should be avoided since it causes a dissolution of carbonates, calcite and dolomite. Calcium can be determined by a titrimetric method using EDTA (ethylenediaminetetracetic acid) or by atomic absorption spectrophotometry.

3.6.4   Magnesium

Magnesium is common in natural waters as Mg^{2+}, and along with calcium, is a main contributor to water hardness (see section 3.3.11). Magnesium arises principally from the weathering of rocks containing ferromagnesium minerals and from some carbonate rocks. Magnesium occurs in many organometallic compounds and in organic matter, since it is an essential element for living organisms. Natural concentrations of magnesium in freshwaters may range from 1 to > 100 mg l^{-1}, depending on the rock types within the catchment. Although magnesium is used in many industrial processes, these contribute relatively little to the total magnesium in surface waters.

Samples for magnesium analysis should be collected in plastic or borosilicate glass containers without preservative. Samples can be analysed using the EDTA titrimetric method or by atomic absorption spectrophotometry. The magnesium concentration in a sample can also be estimated by calculating the difference between the total hardness and the calcium concentration.

3.6.5   Carbonates and bicarbonates

The presence of carbonates (CO_3^{2-}) and bicarbonates (HCO_3^-) influences the hardness and alkalinity of water (see sections 3.3.11 and 3.3.7). The inorganic carbon component (CO_2) arises from the atmosphere (see section 3.3.10) and biological respiration. The weathering of rocks contributes carbonate and bicarbonate salts. In areas of non-carbonate rocks, the HCO_3^- and CO_3^{2-} originate entirely from the atmosphere and soil CO_2, whereas in areas of carbonate rocks, the rock itself contributes approximately 50 per cent of the carbonate and bicarbonate present.

The relative amounts of carbonates, bicarbonates and carbonic acid in pure water are related to the pH as shown in Figure 3.1. As a result of the weathering process, combined with the pH range of surface waters (~6–8.2), bicarbonate is the dominant anion in most surface waters. Carbonate is uncommon in natural surface waters because they rarely exceed pH 9, whereas groundwaters can be more alkaline and may have concentrations of carbonate up to 10 mg l^{-1}. Bicarbonate concentrations in surface waters are usually < 500 mg l^{-1}, and commonly < 25 mg l^{-1}.

The concentration of carbonates and bicarbonates can be calculated from the free and total alkalinity. However, the calculation is valid only for pure water since it assumes that the alkalinity derives only from carbonates and bicarbonates. In some cases, hydroxyl ions are also present, and even unpolluted or mildly polluted waters contain components which affect the calculation.

3.6.6   Chloride

Most chlorine occurs as chloride (Cl^-) in solution. It enters surface waters with the atmospheric deposition of oceanic aerosols, with the weathering of some sedimentary rocks (mostly rock salt deposits) and from industrial and sewage effluents, and agricultural and road run-off. The salting of roads during winter periods can contribute significantly to chloride increases in groundwaters. High concentrations of chloride can make waters unpalatable and, therefore, unfit for drinking or livestock watering.

In pristine freshwaters chloride concentrations are usually lower than 10 mg l^{-1} and sometimes less than 2 mg l^{-1}. Higher concentrations can occur near sewage and other waste outlets, irrigation drains, salt water intrusions, in arid areas and in wet coastal areas. Seasonal fluctuations of chloride concentrations in surface waters can occur where roads are salted in the winter. As chloride is frequently associated with sewage, it is often incorporated into assessments as an indication of possible faecal contamination or as a measure of the extent of the dispersion of sewage discharges in water bodies.

Samples for chloride determination need no preservation or special treatment and can be stored at room temperature. Analysis can be done by standard or potentiometric titration methods. Direct potentiometric determinations can be made with chloride-sensitive electrodes.

3.6.7   Sulphate

Sulphate is naturally present in surface waters as SO_4^{2-}. It arises from the atmospheric deposition of oceanic aerosols and the leaching of sulphur compounds, either sulphate minerals such as gypsum or sulphide minerals such as pyrite, from sedimentary rocks. It is the stable, oxidised form of sulphur and is readily soluble in water (with the exception of lead, barium and strontium sulphates which precipitate). Industrial discharges and atmospheric precipitation can also add significant amounts of sulphate to surface waters. Sulphate can be used as an oxygen source by bacteria which convert it to hydrogen sulphide (H_2S, HS^-) under anaerobic conditions.

Sulphate concentrations in natural waters are usually between 2 and 80 mg l^{-1}, although they may exceed 1,000 mg l^{-1} near industrial discharges or in arid regions where sulphate minerals, such as gypsum, are present. High concentrations (> 400 mg l^{-1}) may make water unpleasant to drink.

Samples collected in plastic or glass containers can be stored in the refrigerator for up to seven days, although when intended for analysis soon after collection they may be stored at room temperature. Prolonged storage should be avoided, particularly if the sample contains polluted water. Sulphate can be determined gravimetrically after precipitation by barium chloride in hot hydrochloric acid. Other methods are available including a titrimetric method.

3.7   Other inorganic variables

3.7.1   Sulphide

Sulphide enters groundwaters as a result of the decomposition of sulphurous minerals and from volcanic gases. Sulphide formation in surface waters is principally through anaerobic, bacterial decay of organic substances in bottom sediments and stratified lakes and reservoirs. Traces of sulphide ion occur in unpolluted bottom sediments from the decay of vegetation, but the presence of high concentrations often indicates the occurrence of sewage or industrial wastes. Under aerobic conditions, the sulphide ion converts rapidly to sulphur and sulphate ions.

Dissolved sulphides exist in water as non-ionised molecules of hydrogen sulphide (H_2S), hydrosulphide (HS^-) and, very rarely, as sulphide (S^{2-}). The equilibrium between these forms is a function of pH (Figure 3.4). Sulphide concentrations need not be considered if the pH is lower than 10. Suspended matter may also contain various metallic sulphides. When appreciable concentrations of sulphide occur, toxicity and the strong odour of the sulphide ion make the water unsuitable for drinking water supplies and other uses.

Sulphide determination should be done immediately after sampling. If this is not possible, the sample should be fixed with cadmium acetate or zinc acetate, after which it can be stored for up to three days in the dark. During sampling, aeration of the sample must be prevented. Total sulphide, dissolved sulphide and free H_2S are the most significant determinations. Variations of pre-treatment (filtration and pH reduction) are used for their speciation. Photometric methods or, at high concentrations, iodometric titration are generally used for sulphide determination.

3.7.2   Silica

Silica is widespread and always present in surface and groundwaters. It exists in water in dissolved, suspended and colloidal states. Dissolved forms are represented mostly by silicic acid, products of its dissociation and association, and organosilicon compounds. Reactive silicon (principally silicic acid but usually recorded as dissolved silica (SiO_2) or sometimes as silicate (H_4SiO_4)) mainly arises from chemical weathering of siliceous minerals. Silica may be discharged into water bodies with wastewaters from industries using siliceous compounds in their processes such as potteries, glass works and abrasive manufacture. Silica is also an essential element for certain aquatic plants (principally diatoms). It is taken up during cell growth and released during decomposition and decay giving rise to seasonal fluctuations in concentrations, particularly in lakes.

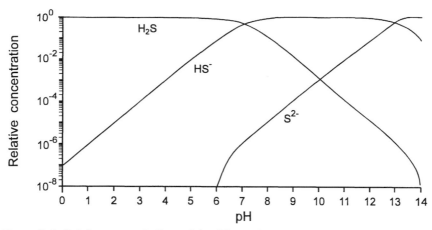

Figure 3.4 Relative concentrations of the different forms of sulphide in relation to the pH of pure freshwaters

The silica content of rivers and lakes usually varies within the range 1–30 mg l^{-1}. Concentrations in ground and volcanic waters are higher, and thermal waters may reach concentrations up to 1 g l^{-1} or more. In the weakly mineralised waters of arctic regions, as well as in marsh and other coloured waters, the reactive silica may account for 50 per cent of the total dissolved solids.

Analytical determinations can be made for dissolved silicic acid (monomeric and dimeric forms), total dissolved silica including polymeric and organic species and suspended or total silica. Plastic containers must be used for samples intended for silica analysis. The samples can be stored without preservation for up to one week, provided they are kept at a low temperature and in the dark. Forms of silicon and total silica are converted to the reactive form prior to analysis using a colourimetric method. Atomic absorption spectrophotometry can also be used.

3.7.3   Fluoride
Fluoride originates from the weathering of fluoride-containing minerals and enters surface waters with run-off and groundwaters through direct contact. Liquid and gas emissions from certain industrial processes (such as metal- and chemical-based manufacturing) can also contribute fluoride ions (F$^-$) to water bodies. Fluoride mobility in water depends, to a large extent, on the Ca^{2+} ion content, since fluoride forms low solubility compounds with divalent cations. Other ions that determine water hardness can also increase F$^-$ solubility.

Fluoride concentrations in natural waters vary from 0.05 to 100 mg l^{-1}, although in most situations they are less than 0.1 mg l^{-1}. Groundwater concentrations are often as high as 10 mg l^{-1}. Very high concentrations of fluoride, far exceeding the WHO guideline value of 1.5 mg l^{-1} (Table 3.4), are encountered in volcanic aquifers and lakes in the East African Rift system and in Hawaii. Localised occurrences of high fluoride in groundwater associated with sedimentary and metamorphic rocks are also reported from Ohio, Sri Lanka, India, Malawi and Tanzania. Fluctuations from year to year are rarely more than two times the base level, or less, for groundwaters.

Measurement of fluoride content is especially important when a water body is used for drinking water supply. At high concentrations fluoride is toxic to humans and animals and can cause bone diseases. However, a slight increase in natural concentrations can help prevent dental caries although, at higher concentrations (above 1.5–2.0 mg l^{-1}), mottling of teeth can occur (WHO, 1984). High fluoride concentrations provide a constraint on the use of groundwaters for potable supply, which may present particular difficulties where there is no practical alternative to groundwater and such values are unlikely to change with time. Where fluoride is known to occur or can be anticipated, it is an essential variable in surveys where community water supplies are being planned but for long-term monitoring it is less important.

Water samples for fluoride determination do not usually require any preservation and can be analysed up to several days following collection. Storage in polyethylene containers is recommended. Determination of the fluoride ion can be made potentiometrically (with a fluoride ion selective electrode) or photometrically. Interference effects from metals in the water can be eliminated by distillation or ion-exchange chromatography.

3.7.4   Boron

Boron is a natural component of freshwaters arising from the weathering of rocks, soil leaching, volcanic action and other natural processes. Industries and municipal wastewaters also contribute boron to surface waters. In addition, agricultural run-off may contain boron, particularly in areas where it is used to improve crop yields or as a pesticide. Boric acid, which does not readily dissociate, is the predominant species in freshwaters.

Despite its widespread occurrence, boron is usually present in natural waters in comparatively low concentrations. Average concentrations in surface waters do not exceed 0.1 mg l^{-1} and only reach 1.5–3 mg l^{-1} in a few areas. Higher concentrations of boron (up to 48 mg l^{-1}) are found in some mineral waters which are sometimes used for special health-related bathing, but not as drinking water. Maximum allowable concentrations of boron in water bodies used for drinking water vary in different countries (Table 3.4). Recommended concentrations of boron in waters used for irrigation vary from 0.5 mg l^{-1} for sensitive crops to 6 mg l^{-1} for short-term irrigation or for tolerant crops.

Containers for samples intended for boron determination must be made of polyethylene or alkali-resistant, boron-free glass. Analysis is normally by photometric methods.

3.7.5 Cyanide

Compounds of cyanide enter freshwaters with wastewaters from industries such as the electroplating industry. Cyanides occur in waters in ionic form or as weakly dissociated hydrocyanic acid. In addition, they may occur as complex compounds with metals. The toxicity of cyanides depends on their speciation; some ionic forms and hydrocyanic acid are highly toxic. The toxicity of complex compounds of cyanide depends on their stability. Weak complexes formed with metals such as zinc, lead and cadmium are extremely toxic. Copper complexes are less toxic, and cobalt and ferrous complexes are only weak toxicants.

Ionic cyanide concentration in water is reduced by carbonic and other acids transforming the ionic form into the volatile hydrocyanic acid. However, the principal mechanism of decreased levels is oxidation, including biochemical oxidation, followed by hydrolysis:

$$2CN^- + O_2 = 2CNO^-; \quad CNO^- + 2H_2O = NH_4^+ + CO_3^{2-}.$$

Strong sunlight and warm seasons favour biochemical oxidation causing a reduction in cyanide concentrations. Cyanides, especially ionic forms, are easily adsorbed by suspended matter and bottom sediments.

Concentrations of cyanides in waters intended for human use, including complex forms (except hexacyanoferrate), are strictly limited because of their high toxicity. The WHO recommends a maximum concentration of 0.07 mg l^{-1} cyanide in drinking water, but many countries apply stricter standards of cyanide concentration both for drinking waters and natural water of importance for fisheries (Table 3.4).

Samples for cyanide determination must be analysed as soon as possible because it is a highly active and unstable variable. If necessary, samples collected in polyethylene bottles can be preserved with sufficient sodium hydroxide to raise the pH to 11 or more and then stored at about 4 °C. A photometric method is normally used for the determination of cyanides in natural waters. A preliminary distillation of cyanides as hydrocyanic acid after acidification, should be made if there are any compounds causing interference in the water, or if the cyanide concentration is too low for direct determination. However, distillation should only be used when really necessary, because the product is very toxic, requiring special safety procedures.

3.8 Metals

3.8.1 General principles

The ability of a water body to support aquatic life, as well as its suitability for other uses, depends on many trace elements. Some metals, such as Mn, Zn and Cu, when present in trace concentrations are important for the physiological functions of living tissue and regulate many biochemical processes. The same metals, however, discharged into natural waters at increased concentrations in sewage, industrial effluents or from mining operations can have severe toxicological effects on humans and the aquatic ecosystem. Water pollution by heavy metals as a result of human activities is causing serious ecological problems in many parts of the world. This situation is aggravated by the lack of natural elimination processes for metals. As a result, metals shift from one compartment within the aquatic environment to another, including the biota, often with detrimental effects. Where sufficient accumulation of the metals in biota occurs through food chain transfer (see Chapter 5), there is also an increasing toxicological risk for humans. As a result of adsorption and accumulation, the concentration of metals in bottom sediments is much higher than in the water above and this sometimes causes secondary pollution problems.

Generally, trace amounts of metals are always present in freshwaters from the weathering of rocks and soils. In addition, particularly in developed countries, industrial wastewater discharges and mining are major sources of metals in freshwaters. Significant amounts also enter surface waters in sewage as well as with atmospheric deposition (e.g. lead). Lead is still widely used as an additive in petroleum for automobiles and is emitted to the atmosphere in their exhaust gases, thereby entering the hydrological cycle.

The toxicity of metals in water depends on the degree of oxidation of a given metal ion together with the forms in which it occurs. For example, the maximum allowable concentration of Cr (VI) in the former USSR was 0.001 mg l^{-1} whereas for Cr (III) it was 0.5 mg l^{-1} (Bestemyanov and Krotov, 1985). As a rule, the ionic form of a metal is the most toxic form. However, the toxicity is reduced if the ions are bound into complexes with, for example, natural organic matter such as fulvic and humic acids. Under certain conditions, metallo-organic, low-molecular compounds formed in natural waters exhibit toxicities greater than the uncombined forms. An example is the highly toxic alkyl-derivatives of mercury (e.g. methylmercury) formed from elemental mercury by aquatic micro-organisms.

Metals in natural waters can exist in truly dissolved, colloidal and suspended forms. The proportion of these forms varies for different metals and for different water bodies. Consequently, the toxicity and sedimentation potential of metals change, depending on their forms.

The assessment of metal pollution is an important aspect of most water quality assessment programmes. The Global Environment Monitoring System (GEMS) programme GEMS/WATER includes ten metals: Al, Cd, Cr, Cu, Fe, Hg, Mn, Ni, Pb, Zn. Arsenic and Se (which are not strictly metals) are also included (Table 3.5). The United States Environmental Protection Agency (US EPA) considers eight trace elements as high priority: As, Cd, Cu, Cr, Pb, Hg, Ni and Zn. Most other countries include the same metals in their priority lists. However, other highly toxic metals such as Be, Tl, V, Sb, Mo should also be monitored where they are likely to occur.

The absence of iron and manganese in some priority lists results from their frequent classification as major elements. The occurrence of iron in aqueous solution is dependent on environmental conditions, especially oxidation and reduction. Flowing surface water, that is fully aerated, should not contain more than a few micrograms per litre of uncomplexed dissolved iron at equilibrium in the pH range 6.6 to 8.5. In groundwater, however, much higher levels can occur. In anoxic groundwaters with a pH of 6 to 8, ferrous iron (Fe^{2+}) concentrations can be as high as 50 mg l^{-1} and concentrations of 1 to 10 mg l^{-1} are common. The iron originates by solution at sites of either reduction of ferric hydroxides or oxidation of ferrous sulphide (Hem, 1989) and the process is strongly influenced by microbiological activity. Reduced groundwater is clear when first brought from a well but becomes cloudy, and then orange in colour, as oxidation immediately occurs with the precipitation of ferric hydroxide. Consequently, obtaining representative samples for iron determination from groundwaters presents special difficulties. High iron concentrations in groundwater are widely reported from developing countries, where iron is often an important water quality issue. Similar problems can be found in anoxic waters for Mn^{2+}, although the concentrations reached are usually ten times less than ferrous iron.

The concentration of different metals in waters varies over a wide range (0.1–0.001 μg l^{-1}) at background sites and can rise to concentrations which are dangerous for human health in some water bodies influenced by human activities. Dissolved metal concentrations are particularly difficult to measure due to possible contamination during sampling, pre-treatment and storage. As a result, large differences may be observed between analyses performed by highly specialised teams. The natural variability of dissolved metals is not yet fully understood. As dissolved metals occur in very low concentrations, it is recommended that metals are measured in the particulate matter, for which there is much more information on variability, reference

Table 3.5   Variables included in the GEMS/WATER monitoring programme at baseline and global flux stations[1]

| Variable | Baseline stations | | Global river flux stations |
	Streams	Headwater lakes	
Basic monitoring			
Water discharge/level	x		x^2
Total suspended solids	x		x
Transparency		x	
Temperature	x	x	x
pH	x	x	x
Conductivity	x	x	x
Dissolved oxygen	x	x	x
Calcium	x	x	x
Magnesium	x	x	x
Sodium	x	x	x
Potassium	x	x	x
Chloride	x	x	x
Sulphate	x	x	x
Alkalinity	x	x	x
Nitrate plus nitrite	x	x	x
Ammonia	x		x
Total phosphorus, unfiltered	x	x	x
Total phosphorus, dissolved	x	x	x
Reactive silica	x	x	x
Chlorophyll *a*	x	x	x
Expanded monitoring			
Total phosphorus, unfiltered			x
Dissolved organic carbon	x	x	x
Particulate organic carbon			x
Dissolved organic nitrogen	x	x	x
Particulate organic nitrogen			x
Aluminium	x^3	x^3	x^4
Iron	x^3	x^3	x^4
Manganese	x^3	x^3	x^4
Arsenic[5]	x^3	x^3	x^4
Cadmium[5]	x^3	x^3	x^4
Chromium			x^4
Copper			x^4
Lead[5]	x^6	x^6	x^4
Mercury[5]	x^6	x^6	x^4
Selenium			x^4
Zinc[5]	x^3	x^3	x^4

Continued

Table 3.5   Continued

Variable	Baseline stations		Global river flux stations
	Streams	Headwater lakes	
Total hydrocarbons			x^7
Total polyaromatic hydrocarbons			x^7
Total chlorinated hydrocarbons			x^7
Dieldrin			x^7
Aldrin			x^7
Sum of DDTs[5]	x	x	x^7
Atrazine	x	x	x^7
Sum of PCBs[5]			x^7
Phenols			x^7

[1]   The selection of variables for trend monitoring is related to different pollution issues
[2]   Continuous monitoring
[3]   Dissolved only
[4]   Dissolved and particulate
[5]   Included as contaminant monitoring at baseline stations
[6]   Total
[7]   Unfiltered water samples

Source: WHO, 1991

background values, etc. (see Tables 4.1 and 4.2).

The variety of metal species is the main methodological difficulty in designing metal-based monitoring programmes. When checking compliance with water quality guidelines, for example, metals should always be determined in the same forms as those for which the guidelines or standards are set. If the quality standards refer to the dissolved forms of metals, only dissolved forms should be monitored. More than 50 per cent of the total metal present (and up to 99.9 per cent) is usually adsorbed onto suspended particles; this is particularly relevant when assessing metal discharge by rivers (see Chapter 4). Consequently, monitoring and assessment programmes such as GEMS/WATER include the determination of both total (unfiltered) and dissolved (filtered through 0.45 µm filter) concentrations of metals when assessing the flux of contaminants into the oceans. More detailed investigations involving the speciation and partition of the metals are rather complicated and should be carried out in special situations only (Hem, 1989).

3.8.2   Sampling and measurement

Samples for metal analysis are usually pre-treated by acidification prior to transportation to the laboratory to suppress hydrolysis, sorption and other processes which affect concentration. However, such preservation techniques destroy the equilibrium of the different forms of the metals, and can be used only for determination of total concentrations. For determination of dissolved metals, it is recommended that the samples are filtered through 0.45 μm pore diameter membrane filters (using ultra-clean equipment in a laminar flow hood). The filtered sample should be acidified for preservation. Removal of the particulate matter by filtration prevents dissolution or desorption of trace metals from the particulate phase to the dissolved phase within the sample. A very high degree of cleanliness in sample handling at all stages of collection and analysis is necessary (such as, use of ultra-pure acids to clean glassware or PTFE (polytetrafluoroethene) utensils, use of a laminar flow hood for sample manipulation and special laboratories with air filtration and purification systems) to avoid contamination and incorrect results.

The low concentrations of metals in natural waters necessitate determination by instrumental methods. Photometric methods, sometimes in combination with extraction, are the oldest and most inexpensive techniques (see various methods handbooks). However, as these have high detection limits, they can only be used for analysis of comparatively polluted waters. Atomic absorption methods are the most widely used. Atomic absorption with flame atomisation is the most simple and available modification of this method, but application for direct determination of metals is possible only if concentrations exceed 50 μg l^{-1}. In other cases, it is necessary to use preconcentration. Atomic absorption with electrothermal atomisation allows direct determination of metals at virtually the full range of concentrations typically found in freshwaters. However, this is a more expensive method requiring specially trained personnel. Even with this method special measures may be needed to eliminate matrix effects.

Atomic emission spectroscopy methods are able to determine a large number of elements simultaneously. Inductively coupled plasma atomic emission spectrometry is becoming popular, especially in the industrially developed countries, due to its high productivity and wide range of quantifiable determinations (despite the high cost of the equipment and the large argon consumption necessary for creating the plasma atmosphere). In contrast, spectrographical analysis is becoming less popular because it is time and labour intensive and has low accuracy.

3.9   Organic contaminants

3.9.1   General principles
Many thousands of individual organic compounds enter water bodies as a result of human activities. These compounds have significantly different physical, chemical and toxicological properties. Monitoring every individual

compound is not feasible. However, it is possible to select priority organic pollutants based on their prevalence, toxicity and other properties. Mineral oil, petroleum products, phenols, pesticides, polychlorinated biphenyls (PCBs) and surfactants are examples of such classes of compounds. However, these compounds are not monitored in all circumstances, because their determination requires sophisticated instrumentation and highly trained personnel. In the future much effort will be needed in monitoring these classes of compounds because they are becoming widespread and have adverse effects on humans and the aquatic environment.

When selecting a list of variables for a survey of organic contaminants, the gross parameters of TOC, COD and BOD should be included. In addition, during preliminary surveys and in emergencies, the whole range of individual organic compounds should be identified. This requires sophisticated instrumental methods, including gas chromatography (GC), liquid chromatography (LC) and gas chromatography/mass spectrometry (GC/MS), in combination with effective pre-concentration. In intensive surveys, the following classes of organic pollutants should be identified: hydrocarbons (including aromatic and polyaromatic), purgeable halocarbons, chlorinated hydrocarbons, different pesticide groups, PCBs, phenols, phthalate esters, nitrosamines, nitroaromatics, haloethers, benzidine derivatives and dioxins. In most cases, analysis for organic contaminants is performed on unfiltered water samples. However, variations observed in samples from turbid rivers may largely reflect variations in total suspended solids. Consequently, it is recommended that analysis of the less soluble organic contaminants (e.g. organochlorine pesticides) is carried out on the particulate material (collected by filtration or centrifugation) in the samples (see Chapter 4).

3.9.2 Mineral oil and petroleum products

Mineral oil and petroleum products are major pollutants responsible for ecological damage especially in inland surface waters. At present, more than 800 individual compounds have been identified in mineral oils. Among them are low- and high-molecular weight aliphatic, aromatic and naphthenic hydrocarbons (or petroleum products), high-molecular unsaturated heterocyclic compounds (resins and asphaltenes) as well as numerous oxygen, nitrogen and sulphur compounds (Table 3.6).

Oil is distributed in water bodies in different forms: dissolved, film, emulsion and sorbed fractions. Interactions between these fractions are complicated and diverse, and depend on the specific gravities, boiling points, surface tensions, viscosities, solubilities and sorption capabilities of the compounds present. In addition, transformation of oil compounds by

Table 3.6   The main components of mineral oils

Component group	Content (%)
Hydrocarbons:	
paraffinic	10–70
naphthenic (mono- and polycyclic)	25–75
aromatic (mono- and polycyclic)	6–40
naphthenon-aromatic	30–70
Unsaturated heterocyclic compounds:	
resins	1–40
asphaltenes	0–80
asphaltenic acids and their anhydrites	0–7

biochemical, microbiological, chemical and photochemical processes occurs simultaneously. Due to the high ecological risk associated with oil extraction, transportation, refining and use, mineral oil is considered a priority pollutant and its determination is important for assessments related to these activities.

The permissible concentration of mineral oil and petroleum products in water depends on the intended use of the water. The recommended maximum concentrations for drinking water supplies and fisheries protection are generally between 0.01 and 0.1 mg l^{-1}. Concentrations of 0.3 mg l^{-1} or more of crude oil can cause toxic effects in freshwater fish.

Since hydrocarbons are the principal component fraction of oils, the definition "petroleum products" applies only to this fraction to ensure comparability of analytical data. The total concentration of dissolved and emulsified oils is the more usual determination and dissolved, emulsified and other fractions should only be determined separately in special cases.

As oil can be biochemically oxidised very easily, it is necessary to extract it from the sample immediately after sampling with carbon tetrachloride or trichlorotrifluoroethane. The extract can then be stored in a cool, dark place for several months. Gravimetric methods of oil determination are the most simple, but are not very sensitive and can give erroneous results due to the loss of volatile components. Ultra violet (UV), infra red (IR) spectrophotometric and luminescent methods are the most popular. Analysis based on column and thin-layer chromatographic separation allows the possibility of the separate determination of volatile and non-volatile polyaromatic hydrocarbons, resins and asphaltenes. Identification and determination of individual oil components is a complicated analytical task which can be undertaken only with the application of capillary gas chromatography, with either mass-spectrometric detection or luminescent spectrometry.

3.9.3   Phenols

Phenols are an important group of pollutants which enter water bodies in the waste discharges of many different industries. They are also formed naturally during the metabolism of aquatic organisms, biochemical decay and transformation of organic matter, in the water column and in bottom sediments.

Phenols are aromatic compounds with one or few hydroxy groups. They are easily biochemically, photochemically or chemically oxidised. As a result, they have detrimental effects on the quality and ecological condition of water bodies through direct effects on living organisms and the significant alteration of biogeneous elements and dissolved gases, principally oxygen.

The presence of phenols causes a marked deterioration in the organoleptic characteristics of water and as a result they are strictly controlled in drinking water and drinking water supplies. Concentrations of phenols in unpolluted waters are usually less than 0.02 mg l^{-1}. However, toxic effects on fish can be observed at concentrations of 0.01 mg l^{-1} and above.

Phenols are usually divided into two groups: steam-distillable phenols (phenol, cresols, xylenes, chlorphenols, etc.) and non-distillable phenols (catechol, hydroquinone, naphthols, etc.). The analytical method used with steam distillation determines only the volatile phenol fractions; these have the worst effects on organoleptic water characteristics. The method does not detect non-volatile phenols which, unfortunately, are often present in greater quantities than the volatile phenols and, furthermore, tend to be highly toxic. Chromatographic determination of individual phenols is more informative, but requires sophisticated instrumentation.

Samples, particularly if required for the determination of volatile phenols, must not be stored for long periods and, ideally, determination should be carried out within four hours. If this is not possible, samples can be preserved with sodium hydroxide and stored for 3–4 days at 2–4 °C.

3.9.4   Pesticides

Pesticides are chemical compounds toxic to certain living organisms, from bacteria and fungi up to higher plants and even mammals. Most pesticides are compounds which do not occur naturally in the environment and, therefore, detectable concentrations indicate pollution. There are approximately 10,000 different pesticides currently available. The most widely used are insecticides (for extermination of insects), herbicides (for extermination of weeds and other undesirable plants) and fungicides (for preventing fungal diseases).

The mode of action of a pesticide is determined by its chemical structure. These structures are similar for the related compounds which comprise separate classes of pesticides such as the organochlorine pesticides, organophosphorus pesticides, the carbamate pesticides, the triazine herbicides and chlorphenolic acids.

The monitoring of pesticides presents considerable difficulties, particularly for groundwaters. There is a wide range of pesticides in common agricultural use, and many of them break down into toxic products. Screening of water samples for all compounds is very expensive; therefore, a preliminary survey of local pesticide use needs to be carried out to reduce the number of target compounds in each specific assessment programme.

Internationally there is considerable variation between, and uncertainty over, guidelines on permissible concentrations of pesticides in drinking water. Nevertheless, the guideline values are at the microgram per litre level which, for the most toxic compounds, are close to the limits of analytical detection. Highly sophisticated analytical procedures are necessary, which normally require a combination of a directly-coupled gas chromatograph (GC) and mass spectrometer (MS). Some pesticides are very polar and cannot be extracted from water for injection into a GC. Furthermore, some other pesticides are too chemically unstable to be heated in a GC. High pressure liquid chromatography is showing promise as a means of isolating compounds in both of these categories. A high degree of cleanliness is necessary for sample handling at all stages, such as the use of ultra-clean solvents to clean glass or stainless steel apparatus.

Organochlorine pesticides
Environmental levels of organochlorine pesticides tend to be higher than other pesticides because of their widespread and prolonged use, combined with their great chemical stability. In the 1950s, DDT was used liberally around the world, but at the beginning of the 1970s most countries severely limited, or even prohibited, its use. However, concentrations of DDT and its metabolites (DDD, DDE) are still high in many environments, especially in arid areas.

Organochlorine pesticides are chlorine derivatives of polynuclear hydrocarbons (e.g. DDT), cycloparaffins (e.g. hexachlorocyclohexane (HCH)), compounds of the diene series (e.g. heptachlor) and aliphatic carbonic acids (e.g. propanide). Most of the compounds are hydrophobic (insoluble in water) but highly soluble in hydrocarbons and fats. They have the ability to accumulate in biological tissues, reaching much higher concentrations in certain aquatic biota than in the surrounding water and sediments (see

Chapter 5). The affinity of pesticides for adsorption onto mineral suspended matter and organic colloids is important for their distribution and mobility in water bodies. Bottom sediments also play a significant role in storage and transformation of organochlorine pesticides.

Where present, concentrations of organochlorine pesticides in water bodies tend to be in the range 10^{-5}–10^{-3} mg l^{-1}. These compounds and their metabolites have been found in sites as distant as the Arctic and Antarctic regions as a result of long-range atmospheric transport. They are sometimes found in groundwaters, where leaching from disposal sites for hazardous substances or from agricultural land usually accounts for their presence. As these compounds are hydrophobic, their occurrence in groundwater may be the result of "solubilisation" in fulvic acid materials.

Due to their toxicity, the maximum allowable concentrations of organochlorine pesticides must be strictly adhered to in waters important for their fish communities or used for drinking water supplies.

Organophosphorus pesticides
Organophosphorus pesticides are complex esters of phosphoric, thiophosphoric and other phosphorus acids. They are widely applied as insecticides, acaricides and defoliants. Their relatively low chemical and biochemical stability is an advantage because many decompose in the environment within a month. Organophosphorus pesticides, like organochlorine pesticides, are readily adsorbed onto suspended matter. Photolysis, as well as hydrolitic, oxidation and enzyme decay processes are the principal mechanisms of decay, resulting in detoxification. When found, the concentrations of organophosphorus pesticides in surface waters range from 10^{-3}–10^{-2} mg l^{-1}.

Unfiltered samples for organochlorine and organophosphorus determination should be collected in glass containers with PTFE caps. Samples can be stored for a short time at low temperature. However, immediate extraction followed by storage at −15 °C is preferable. In this case, samples may be stored for up to three weeks.

3.9.5   Surfactants
Synthetic surfactants are compounds belonging to different chemical classes but containing a weak-polar hydrophobic radical (e.g. alkyl or alkylaryl) and one or more polar groups. They can be classified into anionic (negatively charged), cationic (positively charged) and nonionic (non-ionising). Anionic surfactants are the most widely produced and used, usually as detergents.

Surfactants enter water bodies with industrial and household wastewaters. Atmospheric inputs (originating from atmospheric discharges from surfactant-producing plants) in the form of precipitation are also significant. Surfactants can exist in surface waters in the dissolved or adsorbed states, as well as in the surface film of water bodies, because they have a pronounced ability to concentrate at the air–water or water–sediment interface. Although surfactants are not highly toxic, they can affect aquatic biota. Detergents can impart taste or odour to water at concentrations of 0.4–3 mg l^{-1} and chlorination can increase this effect. Surfactants are responsible for foam formation in surface waters and other pollutants, including pathogens, can become concentrated in the foam. The presence of foam on the water surface makes water aeration difficult, lowering oxygen levels, reducing self-purification processes and adversely affecting aquatic biota. The threshold concentration for foam formation is 0.1–0.5 mg l^{-1} depending on the structure of the surfactant.

In terms of biodegradability, surfactants are divided into highly degradable, intermediate and stable, or non-degradable, with corresponding biochemical oxidation rate constants of > 0.30 day^{-1}, 0.30–0.05 day^{-1} and < 0.05 day^{-1} respectively. In recent years there has been a tendency to substitute non-degradable surfactants for degradable ones. However, this approach to reducing pollution has the drawback of causing a significant decrease in dissolved oxygen concentrations.

The inherent properties of surfactants require special procedures for sample preservation, principally to avoid foam formation and their adsorption onto the walls of the sample containers. Photometric methods are the most widely used for determination of all three types of surfactants and are well documented. Analytical methods are sufficiently simple and convenient for application in any laboratory.

3.10 Microbiological indicators

The most common risk to human health associated with water stems from the presence of disease-causing micro-organisms. Many of these micro-organisms originate from water polluted with human excrement. Human faeces can contain a variety of intestinal pathogens which cause diseases ranging from mild gastro-enteritis to the serious, and possibly fatal, dysentery, cholera and typhoid. Depending on the prevalence of certain other diseases in a community, other viruses and parasites may also be present. Freshwaters also contain indigenous micro-organisms, including bacteria, fungi, protozoa (single-celled organisms) and algae (micro-organisms with photosynthetic pigments), a few of which are known to produce toxins and transmit, or cause, diseases.

Intestinal bacterial pathogens are distributed world-wide, the most common water-borne bacterial pathogens being *Salmonella*, *Shigella*, enterotoxigenic *Escherichia coli*, *Campylobacter*, *Vibrio* and *Yersinia*. Other pathogens occasionally found include *Mycobacterium*, *Pasteurella*, *Leptospira* and *Legionella* and the enteroviruses (poliovirus, echo virus and Coxsackie virus). Adenoviruses, reoviruses, rotaviruses and the hepatitis virus may also occur in water bodies. All viruses are highly infectious. *Salmonella* species, responsible for typhoid, paratyphoid, gastro-enteritis and food poisoning, can be excreted by an apparently healthy person acting as a carrier and they can also be carried by some birds and animals. Therefore, contamination of water bodies by animal or human excrement introduces the risk of infection to those who use the water for drinking, food preparation, personal hygiene and even recreation.

Sewage, agricultural and urban run-off, and domestic wastewaters are widely discharged to water bodies, particularly rivers. Pathogens associated with these discharges subsequently become distributed through the water body presenting a risk to downstream water users. Typical municipal raw sewage can contain 10 to 100 million coliform bacteria (bacteria originating from the gut) per 100 ml, and 1 to 50 million *Escherichia coli* or faecal streptococci per 100 ml. Different levels of wastewater treatment may reduce this by a factor of 10 to 100 and concentrations are reduced further after dilution by the receiving waters.

The practice of land application of wastewaters, particularly poorly treated wastewaters, can lead to pathogen contamination of surface and groundwaters. Surface water contamination is usually as a result of careless spraying or run-off, and groundwater pollution arises from rapid percolation through soils. Other sources of pathogens are run-off and leachates from sanitary landfills and urban solid waste disposal sites which contain domestic animal and human faecal material. The use of water bodies by domestic livestock and wildlife is also a potential source of pathogens.

The survival of microbiological pathogens, once discharged into a water body, is highly variable depending on the quality of the receiving waters, particularly the turbidity, oxygen levels, nutrients and temperature. *Salmonella* bacilli have been reported in excess of 50 miles downstream of the point source, indicating an ability to survive, under the right conditions, for several days. Once in a water body, micro-organisms often become adsorbed onto sand, clay and sediment particles. The settling of these particles results in the accumulation of the organisms in river and lake sediments. The speed at which the settling occurs depends on the velocity and turbulence of the water body. Some removal of micro-organisms from the water column also occurs as a result of predation by filter feeding microzooplankton.

Counts of bacteria of faecal origin in rivers and lakes around the world which suffer little human impact vary from < 1 to 3,000 organisms per 100 ml. However, water bodies in areas of high population density can have counts up to 10 million organisms per 100 ml. Natural groundwaters should contain no faecal bacteria unless contaminated, whereas surface waters, even in remote mountain areas, may contain up to 100 per 100 ml. To avoid human infection, the WHO recommended concentration for drinking water is zero organisms per 100 ml. Detection of pathogens other than faecal bacteria, particularly viruses, is less common partly due to the lack of appropriate, routinely available methodology. *Salmonella* organisms have been recorded at concentrations 10 to 20 times less than the faecal coliform numbers in the same sample. Where faecal coliform bacteria counts are high, viruses may also be detectable, but only in volumes of 20 to 100 litres of water. Enteroviruses occur in raw sewage at very much lower concentrations than bacterial pathogens; measured as plaque forming units (PFUs) they rarely occur at more than 1,000 units per litre.

Monitoring for the presence of pathogenic bacteria is an essential component of any water quality assessment where water use, directly or indirectly, leads to human ingestion. Such uses include drinking, personal hygiene, recreation (e.g. swimming, boating), irrigation of food crops and food washing and processing. Monitoring to detect pathogens can be carried out without accompanying physical and chemical measurements and, therefore, can be very inexpensive.

Prior to using any new drinking water source it should be examined for the presence of faecal bacteria. Sampling localities should be carefully chosen so that the source of the contamination can be identified and removed. Even when drinking water sources have been subjected to treatment and disinfection, it is essential that routine examination of the supply is carried out at weekly, or even daily intervals where the population at risk is large (tens or hundreds of thousands). Where water is used mainly for personal hygiene or recreation, there is still a risk of accidental ingestion of intestinal pathogens as well as a risk of other infections, particularly in the eyes, ears and nose. Less than 10^3 coliforms per 100 ml presents little risk of intestinal diseases although the risk of virus-borne infections always remains.

Where irrigation with wastewater is carried out by spraying food crops, it is advisable to monitor for faecal bacteria as there is a risk of contamination to those eating the crop. This risk is less when irrigation is ceased some time before harvest, as many bacteria do not survive for long periods unless in ideal conditions of temperature and nutrients. The use of contaminated water in any stage of food processing presents a serious risk to human health as food provides an ideal growth medium. All water which may come into contact with food must, therefore, be checked for faecal contamination. Where treated water is temporarily stored in a tank it should be examined immediately prior to use.

Faecal contamination can be measured to indicate the presence of organic pollution of human origin. Other naturally present micro-organisms, such as the algal and protozoan communities may, however, in some situations be more useful to gain an insight into the level of pollution (see Chapter 5).

Methods for detection of the presence of faecal material have been developed which are based on the presence of "indicator" organisms, such as the normal intestinal bacterium *Escherichia coli* (see Chapter 5 for further details). Such methods are cheap and simple to perform and some have been developed into field kits, particularly for use in developing countries (Bartram and Ballance, 1996). Positive identification of the pathogenic bacteria *Salmonella*, *Shigella* or *Vibrio* spp. can be quite complex, requiring several different methods. A special survey may be undertaken if a source of an epidemic is suspected, or if a new drinking water supply is being tested. As these organisms usually occur in very low numbers in water samples, it is necessary to concentrate the samples by a filtration technique prior to the analysis. Although methodologies for identification of viruses are constantly being improved and simplified, they require advanced and expensive laboratory facilities. Local or regional authorities responsible for water quality may be unable to provide such facilities. However, suitably collected and prepared samples can easily be transported, making it feasible to have one national or regional laboratory capable of such analyses. Sample collection kits have been developed for use in such situations.

3.11   Selection of variables

The selection of variables to be included in a water quality assessment must be related to the objectives of the programme (see Chapters 1 and 2). The various types of monitoring operations and their principal uses have been given in Table 2.1. Broadly, assessments can be divided into two categories, use-orientated and impact-orientated. In addition, operational surveillance can be used to check the efficiency of water treatment processes by monitoring the quality of effluents or treated waters, but this is not discussed here.

3.11.1   Selection of variables in relation to water use

Use-orientated assessment tests whether water quality is satisfactory for specific purposes, such as drinking water supply, industrial use or irrigation. Many water uses have specific requirements with respect to physical and chemical variables or contaminants. In some cases, therefore, the required quality of the water has been defined by guidelines, standards or maximum allowable concentrations (see Table 3.4). These consist of recommended (as in the case of guidelines) or mandatory (as in the case of standards) concentrations of selected variables which should not be exceeded for the prescribed water use. For some variables, the defined concentrations vary from country to country (Table 3.4). Existing guidelines and standards define the minimum set of variables for inclusion in assessment programmes. Tables 3.7 and 3.8 suggest variables appropriate to specific water uses and can be used where guidelines are not available. Other variables can also be monitored, if necessary, according to special conditions related to the intended use. Acceptable water quality is also related to water availability. When water is scarce, a lower level of quality may have to be accepted and the variables measured can be kept to a minimum.

Background monitoring

The water quality of unpolluted water bodies is dependent on the local geological, biological and climatological conditions. These conditions control the mineral quality, ion balances and biological cycles of the water body. To preserve the quality of the aquatic environment, the natural balances should be maintained. A knowledge of the background quality is necessary to assess the suitability of water for use and to detect future human impacts. Background quality also serves as a "control" for comparison with conditions at sites presently suffering from anthropogenic impacts. In most assessment programmes, some variables related to background water quality are always included, such as those suggested in Table 3.7.

Aquatic life and fisheries

Individual aquatic organisms have different requirements with respect to the physical and chemical characteristics of a water body. Available oxygen, adequate nutrients or food supply, and the absence of toxic chemicals are essential factors for growth and reproduction. Various guidelines have been proposed for waters important for fisheries or aquatic life (Table 3.4). Detailed information has been prepared by the European Inland Fisheries Advisory Commission (EIFAC) of the Food and Agriculture Organization of the United Nations (FAO) (EIFAC, 1964 onwards). As fish are an essential source of protein for man, it is imperative to avoid accumulation of contaminants in fish or shellfish (see Chapter 5). Suggested variables for inclusion in an assessment programme aimed at protecting aquatic life and fisheries are given in Table 3.7.

Table 3.7   Selection of variables for assessment of water quality in relation to non-industrial water use[1]

	Background monitoring	Aquatic life and fisheries	Drinking water sources	Recreation and health	Agriculture	
					Irrigation	Livestock watering
General variables						
Temperature	xxx	xxx		x		
Colour	xx		xx	xx		
Odour			xx	xx		
Suspended solids	xxx	xxx	xxx	xxx		
Turbidity/transparency	x	xx	xx	xx		
Conductivity	xx	x	x		x	
Total dissolved solids		x	x		xxx	x
pH	xxx	xx	x	x	xx	
Dissolved oxygen	xxx	xxx	x		x	
Hardness		x	xx			
Chlorophyll *a*	x	xx	xx	xx		
Nutrients						
Ammonia	x	xxx	x			
Nitrate/nitrite	xx	x	xxx			xx
Phosphorus or phosphate	xx					
Organic matter						
TOC	xx		x	x		
COD	xx	xx				
BOD	xxx	xxx	xx			
Major ions						
Sodium	x		x		xxx	
Potassium	x					
Calcium	x				x	x
Magnesium	xx		x			
Chloride	xx		x		xxx	
Sulphate	x		x			x
Other inorganic variables						
Fluoride			xx		x	x
Boron					xx	x
Cyanide		x	x			
Trace elements						
Heavy metals		xx	xxx		x	x
Arsenic & selenium		xx	xx		x	x

Continued

Table 3.7   Continued

	Background monitoring	Aquatic life and fisheries	Drinking water sources	Recreation and health	Agriculture Irrigation	Agriculture Livestock watering
Organic contaminants						
Oil and hydrocarbons		x	xx	xx	x	x
Organic solvents		x	xxx^2			x
Phenols		x	xx			x
Pesticides		xx	xx			x
Surfactants		x	x	x		x
Microbiological indicators						
Faecal coliforms			xxx	xxx	xxx	
Total coliforms			xxx	xxx	x	
Pathogens			xxx	xxx	x	xx

TOC   Total organic carbon
BOD   Biochemical oxygen demand
COD   Chemical oxygen demand

x – xxx Low to high likelihood that the concentration of the variable will be affected and the more important it is to include the variable in a monitoring programme.
Variables stipulated in local guidelines or standards for a specific water use should be included when monitoring for that specific use.
The selection of variables should only include those most appropriate to local conditions and it may be necessary to include other variables not indicated under the above headings.

[1]   For industrial uses see Table 3.8
[2]   Extremely important in groundwater

Drinking water sources

In some regions groundwater, or water from rivers and lakes, is used for drinking without treatment. In other areas, it is subjected to treatment and/or disinfection before use. In both cases, the water which is eventually consumed should be monitored for variables which may pose a potential risk to human health. Guidelines for maximum concentrations of such variables in drinking water have been set by WHO (WHO, 1993) and regional and national authorities such as the Commission of the European Communities (CEC, 1980), US EPA (US EPA, 1993) and Environment Canada (Environment Canada, 1987) (Table 3.4). Drinking water sources should also be monitored to establish the required level of water treatment and to detect any contaminants which may not be removed during treatment, or which may interfere with the treatment process. Water ready for distribution and consumption can also be monitored to check the efficiency of the treatment process. Table 3.7 lists variables which should be measured with respect to recommended guidelines and those which are potentially a problem for drinking water sources. The selection of those which would actually be included in an assessment programme depends on the nature of the water source and the level of subsequent treatment. Many groundwater sources require minimal or no treatment and, consequently, need only be monitored infrequently for such variables as pathogens and organic solvents.

Recreation and health

Besides drinking, human populations use water for hygiene purposes (e.g. washing) and recreation (e.g. swimming and boating). Such activities have an associated health risk if the water is of poor quality due to the possibility of ingesting small quantities, or the pathogens directly entering the eyes, nose, ears or open wounds. Most recommended variables with respect to recreation and health are associated with pathogens or the aesthetic quality of the water (see Table 3.7). Guideline values are usually set with respect to use of the water for swimming (e.g. CEC, 1976) and other water-contact sports.

Agricultural use

Irrigation of food crops presents a possible health risk to food consumers if the quality of the irrigation water is inadequate, particularly with respect to pathogens and toxic compounds. The risk is greatest when the water is sprayed directly onto the crop rather than flooded around the base of the plants. The presence of certain inorganic ions can also affect the soil quality and, therefore, the growth potential of the crops. Recommended guidelines have been set for some variables in irrigation water (e.g. Environment Canada, 1987) but higher levels may be tolerated if water is scarce. Suggested variables for the monitoring of irrigation water are included in Table 3.7.

In principle, water for livestock watering should be of high quality to prevent livestock disease, salt imbalance or poisoning from toxic compounds. Nevertheless, higher levels of suspended solids and salinity may be tolerated by certain livestock than by humans. Many of the variables included in monitoring the quality of livestock water are the same as for drinking water sources (see Table 3.7).

Industrial uses

The requirements of industry for water quality are diverse, depending on the nature of the industry and the individual processes using water within that industry. Table 3.8 summarises some of the key variables for some major industrial uses or processes. Although some guidelines have been proposed, they need to be considered in relation to the specific industrial needs and water availability.

Table 3.8   Selection of variables for the assessment of water quality in relation to some key industrial uses

	Heating	Cooling	Power generation	Iron and steel	Pulp and paper	Petrol	Food processing
General variables							
Temperature	xxx	xxx		xxx	x		
Colour	x				x		xx
Odour							xxx
Suspended solids	xxx	xxx	xx	xx	x	xxx	xx
Turbidity	xx				xx		xx
Conductivity	x	x					
Dissolved solids	xx	xx	xxx	xx	xxx	x	xxx
pH	x	xxx	xxx	xx	xx	xxx	xxx
Dissolved oxygen	xxx		x	xxx	x		
Hardness	xxx	xx	xxx	xx	xxx	xxx	xxx
Nutrients							
Ammonia	xxx		x				x
Nitrate						x	xx
Phosphate					x		
Organic matter							
COD		x	xx				
Major ions							
Calcium		xxx	xxx		x	xxx	x
Magnesium			x		x	xxx	x
Carbonate components	xx		xxx		xxx	x	x
Chloride	x	x	xx	xx	x	xxx	xxx
Sulphate		x	xx	xx	xx	x	xxx
Other inorganic variables							
Hydrogen sulphide	xxx	x					xx
Silica	xx	xx	x		x	x	x
Fluoride						x	xx
Trace elements							x
Aluminium		x	x				
Copper		x	x				
Iron	xx	x	x		x	x	xx
Manganese	xx	x	x		x		xx
Zinc			x				
Organic contaminants							
Oil & hydrocarbons	x	x	x	x			x
Organic solvents							x

Continued

Table 3.8   Continued

	Heating	Cooling	Power generation	Iron and steel	Pulp and paper	Petrol	Food processing
Phenols							x
Pesticides							x
Surfactants	x	x	x				x
Microbiological indicators							
Pathogens							xxx

COD   Chemical oxygen demand

x – xxx Low to high likelihood that the concentration of the variable will be affected and the more important it is to include the variable in a monitoring programme.

The precise selection of variables depends on the required quality of the water in the individual industrial processes and any standards or guidelines that are applied.

3.11.2 Selection of variables in relation to pollutant sources

Water quality assessment often examines the effects of specific activities on water quality. Typically, such assessment is undertaken in relation to effluent discharges, urban or land run-off or accidental pollution incidents. The selection of variables is governed by knowledge of the pollution sources and the expected impacts on the receiving water body. It is also desirable to know the quality of the water prior to anthropogenic inputs. This can be obtained, for example, by monitoring upstream in a river or prior to the development of a proposed waste disposal facility. When this cannot be done, background water quality from an adjacent, uncontaminated, water body in the same catchment can be used. Appropriate variables for assessing water quality in relation to several major sources of pollutants are given in Tables 3.9 and 3.10.

Sewage and municipal wastewater

Municipal wastewaters consist of sewage effluents, urban drainage and other collected wastewaters. They usually contain high levels of faecal material and organic matter. Therefore, to assess the impact of such wastewaters it is advisable to measure variables which are indicative of organic waste such as BOD, COD, chloride, ammonia and nitrogen compounds. If the wastes contain sewage, then faecal indicators are also important. Depending on the collection and treatment systems in operation, municipal wastes may contain various other organic and inorganic contaminants of industrial origin. Suggested variables for inclusion in an assessment programme are given in Table 3.9. Effluents from food processing also contain large amounts of organic matter and even pathogens, therefore, variables used to monitor the effects

Table 3.9   Selection of variables for the assessment of water quality in relation to non-industrial pollution sources

	Sewage and municipal wastewater[1]	Urban run-off	Agricultural activities	Waste disposal to land		Long range atmospheric transport
				Solid municipal	Hazardous chemicals	
General variables						
Temperature	x	x	x			
Colour	x	x	x	x		
Odour	x	x	x			
Residues	x	x	xxx	xxx	xx	
Suspended solids	xxx	xx	xxx	xx	xx	
Conductivity	xx	xx	xx	xxx	xxx	xxx
Alkalinity				xx		xxx
pH	x	x	x	xx	xxx	xxx
Eh	x	x	x			
Dissolved oxygen	xxx	xxx	xxx	xxx	xxx	
Hardness	x	x	x		x	x
Nutrients						
Ammonia	xxx	xx	xxx	xx		
Nitrate/nitrite	xxx	xx	xxx	xx		xxx
Organic nitrogen	xxx	xx	xxx	xx		
Phosphorus compounds	xxx	xx	xxx	x		x
Organic matter						
TOC	x	x	x			
COD	xx	xx	x	xxx	xxx	
BOD	xxx	xx	xxx	xxx	xx	
Major ions						
Sodium	xx	xx	xx			
Potassium	x	x	x			
Calcium	x	x	x			
Magnesium	x	x	x			
Carbonate components				x		
Chloride	xxx	xx	xxx	xx	xx	
Sulphate	x	x	x			xxx
Other inorganic variables						
Sulphide	xx	xx	x		x	
Silica	x	x				
Fluoride	x	x				
Boron			x			

Continued

Table 3.9   Continued

	Sewage and municipal wastewater[1]	Urban run-off	Agricultural activities	Waste disposal to land		Long range atmospheric transport
				Solid municipal	Hazardous chemicals	
Trace elements						
Aluminium						xx
Cadmium		x		xxx	xxx	x
Chromium		x		xxx	xx	x
Copper	x	x	xx[2]	xxx	xx	x
Iron	xx	xx		xxx	xx	x
Lead	xx	xxx		xxx	xx	xx
Mercury	x	x	xxx[2]	xxx	xxx	
Zinc		x	xx[2]	xxx	xx	x
Arsenic		x	xxx[2]	xx	xxx	x
Selenium		x	xxx[2]	x	x	
Organic contaminants						
Fats	x	x				
Oil and hydrocarbons	xx	xxx		xx	x	
Organic solvents	x	x		xxx	xxx	
Methane				xxx[3]		
Phenols	x			xx	xx	
Pesticides		x	xxx[4]	xx	xxx	xxx
Surfactants	xx		x		x	
Microbiological indicators						
Faecal coliforms	xxx	xx	xx	xxx		
Other pathogens	xxx		xx	xxx		

TOC   Total organic carbon
COD   Chemical oxygen demand
BOD   Biochemical oxygen demand

x –xxx Low to high likelihood that the concentration of the variable will be affected and the more important it is to include the variable in a monitoring programme.

The final selection of variables is also dependent on the nature of the water body.

[1]   Assumes negligible industrial inputs to the wastewater
[2]   Need only be measured when used locally or occur naturally at high concentrations
[3]   Important only for groundwater in localised industrial areas
[4]   Specific compounds should be measured according to their level of use in the region.

of food processing operations (Table 3.10) are similar to those for sewage and organic wastewater.

Urban run-off

Rivers running through, or lakes adjacent to, large urban developments are inevitably subjected to urban run-off during periods of heavy rain. In some cities, rain water is collected in drains and directed through the sewage collection and treatment facilities before discharge to the river or lake. In other cities, rainwater is channelled directly into the nearest water body. Even where urban run-off is collected in the sewerage system, excessive rainfall can lead to an overload which by-passes the sewage treatment plants. Variables associated with urban run-off are largely the same as those selected for municipal wastewater (Table 3.9). However, water quality problems particularly associated with urban run-off are high levels of oil products and lead (both arising from the use of automobiles), as well as a variety of other metals and contaminants associated with local industrial activity.

Agricultural activities
Impacts relating to agricultural activities principally concern organic and inorganic matter (such as arising from intensive animal rearing and land run-off associated with land clearing) and those chemicals incorporated in fertilisers and pesticides. Irrigation, especially in arid areas, can lead to salinisation of surface and groundwaters and, therefore, inclusion of conductivity, chloride, alkalinity, sulphate, fluoride and sodium is important in water quality assessment programmes in these areas. Suggestions for necessary variables in relation to agricultural activities are given in Table 3.9.

Land disposal of solid municipal and hazardous wastes
In most countries, municipal solid waste is dumped in designated land sites. Similar sites are also used for specific industrial, hazardous wastes which are too toxic to be released to the environment, and are usually sealed in containers. Land disposal sites are often poorly planned and controlled, resulting in the formation of leachates which pose particular risks for groundwaters in the vicinity of the sites (see section 9.4.3). Leachates can contain many contaminants, including pathogens, metals and organic chemicals, depending on the material deposited at the site. Leachates from municipal wastes are usually rich in biodegradable organic matter. Recently, it has been recognised that it is important to include the monitoring of methane in groundwaters close to land waste disposal sites (Table 3.9).

Atmospheric sources
Studies of atmospheric pollutants are constantly increasing the number of variables which need to be included in water quality assessment programmes. Acidic depositions lead to a loss of the acid neutralising capacity or alkalinity, which in turn decreases the pH and affects the normal chemical balance of water bodies. Assessment programmes for lakes in susceptible regions

should include alkalinity, pH, sulphate and nitrate (see Table 3.9). Assessment of atmospheric impact with respect to contaminants depends on local and regional sources of emissions. However, widespread atmospheric transport has been proven for lead, cadmium, arsenic, certain pesticides and other organic compounds.

Industrial effluents and emissions
There are few industries which do not make use of water, either directly as part of the manufactured product or indirectly for cooling, cleaning and circulating. Many of these activities generate liquid effluents which may contain many different chemicals, as well as organic matter, depending on the nature of the industrial processes involved. Assessment programmes for water bodies receiving industrial waste, or close to industrial developments, should include variables selected to indicate background water quality and others selected in relation to the local industrial processes, especially contaminants which may cause harm to the environment or make it unsuitable for other uses. The choice of contaminants should be based on an inventory of the chemicals used and discharged during the industrial processes. Some suggestions of appropriate variables for the principal industrial sectors are given in Table 3.10. When water is polluted as the result of an industrial accident the number of variables to be monitored in the receiving water body may be restricted to those known to be accidentally released (or normally used or produced in the industrial process), together with those physical, chemical and biological variables likely to be affected.

3.12   Summary and recommendations
This chapter provides the basic information necessary to aid selection of appropriate variables for the major types of assessment programmes. These variables should be selected in relation to other methods which may be appropriate, such as analysis of particulate and biological material. The choice of variables will also be influenced by the ability of an organisation to provide the facilities, and suitably trained operators, to enable the selected measurements to be made accurately. Further information on recommended field measurement and laboratory techniques is available in the companion volume to this guidebook by Bartram and Ballance (1996).

The suggested variables for different types of assessment given in Section 3.11 are based on common situations and should be taken as guides only. Full selection of variables must be made in relation to assessment objectives and specific knowledge of each individual situation.

Table 3.10 Selection of variables for the assessment of water quality in relation to some common industrial sources of pollution

	Food processing	Mining	Oil extraction/ refining	Chemical/ pharmaceut.	Pulp and paper	Metallurgy	Machine production	Textiles
General variables								
Temperature	x	x	x	x	x	x	x	x
Colour	x	x	x	x	x	x	x	x
Odour	x	x	x	x	x	x	x	x
Residues	x	x	x	x	x	x	x	x
Suspended solids	x	xxx	xxx	x	xxx	xxx	xxx	xxx
Conductivity	xxx	xxx	xxx	x	xxx	xxx	xxx	xxx
pH	xxx	xxx	x	xxx	x	xxx	x	x
Eh	x	x	x	x	x	x	x	x
Dissolved oxygen	xxx	xxx	xxx	xxx	xxx	x	x	xxx
Hardness	x	x	x	x	x	xx	x	x
Nutrients								
Ammonia	xxx	x	xx	xx	x	x	x	x
Nitrate/nitrite	xx	x		xx	x	x		x
Organic nitrogen	xx			x	x			x
Phosphorus compounds	xx			xx			x	x
Organic matter								
TOC	x	x	x	xx	xxx	x	x	x
COD	x	x	x	xxx	xxx	x	x	x
BOD	xxx	x	xxx	xx	xxx	x	x	xxx

Continued

Table 3.10 Continued

	Food processing	Mining	Oil extraction/ refining	Chemical/ pharmaceut.	Pulp and paper	Metallurgy	Machine production	Textiles
Major ions								
Sodium	x	x	x	x				x
Potassium	x	x	x	x				x
Calcium	x	x	x	x	x	xx	x	x
Magnesium	x	x	x	x	x	x		x
Carbonate components	x	x	x	x				
Chloride	xx	xxx	xx	xx	x	x	x	xxx
Sulphate	x	x	xx	xx	xxx	x	x	x
Other inorganic variables								
Sulphide		x	xxx	xxx	xxx	xxx		x
Silica		x	x	x			x	x
Fluoride		x	x	xx		x		x
Boron		x	x	x	x	x	x	x
Cyanide		x		x		x	x	x
Trace elements								
Heavy metals		xxx	xx	xx	x	xxx	xxx	xx
Arsenic		x		x		x		x
Selenium		x		x		x	x	x
Organic contaminants								
Fats	xx							
Oil and hydrocarbons			xxx	xx		xx	xxx	x

Continued

Table 3.10 Continued

	Food processing	Mining	Oil extraction/ refining	Chemical/ pharmaceut.	Pulp and paper	Metallurgy	Machine production	Textiles
Organic solvents				xxx	xxx		x	x
Phenols	x		xx	xxx	xxx	x	x	x
Pesticides	x			xxx				
Other organics				xxx	xxx	x		
Surfactants	xx		xx	xxx	x	x	x	xx
Microbiological indicators								
Faecal coliforms	xxx							
Other pathogens	xxx							

x – xxx Low to high likelihood that the concentration of the variable will be affected and the more important it is to include the variable in a monitoring programme. The final selection of variables to be monitored depends on the products manufactured or processed together with any compounds present in local industrial effluents. Any standards or guidelines for specific variables should also be taken into consideration.

TOC Total organic carbon
COD Chemical oxygen demand
BOD Biochemical oxygen demand

3.13   References

AOAC 1990 *Official Methods of Analysis of the Association of Official Analytical Chemists*. 15th edition. Association of Official Analytical Chemists, Washington D.C.

APHA 1989 *Standard Methods for the Examination of Water and Wastewater*. 17th edition, American Public Health Association, Washington D.C., 1,268 pp.

Bartram, J. and Ballance, R. [Eds] 1996 *Water Quality Monitoring: A Practical Guide to the Design of Freshwater Quality Studies and Monitoring Programmes*. Chapman & Hall, London.

Bestemyanov, G.P. and Krotov, Ju.G. 1985 *Maximum Allowable Concentrations of Chemicals in the Environment*. Khimiya, Leningrad, [In Russian].

CEC (Commission of European Communities) 1976 Council Directive of 8 December 1975 concerning the quality of bathing water, (76/160/EEC). *Official Journal*, **L/31**, 1-7.

CEC (Commission of European Communities) 1978 Council Directive of 18 July 1978 on the quality of fresh waters needing protection or improvement in order to support fish life, (78/659/EEC). *Official Journal*, **L/222**, 1-10.

CEC (Commission of European Communities) 1980 Council Directive of 15 July 1980 relating to the quality of water intended for human consumption, (80/778/EEC). *Official Journal*, **L/229**, 23.

Committee for Fisheries 1993 *List of Maximum Allowable Concentrations and Approximately Harmless Levels of Impact of Toxic Chemicals on Water Bodies of Fisheries Importance*. Kolos, Moscow.

Dessery, S., Dulac, C., Lawrenceau, J.M. and Meybeck, M. 1984 Evolution du carbone organique particulaire algal et détritique dans trois rivières du Bassin Parisien. *Arch. Hydrobiol.*, **100** (2), 235-260.

EIFAC (European Inland Fisheries Advisory Commission) 1964 onwards Working Party on Water Quality Criteria for European Freshwater Fish. *Water Quality Criteria for European Freshwater Fish,* EIFAC Technical Paper Series (various titles), Food and Agriculture Organization of the United Nations, Rome.

Environment Canada 1979 *Water Quality Source Book. A Guide to Water Quality Parameters*. Environment Canada, Ottawa, 89 pp.

Environment Canada 1987 *Canadian Water Quality Guidelines* [with updates]. Prepared by the Task Force on Water Quality Guidelines of the Canadian Council of Resource Ministers, Environment Canada, Ottawa.

Gray, N.F. 1994 *Drinking Water Quality. Problems and Solutions*. John Wiley and Sons, Chichester, 315 pp.

Hagebro, C., Bang, S. and Somer, E. 1983 Nitrate/load discharge relationships and nitrate load trends in Danish rivers. *IAHS Publ.* No. **141**, 377-386.

Hem, J.D. 1989 *Study and Interpretation of the Chemical Characteristics of Natural Waters*. Water Supply Paper, 2254, 3rd edition, U.S. Geological Survey. Washington D.C., 263 pp.

ISO 1984 *Water Quality — Determination of the sum of calcium and magnesium — EDTA titrimetric method*. International Standard ISO 6059-1984 (E), First edition 1984-06-01, International Organization for Standardization.

Keith, L.H. 1988 *Principles of Environmental Sampling*. American Chemical Society, 458 pp.

Lorenzen, C.J. 1967 Determination of chlorophyll and phaeopigments: Spectrophotometric equations. *Limnol. Oceanogr.*, **12**, 343-346.

NIH 1987-88 *Physico-chemical Analysis of Water and Wastewater*. National Institute of Hydrology, Roorkee - 247667(UP), India.

Roberts, G. and Marsh, T. 1987 The effects of agricultural practices on the nitrate concentrations in the surface water domestic supply sources of western Europe. In: *Water for the Future: Hydrology in Perspective*, IAHS Publ. No. **164**, 365-380.

Semenov, A.D. [Ed.] 1977 *Guidebook on Chemical Analysis of Inland Surface Waters*. Hydrometeoizdat, Leningrad, 542 pp, [In Russian].

Strickland, J.D.H. and Parsons, T.R. 1972 A practical handbook of seawater analysis. 2nd edition. *Bull. Fish. Res. Bd Canada*, **167**, 310 pp.

US EPA 1993 *Drinking Water Regulations and Health Advisories*. Health and Ecological Criteria Division, United States Environmental Protection Agency, Washington D.C.

Velz, C.J. 1984 *Applied Stream Sanitation*. 2nd edition, John Wiley and Sons, New York, 800 pp.

WHO 1984 *Guidelines for Drinking-Water Quality. Volume 2. Health Criteria and Other Supporting Information*. World Health Organization, Geneva, 335 pp.

WHO 1991 *GEMS/WATER 1990-2000 The Challenge Ahead*. WHO/PEP/91.2, World Health Organization, Geneva.

WHO 1992 *GEMS/Water Operational Guide*. Third edition. World Health Organization, Geneva.

WHO 1993 *Guidelines for Drinking-Water Quality. Volume 1. Recommendations*. Second edition, World Health Organization, Geneva, 188 pp.

WMO 1974 *Guide to Hydrological Practices*. Publication No. 168, World Meteorological Organization, Geneva.

WMO 1980 *Manual on Stream Gauging*. Publication No. 519, World Meteorological Organization, Geneva.

Chapter 4*

THE USE OF PARTICULATE MATERIAL

4.1 Introduction

Since the publication of the original version of this guidebook in 1978 (UNESCO/WHO, 1978), much new information has been published on the role of particulates in the uptake, release and transport of pollutants, as well as sediment-bound nutrient and contaminant interactions with water and biota, within the aquatic environment. Assessment of the literature on sediments clearly reveals the prominent role that they play in elemental cycling, and this has been used to great effect in environmental monitoring and assessment. For this reason, a separate chapter is now devoted to this topic to provide the basic background and understanding needed to interpret accurately data derived from sediment sampling programmes. More detailed information is also available in Golterman *et al.* (1983), Häkanson and Jansson (1983) and Salomons and Förstner (1984).

It is common practice to accept, as an operational definition, that particulate matter (PM) refers to particles greater than 0.45 μm. By this definition dissolved matter includes particles finer than 0.45 μm, including colloids. Particulate matter is derived primarily from rock weathering processes, both physical and chemical, and may be further modified by soil-forming processes. Erosion subsequently transfers the sediments or soil particles from their point of origin into freshwater systems. During transport, the sediment is sorted into different size ranges and associated mineral fractions until it is deposited on the bottom of the receiving water body. Sediment may then be resuspended, and transported further afield, by intermittent storm activity until it comes to its ultimate resting point or sink, where active sediment accumulation occurs. Modification of the composition of sediments may occur as a result of the input of autochthonous organic and inorganic particles (e.g. calcite, iron hydroxides) generated in the water column and by chemical alterations, especially during periods of deposition.

Particle size and mineralogy are directly related because individual minerals tend to form within characteristic size ranges. Sediments may thus be described in terms of discrete compositional fractions, the overall characteristics of which are due to the variation in the proportions of these fractions and the consequent changes in particle size. Four major categories of particle pollutants may be defined as follows:

This chapter was prepared by R. Thomas and M. Meybeck

- *Particulate organic matter:* either dissolved organic substances adsorbed from solution onto mineral particles or particulate-sized organic detritus of allochthonous (external) or autochthonous (internal) origin (such as algal cells). Organic matter largely originates from plant detritus although some animal debris may also be present. Microbially mediated decay of the organic matter results in the use of oxygen from the water which can, in extreme cases, cause complete anoxia when all the oxygen has been consumed.

- *Nutrients:* adsorbed nutrient elements required for plant growth (of which the most important are phosphorus and nitrogen) which actively exchange between sediment and water. Sediment-bound nutrients create a reserve pool which, under specific conditions, can be released back to the overlying waters, enhancing nutrient enrichment effects (eutrophication).

- *Toxic inorganic pollutants:* sorbed heavy metals, arsenic, etc., controlled by various processes, such as adsorption and desorption, uptake and recycling, and redox conditions.

- *Toxic organic pollutants:* sorbed organochlorine compounds, hydrocarbons, etc., controlled, for example, by hydrophilic/hydrophobic characteristics and liposolubility.

4.2   Composition of particulate matter

4.2.1 Natural sources of particulate matter

Two major natural sources of sediment to rivers and lakes can be considered: (i) products of continental rock and soil erosion, and (ii) the autochthonous material which is formed within the water body and which usually results from the production of algae and the precipitation of a few minerals, mostly calcite (Campy and Meybeck, 1995).

The mechanical erosion of rock and soil results from the combined effects of various erosion agents, i.e. running water, wind, moving ice, mass movements of material on slopes. Where human activities are negligible, natural erosion is maximum in mountainous areas and in active volcanic regions. In particular, it is enhanced when the climate is characterised by alternating wet and dry seasons as in tropical areas (e.g. monsoon climate of South East Asia). Erosion rates may vary from less than 10 t km^{-2} a^{-1} to more than 10,000 t km^{-2} a^{-1}.

As a result of the combined processes of erosion and river transport, the concentration of river suspended matter (SM) (usually measured after filtration through 0.45 μm or 0.5 μm pore filters and referred to as total suspended solids (TSS)) is one of the most variable characteristics of water quality. The yearly TSS average may range from 1 to > 10,000 mg l^{-1}, and for a given river it may vary over three orders of magnitude. The lowest TSS values are measured in lowland regions where lakes are abundant, as in Amazonia, the Canadian Shield, Finland and Zaire. Highest levels are encountered in semi-arid regions, as in North Africa, South West USA, South Central regions of the former USSR, etc. (Meybeck *et al.*, 1989).

Concentrations of autochthonous material are usually low in rivers not influenced by human activities. However, autochthonous material is a major source of lake sediments. The production of macrophytes (aquatic plants) and phytoplankton (free-floating algae) leads to organic debris that eventually sinks to lake bottom sediments. In hard-water lakes rich in Ca^{2+} and HCO_3^-, increases in pH from algal productivity can cause precipitation of calcite ($CaCO_3$), which sinks to the bottom. A third origin of autochthonous material is the debris of algal diatoms which are very rich in silica.

When allochthonous sediment sources to lakes (dust, river inputs and shoreline erosion) are limited, the sediments may be formed mostly by autochthonous material, i.e. diatomite, organic debris and lacustrine chalk.

4.2.2 Chemical composition of river suspended matter

In regions of very high mechanical erosion, the elemental content of river suspended matter reflects the principal origins. The composition is generally close to the composition of the parent rocks and, depending on the lithological nature of the parent rock, the suspended matter may present some variations in major elements (Table 4.1A). When chemical alteration exceeds mechanical erosion, the most soluble elements are carried in the dissolved phase as ions (Ca^{2+}, Mg^{2+}, Na^+, K^+) and dissolved SiO_2, whereas the least soluble ones (Al, Fe, Ti, Mn) remain in the soil which gradually becomes more enriched. As a result of this relative enrichment, the soil particles, which are eventually eroded during heavy rains, are quite different from the parent rock. This is well documented for major elements. Lowland tropical rivers have higher Al, Fe and Ti concentrations than highland temperate rivers (Table 4.1A).

The organic carbon content of river suspended matter, usually expressed as a percentage of TSS, ranges from 0.5 per cent to 20 per cent and is inversely related to the amount of particulate matter found in the river (Meybeck, 1982). The particulate organic nitrogen (PON) is closely linked to particulate organic carbon (POC) and the POC/PON ratio is very constant: between 7 and 10 g g^{-1} in unpolluted rivers (Table 4.1B).

The natural trace element concentration of river suspended matter is difficult to determine since many rivers are already subject to anthropogenic

Table 4.1 The natural chemical composition of river suspended matter

A. MAJOR ELEMENTS (mg kg⁻¹)[1]

	Si	Al	Fe	Mn	Mg	Ca	Na	K	Ti	P
INFLUENCE OF LITHOLOGY[1]										
Basalt river basin	290,000	78,300	52,600	1,300	17,200	35,400	22,700	19,300	11,700	
Metamorphic rocks basin	388,000	49,800	19,000	235	3,110	<3,000	7,250	23,800	3,400	
Limestone basin	211,000	35,300	17,400	300	8,500	178,000	2,500	8,800	2,000	
WORLD AVERAGES[2]										
World rivers	274,000	91,000	51,800	1,000	11,400	23,600	6,900	20,900	5,800	1,400
Tropical and arid zone basins	264,000	114,000	61,700	890	9,600	7,500	5,100	18,300	7,300	1,600
Cold and temperate zone basins	283,000	75,000	46,600	1,100	12,500	31,500	8,000	23,000	4,900	1,350
World surficial continental rock	275,000	69,300	35,900	720	16,400	45,000	14,200	24,400	3,800	610

B. DISTRIBUTION OF ORGANIC CARBON (POC) AND NITROGEN (PON) IN WORLD RIVERS[3]

Discharge weighted percentage of river water reaching the ocean	10 %	50 %	90 %
Suspended matter (mg l⁻¹)	<20	<150	<1,000
POC (% of TSS)	<10	<1.0	<0.5
PON (% of TSS)	<1.2	<0.12	<0.06

C. AVERAGE CONCENTRATIONS OF TRACE ELEMENTS (mg kg⁻¹)[2,4]

	As	Ba	Cd	Co	Cr	Cu	Ni	Pb	Zn
World rivers	8	600	0.3	20	120	50	80	40	110
World surficial continental rocks	7.9	445	0.2	13	71	32	49	16	127

POC Particulate organic carbon
PON Particulate organic nitrogen
TSS Total suspended solids

[1] Three unpolluted monolithologic watersheds in France; inorganic fraction of particulate matter (Meybeck, unpublished)

[2] Sources: Martin and Meybeck, 1979; Meybeck, 1988

[3] Source: Meybeck, 1982

[4] Source: Elbaz-Poulichet, F. and Seyler, P., Ecole Normale Superior, Paris, pers. comm.

influences, particularly in the Northern temperate regions. World averages can be estimated accurately for a few elements (Table 4.1C). For most of these, the averages are close to the world average surficial rock value and to the average content of trace elements in shales (given in Table 4.2C).

4.2.3 Natural composition of lake sediments

The chemical and mineralogical composition of lake sediments may be greatly influenced by the occurrence of autochthonous material in addition to the allochthonous fraction resulting from basin erosion (Table 4.2A). In Lake Geneva (Lake Léman), for example, the chemical composition of the deepest sediments reflects the combination of its various allochthonous origins (Jaquet *et al.*, 1982). When autochthonous matter is dominant, lake sediments may be either carbonate-rich (e.g. Annecy lake, France) or silica-rich (e.g. Pavin lake, France) due to the accumulation of siliceous diatoms, or they may be mostly organic. In the latter case, the organic carbon content may reach 20 to 25 per cent, but in peat bogs it may be even higher (Campy and Meybeck, 1995).

As a result of these various origins of particulate matter, and of the post-depositional processes (chemical diagenesis), the trace element content of world lake sediments may naturally range over an order of magnitude (Förstner and Whitman, 1981). However, the median values of this distribution (usually log-normal) are very close to the content of average shale (fine detrital sedimentary rock) reflecting, therefore, the major influence of allochthonous inputs in most lake sediments (Table 4.2C).

4.2.4 Anthropogenic chemicals in particulate matter

Natural sediment formed during weathering processes may be modified quite markedly during transportation and deposition by chemicals of anthropogenic origin. Major point or diffuse sources of pollutants to sediments have been described in Chapter 1 and are summarised in Figure 4.1. Firstly, it must be noted that anthropogenic chemicals may be scavenged by fine sediment particles at any point from their origin to the final sink or their deposition. Secondly, to compute a geochemical mass balance for sediment-associated elements, it is imperative to derive, by measurement, a mass balance for the sediment in the system under evaluation. This includes deposition of atmospheric particles, total sediment loadings in rivers, accumulation in lakes, and river output to the marine system. These are discussed further in section 4.3.

4.3 Transport and deposition

As noted previously, sedimentation can be defined in terms of particle size and mineralogical composition, both of which are inter-related. The chemical

Table 4.2 Natural chemical composition of lake surficial sediments

A. MAJOR ELEMENTS (% OF INORGANIC FRACTION AFTER IGNITION AT 550 °C)

	SiO$_2$	Al$_2$O$_3$	Fe$_2$O$_3$	MnO	CaO	MgO	Na$_2$O	K$_2$O	P$_2$O$_5$	TiO$_2$	IL[1]
Annecy lake, France[2]	5.3	1.2	0.45	0.014	54.8	0.43	0.095	0.21	0.042	0.042	39.3
Pavin crater lake, France[3]	89.7	2.8	2.22	0.04	0.95	0.19	0.40	0.42	0.55	0.12	3.05
Lake Geneva[4]	48.0	11.2	4.05	0.325	17.0	3.65	0.86	2.25	0.22	0.62	13.6

B. INORGANIC CARBON (POC % DRY WEIGHT, TOTAL FRACTION)

World lakes   Minimum 0.5 %   Maximum 20 %

C. TRACE ELEMENTS (mg kg^{-1} DRY WEIGHT, TOTAL FRACTION)

	As	Cd	Cr	Co	Cu	Hg	Ni	Pb	Sr	Zn
Average world lake sediments[5]										
minimum		0.1	20	4	20	0.15	30	10	60	50
maximum		1.5	190	40	90	1.5	250	100	750	250
mode(s)			60		60		60	30	60/250[6]	120
Average shale[7]	13	0.3	90	19	45	0.4	68	20	300	95

1 IL = ignition loss between 550 °C and 1,000 °C, mostly attributed to the CO$_2$ of carbonate minerals

2 Carbonate sediment mostly derived from autochthonous precipitation

3 Mostly diatomaceous sediment (allochthonous fraction < 10 %)

4 Mixing of allochthonous fraction resulting from erosion of both crystalline and carbonate rocks plus autochthonous carbonate precipitation

5 Source: Förstner and Whitman, 1981; 87 lake sediments mostly from remote areas

6 The bimodal distribution for Sr reflects its double origin – silicate minerals and carbonate minerals

7 Source: Turekian and Wedepohl, 1961

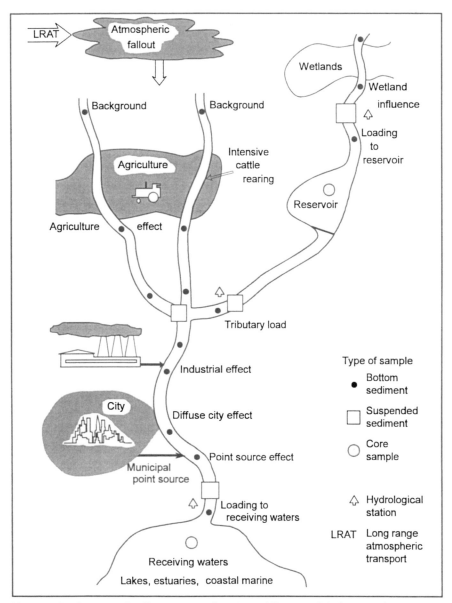

Figure 4.1   Sources of pollutants to sediments and the associated appropriate sampling operations for surveys of particulate pollutants

composition of the sediment at its point of deposition is a product of the composition of the source material, the size of the source material, the sorting during transport, and the physical conditions at the point of deposition.

Transportation occurs in a similar fashion in both rivers and lakes, and is a direct function of water movement. In rivers, water movement is linear, whereas in lakes water movement is mainly orbital or oscillatory, due to the passage of wind-generated waves. In lakes, wind stress also induces major water circulation patterns involving low velocity currents which influence the transport directions of wave-perturbated sediment.

4.3.1 Particle size fractions

The size range (diameter ø) of transported particles ranges upwards from the clay-sized material conventionally defined as 8ø (< 4 µm). This fraction consists mostly of clay minerals such as montmorillonite, kaolinite, etc., but may also include some other fine minerals and organic debris. The silt fraction is medium sized (4ø–8ø; 64–4 µm) and the sand (– 1ø–4ø; 2 mm–64 µm) and gravel (< – 1ø; > 2 mm) make up the coarser size fraction. These limits are only conventional and may slightly change from one scale to another (Krumbein and Pettijohn, 1938). There is a marked relationship between the particle size and its origin (rock minerals, rock fragments, pollutants, etc.) as shown in Figure 4.2.

4.3.2 Transport mechanisms

Erosion, transportation and deposition of sediment is a function of current velocity, particle size, and the water content of the materials. These factors have been integrated into a set of velocity curves (the Hjulstrom curves), which set the threshold velocities for erosion, transport and deposition of various particle sizes (Figure 4.3). Two distinct sediment transport systems are functional under hydraulic conditions. These are defined as transport in suspension and transport by traction along the bottom, often termed bedload. The suspended particles normally consist of finer materials, usually clays and colloids, occasionally with a substantial proportion of silt. Under extreme flow conditions sands, and even gravels, may become suspended. This condition, however, is rare and confined to major storms in high gradient rivers and to the breaker zone of large water bodies. The bedload consists of coarser materials, sands, gravels and larger particles, which move along the bottom by rolling and saltation. Saltation is a process in which a particle is plucked from the bed and moves in a series of bounces in the downstream direction.

Transport brings about a separation by particle size of the material introduced into a moving water body, whether a river or a lake. The resultant separation is: (i) fine grained, geochemically active, suspended material, and (ii) a coarse, geochemically (and relatively) inactive bedload.

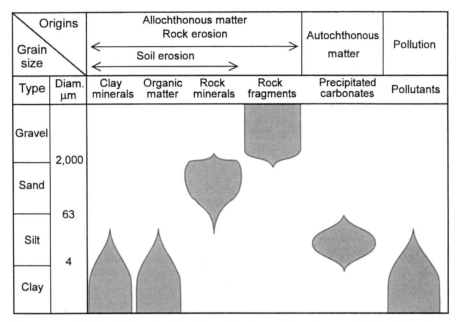

Figure 4.2   Major origins of particulate matter in aquatic systems and their distribution in class sizes

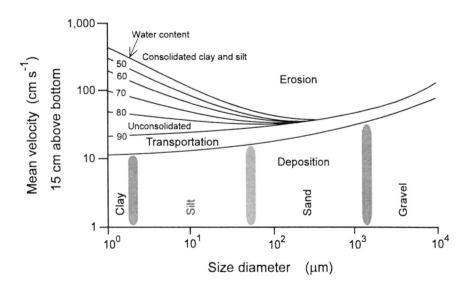

Figure 4.3   Velocity curves defining erosion, transport and deposition of sediments of differing grain size and water content (After Postma, 1967)

4.3.3   River transport and variations in total suspended solids with water discharge

An idealised vertical profile of the proportional composition of sediment in a river would show the clay fraction dominant in the upper water layer, silt in the middle layer and fine to coarse sand near the river bottom. This situation rarely occurs, mainly due to the composition of the source material which generally tends to be deficient in silt. In reality, in many cases, little change in particle size is observed under different flow regimes and there is a clear separation of the sand and clay materials with only a small proportion of silt.

The concentration of total suspended solids varies dramatically with changes in discharge. This is illustrated by Figure 4.4 for the River Exe in England, where a generalised relationship occurs with the peaks in sediment concentration closely approximating to the peaks in the discharge. In most rivers the sediment peaks slightly precede the hydrography peaks, a condition which is known as advanced (see the flood on 25 December in Figure 4.4). Also, the peak concentration of suspended sediment decreases for each of the five consecutive storms measured in the River Exe. Resuspension of fine grained bottom sediment with increasing discharge is the cause of the major increase in suspended solids. When this happens in series, less sediment is available in the river bed to be remobilised in each subsequent event. This removal process is termed sediment exhaustion. However, during calm periods, bed sediment is replaced by deposition from newly eroded and deposited sediment. As a result, a major scatter is often observed in the short-term relationship between sediment concentration and river discharge. This causes a succession of clockwise hysteresis curves in the data. In many rivers, however, the TSS is generally linked to water discharge Q on an annual basis according to the relationship: $TSS = aQ^b$. Where b > 1, this corresponds to a linear variation in a log-log diagram.

Two other points can be illustrated by the data from the River Exe. Firstly, the suspended solids range from about 15 mg l^{-1} to nearly 2,500 mg l^{-1}, i.e. somewhat in excess of two orders of magnitude. This TSS variability far exceeds the variability in concentrations of pollutants measured on the sediment particles. Hence, to compute contaminant loadings in river systems, accurate measurements of discharge and sediment concentrations are absolutely essential. Secondly, Figure 4.4 illustrates the process of sediment storage within the river drainage system, or basin, which is a function of river basin size, slope and water discharge regime. Individual events, or event series, remove a proportion of the stored sediment, including associated pollutants. Extreme storm discharges may flush all of the stored sediment.

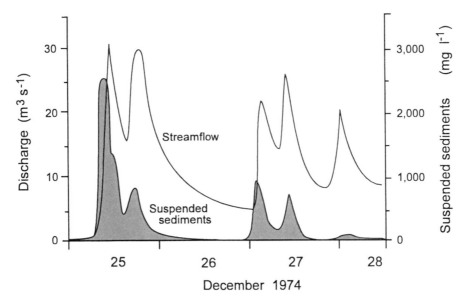

Figure 4.4   The temporal relationship of total suspended solids to the hydrography of the River Exe, UK (After Walling, 1977)

4.3.4 Lake sedimentation
The sediment input to lakes and reservoirs is derived from:
- River input: fine grained suspended load (inorganic and organic particles), coarse traction load.
- Shoreline erosion: sediment of mixed particle size.
- Lake bed erosion: size determined by the strength of the erosional forces.
- Airborne inputs: fine particulate material of inorganic or organic origin (e.g. pollen grains).
- Autochthonous organic matter and autochthonous inorganic precipitates: usually fine particles, but larger algal aggregates and faecal pellets from zooplankton can occur.

In reservoirs the first two sources of sediment input are dominant.

The different particle sizes, in both lakes and reservoirs, are separated by hydraulic transport in a similar manner to that in rivers. Coarse sediment, derived from large river inputs, is deposited first at the river mouth, forming both emerged and submerged deltaic deposits (e.g. the Selenga delta in Lake Baikal, Russia and the Rhône delta in Lake Geneva, Switzerland).

Fine sediment in lakes is transported in suspension by major lake circulatory currents established by the wind stress. During extended periods of calm, suspended sediment will settle, even in very shallow water. An increase in wind leads to resuspension and the particles then continue to be transported. This intermittent transport occurs until the sediment is deposited in an area where water movements are insufficient to resuspend or remobilise it. Fine grained sediment deposits normally define the areas in the lake where active accumulation is taking place. For pollution studies, these depositional basins are of critical importance since they represent the only areas which can be sampled to determine accurately levels of pollutants in lake sediments. Such basins also preserve, with depth, the history of the influence of man on the composition of the sediment.

Lake sedimentation models are relatively simple since they lack tidal currents, and complexity is more related to lake morphology. Four major models of lake sedimentation (with variants) can be conceptualised. Such models are given in Figure 4.5 and described as follows (Thomas, 1988):

A. *Shallow lake: Non-depositional (Figure 4.5A):* The input of fine grained sediment is approximately equal to the output of fine sediment. Coarse sediment normally forms a delta which is progressive and ultimately results in lake infilling. Fine particles may deposit on the open lake bed, but are eventually resuspended under storm conditions. The fine sediment cover remains thin and is intermittently mixed by physical processes. Sediment cores taken in lakes of this type have a thin, modern sediment in which, due to mixing, the profiles of elements are randomly distributed throughout and are, therefore, unfit for pollution assessment.

B. *Shallow lake: Depositional (Figure 4.5B):* In this model, the fine sediment input load is greater than the output and, hence, net accretion or deposition occurs. Deltas or bars may form depending on the sand input, but the lake essentially fills from the bottom upwards. Excess energy derived from storm waves resuspends and mixes the surface of the fine sediment to depths which sometimes exceed 10 cm. Coring of this type of lake gives random element profiles for the top layers, which have been subjected to physical mixing, with smooth profiles below. These profiles reflect the upward movement of averaged concentrations in a mixing zone of constant thickness, analogous to moving averages in smoothing data trends.

C. *Shallow water: Fetch controlled deposition (Figure 4.5C):* Fetch controlled deposition occurs in moderate to large lakes in which wind fetch dominates water depth as the controlling factor bringing about the focused deposition of fine material. The fine material input exceeds output, hence net deposition occurs. Resuspension is less significant, therefore, permanent and continuous sedimentation of fine material can be observed. The sediment texture coarsens downwind with increasing

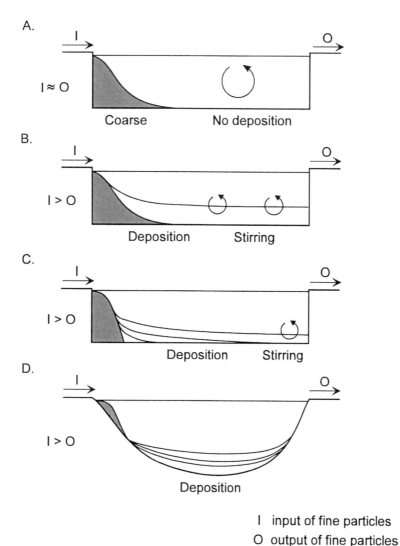

A.

I ≈ O

Coarse   No deposition

B.

I > O

Deposition   Stirring

C.

I > O

Deposition   Stirring

D.

I > O

Deposition

I   input of fine particles
O   output of fine particles

Figure 4.5   Lake sedimentation models with special reference to fine particle
sedimentation (After Thomas, 1988)

wave energy until deposition of fine particles ceases, and erosion and lag
deposit formation occurs. Coring in fine material may provide good elemental profiles in quieter upwind reaches relevant to the prevailing winds.

D. *Deep water model (Figure 4.5D):* This model describes the most common lake condition. Coarse materials occur as bars or deltas and the
shallow water periphery is almost exclusively an erosional or

non-depositional zone. Fine sediments occur in the deeper water and fan outwards from all sides into the deep water basins. These sediments are not subjected to physical mixing and any disruption of the sediment surface is exclusively due to bioturbation. In deep lakes, the accretion rates of sediments (see section 4.8.4) are commonly between 0.1 and 1.0 mm a^{-1}. Coring in such deep water sediments tends to produce elemental profiles which are readily interpreted with respect to lake and basin history. However, care still has to be taken to account for any post depositional sediment stirring, or the occurrence of turbidites, slump deposits and bioturbation in the core. Biological mixing, or bioturbation, involves the physical reworking of sediment by a variety of benthic organisms. In lakes, bioturbation may extend downwards for many centimetres. This depth is normally controlled by the oxygen content of the interstitial waters. Turbidites and slump deposits are coarser material which may reach the deepest parts of some lakes during rare events (usually extreme river floods, slumping on steep slopes, etc.). These layers are unfit for determining the pollution history.

4.4   Environmental control of particulate matter quality

4.4.1 Grain-size influence

The specific surface area is a key particle property which controls adsorption capacity. It is inversely proportional to particle size and decreases over three orders of magnitude from clay-sized particles (10 m^2 g^{-1}) to sand grains (0.01 m^2 g^{-1}). Therefore, the finest particles are generally the richest in trace elements. This effect is particularly evident when separate chemical analyses are made on different size fractions as shown for Cu and particulate matter in the Fly River Basin, Papua New Guinea (Figure 4.6). When total particulate matter is considered, the trace element content is usually directly proportional to the amount of the finest fraction as shown in the Rhine river for the < 16 μm fraction (Salomons and De Groot, 1977).

4.4.2 The form of pollutants bound to particulate matter

Particulate pollutants and nutrients can be partitioned into different forms or phases (speciations), likely to occur in suspended or deposited sediments. These forms depend on the origin of the substances bound to the particulate matter and on the environmental conditions, such as pH, redox potential, etc. The major forms in which pollutants and nutrients occur in the particulate matter are as follows (approximately ranked from the most reactive to the least reactive):

(i)   adsorbed (electrostatically or specifically) onto mineral particles;

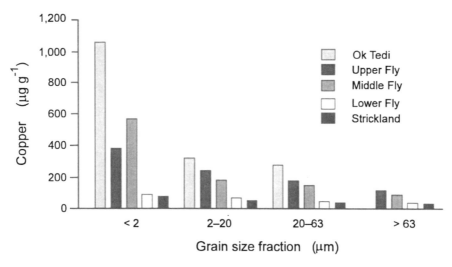

Figure 4.6   Copper in various grain-size fractions in the Fly River basin, Papua New Guinea. Ok Tedi tributary and the Middle Fly River are reaches influenced by copper mining operations. The Lower Fly reach has concentrations close to the background concentrations which are observed in the Strickland River (After Salomons *et al.*, 1988)

(ii) bound to the organic material, which consists mainly of organic debris and humic substances;

(iii) bound to carbonates;

(iv) bound to sulphides;

(v) occluded in Fe and Mn oxides, which occur commonly as coatings on particles;

(vi) within the mineral lattice (e.g. apatite or calcium phosphate for phosphorus; copper oxide or sulphide for Cu); and

(vii)in silicates and other non-alterable minerals.

In unpolluted conditions, the majority of the inorganic compounds (i.e. trace elements, phosphorus) are found in the last three categories. In polluted environments, the additional inputs are mainly found adsorbed onto particles and bound to organic material. The great majority of synthetic organic compounds are found in the adsorbed fraction. Particulate organic matter (terrestrial or aquatic organic detritus) has a very high specific area and consequently a high adsorption capacity. As a result, the concentration of pollutants in the particulate matter may also be proportional to the amount of organic particulates or to the amount of carbon adsorbed on mineral surfaces.

The determination of the chemical phases of trace elements is a tedious task, undertaken by successive chemical extractions, which can only give an operational definition of the actual speciation. The analytical procedures are numerous (Salomons and Förstner, 1984). Some of the most commonly used are described by Tessier *et al.* (1979) for trace elements and by Williams *et al.* (1976) for phosphorus. Most procedures differentiate up to five main phases: sorbed, organic-bound, carbonate-bound, hydroxide-bound and detritus. Criticisms of these methods include the non-selectivity (i.e. some extraction steps may release portions of other forms), the difficulty of inter-comparison of results obtained in various environments (Martin *et al.*, 1987), the time involved (only a dozen samples treated, per week, per person) and the need for highly trained analysts.

Of the chemical analysis techniques currently used the simplest is total digestion (di- or tri-acid attack) which solubilises all material present. However, under natural conditions, only part of the total trace element content is actually reactive to changes in environmental conditions or available for accumulation by biota, since the elements are strongly bound to the minerals or even incorporated within them. Some workers advocate a strong acid attack method which solubilises most specific forms, including oxide coatings, but excluding the lattice-bound elements in aluminosilicates or mineral oxides. Partial leaching at moderate pH values (around pH 2) is often used as an estimate of the maximum content of reactive and available elements. Even though partial leaching is difficult to standardise, it provides very useful information when applied to comparable environmental conditions (e.g. within a lake or river basin). The complete determination of four to six chemical forms should only be undertaken within research programmes.

4.4.3   Effects of changing environmental conditions

As environmental conditions change, the various phases of elements, nutrients, etc., found in particulate matter are altered, and various amounts of these substances may be released into solution. Various forms of organic matter, such as detritus and organic coatings on mineral particles, can be degraded under oxidising conditions, leading to the release of bound substances into solution. The solubility of metals is primarily a function of the oxidation state. For example, reduced forms of iron and manganese (Fe^{2+} and Mn^{2+}) are highly soluble under anoxic conditions and, as a result, are released from particulate matter into solution. Particulate phosphates, in the form of Al-phosphates, Fe-phosphates and Ca-phosphates are more soluble at low pH. In general, acidification (pH < 5) results in the solubilisation of Fe, Mn, Al and other metals from most minerals. In contrast, some elements, like Pb, form insoluble sulphides under low pH and redox conditions. The solubility of sulphides is inversely related to the pH.

The adsorption of trace elements, hydrocarbons, organochlorines, as well as of some forms of nutrients (PO_4^{3-}, NH_4^+, etc.), onto particulate material has been clearly established. When salinity increases, as in estuarine waters, the major cations cause the release of some of the above substances because the cations have a stronger bonding to adsorption sites. Particulate pollutants may also become soluble within the digestive tract of organisms due to the acidic conditions. As a result, the pollutants become more readily available to the organisms and bioaccumulation in body tissues may occur (see Chapter 5). As noted above, trace elements may also exist in the crystalline matrix of minerals (e.g. silicates). Such trace elements are seldom released into solution under the conditions normally encountered in the aquatic environment.

4.4.4 Internal recycling

As a result of changing environmental conditions there is an internal recycling of pollutants in the aquatic environment which is not yet fully understood. These processes are complex and require specific conditions within a multivariate system. The most studied and best understood elements are mercury and phosphorus. In the case of mercury, the transfer from sediment is mediated by bacteria which convert sediment-bound mercury to soluble, mono-methylmercury or to volatile di-methylmercury, depending on the pH. This methylation process, together with its impact on water quality and aquatic organisms, has been very well described in the English Wabigon river-lake system in north western Ontario, Canada (Jackson, 1980) (see section 6.6.1).

Many studies have been carried out on the recycling, or internal loading, of phosphorus from lake sediments to water. This process is particularly important since it amplifies trophic levels in eutrophic (nutrient rich) lakes by producing significant release of phosphorus to the hypolimnion waters, with subsequent mixing during overturn. The process makes phosphorus directly available to plankton in shallow lakes, and slows the rate of reversal of nutrient enrichment when management action is taken to reduce phosphorus loadings. Many environmental and physical processes are involved in the release of phosphorus. The most common is the release of phosphorus bound to iron oxide under the reducing conditions which occur in the interstitial waters of lake sediments. When bottom waters are oxygenated, the phosphorus release is stopped at the sediment–water interface. However, when bottom waters are anoxic, the redox barrier is no longer effective and interstitial phosphate diffuses to the overlying water, accelerating eutrophication.

4.5   Sampling of particulate matter

Specific systems deployed for the sampling of sediment in rivers and lakes may be sub-divided into two categories, those for suspended sediment and those for bed sediments. Different systems are appropriate for rivers and lakes and for various environmental conditions. The following discussion of equipment is based on the summary in Table 4.3.

Bottom sampling devices

Commonly used sampling equipment for lakes includes grab samplers and simple gravity coring devices (Table 4.4). A complete description of samples and sampling operations can be found in Golterman *et al.* (1983) and in Häkanson and Jansson (1983).

The grab samplers used for sampling the beds of large rivers are the same as those used in the sampling of lakes. This equipment must be used from a boat of adequate dimensions to ensure safety. Small, shallow rivers may be sampled by wading into the water and scooping sediment into an appropriate container. In deeper waters, a container attached to a pole may be used. Appropriate bank deposits can be sampled directly below the water surface and should represent recent deposits from the river system. Finer bed deposits can be found behind structures which create backeddies, or in still water conditions and in slack water on the downstream, inside banks of river curves.

River sampling for total suspended solids

Measurement of TSS is now widely employed in river monitoring. Ideally, individual samples should be taken from three to five depths along three to eight vertical profiles at the river station. These samples are then united proportionally to the measured velocity at each depth. When velocities are not measured, special depth integrating samplers can be used: they provide a velocity-averaged water sample for each vertical profile. Once the composite water sample is obtained, it is filtered through a 0.45 μm filter. A full description of these procedures can be found in WMO (1981).

Sampling TSS for chemical analysis requires more precautions in order to avoid contamination. These samples are generally taken at mid-depth in the middle of rivers assumed to be representative of the average quality of river particulates, or with depth integrating samplers. For eventual chemical analysis the TSS samples and filtration kits must be treated in the same manner as laboratory glassware for the same categories of pollutants. For trace metal analysis they must be pre-cleaned with high quality, dilute acids and for trace organics with purified solvents, etc. During field operations, great care must be taken to avoid any contact with rusted devices, greasy wires, etc.

Table 4.3   Sampling methodology for particulate matter in lakes and rivers

Water body	Bottom sediment	Suspended material
Lake	Grab samplers Coring devices	Sediment traps Water sampling followed by: filtration or centrifugation
River	Grab samplers Bank sampling by hand	Water sampling followed by: filtration or centrifugation

Table 4.4   Suitability of bottom sediment samplers

Sampler type	Soft mud	Silty clay	Sand	Trigger reliability	Jaw cut	Sample preserv-ation	Sampler stability	Biological samples	Oper-ation
Corers									
Benthos	xx	x	o	na	na	excellent		xx	EW
Alpine	x	x	o	na	na	poor		o	EW
Phleger	x	xx	x	na	na	fair		o	EW,M
Grab samplers									
Franklin-Anderson	x	x	x	good	poor	fair	fair	o	W
Dietz-Lafond	x	x	o	poor	poor	fair	poor	o	W
Birge-Ekman	xx	x	o	good	excellent	good	fair	xx	M
Peterson	x	x	x	good	poor	good	good	o	EW
Shipek	xx	x	x	good	excellent	excellent	excellent	x	EW
Ponar	xx	xx	x	good	excellent	good	excellent	xx	EW

x	Good	EW	Electric winch
xx	Excellent	M	Manual
o	Not recommended	W	Winch
na	Not applicable		

Water may also be collected and transported to a laboratory centrifuge or processed in the field by pumping at a controlled flow rate through a high capacity centrifuge. The Westfalia and Alfa Laval commercial high flow separators are examples of such centrifuges (Burrus *et al.*, 1988; Horowitz *et al.*, 1989). When used at flow rates of up to 6 litres per minute, the efficiency of recovery exceeds 95 per cent of the total solids finer than 0.45 μm. This technique has become progressively more popular as it provides sufficient material for complete particulate matter analysis. When pollutant loads are required, it is essential to link the sampling operations for chemical analysis to the TSS measurement operations and discharge measurements.

Lake sampling devices for total suspended solids
Most lakes are characterised by very low TSS values (generally < 1 ppm). Therefore, TSS sampling is a particularly difficult task. Recovery of sufficient quantity of material to accomplish a wide range of analyses requires either long time period sampling or the processing of large volumes of water. Another strategy is the deployment of sediment traps (Bloesch and Burns, 1980). Most traps consist of vertical tubes with open tops exposed to settling particles. Traps are attached to fixed vertical lines, anchored at the bottom, and attached at the top to buoys. The traps are deployed in the lake for periods of two weeks or one month to allow the capture of sufficient sediment particles without excess decomposition of the organic matter. Another method for collecting suspended solids is filtration of water samples using a 0.45 μm filter. However, only small quantities of sediment can be collected due to filter clogging. As a result, the sediment mass is usually low, permitting only a few chemical analyses. Sediment samples can also be obtained by continuous pumping from the appropriate water depth and processing the water with a continuous flow centrifuge.

4.6   Analysis of particulate matter
Details of sediment analysis procedures are not given here but full details are available in the appropriate texts (e.g. Salomons and Förstner, 1984). Desirable analyses are outlined in the following sub-sections in three levels, ranging from a minimum of simple, and essential, variables to a complete analytical scheme which can be carried out by the most sophisticated laboratories. The monitoring of sediment chemistry is expensive. Therefore, such work must only be undertaken based on clear programme objectives and on a specific list of chemicals chosen to meet those objectives. Examples of such lists have been discussed in Chapters 1 and 3 and are summarised in Tables 1.3, 1.4, 3.9 and 3.10. Care must be taken to avoid comprehensive analytical procedures of organic and inorganic pollutants where many of the results are not actually required and merely increase the expense. Such comprehensive analyses are only valid for special investigations and for preliminary or basic surveys of river basin and lake systems.

4.6.1   Chemical analysis schemes for sediment
Table 4.5 shows a scheme for three levels of sophistication (levels A, B and C) of sediment analyses for bottom sediment and total suspended solids in

Table 4.5 Suggested sediment analyses for three levels of assessment with increasing complexity

	Assessment level[1]			
Analyses	A	B	C	Comments
Particle size				
% sand, silt, clay	x	x	x	Sieve at 63 μm and 4 μm
Full spectrum analysis:				
settling		x	x	Pipette, Hydrometer
instrumentation			x	Coulter, Laser, X-ray
Mineralogy				
Microscope		x	x	
X-ray			x	X-ray, Diffraction
Major elements				
Al only		x		Colorimetry
Total			x	X-ray spectrometry, ICPS etc.
Nutrients				
Total P	x	x	x	Colorimetry
Forms of P			x	Chemical fractionation
N		x	x	Kjeldahl
Carbon				
Loss of ignition	x	x	x	Ignition
Organic C		x	x	Combustion, TOC analyser
Inorganic C		x	x	Acid CO_2, Evolution
Trace elements				
Total or strong acid extractable		x	x	Colorimetry, AAS, ICPS, X-ray
Fractionation			x	Chemical fractionation
Organic micropollutants				
Organochlorine compounds			x	Gas chromatography (GC)
Other micropollutants			x	GC/MS (Mass spectrometry)

TOC Total organic carbon
AAS Atomic absorption spectrophotometry
ICPS Inductively coupled plasma spectroscopy

[1] Level A: basic equipment required
Level B: some specific equipment necessary
Level C: sophisticated equipment necessary

rivers and lakes. For most trace elements, the total or strong acid extractable forms are determined. However, for nutrients and metals sequential chemical extraction techniques are available which provide insight into which of the chemical phases the substances are bound (see section 4.4.2).

A comprehensive outline for the analysis of the inorganic component of sediment is given for the most sophisticated level of analysis in Figure 4.7. Steps may be omitted to provide the analyses which are necessary at levels A and B defined in Table 4.5.

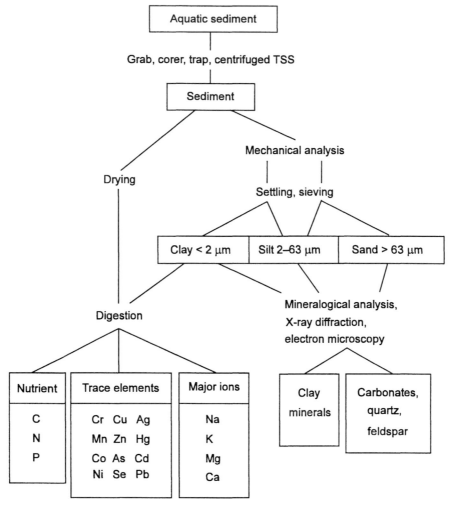

Figure 4.7   A system for the complete analysis of the inorganic components of sediments (Modified from Häkanson and Jansson, 1983)

4.6.2 Core dating

More advanced analyses (levels B and C in Table 4.5) may require additional procedures when core samples are being investigated. These may include stratigraphic analyses of the core to investigate internal structure for slumps, turbidites, general homogeneity and bioturbation. To establish an historical record, the cores must be accurately sub-sampled into appropriate depth increments, usually centimetre intervals (as in Figure 7.12), and analysed for the pollutants of interest (nutrients, trace elements or organic pollutants). If

Table 4.6   Methods used for dating lake sediment cores

Methods based on events	Stratigraphic methods	Radiochemical methods
Ash bands	Magnetostratigraphy	^{14}C
Slumps	Fossil assemblages	^{210}Pb
Turbidites	Chemical	(^{137}Cs)
Hydraulic regime	Textural	
^{137}Cs		
Faunal change		
Anthropogenic materials		

Source: Thomas, 1988

possible, the same increments can be dated (see Table 4.6) to provide a chronological interpretation. Any datable event observed in the stratigraphic assessment can be used to provide a sediment accretion rate for the core (see Table 4.6). The most sophisticated determinations include analyses of ^{137}Cs and ^{210}Pb to establish an accurate chronology of sedimentation (Krishnaswamy and Lal, 1978).

4.6.3   Analytical compensation for grain size effect

As already discussed, the relationship between concentration of a pollutant and sediment grain size leads to a "grain size effect" which must be eliminated to allow a reasonable inter-comparison between samples either spatially or vertically within a core. This can be carried out in two ways: analysis of the same grain size fraction in all samples, or normalisation procedures.

Analysis of the same grain size fraction in all samples
For chemical analysis, the most commonly used fraction is the silt and clay fraction (less than 50 or 64 μm grain size), obtained by wet sieving of the collected sample. Despite giving improved inter-comparative results, this approach suffers from variations in the relative proportions of silt and clay and in the probability that, in many lake samples, the silt may contain significant quantities of calcite which may dilute the pollutant concentrations.

Normalisation procedures
These include taking the ratio of the concentration of the variable of interest to some other sediment element or component that quantifies the geochemically active and/or geochemically inactive sites. Such ratios can be made using

sand (quartz), clay, organic carbon, aluminium or other major, or trace, elements lattice-bound in clay (e.g. scandium, K, Ti). An example for aluminium is given in section 4.8.3.

4.7   Development of a programme for assessing particulate matter quality

4.7.1 Objectives
The objectives of an assessment programme for particulate matter quality can be numerous as indicated below:

- To assess the present concentrations of substances (including pollutants) found in the particulate matter and their variations in time and in space (basic surveys), particularly when pollution cannot be accurately and definitely shown from water analysis.
- To estimate past pollution levels and events (e.g. for the last 100 years) from the analysis of deposited sediments (environmental archive).
- To determine the direct or potential bioavailability of substances or pollutants during the transport of particulate matter through rivers, lakes and reservoirs (bioavailability assessment).
- To determine the fluxes of substances and pollutants to major water bodies (i.e. lakes, reservoirs, regional seas, oceans) (flux monitoring).
- To establish the trends in concentrations and fluxes of substances and pollutants (trend monitoring).

The objectives are listed above in increasing order of complexity, with each step requiring more sampling and measurement effort. The type of information obtained through the study of particulate matter (Table 4.7) is highly variable, depending mainly on the types of studies carried out.

4.7.2 Preliminary surveys
Before establishing a new monitoring programme or extending an existing one, preliminary surveys are recommended to collect information on the present characteristics of the water bodies of interest. These surveys are needed for the selection of sampling sites and devices, establishing sampling periods, and to aid interpretation of results. Table 4.8 summarises the information obtained from different types of appropriate surveys.

4.7.3 Sampling design
Sampling design mostly depends on: (i) the objectives, (ii) the available funds and materials, both in the field and at the laboratory, and (iii) knowledge obtained from preliminary surveys. Some examples of good design are given in section 4.9. A tentative list of possible assessments of the quality of

Table 4.7   Information obtained from chemical analysis of particulate matter in relation to specific assessment objectives

Rivers	Objectives	Lakes and reservoirs	Objectives
Suspended matter			
Present concentrations of substances and pollutants	a, c	Present concentrations of substances and pollutants	a
Pollutant and nutrient fluxes to seas or lakes	d, e	Present nutrient concentrations and associated eutrophication	d
		Present rate of vertical settling of pollutants and nutrients	c, d
Bottom deposits			
Present concentrations of pollutants	a, c	Present concentrations of sediment pollutants	a, c
Past concentrations of pollutants in some cases	b, c	Past concentrations of pollutants e.g. since the beginning of industrialisation	b

Objectives
a — basic surveys   d — flux monitoring
b — environmental archives   e — trend monitoring
c — bioavailability assessment

the aquatic environment through the study of particulate material is given in Table 4.9 for the three levels of monitoring discussed earlier.

Sediment sampling strategies have been discussed in considerable detail by Golterman *et al.* (1983) and a full discussion is beyond the scope of this guidebook. However, some observations can be made which emphasise certain aspects of river and lake sampling.

Rivers
To establish background levels of particulate matter composition, samples of bottom sediment should be taken in the upper reaches of the river basin. The effects of tributaries on the main river should be covered by sampling tributaries close to their junction with the main river. The possible effects of point sources can be estimated from a sample taken from the point source (effluent or tributary), whereas the impact on the river is determined by taking samples immediately upstream and downstream of the source. These samples must be taken from the same side of the river as the effluent input, since the river flow will maintain an influx to the bank of origin for many kilometres downstream. The impact of land-use (diffuse sources) and the influence of a city

Table 4.8   Preliminary surveys pertinent to particulate matter quality assessments

Water bodies	Type of survey	Information obtained
Rivers	Water discharge Q	River regime Extreme discharge statistics
	Suspended sediment (TSS)	TSS variability Relationship TSS = f Q Annual sediment discharge
	Inventory of major pollutant sources	Location of pollutant sources Types of pollutants Estimated quantities discharged
Lakes and reservoirs	Bathymetric survey	Volume Hypsometric curve Deepest points
	Temperature and O_2 profiles	Thermal structure Turnover period Intensity of vertical mixing
	Chlorophyll and transparency	Periods of algal production Resuspension of sediments
	Sedimentological survey (grain-size)	Area of deposition Occurrence of fine deposits
	Inventory of major pollutant sources	As for rivers

TSS   Total suspended solids

should be covered by sampling both upstream and downstream of the city or land-use area. Single bottom sediment samples are adequate provided the objective is to assess only the qualitative impact on the composition of the sediment. This sampling regime is summarised schematically in Figure 4.1.

Lakes
Lakes represent more static conditions than those observed in rivers and, therefore, the sampling intensity is related to the purpose of the study. For example, historical trends can be determined easily by the accurate analysis of a single sediment core recovered from an active depositional basin of fine grained sediment, generally at the deepest point of the lake. For extensive monitoring, surface sediment can be collected from an appropriate grid which is related to the size and shape of the lake, or to a particular region of the lake which is important because of a specific use. Some examples are given in Figure 4.8. To provide mean concentration values, at least five, and preferably ten, samples should be taken for each sediment type observed in the lake. Subsequent

Table 4.9   Development of particulate matter quality assessment in relation to increasing
levels of monitoring sophistication

| | Monitoring level[1] | | |
	Level A	Level B	Level C
Rivers			
Suspended matter (SM)	Survey of SM quantity throughout flood stage (mostly when rising)	Survey of SM quality at high flow (filtration or concentration)	Full cover of SM quality throughout flood stage
Deposited matter	Grab sample at station (end of low flow period)	Longitudinal profiles of grab samples (end of low flow period)	Cores at selected sites where continuous sedimentation may have occurred
Lakes			
Suspended matter	Transparency measurements	Survey of total phosphorus inputs from tributaries (for eutrophication assessment)	Sediment trap
Deposited matter	Grab surface sample at deepest points	Coring at deepest point	Complete surface sediment mapping; Longitudinal profiles of cores

[1] Level A: simple monitoring, no requirement for special field and laboratory equipment
      Level B: more advanced monitoring requiring special equipment and more manpower
      Level C: specialised monitoring which can only be undertaken by fully trained and equipped teams
                    of personnel

sampling episodes should take samples from the same locations.

More advanced analysis of TSS in lakes should only be undertaken by laboratories which are capable of strong field sampling support. Sediment traps should be deployed in offshore regions, with one chain in a small lake and additional chains in lakes of increasing dimensions.

4.7.4 Sampling frequency

Lakes and reservoirs
The sampling frequency for sediments varies according to the monitoring level and type. As the velocity of sediment accretion is generally low $(0.1 \text{ mm a}^{-1}$ to $> 1 \text{ cm a}^{-1})$, lake deposited sediments need only be analysed occasionally. An average frequency of once every five years is usually sufficient. However, in water bodies with high sedimentation rates, such as reservoirs, it may be appropriate to carry out coring operations more often.

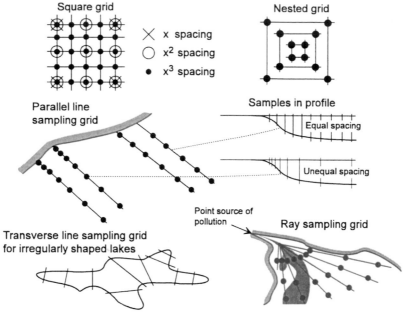

Figure 4.8 Examples of bottom sediment sampling grids in lakes (After Golterman et al., 1983)

Sediment traps in lakes should be operated at least twice a year during periods of minimum and maximum algal productivity. If these periods do not coincide with those of high input of sediment from rivers, additional samples might be needed. The exposure time of the trap should not exceed two weeks at a time in order to avoid excessive algal development and organic matter decomposition within the trap.

Rivers
The optimum sampling frequency for rivers varies according to the objectives of the assessment. If the identification of the peak pollution level of particulate matter is required, two situations must be considered. Whenever pollutants originate from point-sources such as sewers, sampling should be done during low-flow periods, when suspended matter is usually low, and when pollution inputs are less diluted by land erosion products. When pollutants originate from diffuse sources (such as agricultural run-off for nitrogen or urban run-off for lead), sampling may be focused on the flood periods during which the pollutant is washed from the soil.

If mass budgets and weighted average concentrations of particulate pollutants are needed, then the emphasis should be placed on high-flow sampling. The sampling frequency should then depend on the size and regime of the river. For the largest rivers, weekly or bi-weekly TSS samples during floods are convenient, while for smaller rivers daily TSS measurements are needed. The frequency of chemical analysis should be adapted to the variability of the considered elements. For example, the data of Cossa *et al.* (1990) from the St Lawrence river suggest 6 to 12 samples a year are appropriate for Fe and Mn respectively.

When sophisticated chemical analyses are required (e.g. sequential extractions) 12 analyses a year should be considered as a starting point to avoid overloading the analytical laboratory. To make results more representative analyses can be carried out on composite samples prepared by mixing aliquots from several discrete samples.

4.8   Data evaluation

4.8.1   Data reporting
Data reporting should include the following information:
- The full description of sample collection procedures including the location, type of sampler, quantity sampled, number of samples, types of filtration (or centrifugation) apparatus and filters.
- A full description of the sample pre-treatments (acid digestion, partial leaching, organic solvent extraction, etc.).
- The analytical method used.

All concentrations should be reported as mass of pollutant per dry mass of suspended or deposited particulate matter (i.e. mg g^{-1}, µg g^{-1}, ng g^{-1}). When analysing a sediment core, each level should be considered as an individual sample and reported on a different reporting form. Major elements can either be reported as elemental contents (Table 4.1) or as oxides (Table 4.2). In the latter, the sum of contents should reach 100 per cent. Specific forms of pollutants and nutrients can be represented graphically, either as circles of various sizes (Figure 4.9A) or as bars (Figure 4.9B).

4.8.2   Correction of results
Interpretation and presentation of laboratory results must take into account various factors, such as the dilution of pollutants by uncontaminated material (e.g. quartz or carbonates), correction for size fraction and evaluation of natural background values (in the case of naturally occurring elements).

Effects of particle size distribution
Coarser material generally dilutes the pollutants (the matrix effect). Therefore, it is common to remove the sediment fraction larger than 175 or 63 µm prior to analysis (note, however, that the coarse floating material (usually

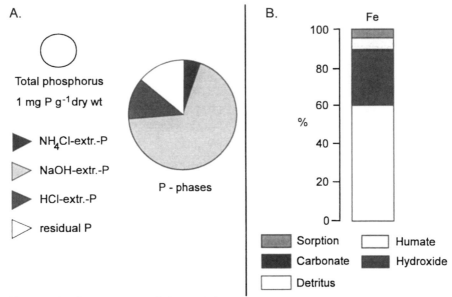

Figure 4.9   Data reporting of chemical phase analysis. **A.** Phosphorus phases in Swedish lakes. The circle diameter can be made proportional to the total phosphorus concentration in the sample (After Boström *et al.*, 1982, in Häkanson and Jansson, 1983). **B.** Phases of Fe in sediment from the River Rhine (After Förstner, 1977)

organic) may be highly contaminated). Nevertheless, the remaining fraction still contains appreciable quantities of inert particles in the sand fraction (usually quartz) or even in the silt. Whenever possible, the analysis should be made on homogeneous size-fractions (as in Figure 4.6), particularly the clay-size fraction. However, separation into individual size fractions is time consuming and, therefore, it is recommended that when necessary complete determinations be performed (preferentially) on the fraction less than 63 μm. This distinction is mostly appropriate for the river-deposited sediments which usually consist of gravels and sands in addition to silts and clays. Other types of particulate matter, such as river suspended matter and deep lake sediments, consist primarily of clay and silt fractions (i.e. below 63 μm). Some workers have even used the fraction less than 16 μm (e.g. Salomons and De Groot, 1977).

Quartz correction
Some minerals such as quartz do not accumulate pollutants. Therefore, when in high proportions, they can produce a dilution of the pollutants present which is known as a "matrix effect". Even when the sand fraction has been removed prior to chemical analysis, some quartz grains may remain in the silt

fraction. Therefore, a quartz correction can be applied to the observed concentration Co (Thomas, 1972): $Cc = (Co \times 100) / (100 - qz)$ where Cc is the corrected concentration and qz the percentage quartz content. Quartz content is determined by X-ray diffraction. In most cases it can be replaced by the sand and silt fraction of the sediment determined after sieving, although in some cases the fine fraction (clay size) may also contain very small quartz particles. A similar correction can also be made for carbonates when they occur in appreciable quantities (e.g. in marine coastal sediments).

Normalisation of results to aluminium
The effect of variable amounts of clay minerals in samples can be minimised by normalising the content of trace elements of concern to the aluminium content. The aluminium content is related to the amount of clay materials and, although it is also part of other minerals, aluminium is normally inert in the aquatic environment. This correction is only valid for trace elements which have a linear relationship with the aluminium content. Results are expressed as the ratio of the concentration of trace elements to the concentration of aluminium in the sample (see below).

4.8.3 Assessment of pollutant concentrations

Estimate of background concentrations
An important problem when interpreting analytical results is the evaluation of the natural background concentrations of substances. This is the case for organic carbon, nutrients, the heavy metals and arsenic, but not for man-made organic micropollutants. River and lake sediments deposited before the industrial era are commonly used for the assessment of natural background concentrations. While post-depositional migrations of trace elements and nutrients are possible in the sediments, the bottom deposits generally provide valuable records of past contamination levels (i.e. they act as environmental archives). For example, an increase of nitrogen and phosphorus in the upper part of the bottom sediments has been clearly verified in many lakes and has been related to accelerated eutrophication (see Chapter 7). Higher levels of trace elements and organic micropollutants are commonly observed in surface sediments of polluted lakes, sometimes as a result of atmospheric deposition. Determination of background concentrations in cores should be done together with core dating.

For rivers, samples from small tributary streams often provide reasonable background concentrations for comparison. However, the concentrations of trace elements in most rivers of the world, even in pristine headwaters, are probably elevated above historic background concentrations owing to atmospheric deposition. The results of chemical analysis of the particulate matter can also be compared with the average composition of the rocks in the basin, if their chemical composition is known. If comparisons with headwater samples or basin rocks are not possible, the world average content of shales, or of river particulate matter (Table 4.2), can be used for the comparison.

The sediment enrichment factor
The sediment enrichment factor (SEF) (Kemp and Thomas, 1976) is one example of the approaches available for evaluating the concentrations of chemicals measured on particulate matter. The concentrations of trace elements in bottom sediments is given by the SEF as follows:

$$SEF = \frac{(C_z/Al_z - C_b/Al_b)}{C_b/Al_b}$$

where: C_z = concentration of the element in the layer or sample z
C_b = concentration of the element in the bottom sediment layers, or background concentration (corresponding to pre-industrial age)
Al_z = concentration of the aluminium in the layer z
Al_b = concentration of the aluminium in the bottom layers.

Whenever possible the SEF, which takes into account the possible variations of sediment grain-size, should be used in preference to the more simple contamination factor which is equivalent to C_z/C_b (Häkanson, 1977). The SEF can also be used for mapping particulate matter pollution.

4.8.4 Mass transfer of pollutants in rivers and lakes

Pollution fluxes in rivers
Determination of pollutant and nutrient fluxes is necessary for the assessment of inputs to lakes, regional seas or oceans and when studying pollutant mass balances within a drainage basin. The flux is dependent on the variations in the total suspended matter content (TSS expressed in mg l^{-1}) and the content Cs of the chemical x in the particulate matter (Cs_x expressed in g kg^{-1} or mg kg^{-1}). The amount of chemical per unit volume of unfiltered water (Cv_x in mg l^{-1} or µg l^{-1}) is easily obtained as: Cv_x = TSS.Cs_x. On unfiltered samples (e.g. for total phosphorus, total Kjeldahl nitrogen or total lead), Cv_x is obtained directly from the analysis.

The sampling frequency is a crucial factor in determining pollutant fluxes. In most surveys, the analyses of suspended material will not be carried out more than 12 times a year. Water discharge Q, the concentration of suspended matter (i.e. TSS) and the chemical content of the TSS Cs_x, are all likely to vary between sampling periods. Therefore, either special high flow sampling or data interpolation is required. When high flow sampling cannot be carried out two types of interpolation can be used:

1. *Constant flux assumption:* The flux Qs_{xi} of pollutant x discharged by the river with the particulate matter is assumed constant during a representative time interval t_i around the time i of sampling. The total mass of pollutant discharged M_x during the entire study period, $T = \Sigma t_i$ is:

$$M_x = \Sigma\, Qs_{xi} \cdot t_i$$

where: Qs_{xi} = $TSS_i \cdot Cs_{xi} \cdot Q_i$ (g s^{-1});

and TSS_i = total suspended matter at time of sampling (mg l^{-1} or g m^{-3})

Cs_{xi} = concentration of pollutant x in the particulate matter at the time of sampling (g g^{-1})

Q_i = the water discharge at the time of sampling (m^3 s^{-1}).

Qs_{xi} is computed for each sample. The length of the representative period t_i can be variable according to the water discharge variations. The basic assumption is particularly valid for point sources releasing a relatively constant flux of pollutants.

On unfiltered samples, the total concentration of chemical compound per unit volume Cv_{xi} is obtained directly as: Total mass $M_x = \Sigma Qs_{xi} \cdot t_i$ where: $Qs_{xi} = Cv_{xi} \cdot Q_i$. This method can also be applied to mass budgets of dissolved compounds, particularly for those which present a marked concentration decrease with discharge (e.g. the dilution of a sewage outfall by a river).

2. *Constant concentration assumption*: The concentration Cs_{xi} is assumed constant during a given time interval t_i around the time of sampling. The amount of suspended matter Ms_i discharged during this period should be measured with maximum accuracy, for example, by daily measurements of suspended matter TSS_j. The total mass M_x of pollutant x discharged is:

$M_x = \Sigma_j\, Cs_{xi} \cdot Ms_i$ where: $Ms_i = t_i \cdot \Sigma_j\, TSS_j \cdot Q_j$.

The constant concentration assumption method takes into consideration the variations in river water discharge and total suspended matter. These variations may be two, or even three, orders of magnitude in rivers and are much more than the variations in pollutant level in the particulate matter Cs_x (which are usually within one order of magnitude). For this reason, the constant concentration assumption method should be the first to be considered for application to available data.

It should be noted that the constant concentration assumption can also be applied to river budgets of dissolved substances. The total mass M_x during a given time interval t_i is: $M_x = Cv_{xi}.V_i$, where V_i is the total volume of water discharged during time t_i and Cv_{xi} is the chemical concentration x per unit volume. This method particularly applies to chemicals that show no dilution during high flows.

Further improvements in data interpretation are possible if relationships can be developed between the pollutant flux Qs_{xi} and the water discharge Q_i, or between the contamination level Cs_{xi} and the amount of suspended material TSS_i, etc. If such relationships exist they permit estimation of Qs_x and Cs_x to be made between two consecutive sampling periods.

Lake sedimentation

In sediment traps, the settling rate, SR, defined as a mass of dry material deposited per unit area per unit time, is expressed in mg m^{-2} day^{-1} or g m^{-2} a^{-1}. The average settling rate during the exposure time T becomes:

$$SR = \frac{m.Cs_x}{A.T}$$

where: Cs_x =   the concentration of the chemical x,
          m   =   the dry mass of the whole sample, and
          A   =   the trap collecting area.

The velocity of sediment accretion, SA, measured by radionuclide dating of a sediment core or estimated by other methods, is the thickness of sediments deposited during a given period (expressed as mm a^{-1}). This result must be transformed to give the sedimentation rate: $SR = SA . \lambda$, where λ is the mass of dry sediment per unit volume, after the sediment is retrieved from the core. To determine λ the mass of dry material recovered from a known volume of wet sediment is measured. If Cs_x is the concentration of chemical x in the sediment (generally reported in mg kg^{-1} dry weight) the sedimentation rate of the chemical becomes: $SR_x = Cs_x . \lambda . SA$.

4.9   The use of particulate material in water quality assessments: case studies

The following examples illustrate the great variability in studies of particulate matter quality depending on the type of water body (rivers, lakes and reservoirs), the type of chemical (organic matter, nutrients, trace elements, organic micropollutants), and the programme objectives (inter-comparison of stations, time series, flux determination). More case studies are reported in Alderton (1985) mostly for lake cores, and in Salomons and Förstner (1984) for trace elements in rivers, lakes and reservoirs.

4.9.1 Preliminary studies

A thorough survey of pollutant sources should always be carried out before beginning an assessment programme, as was done for Lake Vättern (Häkanson, 1977), where major cities, cultivated grounds and industries were mapped in the lake basin (see Figure 2.1). A detailed bathymetric chart must be set up for lakes (see Figure 7.16), particularly when multiple basins may occur. The sampling grid for the lake should be densest near the pollution sources, as illustrated by the Lake Vättern study (Figure 4.10), which combined 20 cores and more than 90 grab samples.

The determination of sediment grain size is often a key operation in the preliminary survey for lake quality assessment. In Lake Ontario, hundreds of determinations have been made to produce a map of grain size distribution (Figure 4.11). Only medium and coarse sands settle along shores which are exposed to strong wave action. Therefore, the most appropriate locations for coring are the three deep basins where clay sized materials are predominant and, consequently, where most pollutants and nutrients have accumulated.

4.9.2 River studies

TSS and organic matter: seasonal variations in the River Seine
Figure 4.12 shows suspended matter in the River Seine ranging from 4 to 150 mg l^{-1} with a maximum during the first flooding stage of the hydrological year (i.e. January). Numerous peaks of TSS occur in this medium-sized river (44,000 km^2 at the Paris monitoring station). The particulate organic carbon (POC) content of the suspended matter is inversely related to the river TSS. This is generally true for most rivers (Meybeck, 1982). A weekly sampling frequency was most appropriate at this station (i.e. monitoring levels A and B as indicated in Table 4.9). This monitoring programme was also carried out in connection with sampling for PCB concentrations (see Figure 6.12).

Phosphorus speciation and fluxes in the Alpine Rhône, Switzerland
The Alpine Rhône is the major tributary of Lake Geneva and has a maximum water discharge from May to July due to glacier melt (see also Figure 4.14). Six analyses of three phosphorus phases were made on suspended particles to differentiate organic phosphorus (PO), apatitic phosphorus (PA), and non-apatitic inorganic phosphorus (PINA) (Figure 4.13). The most abundant form was PA and its concentration in the TSS remained very stable throughout the year, whereas PO and PINA concentrations were maximum at the low water stage when TSS was at a minimum (Figure 4.13). The flux of particulate matter, in kg s^{-1}, was maximum during the high water stage (26 May–6 October)

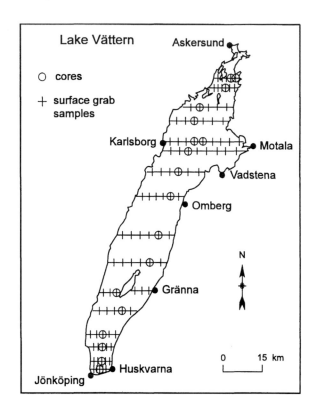

Figure 4.10 The sampling grid of surficial sediments in Lake Vättern, Sweden (After Häkanson, 1977)

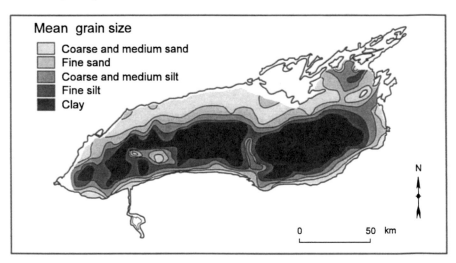

Figure 4.11 Grain-size distribution of bottom sediments in Lake Ontario, reflecting wind action

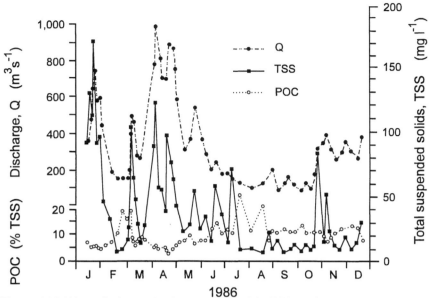

Figure 4.12 Water discharge, total suspended solids (TSS) and suspended particulate organic carbon (POC expressed as % TSS) in the River Seine in Paris during 1986 (After Chevreuil *et al.*, 1988)

and negligible during the rest of the year. In this river, the sampling frequency for suspended matter should be at least weekly, but chemical analysis of particulate matter every two months is acceptable. As a result of the TSS variation, the PO and PINA fluxes were minor, despite their relative abundance from October to May. The arrows (Figure 4.13) indicate the time interval attributed to a given sample when annual fluxes were computed, using an assumption of constant concentration, which is a valid hypothesis for PA. For PO and PINA, this is no longer valid and the inverse relationship between TSS, PO and PINA levels should be considered when computing fluxes. This design corresponds to monitoring level C in Table 4.9.

Mercury variations in the suspended matter of the Alpine Rhône
Mercury concentrations in TSS were analysed in the Alpine Rhône six times a year in 1987 and 1988 (Figure 4.14). Mercury concentrations ranged from < 50–500 µg kg^{-1} with the minimum occurring when the TSS content of the river was maximum, i.e. during the summer. As with PO and PINA above, the flux computation for mercury should account for variabilities in both the suspended matter and mercury. The water discharge showed weekly variations due to the operation of major dams. This design corresponds to monitoring level B in Table 4.9.

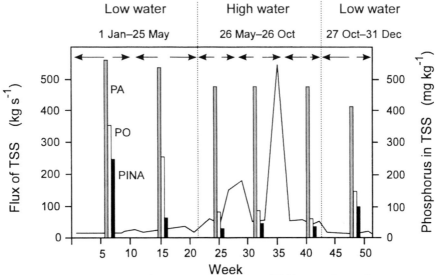

Figure 4.13 Annual pattern of total suspended solids (TSS) and different forms of phosphorus in the Alpine Rhône River, Switzerland for 1987. The flux of TSS is based on 26 samples per year. Horizontal arrows correspond to the period for which time integrated TSS samples were collected for phosphorus analysis. Vertical bars indicate the proportions of the three major forms of particulate phosphorus: PA apatitic P; PO organic-bound P; PINA non-apatitic inorganic P (After Favarger and Vernet, 1988)

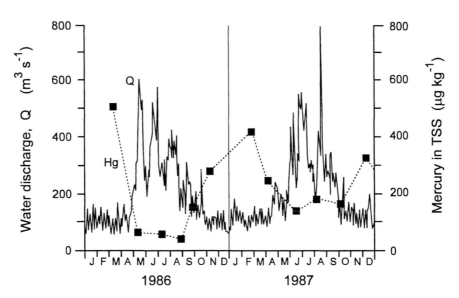

Figure 4.14 Water discharge Q and changes in mercury concentrations in the suspended matter during 1986 and 1987 in the Alpine Rhône River, Switzerland (After Favarger and Vernet, 1989)

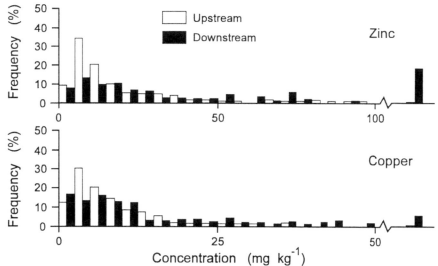

Figure 4.15 Frequency distribution of 0.5N HCl soluble metals (Zn, Cu) in bottom muds of Japanese rivers (After Tada *et al.*, 1983)

Statistical distribution of trace metals in sediments of rivers in Japan
In the sediments of Japanese rivers, Zn, Cu, Pb and Cd were extracted from mud samples with a 0.5 N HCl digestion (Tada *et al.*, 1983). Two river stretches were considered: the upstream section which is supposedly less polluted, and the downstream section (Figure 4.15). The maximum values in the downstream stations were obviously outside of the statistical distribution of the upstream values, suggesting an anthropogenic impact, although the modal values were similar. These distributions were of the log-normal type, characteristic of most trace elements. This design corresponds to monitoring levels A and B in Table 4.9.

Trends in the longitudinal profiles of mercury in sediments of the Alpine Rhône, Switzerland
Four longitudinal profiles of river sediments, at approximately 5 km intervals, were made annually in the Alpine Rhône from 1970 to 1986 (Figure 4.16). The profiles enabled the exact location of pollutant sources (major chemical industries) to be detected downstream of the city of Brigue. This source of pollution resulted in sediment concentrations of mercury which gradually decreased from a peak value of 25 µg g[-1] in 1980 (a value similar to those found in Minimata Bay, Japan). The downstream dilution of polluted sediment by unpolluted tributaries is quite effective. This design corresponds to monitoring level B in Table 4.9.

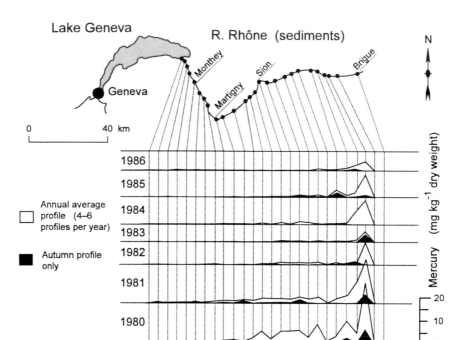

Figure 4.16 Longitudinal profiles of mercury in sediments in the Alpine Rhône River, Switzerland from 1980 to 1986 (After Favarger and Vernet, 1989)

The impact of mine development on metal pollution of the Fly River, Papua New Guinea

Suspended matter at seven stations on the Fly River system (78,000 km^2 and 7,500 m^3 s^{-1} at the mouth) was analysed for copper between 1982 and 1988 (Figure 4.17). The pre-mining data (prior to 1981) served as background values. The copper–gold mine was established in 1982/84, with gold-processing only occurring during the period 1982/86. The extraction of copper was eventually added in 1986/88. The Ok Tedi, Strickland and Upper Fly stations were located upstream of the mine or on another river branch and thus were not affected by the mining. These all showed low, and very stable, copper contents similar to world background values (Tables 4.1 and 4.2). The Ningerum and Kuambit stations, downstream of the mine, showed the effects of mining operations with more than ten-fold increases in copper concentrations. Stations Obo and Ogwa, located far downstream of the mine, showed evidence of recent, but only moderate, pollution due to dilution by the very high sediment load ($> 100 \times 10^6$ t a^{-1}) of the river. Most of the copper from the mining operation occurred in the clay and fine silt fractions of the river sediments (see Figure 4.6). The design of this study corresponds to monitoring level B in Table 4.9.

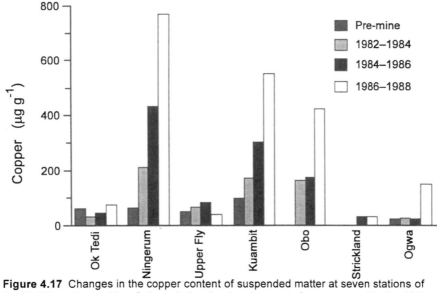

Figure 4.17 Changes in the copper content of suspended matter at seven stations of the Fly River, Papua New Guinea during the development of a copper–gold mine (After Salomons *et al.*, 1988)

4.9.3 Lake studies

Zinc mapping in Lake Vättern, Sweden

The zinc content of the surficial sediments (0–1 cm) of Lake Vättern was analysed in approximately 90 samples taken from a very dense sampling grid (see Figure 4.10). The results (Figure 4.18) showed slight enrichment in the deepest parts of the lake (compare with the bathymetric map in Figure 7.16). The major source of zinc was a small bay located in the very northern part of the lake where concentrations reached 23,700 µg g^{-1}, i.e. about 100 times the published natural concentrations (see Tables 4.1 and 4.2). However, the general concentration in central sediments was less than 800 µg g^{-1}. This industrial point source was well documented in the pollutant source inventory made during the preliminary survey (see Figure 2.1). This regime is equivalent to monitoring level A in Table 4.9.

Mapping of DDT degradation products in Lake Geneva

A DDT degradation product, pp'DDE, has been mapped in 115 samples from surficial sediments in Lake Geneva. Instead of iso-concentration contours, an indication of concentrations is given in Figure 4.19 by the sizes of the symbols. Contamination still exists despite the ban on the use of DDT in Switzerland since the 1950s, thereby illustrating the time-lag between

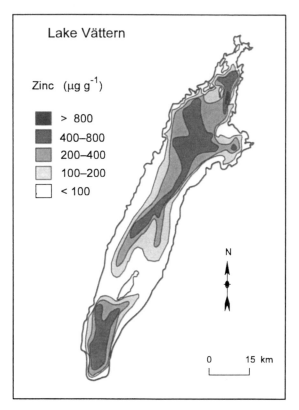

Figure 4.18 Map of zinc concentrations in surface sediments of Lake Vättern, Sweden in the early 1970s (After Häkanson, 1977)

environmental protection measures and actual improvements. The higher concentrations in the west central part of the lake, where sedimentation rates are moderate, illustrate that agricultural sources were likely. The design of this study is equivalent to monitoring level B in Table 4.9.

Comparison of historical copper contamination in southern Wisconsin lakes, USA
Four cores of similar length (60 to 80 cm) were taken in four southern Wisconsin lakes (Iskandar and Keeney, 1974). The velocities of sediment accretion were similar since the depth scales and time scales were the same for all cores (Figure 4.20). Mendota Lake showed no variation in Cu, whereas the Waubesa, Monona and Kegonsa profiles showed major peaks near the turn of the century, around 1930, and from 1930 to World War II, respectively. It is important to note that the copper levels near 1850, and earlier, were not exactly equal for all the lakes. Local conditions (grain-size, lithology of lake basin, etc.) were probably responsible for the observed four-fold

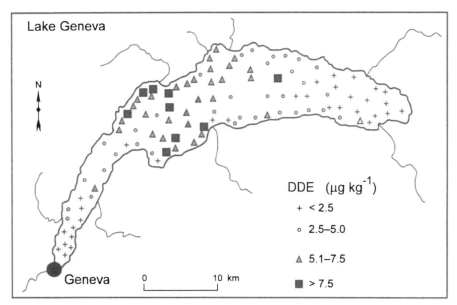

Figure 4.19 Map of DDE concentrations in the surficial sediments of Lake Geneva, Switzerland in the early 1980s (After Corvi *et al.*, 1986)

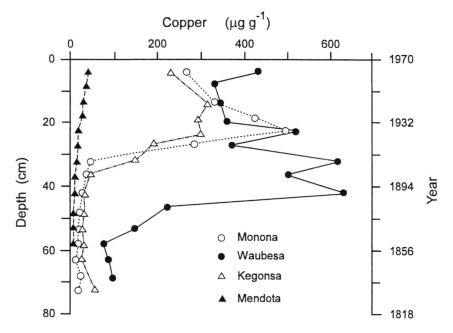

Figure 4.20 A comparison of the history of copper concentrations in four lake cores from Wisconsin lakes, USA (After Iskandar and Keeney, 1974, in Alderton, 1985)

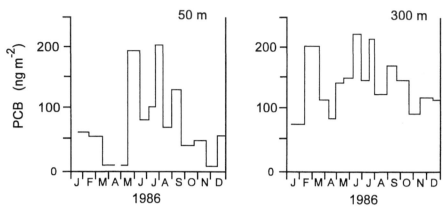

Figure 4.21 Settling rate of particulate PCBs in sediment traps at 50 m and 300 m depths in Lake Geneva, Switzerland during 1986 (After Gandais, 1989)

difference. The Mendota core was used to check for regional atmospheric pollution but, because the Cu levels were very similar throughout the core, this pollutant pathway seemed unlikely. Therefore, local sources (sewage, urban run-off and agriculture) were held responsible for the levels in the other three lakes. The earlier copper maximum found in Lake Waubesa was actually interpreted as an artefact due to sediment reworking (Alderton, 1985). This study corresponds to monitoring level B in Table 4.9.

Settling rate of total particles and PCBs in Lake Geneva
Figure 4.21 shows the results from two sediment traps in Lake Geneva deployed at 60 m and 300 m depths, and retrieved 15 times during 1986. The deepest trap was representative of the amount and quality of material reaching the deepest part of the lake (309 m) and, supposedly, was not affected by possible resuspension of the sediment by turbidity currents. The total settling rate of particles varied from 1–8 g cm^{-2} day^{-1}, with the minimum in November and maximum in the summer. The settling pattern of PCBs at 60 m indicated a peak rate in late May, whereas at 300 m it showed marked variations with a notable peak in early March. The annual rate of settling for PCBs was greater at the 300 m location. This can be attributed to a lateral input of particles, possibly from the Alpine Rhône river, which is by far the largest sediment input to the lake. This survey, carried out for the International Surveillance Commission of Lake Geneva (CIPEL), is a good example of the sophistication of some surveys of particulate matter at monitoring level C (see Table 4.9).

4.10 Conclusions and future developments

The importance of particles in transporting nutrients and anthropogenic substances has been well documented in the scientific literature of the past 20 years. No assessment programme concerned with the aquatic environment should ignore this fact. Assessment programmes must include sediment sampling and analysis to the maximum extent allowed by funding and the capability of personnel. A substantive programme requires, together with an environmental chemist, a trained sedimentary scientist to optimise design, implementation, interpretation and reporting of the resultant data.

Although a three tier system of sophistication for analysis of sediment has been discussed here, these are not strict guidelines. Even for the most sophisticated level of water quality assessment, the simplest, most reliable, robust and well-tried systems are first recommended. The complexity of sediment chemistry studies has resulted in an array of sophisticated analytical techniques. These techniques are often very tempting for inexperienced sedimentary geochemists and may lead to over-commitment of resources and support systems with over-complicated analytical requirements. Simple measures must not be overlooked. As a general rule the simplest approach for accomplishing the objectives of the assessment should form the basis for the design and implementation of a monitoring programme. Any necessary research is best left to the appropriate agencies and to the universities.

Much remains to be done by the research community, both in the field and in the laboratory, to improve sediment monitoring and routine sediment survey programmes. Simple techniques for carrying out analyses in remote, and occasionally hostile, environments must be developed. Fractionation procedures should also be improved to provide greater resolution and understanding of the bonding of nutrients, trace elements and organic compounds to the various mineral compartments.

It has been demonstrated that highly polluted sediment (through resuspension and mobilisation) may continue to pollute a localised environment, even though the source of pollution has been removed. A better understanding is required of the mechanisms involved in remobilisation of sedimentary pollutants in order to improve the ability to remedy this problem of *"in-situ"* pollutants.

Sediment systems, as they interact with water and biology, are highly dependent on the physical and chemical condition of the water body. The interactive effects of water, sediment and biota require urgent research. The toxic effects of sediment pollutants must be defined more clearly and standardised biological tests must be developed to assess sediment toxicity. Not only must these be laboratory based, but they must also be designed to assess ambient conditions in the range of aquatic habitats observed under different global climate conditions.

Finally, there is a need to define more clearly the atmospheric deposition of particulates and the role that these play in the major global transport of nutrients and toxic contaminants, particularly to lakes. This topic will become more important as the true ramifications of the global distribution of pollutants by the atmosphere is understood.

4.11   References

Alderton, D.H.M. 1985 Sediments. In: *Historical Monitoring*, MARC Report No. 31, Monitoring and Assessment Research Centre, King's College London, University of London, 1-95.

Bloesch, J. and Burns, N.M. 1980 A critical review of sedimentation trap technique. *Schwiz. Z. Hydrol.*, **42**, 15-55.

Boström, B., Jansson, M. and Forsberg, C. 1982 Phosphorus release from lake sediments. *Arch. Hydrobiol. Beih Ergebn Limnol.*, **18**, 5-59.

Burrus, D., Thomas, R.L., Dominik, J. and Vernet, J.P. 1988 Recovery and concentration of suspended solids in the Upper Rhone River by continuous flow centrifugation. *J. Hydrolog. Processes*, **3**, 65-74.

Campy, M. and Meybeck, M. 1995 Les sédiments lacustres. In: R. Pourriot and M. Meybeck [Eds] *Limnologie Générale*. Masson, Paris, 185-226.

Chevreuil, M., Chesterikoff, A. and Létolle, R. 1988 Modalités du transport des PCB dans la rivière Seine (France). *Sciences de l'Eau,* **1**, 321-337.

Corvi, C., Majeux, C. and Vogel, J. 1986 Les polychlorobiphényles et le DDE dans les sédiments superficiels du Léman et de ses affluents. In: *Rapports sur les Etudes et Recherches Entreprises dans le Bassin Lémanique, 1985*. Commission Int. Protection Eaux du Léman, Lausanne, 206-216.

Cossa, D., Tremblay, G.H. and Gobeil, C. 1990 Seasonality of iron and manganese concentrations of the St Lawrence river. *Sci. Total Environ.*, **97/98**, 185-190.

Favarger, P.Y. and Vernet, J.P. 1988 Flux particulaires de quelques nutriments et métaux dans les suspensions du Rhône à la Porte du Scex. In: *Rapports sur les Etudes et Recherches Entreprises dans le Bassin Lémanique, 1987*. Commission Int. Protection Eaux du Léman, Lausanne, 90-96.

Favarger, P.Y. and Vernet, J.P. 1989 Pollution du Rhône par le mercure: un suivi de 17 ans. *Cahiers de la Faculté des Sciences, Genève*, **19**, 35-44.

Förstner, U. 1977 Metal concentrations in freshwater sediments. Natural background and cultural effects. In: H.L. Golterman [Ed.] *Interactions between Freshwater and Sediments*, Junk, The Hague, 94-103.

Förstner, U. and Whitman, G.T.W. 1981 *Metal Pollution in the Aquatic Environment.* Springer Verlag, Berlin, 486 pp.

Gandais, V. 1989 Un exemple d'évolution spatio-temporelle des flux de matières particulaires au centre du Léman. *Cahiers Faculté Sciences Genève,* **19**, 75-82.

Golterman, H.L., Sly, P.G. and Thomas, R.L. 1983 *Study of the Relationship Between Water Quality and Sediment Transport: A Guide for the Collection and Interpretation of Sediment Quality Data.* United Nations Educational Scientific and Cultural Organization, Paris, 231 pp.

Häkanson, L. 1977 *Sediments as Indicators of Contamination. Investigation in the Four Largest Swedish Lakes.* Naturvarsverkets Limnologiska Undersökning Report 92, Uppsala, 159 pp.

Häkanson, L. and Jansson, M. 1983 *Principles of Lake Sedimentology.* Springer Verlag, New York, 316 pp.

Horowitz, A.J., Elrick, K.A. and Hooper, R.C. 1989 A comparison of instrumental dewatering methods for the separation and concentration of suspended sediment for subsequent trace element analysis. *J. Hydrolog. Processes,* **2**, 163-184.

Iskandar, I.K. and Keeney, D.R. 1974 Concentration of heavy metals in sediment cores from selected Wisconsin lakes. *Environ. Sci. Technol.,* **8**, 165-170.

Jackson, T. 1980 *Mercury Speciation and Distribution in a Polluted River-lake System as Related to the Problem of Lake Restoration.* Proc. Internat. Symposium for Inland Waters and Lake Restoration. US EPA/OECD, Portland, Maine, 93-101.

Jaquet, J.M., Davaud, E., Rapin, F. and Vernet, J.P. 1982 Basic concepts and associated statistical methodology in the geochemical study of lake sediments. *Hydrobiologia,* **91**, 139-146.

Kemp, A.L.W. and Thomas, R.L. 1976 Cultural impact on the geochemistry of the sediments of Lake Ontario. *Geoscience Canada,* **3**, 191-207.

Krishnaswamy, S. and Lal, D. 1978 Radionuclide limno-chronology. In: A. Lerman [Ed.] *Lakes, Chemistry, Geology, Physics,* Springer-Verlag, New York, 153-177.

Krumbein, W.C. and Pettijohn, F.J. 1938 *Manual of Sedimentary Petrography.* Appleton-Century-Crofts Inc., New York, 549 pp.

Martin J.M. and Meybeck M., 1979 Elemental mass-balance of material carried by world major rivers. *Mar. Chem.,* **7**, 173-206.

Martin, J.M., Nirel, P. and Thomas, A.J. 1987 Sequential extraction techniques: promises and problems. *Mar. Chem.,* **22**, 313-341.

Meybeck M., 1982, Carbon, nitrogen and phosphorus transport by world rivers. *Amer. J. Sci.,* **282**, 401-450.

Meybeck M., 1988 How to establish and use world budgets of riverine materials. In: A. Lerman and M. Meybeck [Eds] *Physical and Chemical Weathering in*

Geochemical Cycles, Kluver, Dordrecht, 247-272.

Meybeck, M., Chapman, D. and Helmer, R. [Eds] 1989 *Global Freshwater Quality: A First Assessment*. Blackwell Reference, Oxford, 306 pp.

Postma, H. 1967 Sediment transport and sedimentation in the estuarine environment. In: G.H. Lauff [Ed.] *Estuaries*, Publ. Amer. Assoc. Adv. Sci., **83**, 158-170.

Salomons, W. and De Groot, A.J. 1977 *Pollution History of Trace Metals in Sediments as Affected by the Rhine River*. 3rd Int. Symp. Environmental Bio-geochemistry, Wolfenbüttel, March 1977, Inst. Soil Fertility, Haren (ND), publ. 184, 20 pp.

Salomons, W., Eagle, M., Schwedhelm, E. Allerma, E., Bril, J. and Mook, W.G. 1988 Copper in the Fly river system (Papua New Guinea) as influenced by discharges of mine residue: overview of the study and preliminary findings. *Environ. Technol. Letters*, **9**, 931-940.

Salomons, W. and Förstner, U. 1984 *Metals in the Hydrological Cycle*. Springer-Verlag, New York, 350 pp.

Tada, F., Nishida, H., Miyai, M. and Suzuki, S. 1983 Classification of Japanese rivers by heavy metals in bottom mud. *Environm. Geology*, **4**, 217-222.

Tessier, A. Campbell, P.G.C. and Bisson, M. 1979 Sequential extraction procedure for the speciation of particulate trace metals. *Analyt. Chem.*, **51**, 844-851.

Thomas, R.L. 1972 The distribution of mercury in the sediments of Lake Ontario. *Can. J. Earth Sci.*, **9**, 636-651.

Thomas, R.L. 1988 Lake sediments as indicators of changes in land erosion rates. In: A. Lerman and M. Meybeck [Eds] *Physical and Chemical Weathering in Geochemical Cycles*, Kluver, Dordrecht, 143-164.

Turekian, K.K. and Wedepohl, K.H. 1961 Distribution of elements in some major units of the earth's crust. *Bull. Geol. Soc. Amer.*, **72**, 175-192.

UNESCO/WHO, 1978 *Water Quality Surveys: A Guide for the Collection and Interpretation of Water Quality Data*. Studies and Reports in Hydrology 23, United Nations Educational Scientific and Cultural Organization, Paris and World Health Organization, Geneva, 350 pp.

Walling, D.E. 1977 Suspended sediments and solute response characteristics of the River Exe, Devon, England. In: R. Davidson-Arnott and W. Nickling [Ed.] *Research in Fluvial Geomorphology*. Geo Abstracts, Norwich, 169-197.

Williams, J.D.H., Jaquet, J.M. and Thomas, R.L. 1976 Forms of phosphorus in the surficial sediments of Lake Erie. *J. Fish. Res. Bd. Can.*, **33**, 413-429.

WMO 1981 *Measurement of River Sediments*. WMO Operational Hydrology Report 16, World Meteorological Organization, Geneva, 61 pp.

Chapter 5*

THE USE OF BIOLOGICAL MATERIAL

5.1 Introduction

Natural events and anthropogenic influences can affect the aquatic environment in many ways (see Chapter 2): synthetic substances may be added to the water, the hydrological regime may be altered or the physical or chemical nature of the water may be changed. Most organisms living in a water body are sensitive to any changes in their environment, whether natural (such as increased turbidity during floods) or unnatural (such as chemical contamination or decreased dissolved oxygen arising from sewage inputs). Different organisms respond in different ways. The most extreme responses include death or migration to another habitat. Less obvious responses include reduced reproductive capacity and inhibition of certain enzyme systems necessary for normal metabolism. Once the responses of particular aquatic organisms to any given changes have been identified, they may be used to determine the quality of water with respect to its suitability for aquatic life.

Organisms studied *in situ* can show the integrated effects of all impacts on the water body, and can be used to compare relative changes in water quality from site to site, or over a period of time. Alternatively, aquatic organisms can be studied in the laboratory (or occasionally in the field) using standardised systems and methods, together with samples of water taken from a water body or effluent. These tests, sometimes known as biotests, can be used to provide information on the intensity of adverse effects resulting from specific anthropogenic influences, or to aid in the evaluation of the potential environmental impact of substances or effluents discharged into surface or groundwater systems. Most kinds of biological analysis can be used alone or as part of an integrated assessment system where data from biological methods are considered together with data from chemical analyses and sediment studies. A full appreciation of natural changes and anthropogenic influences in a water body can only be achieved by means of a combination of ecological methods and biotests. Sometimes these studies need to be carried out over a period of many years in order to determine the normal variation in biological variables as well as whether any changes (natural or unnatural) have occurred or are occurring. An example of a continuous programme of biological assessment using a variety of methods is that carried out in Lake Baikal, Russia (Kozhova and Beim, 1993).

*This chapter was prepared by G. Friedrich, D. Chapman and A. Beim

It is not possible to describe in this chapter, in detail, all the methods and variations that exist for biological analysis of water quality. There are several comprehensive texts and reviews which cover this subject (e.g. Ravera, 1978; OECD, 1987; Newman, 1988; Abel, 1989) and the details of many of the methods are published in appropriate reports and journals. Since many biological methods have been developed for local use and are based on specific species, an attempt has been made in this chapter to give only the basic principles behind the methods. With the help of such information it should be possible to decide whether such methods are applicable to the water quality assessment objectives in question. In many cases, when such methods are chosen, it will be necessary to adapt the basic principle to the local hydrobiological conditions, including the flora and fauna.

5.2   Factors affecting biological systems in the aquatic environment

5.2.1   Natural features of aquatic environments
The flora and fauna present in specific aquatic systems are a function of the combined effects of various hydrological, physical and chemical factors. Two of these factors specific to water bodies are:

- The density of the water, which allows organisms to live in suspension. Organisms which exploit this are called plankton, and consist of photosynthetic algae (phytoplankton), small animals (zooplankton) which feed on other planktonic organisms and some fish species which feed on other plankton and/or fish. The development of a rich planktonic community depends on the residence time (or retention time) of the water in the water body (see sections 6.4.2 and 7.2.5). Fast flowing water tends to carry away organisms before they have time to breed and to establish populations and, therefore, planktonic communities are more usually associated with standing waters such as lakes and reservoirs. As many fish are strong swimmers they are able to live in rivers, provided there are suitable breeding grounds present (see sections 6.4.1 and 6.4.2). Organisms living permanently in fast flowing waters, require specific adaptations to their body shape and behaviour (see section 6.4.1).
- The abundance of dissolved and particulate nutrients in the water. The constant supply of these often allows diverse and rich communities of planktonic and benthic (those living in or on the bottom) organisms to develop. An abundance of dissolved nutrients in shallow, slow flowing or standing waters allows the growth of larger aquatic plants (macrophytes), which in turn provide food, shelter and breeding grounds for other organisms.

The photosynthetic organisms which depend on the dissolved nutrients and sunlight for their own carbon production are termed the primary producers. These organisms are the food source of the zooplankton and small fish (secondary producers), which in turn are the food source of other fish (tertiary producers). This simplified system is known as the food chain and, together with the processes of decay and decomposition, is responsible for carbon transfer within the aquatic environment. In practice, the interactions between different groups are more complex and may be referred to as the food web. For more detailed information on the fundamentals of biological systems in water bodies see Hutchinson (1967), Hynes (1970), Wetzel (1975), Whitton (1975) and Moss (1980).

5.2.2   Anthropogenic influences on water bodies
In addition to natural features, biological communities are often affected directly by human activities (such as inputs of toxins, increased suspended solids, habitat modification or oxygen depletion) or indirectly by processes influenced by anthropogenic activities (e.g. chelating capacity).

The variety of effects that can be observed on different aquatic organisms as a result of anthropogenic influences can be demonstrated by the example of domestic sewage. Purely domestic sewage, without the input of modern, synthetic, harmful substances, such as chlorinated hydrocarbons, detergents etc., is characterised by high concentrations of easily biodegradable organic matter. It also contains high concentrations of bacteria, viruses and other pathogens from which water-borne diseases may arise. During the process of biodegradation of sewage in a river there is an initial rapid decline in oxygen concentration in the water resulting from microbial respiration during self-purification. However, microbial activity also leads to an increase in nutrient content and sometimes other harmful substances are formed such as hydrogen sulphide or ionised ammonia (Figure 5.1). Hydrogen sulphide is very toxic, but ionised ammonia is a nutrient which is more easily assimilated than nitrate. However, if the pH value exceeds 8.5, a rapid increase in unionised ammonia occurs (see Figure 3.2) which is very toxic to fish. Phosphate also becomes available following the biological decomposition of domestic sewage. These changes in the chemical composition of the water are followed by significant changes in the structure of the biota, some of which exploit the increased nutrients and others which can tolerate reduced oxygen concentrations (Figure 5.1). Such changes form the basis of water quality assessments using biota as indicators of the intensity of pollution.

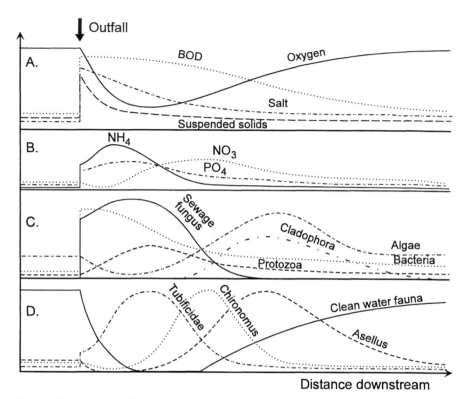

Figure 5.1 Typical effects on water quality and the associated biota which may be observed downstream of a sewage outlet. **A** and **B**. Physical and chemical changes; **C**. Changes in micro-organism populations; **D**. Changes in invertebrate populations (After Hynes, 1960)

5.2.3   Physical alterations in the aquatic environment

The presence or absence of specific aquatic organisms depends on the physical environment and its associated habitats, such as fast flowing water with large stones or boulders or still waters with fine deposited sediments. Although these environments can easily be modified by human activities, including river damming, canalisation and drainage schemes, natural changes can occur in the physical environment due to local climatological and geographical conditions. Events such as torrential rain storms or prolonged droughts can lead to sudden or gradual modifications of a natural habitat, e.g. by increased siltation or scouring of river beds, which in turn lead to changes in the flora and fauna of the water body. These changes can be quite dramatic, including short term or permanent loss of species. It is important to understand the hydrological regime of water bodies when designing biological assessment programmes so that effects due to natural changes in the environment can be separated from those that may be caused by human activities.

5.2.4   Dissolved oxygen

Oxygen is an important factor for aquatic life and the chemical characteristics of the environment. Concentrations less than 100 per cent saturation can occur normally under certain circumstances, e.g. at the bottom of nutrient rich lakes (see Figure 7.8), or at night in slow flowing rivers (see Figures 6.19 and 6.20). In such locations, species may be found which are adapted to low concentrations of oxygen. Under normal conditions these species would be rare, but they can become widespread in association with pollution or nutrient enrichment. However, many species are able to survive a potentially harmful lack of oxygen for a short time, but rarely for days or many hours. The ability of organisms to survive different levels of oxygen depletion in water forms the basis of some biotic indices and water quality assessment methods. Tolerance of low concentrations of oxygen varies from species to species, even within the same genus and, therefore, it is more appropriate to work at the species level for some biological assessment methods. Further details are given in section 5.4.

5.2.5   The duration of exposure

The duration of exposure, or influence on the organism, is generally the period of effective concentration of the contaminant or other variable of interest in the environment, or in a laboratory test system. In a biological sense this is the duration of actual exposure of an organism to a harmful concentration, or an effective concentration of, for example, a substance which can be bioaccumulated (see section 5.8). In some field situations the actual period of influence may be longer than the measured duration of unusually high contaminant concentrations. Alternatively, it may be shorter as a result of incomplete mixing, such as occurs at the beginning of the "toxic wave" arising from a point source input to a river. Sometimes, as a result of incomplete mixing, high concentrations may occur only on one side of a river or lake, or near the bottom. Adverse effects may be felt by organisms which are free living in the water or which live in, or on, the substrate and are not able to escape from the dangerous area. In severe cases of toxic pollution or deoxygenation a "fish-kill" may occur. In many cases, long after the fish-kill is over, the continuing absence of organisms which would otherwise be present (together with the benthic fauna which colonise the surfaces of the bottom of the river, i.e. stones, gravel, sand or mud) enables the investigator to establish the severity of the event, and the length of the affected stretch of the river.

The body of an organism takes some time (seconds or longer) to absorb a toxin and then react. Nevertheless, many aquatic organisms react very rapidly, especially against toxic substances, and this can be an advantage for the development of biomonitoring techniques. Alternatively, toxic substances which are accumulated gradually until they reach harmful concentrations which produce sudden or very subtle effects in the organisms, present a particular problem. Nutrient absorption by aquatic organisms is usually rapid but their subsequent growth takes time. Consequently, the effect of nutrient enrichment (eutrophication) in a water body is a long-term effect. In many rivers, the effects of eutrophication may occur some distance downstream of the source of nutrients. In lakes, the effects of nutrient enrichment also occur sometime after increased nutrient levels begin (for details see section 7.3.1).

5.2.6   Concentration

The physiological or behavioural reactions of aquatic organisms depend on the concentration of natural substances and pollutants in the environment, and the time required for these substances to affect the internal systems of the organisms. The actual environmental concentration of a substance or compound which produces toxic effects in an organism can also be influenced by many other environmental factors (e.g. presence of other toxins, inadequate food supply, and physical factors such as habitat alteration, sedimentation, drought or oxygen depletion). An organism under stress will not be able to survive the same concentration of a contaminant as when its environmental conditions are optimal. Consequently, the toxic effects determined by laboratory tests may vary with different experimental circumstances. Many substances also have significant differences in their toxicity to different species. Therefore, to determine environmental effects fully, it is necessary to use a set of tests under standardised circumstances.

The reactions of organisms are often not very specific in relation to any given concentration, but can be observed mainly in relation to exceeding, or remaining below, a "no observed effect concentration" (NOEC). The NOEC plays an important role in international discussions on water quality management, as it is important in determining the toxicity of substances and setting priorities for control measures for effluents discharged to freshwaters. This specialised topic is beyond the scope of this guidebook.

5.2.7   Chelating capacity

Chelation is the ability of organic compounds to bind metal ions and maintain them in solution. Examples of chelating agents are humic and fulvic acids and compounds such as EDTA (ethylenediaminetetraacetic acid). These compounds can also slowly release bound metal ions back into the water. The chelating capacity of the water, therefore, depends on the content of humic acids and other ligands, as well as on the hardness of the water (see section 3.3.11). Hardness plays an important role in the distribution of aquatic biota and many species can be distinguished as indicators for hard or soft water. Organisms with shells which are composed of calcium carbonate need high concentrations of calcium in the water, whereas stoneflies and some triclad worms are characteristic of soft water. Different requirements can be found within the same family of organisms. The microcrustacea *Gammarus pulex* and *G. roeseli* have a preference for hard water and can survive some depletion of oxygen, whereas *G. fossarum* is more sensitive to organic pollution and oxygen depletion but can survive in less hard water. However, *G. fossarum* cannot tolerate very soft water although the closely related genus *Niphargus* lives in soft water wells and clean mountain rivers which are very poor in calcium.

The toxicity of trace elements to a given species may also vary according to the water hardness. For example, the toxicity of copper and zinc varies over a wide range depending on the concentration of calcium in the water. The higher the concentration of calcium, the lower the toxicity of both metals. As a result the European Union directive for the protection of water as a habitat and spawning ground for fish (CEC, 1978) gives different concentration limits for zinc and copper for different degrees of water hardness (Table 5.1). However, zinc is much more toxic to bacteria than to all other organisms, including man, and bacteria are the main organisms responsible for self-purification in freshwaters. Therefore, with respect to the total environment, it is necessary to look for the most sensitive components of the system in order to establish permissible limits for toxic compounds. The toxicity of metals may be reduced in waters high in humic acids (often known as brown waters, e.g. Rio Negro, Brazil) as a result of their chelating capacity. Examples from the temperate zone include rivers and lakes in peat bog areas, both of which have specific communities of plants and animals.

5.2.8 Acidity

Some organisms are sensitive to the acidity or alkalinity of water. Aquifers, rivers and lakes situated in catchment areas consisting of acid rocks or pure quartz have waters poor in calcium and magnesium with a low buffering capacity. Such water bodies are widely distributed in North America (the Laurentian shield), Scandinavia and specific areas of the arctic, temperate and tropical zones. In these water bodies, additional acid input from "acid rain" produces a drop in the pH of the water and may result in increased

Table 5.1   Permissible concentrations of copper and zinc in freshwaters of different degrees of hardness according to the European Union directive for the support of fish life in freshwaters

	Water hardness (mg I^{-1} CaCO$_3$)			
	10	50	100	500
Total zinc (mg I^{-1})				
Salmonid waters	0.03	0.2	0.3	0.5
Cyprinid waters	0.3	0.7	1.0	2.0
Dissolved copper (mg I^{-1})				
Salmonid and				
Cyprinid waters	0.005	0.022	0.04	0.112

Source: CEC, 1978

concentrations of reactive aluminium species released from the soil of the surrounding watershed (Meybeck *et al.*, 1989). Both low pH (lower than 5.5) and increased aluminium content are toxic to many invertebrates and fish. Consequently, "acid rain" has led to a reduction of fish populations and other organisms in lakes and rivers which had previously been excellent fishing grounds in North America, Scandinavia and other parts of the world (Meybeck *et al.*, 1989). Sudden effects due to acidification of waters can also occur after heavy rain and snow melt. Although acid waters occur naturally, evidence now suggests that acidification arising from "acid rain" is also occurring in some tropical countries, e.g. Brazil and parts of South East Asia. Detailed information is available in specialised texts, e.g. Rodhe and Herrera (1988) and Bhatti *et al.* (1989). Other adverse effects associated with acidification arise from the mobilisation of mercury and cadmium, both of which are highly toxic and can be accumulated by fish in their body tissues (see section 5.8).

5.3   Uses and benefits of biological methods

5.3.1   Biological effects used for assessment of the aquatic environment

A variety of effects can be produced on aquatic organisms by the presence of harmful substances or natural substances in excess, the changes in the aquatic environment that result from them, or by physical alteration of the habitat. Some of the most common effects on aquatic organisms are:

- changes in the species composition of aquatic communities,
- changes in the dominant groups of organisms in a habitat,
- impoverishment of species,
- high mortality of sensitive life stages, e.g. eggs, larvae,
- mortality in the whole population,
- changes in behaviour of the organisms,
- changes in physiological metabolism, and
- histological changes and morphological deformities.

As all of these effects are produced by a change in the quality of the aquatic environment, they can be incorporated into biological methods of monitoring and assessment to provide information on a diverse range of water quality issues and problems, such as:

- the general effects of anthropogenic activities on ecosystems,
- the presence and effects of common pollution issues (e.g. eutrophication, toxic metals, toxic organic chemicals, industrial inputs),
- the common features of deleterious changes in aquatic communities,
- pollutant transformations in the water and in the organisms,
- the long-term effects of substances in water bodies (e.g. bioaccumulation and biomagnification, see section 5.8),
- the conditions resulting from waste disposal and of the character and dispersion of wastewaters,
- the dispersion of atmospheric pollution (e.g. acidification arising from wet and dry deposition of acid-forming compounds),
- the effects of hydrological control regimes, e.g. impoundment,
- the effectiveness of environmental protection measures, and
- the toxicity of substances under controlled, defined, laboratory conditions, i.e. acute or chronic toxicity, genotoxicity or mutagenicity (see section 5.7),

Biological methods can also be useful for:

- providing systematic information on water quality (as indicated by aquatic communities),
- managing fishery resources,
- defining clean waters by means of biological standards or standardised methods,
- providing an early warning mechanism, e.g. for detection of accidental pollution, and
- assessing water quality with respect to ecological, economic and political implications.

Examples of some of these uses are mentioned in the following sections, and in Chapters 6, 7 and 8 in relation to rivers, lakes and reservoirs.

5.3.2   Advantages of biological methods

Biological assessment is often able to indicate whether there is an effect upon an ecosystem arising from a particular use of the water body. It can also help to determine the extent of ecological damage. Some kinds of damage may be clearly visible, such as an unusual colour in the water, increased turbidity or the presence of dead fish. However, many forms of damage cannot be seen or detected without detailed examination of the aquatic biota.

Aquatic organisms integrate effects on their specific environment throughout their lifetime (or in the case of laboratory tests, during the period of exposure used in the test). Therefore, they can reflect earlier situations when conditions may have been worse. This enables the biologist to give an assessment of the past state of the environment as well as the present state. The length of past time that can be assessed depends on the lifetime of the organisms living in the water under investigation. Micro-organisms, such as ciliated protozoa, periphytic algae or bacteria, reflect the water quality of only one or two weeks prior to their sampling and analysis, whereas insect larvae, worms, snails, and other macroinvertebrate organisms reflect more than a month, and possibly several years.

When biological methods are carried out by trained personnel they can be very quick and cheap, and integrated into other studies. Compared with physico-chemical analysis, much less equipment is necessary and a large area can be surveyed very intensively in a short time, resulting in a large amount of information suitable for later assessment. Recent developments in water quality assessment, especially for the purpose of effluent control, have begun to include bioindicators and tests such as bioassays (as in Germany under the "Waste Water Levy Act"). The costs of chemical analytical equipment, trained personnel and materials, repairs and energy consumption are enormous due to the number of different polluting substances that now have to be legislated and controlled. In some situations biological methods can offer a cheaper option. The advantages of biological methods, however, do not eliminate the need for chemical analysis of water samples. Agencies and individuals responsible for establishing assessment programmes must integrate both methods to provide a system which is not too expensive and which provides the necessary information with maximum efficiency.

Acute toxicity testing (see section 5.7.1) is particularly useful in cases of emergency and accidental pollution where it can minimise the amount of chemical analysis required. When investigating a fish-kill, samples of the water are usually taken for analysis in order to determine the cause. However, if a toxicity test (using an aquatic organism in samples of the contaminated water) is conducted immediately in parallel to the chemical analyses it is possible to ascertain whether toxic concentrations are present in one, or all, of the samples taken. This initial "screening" enables the chemical laboratory to focus their efforts on the most toxic water samples and helps the water quality managers and decision-makers prepare (or stop) further action. Immediate remedial action may, therefore, be possible although the reaction of the biota in the test does not necessarily give specific information about the type of substance causing the toxicity, or an indication of the concentration present.

5.3.3   Types of biological assessment

Biological assessment of water, water bodies and effluents is based on five main approaches.

(i)   Ecological methods:
- analysis of the biological communities (biocenoses) of the water body,
- analysis of the biocenoses on artificial substrates placed in a water body, and
- presence or absence of specific species.

(ii)   Physiological and biochemical methods:
- oxygen production and consumption, stimulation or inhibition,
- respiration and growth of organisms suspended in the water, and
- studies of the effects on enzymes.

(iii) The use of organisms in controlled environments:
- assessment of the toxic (or even beneficial) effects of samples on organisms under defined laboratory conditions (toxicity tests or bioassays), and
- assessing the effects on defined organisms (e.g. behavioural effects) of waters and effluents *in situ*, or on-site, under controlled situations (continuous, field or "dynamic" tests).

(iv)   Biological accumulation:
- studies of the bioaccumulation of substances by organisms living in the environment (passive monitoring), and
- studies of the bioaccumulation of substances by organisms deliberately exposed in the environment (active monitoring).

(v)   Histological and morphological methods:
- observation of histological and morphological changes, and
- embryological development or early life-stage tests.

Some of these methods are widely used in freshwaters although other methods have been developed for use in specific environments, or in relation to particular environmental impacts. The principal advantages and disadvantages of the major methods for freshwater quality assessment are given in Table 5.2.

Biological assessment is of increasing importance in many places around the world, and many methods have been developed which are being used on a national or local basis. In the following sections only the basic principles, and selected methods which are in more general use, are described. More details are available in specialised texts which summarise the literature on the topic (e.g. Alabaster, 1977; Burton, 1986; Hellawell, 1986; De Kruijf *et al.*, 1988; Yasuno and Whitton, 1988; Abel, 1989; Samiullah, 1990).

5.4   Ecological methods

Aquatic organisms have preferred habitats which are defined by physical, chemical and other biological features. Variation in one or more of these can lead to stress on individuals and possibly a reduction in the total numbers of species or organisms that are present. In extreme situations of environmental change, perhaps due to contamination, certain species will be unable to tolerate the changes in their environment and will disappear completely from the area concerned, either as a result of death or migration. Thus the presence or absence of certain species or family groups, or the total species number and abundance, have been exploited as a means of measuring environmental degradation (see the example of the sewage input into a river in Figure 5.1 and section 5.2.2). Two main approaches have been used: methods based on community structure and methods based on "indicator" organisms. An indicator organism is a species selected for its sensitivity or tolerance (more frequently sensitivity) to various kinds of pollution or its effects, e.g. metal pollution or oxygen depletion. Some groups of organisms, such as benthic invertebrates, have been exploited more than others in the development of ecological methods and this is due to a combination of the specific role of the organisms within the aquatic environment, their lifestyle and the degree of information available to hydrobiologists (Table 5.3). Principal approaches based on the commonly used groups of organisms are described below. For further details see the specialised literature quoted.

5.4.1   Indices based on selected species or groups of organisms

Biological indices are usually specific for certain types of pollution since they are based on the presence or absence of indicator organisms (bioindicators), which are unlikely to be equally sensitive to all types of pollution. Such indices often use macroinvertebrate populations because they can be more easily and reliably collected, handled and identified. In addition, there is often more ecological information available for such taxonomic groups.

Table 5.2 Principal biological approaches to water quality assessment; their uses, advantages and disadvantages

	Ecological methods		Microbiological methods	Physiological and biochemical methods	Bioassays and toxicity tests	Chemical analysis of biota	Histological and morphological studies
	Indicator species[1]	Community studies[2]					
Principal organisms used	Invertebrates, plants and algae	Invertebrates	Bacteria	Invertebrates, algae, fish	Invertebrates, fish	Fish, shellfish, plants	Fish, invertebrates
Major assessment uses	Basic surveys, impact surveys, trend monitoring	Impact surveys, trend monitoring	Operational surveillance, impact surveys	Early warning monitoring, impact surveys	Operational surveillance, early warning monitoring, impact surveys	Impact surveys, trend monitoring	Impact surveys, early warning monitoring, basic surveys
Appropriate pollution sources or effects	Organic matter pollution, nutrient enrichment, acidification	Organic matter pollution, toxic wastes, nutrient enrichment	Human health risks (domestic and animal faecal waste), organic matter pollution	Organic matter pollution, nutrient enrichment, toxic wastes	Toxic wastes, pesticide pollution, organic matter pollution	Toxic wastes, pesticide pollution, human health risks (toxic contaminants)	Toxic wastes, organic matter pollution, pesticide pollution
Advantages	Simple to perform. Relatively cheap. No special equipment or facilities needed	Simple to perform. Relatively cheap. No special equipment or facilities needed. Minimal biological expertise required	Relevant to human health. Simple to perform. Relatively cheap. Very little special equipment required	Usually very sensitive. From simple to complex methods available. Cheap or expensive options. Some methods allow continuous monitoring	Most methods simple to perform. No special equipment or facilities needed for basic methods. Fast results. Relatively cheap. Some continuous monitoring possible	Relevant to human health. Requires less advanced equipment than for the chemical analysis of water samples	Some methods very sensitive. From simple to complex methods available. Cheap or expensive options
Disadvantages	Localised use. Knowledge of taxonomy required. Susceptible to natural changes in aquatic environment	Relevance of some methods to aquatic systems not always tested. Susceptible to natural changes in aquatic environment	Organisms easily transported, therefore, may give false positive results away from source	Specialised knowledge and techniques required for some methods	Laboratory based tests not always indicative of field conditions	Analytical equipment and well trained personnel necessary. Expensive	Specialised knowledge required. Some special equipment needed for certain methods

1 e.g. biotic indices
2 e.g. diversity or similarity indices

Table 5.3   Advantages and disadvantages of different groups of organisms as indicators of water quality

Organisms	Advantages	Disadvantages
Bacteria	Routine methodology well developed. Rapid response to changes, including pollution. Indicators of faecal pollution. Ease of sampling.	Cells may not have originated from sampling point. Populations recover rapidly from intermittent pollution. Some special equipment necessary.
Protozoa	Saprobic values well known. Rapid responses to changes. Ease of sampling.	Good facilities and taxonomic ability required. Cells may not have originated from sampling point Indicator species also tend to occur in normal environments.
Algae	Pollution tolerances well documented. Useful indicators of eutrophication and increases in turbidity.	Taxonomic expertise required. Not very useful for severe organic or faecal pollution. Some sampling and enumeration problems with certain groups.
Macroinvertebrates	Diversity of forms and habits. Many sedentary species can indicate effects at site of sampling. Whole communities can respond to change. Long-lived species can indicate integrated pollution effects over time. Qualitative sampling easy. Simple sampling equipment. Good taxonomic keys.	Quantitative sampling difficult. Substrate type important when sampling. Species may drift in moving waters. Knowledge of life cycles necessary to interpret absence of species. Some groups difficult to identify.
Macrophytes	Species usually attached, easy to see and identify. Good indicators of suspended solids and nutrient enrichment.	Responses to pollution not well documented. Often tolerant of intermittent pollution. Mostly seasonal occurrence.
Fish	Methods well developed. Immediate physiological effects can be obvious. Can indicate food chain effects. Ease of identification.	Species may migrate to avoid pollution.

Source: Based on Hellawell, 1977

It is frequently argued that the indicator organisms incorporated into biotic indices should be distributed world-wide. However, few animal and plant species have true global distributions apart from ciliated protozoa which are difficult to collect, preserve and identify. Those species which do occur world-wide probably have broad ecological requirements and are, therefore, generally not suitable as indicators. Recent developments in biological monitoring methods have favoured the use of families of organisms instead of species in order to limit the time and effort required in identifying organisms to the species level. Although the ecological requirements of families of organisms are often so broad that their use in biological indices is limited (Friedrich, 1992), the basic principles can be applied to develop regionally suitable methods. Recent experience in Brazil has shown that it can be useful, when establishing a new monitoring system based on bioindicators, to use a provisional form of taxonomy such as "morphological type" which can be observed at species level (Friedrich *et al.*, 1990). A general problem associated with establishing such systems is the lack of basic data from physico-chemical analyses. Statistical correlations of species alone are not satisfactory.

The basic principles of some indices are described below. It must be stressed that in most cases the indices only work well for the water bodies in the regions in which they were developed, and that they may give anomalous results in other types of water body, largely due to natural variations in species distributions. When applying biological indices in other regions careful selection of the appropriate method must be made to suit local conditions (Tolkamp, 1985) and caution must be applied in interpreting the results. Alternatively, the indices can be modified to suit local conditions. Reviews of some of the widely used biotic indices for water quality assessment are available in Hellawell (1978) and Newman *et al.* (1992).

The Saprobic system and Saprobic Indices

At the very beginning of the twentieth century the effect of point source pollution from sewage discharges on aquatic fauna and flora downstream of urbanised areas became evident. Kolkwitz and Marsson (1902, 1908, 1909) were the first to exploit these effects and present a practical system for water quality assessment using biota. Their system, known as the Saprobic system, has been used mainly in Central Europe. It is based on the observation that downstream of a major source of organic matter pollution a change in biota occurs. As self-purification takes place further ecosystem changes can be observed, principally in the components of the biotic communities. Odour and other chemical variants in the water also change. Kolkwitz and Marsson were principally concerned with an ecological approach, dealing with biological communities and not purely with indicator species. Since the taxonomy of aquatic organisms in Central Europe is well developed, it is possible to use the species level (which is the most precise) in the Saprobic system developed in that region.

The Saprobic system is based on four zones of gradual self-purification: the polysaprobic zone, the α-mesosaprobic zone, the ß-mesosaprobic zone, and the oligosaprobic zone. These zones are characterised by indicator species, certain chemical conditions and the general nature of the bottom of the water body and of the water itself, as described below:

- *Polysaprobic zone* (extremely severe pollution): Rapid degradation processes and predominantly anaerobic conditions. Protein degradation products, peptones and peptides, present. Hydrogen sulphide (H_2S), ammonia (NH_3) and carbon dioxide (CO_2) are produced as the end products of degradation. Polysaprobic waters are usually dirty grey in colour with a faecal or rotten smell, and highly turbid due to the enormous quantities of bacteria and colloids. In many cases, the bottom of the watercourse is silty (black sludge) and the undersides of stones are coloured black by a coating of iron sulphide (FeS). Such waters are characterised by the absence of most autotrophic organisms and a dominance of bacteria, particularly thiobacteria which are particularly well adapted to the presence of H_2S. Various blue-green algae, rhizopods, zooflagellates and ciliated protozoa are also typical organisms in polysaprobic waters. The few invertebrates that can live in the polysaprobic zone often have the blood pigment, haemoglobin, (e.g. *Tubifex*, *Chironomus thummi*) or organs for the intake of atmospheric air (e.g. *Eristalis*). Fish are not usually present.

- *α-mesosaprobic zone* (severe pollution): Amino acids and their degradation products, mainly the fatty acids, are present. The presence of free oxygen causes a decline in reduction processes. The water is usually dark grey in colour and smells rotten or unpleasant due to H_2S or the residues of protein and carbohydrate fermentation. This zone is characterised by "sewage fungus", a mixture of organisms dominated by the bacterium *Sphaerotilus natans*. The mass of organisms, which form long strands, can become detached from the bottom by the gas generated during respiration and decomposition processes, and then drift in the water column as dirty-grey masses. Frequently, they form a mat over the entire surface of the stream bed. Sewage fungus is particularly common in waters containing wastes rich in carbohydrates such as sewage and effluents from sugar and wood processing factories.

- *ß-mesosaprobic zone* (moderate pollution): Aerobic conditions normally aided by photosynthetic aeration. Oxygen super-saturation may occur during the day in eutrophic waters. Reduction processes are virtually complete and protein degradation products such as amino acids, fatty acids and ammonia are found in low concentrations only. The water is usually transparent or slightly turbid, odour-free and generally not coloured. The surface waters are characterised by a rich submerged vegetation, abundant macrozoobenthos (particularly Mollusca, Insecta, Hirudinae, Entomostraca) and coarse fish (Cyprinidae).
- *Oligosaprobic zone* (no pollution or very slight pollution): Oxygen saturation is common. Mineralisation results in the formation of inorganic or stable organic residues (e.g. humic substances). More sensitive species such as aquatic mosses, planaria and insect larvae can be found. The predominant fish are Salmonid species.

Each of the four zones can be characterised by indicator species which live almost exclusively in those particular zones. Therefore, comparison of the species list from a specific sampling point with the list of indicator species for the four zones enables surface waters to be classified into quality categories, particularly when combined with other important and often characteristic details (e.g. generation of gas in sediments, development of froth, iron sulphide on the undersides of stones, etc.).

The above classification system has been used to design Saprobic Indices particularly for data treatment, assessment and interpretation in relation to decision making and management. The first Saprobic Index designed by Pantle and Buck (1955) has been modified by Liebmann (1962). The frequency of occurrence of each species at the sampling point, as well as the saprobic value of that indicator species are expressed numerically. The frequency ratings or abundance, a, are:

random occurrence	$a = 1$
frequent occurrence	$a = 3$
massive development	$a = 5$

and the preferred saprobic zones of the species are indicated by the numerical values, s, as follows:

oligosaprobic	$s = 1$
ß-mesosaprobic	$s = 2$
α-mesosaprobic	$s = 3$
polysaprobic	$s = 4$

For any given species i the product of abundance a_i and saprobic zone preference s_i expresses the saprobic value S_i for that species, i.e. $S_i = a_i s_i$.

The sum of saprobic values for all the indicator species determined at the sampling point divided by the sum of all the frequency values for the indicator species gives the Saprobic Index (S) which can be calculated from the following formula:

$$S = \sum_{i=1}^{n}(s_i \cdot a_i) \Big/ \sum_{i=1}^{n}(a_i)$$

The Saprobic Index S, a number between 1 and 4, is the "weighted mean" of all individual indices and indicates the saprobic zone as follows:

S = 1.0 – < 1.5	oligosaprobic
S = 1.5 – < 2.5	ß-mesosaprobic
S = 2.5 – < 3.5	α-mesosaprobic
S = 3.5 – 4.0	polysaprobic

The Saprobic Index can be plotted directly against the distance along a river as shown in Figure 5.2. This is an example of an early use of biological data for engineers and decision makers. Note, however, that use of the index requires the organisms normally occurring in each of the river classification zones for a particular region to be known so that they can be assigned to a preferred saprobic zone during the calculation of the index. This information can only be obtained by detailed studies of the river systems, including precise identification of the individual species.

A comprehensive revision of the Saprobic system, carried out in 1973 by Sládecek (1973), has been adopted for widespread use in Central and Eastern Europe (LAWA, 1976; Breitig and Von Tümpling, 1982). Based on many years of practical experience and a large amount of data (especially physico-chemical data from water analysis) the system has again been revised by a group of experts in Germany (Friedrich, 1990). The main revisions were:

- The species used must be benthic in order to reflect the situation at the place where they are found, since planktonic species reflect the situation at an unknown place upstream.
- Photoautotrophic species are no longer included to avoid interactions between indicating saprobity and indicating trophic status.
- The species selected must be able to be identified without doubt by trained biologists and not only by specialists in taxonomy.
- The species used must be distributed over most of Central Europe.
- All species included must be well known with respect to their ecological requirements.

In this revision, the organisms were assigned a saprobic value (s) between 1 and 20 for a more precise description of the ecological range of the species. The Saprobic Index is calculated using the formula of Zelinka and Marvan

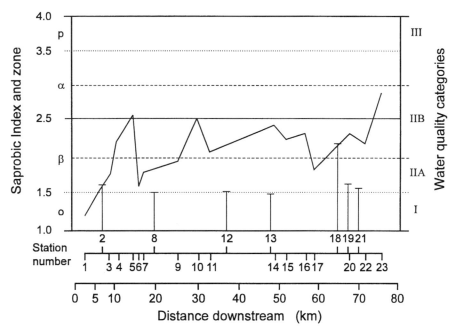

Figure 5.2 Changes in the Saprobic Index of Pantle and Buck (1955) along the Werra river, Germany (After UNESCO/WHO, 1978)

(1961) which takes account of the fact that very few species occur only in one saprobic zone. Using many years' experience in Germany, species with very narrow ecological ranges have been distinguished from less sensitive ones and a weighting factor, g (a value of 1, 2, 4, 8 or 16) has been assigned to each organism and incorporated into the revised formula as follows:

$$S = \sum_{i=1}^{n} (s_i . a_i . g_i) \bigg/ \sum_{i=1}^{n} (a_i . g_i)$$

Further details and the formula for calculating the uncertainty of this index are available in Friedrich (1990). This revised system has been designated as a German Standard Method (DIN 38410 T.2) and forms part of the basis of an integrated system of water quality classification, which includes biological and chemical variables.

The sampling procedure to be employed for collecting organisms for determination of the Saprobic and other biotic indices has been standardised at the national and international level (e.g. ISO, 1985).

Biotic indices
Alternative approaches to the Saprobic Index have been developed by Cairns *et al.* (1968), Woodiwiss (1964), Chandler (1970) and others. These methods are based on the presence or absence of certain "indicator" groups (e.g. Plecoptera, Ephemeroptera, Gammaridae), and/or "indicator" species, at the sampling point. As with the Saprobic Index, they are best suited to use in waters polluted with organic matter, particularly sewage, since the indicator organisms are usually sensitive to decreases in oxygen concentrations. However, a similar approach has recently been developed for the biological monitoring of acidification in streams and lakes using an "Acidification Index" based on the tolerance of invertebrates to acidity (Raddum *et al.*, 1988; Fjellheim and Raddum, 1990).

The Trent Biotic Index was originally developed by Woodiwiss (1964) for assessing pollution in the River Trent in England and forms the basis for many similar types of index. The index is based on the number of defined taxa of benthic invertebrates in relation to the presence of six key organisms found in the fauna of the sample site. Depending on the number of taxonomic groups present and the key organisms found in the fauna, the index ranges from ten, for clean water, to zero for polluted water (Table 5.4). This index has been found to be rather insensitive and has been largely replaced by further developments of the basic principle which have become widely used in Europe. One variation, used in the UK, is the Chandler Biotic Index (Chandler, 1970). To derive the index for a particular river station the invertebrate fauna, collected according to a standard procedure, are identified and then counted. Each group is given a score according to its abundance as shown in Table 5.5. The total score represents the index and the higher the score the cleaner the water. Calculation of the Trent Biotic Index and Chandler Biotic Index using invertebrate samples collected from an upland stretch of a lowland river in England is illustrated in Table 5.6.

In order to limit the taxonomic requirement of earlier biotic indices to identify organisms to species level, some alternative indices have been developed which use only the family level of identification (Hellawell, 1986; Abel, 1989). An example is the Biological Monitoring Working Party-score (BMWP) which has been published as a standard method by an international panel (ISO-BMWP, 1979). This score was devised in the UK but was not specific to any single river catchment or geographical area. Invertebrates are collected from different habitats (e.g. gravel, silt, weed beds) at representative sites of river stretches. The organisms are identified to the family level and then each family is allocated a score between one and ten. The most

Table 5.4   Use of invertebrates to calculate the Trent Biotic Index

Organisms in order of tendency to disappear as degree of pollution increases		Total number of groups[1] present				
		0–1	2–5	6–10	11–15	16 +
Clean				Biotic index		
Plecoptera larvae present	More than one species		7	8	9	10
	One species only		6	7	8	9
Ephemeroptera larvae present	More than one species[2]		6	7	8	9
	One species only[2]		5	6	7	8
Trichoptera larvae present	More than one species[3]		5	6	7	8
	One species only[3]	4	4	5	6	7
Gammarus present	All above species absent	3	4	5	6	7
Asellus present	All above species absent	2	3	4	5	6
Tubificid worm and/or red chironomid larvae present	All above species absent	1	2	3	4	
All above types absent	Some organisms such as *Eristalis tenax* not requiring dissolved oxygen may be present	0	1	2		
Polluted						

The boxed index number represents the result of calculating the Trent Biotic Index using the species list in Table 5.6.

[1] The term "group" means any one of the species included in the list of organisms below (without requiring detailed identification)

[2] *Baetis rhodani* excluded

[3] *Baetis rhodani* (Ephemeroptera) is counted in this section for the purpose of classification

Groups:
Each known species of Platyhelminthes (flat-worms)
Annelida (worms excluding genus *Nais*)
Genus *Nais* (worms)
Each known species of Hirudinae (leeches)
Each known species of Mollusca (snails)
Each known species of Crustacea (hog louse, shrimps)
Each known species of Plecoptera (stone-fly)
Each known genus of Ephemeroptera (may-fly, excluding *Baetis rhodani*)
Baetis rhodani (may-fly)
Each family of Trichoptera (caddis-fly)
Each species of Megaloptera larvae (alder-fly)
Family Chironomidae (midge larvae except *Chironomus riparius*)
Chironomus riparius (blood worms)
Family Simulidae (black-fly larvae)
Each known species of other fly larvae
Each known species of Coleoptera (Beetles and beetle larvae)
Each known species of Hydracarina (water mites)

Source: After Mason, 1981

Table 5.5 The use of invertebrates to calculate the Chandler Biotic Index

Row	Groups present in sample		Abundance in sample				
			Present 1–2	Few 3–10	Common 11–50	Abundant 51–100	Very abundant 100 +
					Points scored		
1	Each species of	Crenobia alpina, Taenopterygidae, Perlidae	90	94	98	99	100
2	Each species of	Perlodidae, Isoperlidae, Chloroperlidae, Leuctridae, Capniidae, Nemouridae (excluding Amphinemura)	84	89	94	97	98
3	Each species of	Ephemeroptera (excluding Baetis)	79	84	90	94	97
4	Each species of	Cased caddis, Megaloptera	75	80	86	91	94
5	Each species of	Ancylus	70	75	82	87	91
6	Genera	Rhyacophila (Tricoptera)	65	70	77	83	88
7	Genus	Dicranota, Limnophora	60	65	72	78	84
8	Genus	Simulium	56	61	67	73	75
9	Genera of	Coleoptera, Nematoda	51	55	61	66	72
10		Amphinemoura (Plecoptera)	47	50	54	58	63
11		Baetis (Ephemeroptera)	44	46	48	50	52
12		Gammarus	40	40	40	40	40
13	Each species of	Uncased caddis (excl. Rhyacophila)	38	36	35	33	31
14	Each species of	Tricladida (excluding C. alpina)	35	33	31	29	25
15	Genera of	Hydracarina	32	30	28	25	21
16	Each species of	Mollusca (excluding Ancylus)	30	28	25	22	18
17		Chironomids (excl. C. riparius)	28	25	21	18	15
18	Each species of	Glossiphonia	26	23	20	16	13
19	Each species of	Asellus	25	22	18	14	10
20	Each species of	Leech (excl. Glossiphonia, Haemopsis)	24	20	16	12	8
21		Haemopsis	23	19	15	10	7
22		Tubifex sp.	22	18	13	12	9
23		Chironomus riparius	21	17	12	7	4
24		Nais	20	16	10	6	2
25	Each species of	air breathing species	19	15	9	5	1
26		No animal life			0		

Source: After Mason, 1981

Table 5.6 Examples of the calculation of the Trent and Chandler Biotic Indices using invertebrates identified in a benthos sample taken from the upper stretch of a river in lowland England[1]

Phylum	Class/Order	Family	Species	Number in sample
Platyhelminthes	Turbellaria	Planariidae	*Polycelis tenuis*	3
Annelida	Oligochaeta	Tubificidae		52
		Naididae	*Nais elinguis*	31
	Hirudinea	Glossiphoniidae	*Glossiphonia companata*	12
			Helobdella stagnalis	9
		Erpobdellidae	*Erpobdella octoculata*	4
Mollusca	Gastropoda	Valvatidae	*Valvata piscinalis*	14
		Hydrobiidae	*Bithynia tentaculata*	1
		Lymnaeidae	*Lymnaea pereger*	11
		Planorbidae	*Planorbis vortex*	9
	Bivalvia	Sphaeriidae	*Sphaerium sp.*	22
			Pisidium sp.	45
Arthropoda	Crustacea	Asellidae	*Asellus aquaticus*	102
		Gammaridae	*Gammarus pulex*	61
	Hydracarina	Elayidae	*Eylais hamata*	37
Insecta	Ephemeroptera	Baetidae	*Baetis rhodani*	22
		Caenidae	*Caenis robusta*	2
		Ephemeridae	*Ephemera danica*	3
	Odonata	Coenagriidae	*Enallagma cyathigerum*	1
	Hemiptera	Corixidae	*Sigara falleni*	13
	Coleoptera	Elminthidae	*Elmis aenia*	7
	Trichoptera	Hydropsychidae	*Hydropsychae angustipennis*	33
		Polycentropidae	*Cymus trimaculatus*	5
		Limnephilidae		3
	Megaloptera	Sialidae	*Sialis lutaria*	3
	Diptera	Chironomidae	Chironominae "red chironomids"	125
			Orthocladinae "green chironomids"	56
		Simulidae	*Simulium sp.*	72
			Total number present	758

Continued

Table 5.6   Continued

Calculation of Trent Biotic Index

No. of groups (see notes to Table 5.4) present in sample = 26 (i.e. > 16), therefore, use final column of Table 5.4. No Plecoptera, but two species of Ephemeroptera present (excluding *Baetis rhodani*), therefore, the Index is given in the third row, final column of Table 5.4. Trent Biotic Index = 9, i.e. unpolluted.

Calculation of Chandler Biotic Index

No *Crenobia alpina* or stoneflies present, therefore, enter Table 5.5 at Row 3 (Ephemeroptera) and score as indicated below:

Row in Table 5.5	Number of taxa	Abundance class(es)	Score
3	2	2 + 3	79 + 84
4	2	1 + 1	75 + 75
5	0		
6	0		
7	0		
8	1	4	73
9	1	2	55
10	0		
11	1	3	48
12	1	4	40
13	2	2 + 3	36 + 35
14	1	2	33
15	1	3	28
16	4	1 + 2 + 3 + 3	30 + 28 + 25 + 25
17	1	4	18
18	1	3	20
19	1	5	10
20	2	2 + 3	20 + 16
21	0		
22	1	4	12
23	1	5	4
24	1	3	10
25	0		
		Total score	879

Chandler Biotic Index = 879, i.e. pollution free

[1] Combination of five two-minute kick samples collected in October

Source: Data from Mason, 1981

Table 5.7   The biological scores allocated to groups of organisms by the Biological Monitoring
Working Party (BMWP) Score

Score	Groups of organisms
10	Siphlonuridae, Heptageniidae, Leptophlebiidae, Ephemerellidae, Potamanthidae, Ephemeridae Taeniopterygidae Leuctridae, Capniidae, Perlodidae, Perlidae, Chloroperlidae Aphelocheiridae Phryganeidae, Molannidae, Beraeidae, Odontoceridae, Leptoceridae, Goeridae, Lepidostomatidae, Brachycentridae, Sericostomatidae
8	Astacidae Lestidae, Agriidae, Gomphidae, Cordulegasteridae, Aeshnidae, Corduliidae, Libellulidae Psychomyiidae (Ecnomidae), Phylopotamidae
7	Caenidae Nemouridae Rhyacophilidae (Glossosomatidae), Polycentropodidae, Limnephilidae
6	Neritidae, Viviparidae, Ancylidae (Acroloxidae) Hydroptilidae Unionidae Corophiidae, Gammaridae (Crangonyctidae) Platycnemididae, Coenagriidae
5	Mesovelidae, Hydrometridae, Gerridae, Nepidae, Naucoridae, Notonectidae, Pleidae, Corixidae Haliplidae, Hygrobiidae, Dytiscidae (Noteridae), Gyrinidae, Hydrophilidae (Hydraenidae), Clambidae, Scirtidae, Dryopidae, Elmidae Hydropsychidae Tipulidae, Simuliidae Planariidae (Dogesiidae), Dendrocoelidae
4	Baetidae Sialidae Pisicolidae
3	Valvatidae, Hydrobiidae (Bithyniidae), Lymnaeidae, Physidae, Planorbidae, Sphaeriidae Glossiphoniidae, Hirudinidae, Erpobdellidae Asellidae
2	Chironomidae
1	Oligochaeta

Groups in brackets are new groups of organisms that were previously contained in the group immediately before it in the list. These new groups are the result of developments in the taxonomic
system since the BMWP score was originally prepared.

sensitive organisms, such as mayfly nymphs score ten, molluscs score three
and the least sensitive worms score one (Table 5.7). The BMWP score is calculated by summing the scores for each family represented in the sample. The
number of taxa gives an indication of the diversity of the community (high
diversity usually indicates a healthy environment, see next section). The

average sensitivity of the families of the organisms present is known as the Average Score Per Taxon (ASPT) and can be determined by dividing the BMWP score by the number of taxa present. A BMWP score greater than 100 with an ASPT value greater than 4 generally indicates good water quality. An evaluation of the performance of the BMWP score in relation to a range of water quality variables is described by Armitage *et al.* (1983).

There are many variations on the biotic index widely employed in temperate zones. Indices have been developed by Tuffery and Verneaux (1967) in France and Gardeniers and Tolkamp (1976) in the Netherlands. Alternative approaches also include the Total Biotic Index according to Coste (1978) in France and modifications of the Woodiwiss (1964) and Tuffery and Verneaux (1967) approach known as the Belgian Biotic Index (Pauw and Vanhooven, 1983), and another by Andersen *et al.* (1984) in Denmark. These indices are used generally in conjunction with chemical monitoring to define water quality classifications. However, it is important that all biotic indices are not used in isolation, but together with all other data available to ensure correct interpretation of the biotic index. Further details, together with discussions of the advantages and disadvantages of biotic indices, are available in Washington (1984).

Predictive models
Knowledge of macroinvertebrate communities in rivers within geographical regions has now reached a sufficiently advanced level that computer-based systems are being developed for predicting the biological communities of unpolluted rivers and streams in relation to the natural features of each specific site. Comparison of the computer predicted communities with the actual fauna sampled and identified produces an indication of the "condition" or degree of pollution for the specific site.

The computer-based system developed in the UK known as RIVPACS (River InVertebrate Prediction and Classification System) (Wright *et al.*, 1993) is being incorporated in the national assessment and classification of river water quality. The ratio of the observed to predicted community status is expressed as an Ecological Quality Index (EQI) which is then used to assign biological grades to rivers and canals (National Rivers Authority, 1994). These grades can be used to assess water quality changes from one annual survey to the next (e.g. in the form of colour maps). Anticipated future uses of RIVPACS include river management, conservation and environmental impact assessment.

5.4.2   Community structure indices

The community structure approach examines the numerical abundance of each species in a community. The methods most widely used to assess aquatic pollution are based on either a diversity index or a similarity index. A diversity index attempts to combine the data on species abundance in a community into a single number. A similarity index is obtained by comparing two samples, one of which is often a control. These methods have the advantage that a knowledge of biology or ecology is not required in order to interpret whether a situation is getting better or worse, since this can be indicated by the numeric value of the index (nevertheless, a knowledge of taxonomy is required to analyse the samples). It is also possible to compare the index of a particular site (e.g. downstream of a sewage outfall) with that of an unpolluted site, although it is essential that the two sites being compared are similar with respect to their natural physical and chemical features.

The theory behind the design of such indices is beyond the scope of this chapter, but a general review, with respect to aquatic ecosystems is available in Washington (1984). The most common diversity indices in use are those based on information theory, such as the Shannon-Wiener Index (H'). Although they are applicable to a wide variety of aquatic situations they have not been thoroughly tested for biological relevance. Nevertheless, such indices can be used until other systems have been adequately field tested or developed.

Abundance and diversity indices
Stable ecosystems are characterised by a great diversity of species, most of which are represented by relatively few individuals. However, where the range of habitats or niches is restricted by physical or chemical factors, high numbers of individuals of only a few species occur. Such reduced diversity can be brought about by the inflow of wastewater into a watercourse, where the increase of nutrients and reduced competition from other species enables a few species to develop to high population densities. As long as all natural factors (e.g. current velocity, temperature, light intensity, sediment structure and stability) are comparable from one sampling point to another, differences in diversity of species can be used to detect changes in water quality or changes over time at the same site. It is important to realise, however, that species diversity can also increase as a result of slight pollution causing nutrient enrichment, although this may not be considered ecologically desirable. Diversity can also be very low where it is naturally limited by the conditions of the habitat, such as in small springs and headwaters. Diversity indices are probably best applied to situations of toxic or physical pollution which impose general stress (Hawkes, 1977).

Abundance or diversity indices are most suitable for use with benthic organisms since plankton are mobile and may reflect the situation elsewhere in the water body rather than at the monitoring site. Samples collected by a standard method (see section 5.10) are sorted into species and counted. It is not usually necessary to identify the species. Table 5.8 shows the calculations involved in some currently used diversity indices. More details are available in Washington (1984) and Abel (1989).

Similarity indices
Similarity indices are based on the comparison of the community structure in two samples, one of which is often a control. Different indices compare abundance in particular species, or abundance in any species, found at the sampling area. Their use is limited by the necessity for a clean water sampling station of similar nature for comparison purposes. Therefore, they are most suitable for point sources of pollution in a river where samples can be taken upstream and downstream of the input. There are many indices, some of which are described in detail by Washington (1984). A typical, simple index is that of Jaccard (1908):

$$\text{Jaccard Index} = \frac{n_c}{n_i + n_j}$$

where: n_c = the number of species common to samples i and j
n_i = number of species in sample i
n_j = number of species in sample j

At present, similarity indices have not been widely used in aquatic systems; they are more frequently applied to terrestrial situations.

Suitability of community structure and biotic indices
When interpreting results from programmes incorporating the use of community structure and biotic indices, it is important to realise that the absence of a species does not always indicate contamination. Absence may be caused by unfavourable changes in the environment which occur naturally. For example, severe storms can lead to changes in the substrate of rivers carrying flood waters or to changes in the vegetation of an area causing a subsequent loss of species depending on these specific habitats.

Attempts have been made to compare the suitability of saprobic, biotic, diversity and similarity indices in the field but these, by necessity, have usually concentrated on a few indices or only one river stretch (Balloch *et al.*, 1976; Ghetti and Bonazzi, 1977; Tolkamp, 1985). Since most indices are based on the natural occurrence of certain organisms in a river system, and all are susceptible to changes induced by natural events within the ecosystem, it is not

Table 5.8   Some currently used diversity indices

Index	Calculation
Simpson Index D	$D = \dfrac{\sum\limits_{i=1}^{s} n_i(n_i - 1)}{n(n-1)}$
Species deficit according to Kothé	$\dfrac{A_1 - A_x}{A_1} \times 100$
Margalef Index D	$D = \dfrac{S-1}{\ln N}$
Shannon and Wiener: Shannon Index H'	$H' = \sum\limits_{i=1}^{s} \dfrac{n_i}{n} \ln \dfrac{n_i}{n}$
Eveness E	$E = \dfrac{H'}{H'max}$

S The number of species in either a sample or a population
A_1 The number of species in a control sample
A_x The number of species in the sample of interest
N The number of individuals in a population or community
n The number of individuals in a sample from a population
n_i The number of individuals of species i in a sample from a population

Further information is available in Washington, 1984

possible to recommend any one method. Each method has disadvantages and advantages in relation to a particular water body, and only a full under-standing of the aquatic systems of a region (perhaps combined with testing the methods) can enable suitable methods to be selected and used effectively. A thorough review of the different methods available in relation to aquatic ecosystems is available by Washington (1984).

5.4.3   Plants as indicators of water quality
Most plant groups, such as macrophytes, filamentous algae, mosses, peri-phyton and phytoplankton, can be used as indicators of water pollution and eutrophication. Examples of their use in water pollution studies are given in Patrick and Hohn (1956), Fjerdingstad (1971), Coste (1978), Lange-Bertalot (1979) and Watanabe *et al.* (1988). Plants are also particularly useful for de-tecting metal pollution (Burton, 1986; Whitton, 1988) (see section 5.8). Problems of taxonomy and laboratory handling have restricted the use of

microscopic algae in water quality studies, although a few good examples exist (e.g. Backhaus, 1968a,b; Fjerdingstad, 1971; Friedrich, 1973). A recent review of the use of algae for monitoring rivers is available in Whitton *et al.* (1991).

Macrophytes

The species composition and density of macrophytes have both been used as indicators of eutrophication. However, the use of macrophytes as indicators of general pollution is more difficult. Nevertheless, they were incorporated into the original Saprobic system. In situations of very severe pollution, autotrophic plants cannot survive, but in less heavily polluted rivers, nutrient enrichment effects may be observed which positively influence macrophyte growth. Therefore, during the revision of the Saprobic system in Germany (Friedrich, 1990) macrophytes were excluded. In tropical rivers with low current velocities, the Water Hyacinth *(Eichhornia crassipes)* may grow extremely well despite the presence of severe pollution and a total depletion of oxygen in the water body. In order to develop systems based on macrophytes a thorough understanding of the local species and their preferred aquatic environments, including physical, chemical and hydrological features, is necessary.

Microphytes and phytoplankton

Attempts have been made to use phytoplankton species diversity and composition as indicators of the trophic status of lakes. These have not been highly successful because the presence or absence of algal species is not always directly related to trophic status and because such relationships have been poorly defined.

Fjerdingstad (1964, 1965) used the microphytobenthos, i.e. the sessile microphytic algae and bacteria, to establish a classification for Danish waters. Other workers, such as Lange-Bertalot (1979), Coste (1978) and Watanabe *et al.* (1988), developed new approaches using diatoms as bioindicators of pollution in running waters. Artificial substrates are particularly suitable for sampling in these situations. Overviews of these techniques are available in Round (1991) and Rott (1991) and an overview of the use of algae in monitoring rivers in Europe is available in Whitton *et al.* (1991).

Some phytoplankton and periphytic algae have been shown to have very narrow tolerance ranges of pH. The recent use of diatoms to indicate acidification in rivers is described by Coste *et al.* (1991). Acid lakes are generally dominated by 5 to 10 acid-tolerant algal species. This has been exploited to assess the past history of acidified lakes using fossilised diatom assemblages (Renberg and Hellberg, 1982; Battarbee and Charles, 1986; NRC, 1986; Smol *et al.*, 1986).

5.5   Microbiological methods

The natural bacterial communities of freshwaters are largely responsible for the self-purification processes which biodegrade organic matter. They are particularly important with respect to the decomposition of sewage effluents and can be indicative of the presence of very high levels of organic matter by forming the characteristic "sewage fungus" community (see section 5.4.1). However, domestic sewage effluents also add to water bodies large numbers of certain bacterial species which arise from the human intestine. These bacteria (in particular *Escherichia coli*) can be used as indicators of the presence of human faecal matter and other pathogens possibly associated with it. Since the presence of human faecal matter in water bodies presents significant health risks when the water is used for drinking, personal hygiene, contact recreation or food processing, it is often the most basic and important reason for water quality assessment (see section 3.10), especially in countries where sewage treatment is negligible or inadequate. Therefore, simple and cheap methods have been developed for detecting the presence of faecal bacteria, some of which (through the use of kits) can be carried out in the field. Further details of field methods are available in Bartram and Ballance (1996).

5.5.1   Indicators of faecal contamination

For the microbiological analysis of water samples in relation to human health it is necessary to determine principally the pathogenic organisms. Detection of all possible pathogens would be a costly and very time consuming process. Methods have, therefore, been developed which detect organisms which are indicative of the presence of faecal pollution, such as the normal intestinal bacteria. If evidence for faecal material is found in a water sample it can be assumed that faecal pathogens may be present and if no evidence is found it is likely, although not totally certain, that the water is safe for human use.

Examination of water samples for the presence of faecal bacteria is a sensitive technique indicating recent faecal contamination. The organisms most commonly used as indicators of faecal pollution are the coliform bacteria, particularly *Escherichia coli* and other faecal coliforms. A count of total viable bacteria in a freshwater sample can distinguish between freshwater species and those from human and animal faeces by their different optimal growth temperatures. Water bacteria show optimal growth at 15 °C to 25 °C (i.e. incubation at 22 °C) and faecal bacteria at 37 °C. Careful sample handling and processing methods are necessary to ensure that there is no contamination from other sources and helps to prevent excess growth of any bacteria present in a water sample. Absence of faecal bacteria in any single sample does not guarantee the absence of faecal contamination.

Samples for bacteriological analysis must be taken in sterile glass or non-toxic plastic bottles with caps. All possible sources of contamination should be avoided such as river banks, stagnant areas, pipes, taps and the hands of the sample collector. All subsequent handling of the samples must be done under sterile conditions to avoid contamination after collection. Ideally samples should be analysed immediately (i.e. within an hour of collection) and before either death, growth or predation can occur within the sample. The rate at which these occur will depend on the temperature and nutrient status of the sample. If absolutely necessary, samples should be cooled to 4 °C and stored at that temperature for up to 30 hours. Samples should never be frozen and any disinfectant present should be neutralised.

Bacteria can be counted by: (i) the growth of colonies directly on a suitable medium, (ii) the increase in turbidity in a liquid medium, (iii) the growth of colonies on a filter and medium, or (iv) the evolution of gas after incubation in a special growth medium. By careful choice of the medium, an inhibitor, or a marker for specific metabolic products it is possible to enumerate specific groups of bacteria, or all bacteria. The method chosen depends on the organisms of interest, the nature of the water samples and the availability of equipment and skilled help. For comparability over time, however, it is essential that the chosen method is always adhered to.

Coliform bacteria
Coliform bacteria occur in high numbers in human faeces, and can be detected at occurrences as low as one bacterium per 100 ml. Therefore, they are a sensitive indicator of faecal pollution. Estimation of the numbers of *Escherichia coli* in a water sample are usually made using the multiple tube or membrane filter technique (WHO, 1984, 1992; Bartram and Ballance, 1996). The multiple tube method gives an estimate of the number of organisms in a given volume of water based on the inoculation of that volume into a number of tubes of growth medium. After incubation the most probable number (MPN) of organisms in the original sample can be statistically estimated from the number of tubes with a positive reaction. The method is cheap and simple but is subject to considerable error. The membrane filtration method determines the number of organisms in a measured volume of sample which is filtered through a 0.45 µm pore diameter membrane filter and incubated face upwards on a selective medium. The number of colonies formed is counted. This technique gives results comparable to the multiple tube method and has the advantage of providing rapid results. However, it is not suitable for highly turbid waters, or where there are many other organisms capable of growing on the medium which may interfere with the coliforms.

The total number of *E. coli*, together with other variables, are included in some water quality indices.

Other intestinal bacteria
When other coliform organisms are found without the presence of faecal coliforms and *Escherichia coli*, other indicator organisms can be used to confirm the presence of faecal contamination. The two most common indicators are faecal streptococci and the sulphite-reducing *Clostridia perfringens*. Faecal streptococci rarely multiply in polluted water and are more resistant to disinfection than coliform organisms. Clostridial spores are also able to survive in water longer than coliform organisms and resist disinfection when inadequately carried out. They are not suitable organisms for routine monitoring because their resistance enables them to survive for long periods and they can be transported long distances after the initial infection. Multiple-tube and membrane-filtration techniques (see above) can be used with appropriate media and growth conditions.

Pathogenic bacteria
Examination for pathogenic bacteria is not normally included in common assessment operations. Positive identification of *Salmonella*, *Shigella* or *Vibrio* spp. can be quite complex, requiring several different methods. A special survey may be undertaken if a source of an epidemic is suspected, or if a new drinking water supply is being tested. As these organisms usually occur in relatively low numbers, it is necessary to concentrate the samples by a filtration technique.

Viruses
Viruses occur in very low numbers in water samples requiring concentration of the sample prior to any analysis. Although the methodology for identification of viruses is constantly being improved and simplified, all methods require advanced and expensive laboratory facilities. Many local or regional authorities responsible for water quality may not be able to provide such facilities. However, suitably collected and prepared samples can easily be transported, making it feasible to have one national or regional laboratory capable of such analyses. Sample collection kits have been developed for use in such situations.

5.6 Physiological and biochemical methods
The physiological response of an organism to change in its environment can also be used for biological assessment of water quality. Many of the methods

which have been developed are most suitable for occasional assessment operations or basic surveys (as defined in Table 2.1). Very few are suitable for routine monitoring and assessment because they are often developed in relation to specific water bodies, are highly qualitative or rather complex and are expensive to perform. The latter includes the more advanced tests, such as determination of sugar levels in the blood and glycogen in the liver and muscles of fish as an indicator of stress, or measurement of specific enzymes (e.g. cholinesterase) in tissues of aquatic organisms as an early indicator of stress due to oxygen deficiency or presence of organic chemicals (Beim, 1986a,b,c; Kotelevtsev et al., 1986). Some tests, such as the inhibition of cholinesterase, are in use to a limited extent world-wide. A summary of enzyme techniques is available in Obst and Holzapfel-Pschorn (1988). A few of the simpler and cheaper methods which can be modified to suit local conditions are described in more detail below.

5.6.1   Growth rate of algae and bacteria

Bringmann and Kuehn (1980) have developed tests to estimate the growth rate of *Scenedesmus quadricauda* (green alga), *Nostoc* sp. (cyanobacteria) and *Escherichia coli* (bacteria) in water samples under standardised conditions. The set of tests can be used to determine the quantity of biodegradable organic compounds (using *E. coli*) and the tendency for eutrophication (using the growth of *Scenedesmus* and *Nostoc*), particularly in very slow-flowing waters. The tests were developed for Berlin rivers in Germany where they are still used for water quality assessment. Since such tests were first developed in the 1960s many better tests have now been developed to determine eutrophication capacity, for example the "Algal Assay Procedure Bottle Test" of the National Eutrophication Research Programme (US EPA, 1978). Some of these tests, based on algal growth rate, are now included in national and international standard methods.

5.6.2   Oxygen Production Potential

The measurement of Oxygen Production Potential (OPP) has been developed as a standardised method in Germany (DIN 38412 T.13 and T.14). Water samples containing the native plankton are incubated in light and dark bottles for a given time (usually 24 hours). This procedure can be carried out in the laboratory, using incubators at 20 °C and standardised illumination, or in the river itself. The net production of oxygen is an indicator of the activity of the phytoplankton and also of toxic inhibition, especially when correlated with the concentration of chlorophyll pigment in the sample. In order to shorten the time of incubation required to give a satisfactory result when conducting

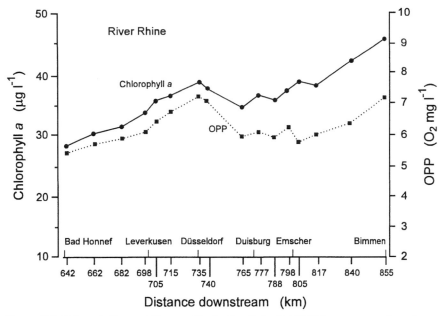

Figure 5.3 Chlorophyll a and Oxygen Production Potential (OPP) measured along the lower Rhine in 1982 (After Friedrich and Viehweg, 1984)

surveys far from a laboratory, the method can be modified by adding algae to water which is otherwise poor in phytoplankton.

An example using this test is given in Figure 5.3, showing a stretch of the lower Rhine River with a slight increase in phytoplankton density (as indicated by chlorophyll *a*) and the parallel activity expressed as OPP. A decline in both chlorophyll *a* and OPP occurred at Duisburg, where toxic inhibition took place as a result of effluent inputs from the metallurgical industry. Once dilution and sedimentation of the metals in the river had taken place, the phytoplankton species were able to increase in biomass, although their OPP was lower than before (i.e. for the equivalent chlorophyll concentrations). The same results can be illustrated by comparing the quantity of chlorophyll required to produce a unit (10 mg l^{-1}) of oxygen (Table 5.9). The toxic effluent from the outfall at Duisburg was subsequently controlled and the plankton was able to recover.

5.6.3   Additional consumption method
Another method using samples taken *in situ* has been developed by Knöpp (1968). It is known as the Zusätzliche Zehrung (ZZ) — additional consumption method. Substrates which stimulate bacterial growth (i.e. peptone or

Table 5.9   Chlorophyll concentrations required to produce 10 mg l^{-1} oxygen in the Oxygen
Production Potential (OPP) test carried out on samples from the Lower Rhine
river in 1982

Site	Distance downstream	Chlorophyll concentration[1]
Bad Honnef	640 km	59.0 µg l^{-1}
Leverkusen	699 km	67.5 µg l^{-1}
Düsseldorf	729 km	63.0 µg l^{-1}
Duisburg	776 km	87.5 µg l^{-1}
Kleve-Bimmen	865 km	82.5 µg l^{-1}

[1]   Median values from 51 weekly measurements

Source: Friedrich and Viehweg, 1984

glucose) are added to the water sample. If the bacterial activity is normal, the respiration associated with the reduction of the additional substrate leads to an increased oxygen consumption. This can be measured as greater oxygen depletion in the bottle at the end of the exposure period. If the bacteria are inhibited by a toxic agent in the sample, oxygen consumption ceases, or is very low. An example of the use of the ZZ method along a stretch of the River Rhine is given in Figure 5.4. The oxygen consumption declined in the samples to which peptone had been added as the BOD (biochemical oxygen demand, see section 3.5.3) of the river increased along the highly populated and industrialised section of the river.

Physiological tests such as the OPP and ZZ methods are limited by the bacteria and plankton organisms present in the water samples. Therefore, the results can vary from day to day and are only relative. However, better results are obtained by doing the test several times. These methods are particularly useful for effluent control where the changes might be demonstrated upstream and downstream of an effluent. They are also useful for monitoring large areas, or for short duration studies.

5.6.4   Chlorophyll fluorescence

Measurement of chlorophyll pigments provides an approximate indication of algal biomass. In addition to the traditional extraction methods for chlorophyll often included in lake assessments, the measurement of chlorophyll fluorescence (see section 3.3.12) has become a routine operation in some survey and monitoring programmes (Nusch, 1980, 1989; Nusch and Koppe, 1981). However, the fluorescence of chlorophyll *a* may be altered, or inhibited, by stress such as that induced by the presence of toxic chemicals. Therefore, comparisons of natural fluorescence with that produced in the

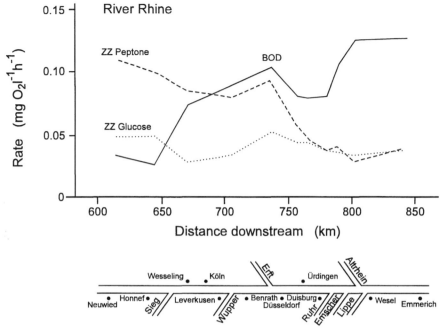

Figure 5.4 Use of the Additional Consumption Method (ZZ) according to Knöpp (1968) in the Lower Rhine. Increases in BOD are accompanied by decreased oxygen consumption by bacteria in the water samples to which additional substrates had been added (indicated as $ZZ_{peptone}$ and $ZZ_{glucose}$) (After UNESCO/WHO, 1978)

presence of possible contamination is a useful indicator of water quality. As it is difficult to produce a chlorophyll standard, the fluorescence measured in water samples must be correlated to other variables such as cell number or extracted chlorophyll. Fluorescence methods have the advantage of being rapid to carry out in the field: using samples taken directly from a water body, or *in situ* using apparatus specially designed to be used under water. The technique is particularly useful in lakes, reservoirs and impounded rivers which are rich in plankton (see Figure 6.18). As the techniques are developed further it is expected that they will become more widely used in water quality assessments, especially in relation to eutrophication.

5.7   Methods for assessing toxic pollution in controlled environments

It is important to know what kind of impact an effluent may have on a receiving ecosystem and the associated organisms. To avoid extensive laboratory simulations, some standardised laboratory procedures have been designed to test the toxic effects of chemical compounds or effluents on selected aquatic organisms. Such tests can also be helpful in determining the toxicity of water from a water body, for example, following an accidental pollution discharge.

Toxicity occurs in two forms: acute or chronic. Acute toxicity is usually caused by exposure to a large dose of a toxic compound for a short period of time. A rapid effect is produced on the organisms, usually death, and this may be used to determine the lethal concentration of a compound or effluent over a given period of time. For example, the concentration which kills 50 per cent of all the organisms in a test within 48 hours is described as the "48 hour LC_{50}". When an effect other than death is used for a similar test it is described in terms of the effective concentration (e.g. 48 hour EC_{50}). Chronic toxicity is caused by very low doses of a toxic compound or effluent over a long period of time and may be either lethal or sub-lethal (not sufficient to cause death). Sub-lethal effects can occur at the biochemical, physiological or behavioural level, including mutagenicity and genotoxicity and interference with the normal life cycle of an organism. There is currently great interest in developing methods for detecting sub-lethal toxicity as a means of early warning for environmental damage.

The testing of samples and compounds for toxicity under controlled laboratory conditions is widely used for pollution monitoring and control. Many methods have been standardised, allowing the results obtained for specific compounds to be compared all over the world. Depending on the organisms used, toxicity tests can indicate the possible effects of water samples, or compounds, on compartments of the natural environment. Tests carried out using effluents alone do not indicate the effects that may occur in the receiving water body due to interactions between the compound (or effluent) tested, with other compounds present in the receiving water body.

The type of test and the species used (Table 5.10) must be chosen in relation to the objectives of the test and to the specific compounds and their expected effects. Depending on the level of information required for any harmful, synthetic substance, different approaches can be applied, such as:

- studies of the toxic effects of pollutants on specific organisms,
- studies of the toxic effects of pollutants on populations and biocenoses,
- the establishment of maximum allowable toxic concentrations (MATC) of toxic compounds (these values can be used to set suitable concentrations for discharge into water bodies and to indicate recommendations for the purification of effluents prior to discharge),
- studies which provide the information to enable the control or suppression of harmful organisms in water, i.e. those which negatively influence the quality of water supplies, and
- studies of modes of action and transformation of toxic substances in organisms.

Table 5.10   Major methods for assessing toxicity in aquatic environments

Aquatic organisms affected		Methods for assessing toxicity	
Trophic level	Principal organisms	Principal methods	Additional methods
Decomposers	Bacteria, fungi, protozoa	Biological oxygen consumption; nitrification	Decomposition of cellulose, lignin, petroleum and other organic matter
Primary producers	Algae, macrophytes	Growth rates; reproductive capacity; oxygen consumption; chlorophyll fluorescence	Photosynthesis and respiration rates; chlorophyll concentrations; morphology, histology and growth
Secondary producers	Invertebrates, some fish species	Survival rate; reproductive capacity; survival of progeny	Feeding, growth and respiration rates; morphology.
Secondary consumers	Most fish and some invertebrates	Fertilisation rate; embryological development; larval survival; life cycle success	Feeding, growth and respiration rates; biochemical analyses e.g. hormones, haemoglobin; morphology and histology

Source: Based on Spynu and Beim, 1986

The variation in harmful effects of excess trace elements or synthetic substances on aquatic organisms is very broad, and potential effects are not entirely predictable. It may be necessary to carry out a series of tests with a potentially toxic sample in order to determine the group(s) of aquatic organisms which may be affected. It is then necessary to choose the species which is most sensitive to the compound in question for further toxicity tests, especially in relation to any management actions. For example, zinc is much more toxic to bacteria than to fish or man. Therefore, the control of zinc concentrations in a water body should be determined from the toxicity of zinc to the bacteria which are responsible for the self-purification of the system, rather than from the limits imposed for drinking water supplies. A large amount of information has been collected about the toxicity to aquatic organisms of natural and synthetic substances and stored in national and international databases, many of which are computerised and quick to search (e.g. the International Register of Potentially Toxic Chemicals (IRPTC) organised by the United Nations Environment Programme (UNEP) in Geneva). Whenever possible, this information should be consulted before beginning a detailed toxicity testing programme, as much unnecessary effort could be saved.

The scientific and technical development of toxicity testing is intensive, particularly with respect to the design of automatic tests and those using very sensitive reactions in organisms. Examples include early life stage tests with fish and behavioural changes, such as the measurement of shell movement in the freshwater clam, *Dreissena polymorpha*. Further information on the variety of tests available for pollution management and control is available in EIFAC (1975), US EPA (1985), Beim (1986a,b), Lillelund *et al.* (1987), OECD (1987), De Kruijf *et al.* (1988), Abel (1989) and many others. Some examples of standardised tests are given in Table 5.11.

5.7.1   Acute and sub-lethal toxicity

Toxicity tests generally use whole organisms because it is not usually known how, and on what part of the organism, the harmful substance may act. Such information cannot be obtained quickly. The use of cell cultures or enzymes from certain organisms is very restricted, partly because the significance of the observed effects for the environment is unknown. Substances which can affect enzymes may never have significant environmental impact as they may not penetrate the protective barriers of the living organisms. In addition, particular enzyme or cell types may be affected only by specific compounds. The use of total organisms ensures that all potential effects on the organism can be detected and the significance for the environment of the test compound is established. Therefore, for waters or effluents containing a mixture of compounds of unknown concentration, whole organisms are essential; whereas for checking the effects of specific compounds, enzymes, cell tissues or organelles provide a more sophisticated method of analysis. Such tests may be developed further in the future to avoid the increasing costs of chemical analysis at very low concentrations.

Acute toxicity is manifest by a severe effect (often death) upon the test organism within a short time of exposure. All substances are capable of being toxic, depending on their concentration, the organism exposed and the environmental conditions of the exposure. Toxicity is, therefore, relative and the term "toxic", as opposed to "harmful", should be used according to clearly described and standardised tests. There are many examples of toxicity tests in use around the world, employing different species and test procedures, some of which have been standardised for national or international use (Table 5.11).

Acute toxicity tests which give rapid results can be very useful in emergencies and cases of accidental pollution even though the reaction of the biota in

Table 5.11 Examples of standardised bioassays and acute toxicity tests

Organism	Exposure time	Reaction expected
Fish		
Brachydanio rerio	1–2 days	Death
Leuciscus idus	2 days	Death
Poecilia reticulata	4 days	Death
Salmo gairdnerii	4 days	Death
Invertebrates		
Daphnia magna	24 hours	Immobilisation
Daphnia similis	24 hours	Immobilisation
Asellus aquaticus	2 days	Immobility, death
Tubifex tubifex	4 days	Oxygen demand
Algae		
Scenedesmus subspicatus	4 days	Growth rate
Ankistrodesmus falcatus	10 days	Reproduction
Haematococcus pluvialis	24 days	Reproduction
Phormidium autumnale	24 hours	Movement
Bacteria		
Pseudomonas putida	30 minutes	Inhibition of oxygen consumption
Phytobacterium phosphoreum[1]	Minutes	Inhibition of luminescence
Protozoa		
Paramecium caudatum	2 days	Death, morphology, movement
Higher plants		
Sinapsis alba	5 days	Germination
Sinapsis alba	10–15 days	Growth
Avena sativa	10–15 days	Growth

[1] Now renamed *Vibrio fischeri*

the test does not give specific information about the type of substance causing the toxicity (see also section 5.3.2). For example, when investigating a
fish-kill, several water samples are usually taken for chemical analysis in order to determine the cause. If parallel fish toxicity tests are conducted
immediately, it may be possible to ascertain whether there is a toxic concentration in one, or all, of the water samples collected in the field. This

screening helps the chemical laboratory focus their efforts and gives information rapidly to water quality managers so that they can prepare for, or stop,
further action. Many tests for acute toxicity have been standardised by national and international organisations and are described in detail in national
water quality assessment guidebooks (e.g. National Committee of the USSR,
1987, 1989; APHA, 1989) or special toxicological methods texts.

Sub-lethal toxicity can be manifest in many ways depending on the organisms. For example, bivalve molluscs may show movement of the shells, fish
may exhibit inhibited gill rhythms or swimming behaviour and phytoplankton species may show reduced chlorophyll fluorescence. Many of these
effects (see also section 5.6) are being developed into bioassay methods to
give relative indications of the toxicity of effluents or water samples. In many
cases, concentrations of substances lower than the acute limit may have an
inhibitive effect on the reproductive capacity of an organism (and, therefore,
its life cycle) and this is also being exploited to develop bioassay methods
(see Table 5.11). At present, few of such methods have been standardised, although the technique based on the life cycle of the crustacean *Daphnia* sp. is
now well developed and becoming widely used (Mount and Norberg, 1984).

5.7.2   Tests for carcinogenic, mutagenic, or teratogenic capacity

Tests to establish the sub-lethal carcinogenic, teratogenic or mutagenic
capacity of chemicals in the aquatic environment are at present poorly developed. However, with the production and use of thousands of new chemical
substances each year (many of which may eventually be discharged to freshwaters or be transported by air or land run-off) such tests are becoming of
increasing importance, particularly with respect to human health and food
quality. Some countries, like Japan, prohibit products containing chemicals
until extensive tests are conducted on the behaviour of the chemicals, or
products, in the environment. As most of these types of test require a high degree of training or expertise and good laboratory facilities, it is unlikely that
they will form part of routine biological monitoring programmes for many
years to come, unless developed into simple kit-based techniques.

An example of a mutagenic test, which has been standardised, is the reverse mutation assay with *Salmonella typhimurium* or the "AMES" test
(Ames *et al.*, 1975; Maron and Ames, 1983; Levi *et al.*, 1989). This test
(based on the growth of the bacterium in special media) has been standardised by the OECD (Organisation for Economic Co-operation and
Development) (OECD, 1983) and is included in some international regulations. Since the test is not particularly complicated, it can be carried out by
most well-equipped laboratories. However, in some cases natural water

samples must be concentrated many times to get a measurable effect, making the results difficult to relate to the natural environment. As long-term toxicity is extremely important, improved tests will probably be available in the near future. At present such tests are principally used in the assessment of effluents from large chemical industries because they reduce the necessity for the highly sophisticated instruments and trained staff which are otherwise needed for effective chemical monitoring.

5.7.3   Continuous or dynamic toxicity tests

Automated biomonitoring can be achieved by determining various forms of toxic effects in continuous or "dynamic" tests. These tests are useful as early warning mechanisms for drinking water supplies, or as a general safeguard for the ecosystem. Typically, a continuous flow of water from the water body is routed through a specially constructed apparatus containing the test organisms. A dynamic fish test has been used on the River Rhine for many years (Besch and Juhnke, 1971; Juhnke and Besch, 1971) and, more recently, a *Daphnia* (crustacean) test has been developed (Knie, 1978). In the dynamic fish test illustrated in Figure 5.5 the presence of a toxic substance in the water passing through the apparatus (which may be directly diverted from a river, for example) is indicated by the inhibition of movement in the fish, which in turn activates an alarm signal and initiates automatic water sampling.

A system of continuous biotests is being investigated for future installation at selected monitoring sites on the River Rhine. The chosen tests will provide a biological warning system for contaminants in the river. Commercially available biomonitors, and test systems designed for the Rhine, have been evaluated for different trophic levels (bacteria, algae, mussels, water fleas and fish) (Schmitz *et al.*, 1994). Development of dynamic tests is an expanding field, especially in relation to monitoring the effects of effluents and sewage discharges.

5.8   The use of aquatic organisms in chemical monitoring

Many organisms have been found to accumulate certain contaminants in their body tissues and this is known as bioaccumulation. Some organisms may do this throughout their whole life time without detectable adverse effects on their normal physiological functions. These species may have detoxifying mechanisms which bind the contaminant in certain sites in the body and render them harmless. Other organisms accumulate contaminants over a period of time and only suffer adverse effects when critical levels are reached in their body tissues, or a change in their metabolic pattern re-releases the contaminants within their body.

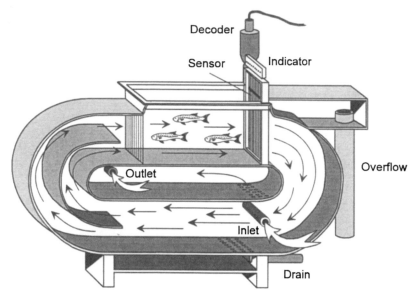

Figure 5.5 An example of a dynamic fish test which is used for continuous monitoring of toxicity in water directed from a water body through the test apparatus

Contaminants may be ingested together with the normal food source, as in the case of filter feeding organisms which have limited ability to select particulate material. Many contaminants are bound to fine particulate material (see section 4.2) and filter feeders are particularly vulnerable to this source of pollution. Other contaminants may be passively absorbed through the body surface, such as with the continuous flow of water through the gills of a fish. When an organism which has accumulated a contaminant is eaten by another organism, which in turn accumulates the contaminant from the tissues of its food source, the process is known as "food chain transfer" of the contaminant. The organisms higher up the food chain may, in this way, accumulate the contaminants to concentrations much higher than those occurring in the water or particulate material itself (see Figure 7.9). This process is known as biomagnification.

If the correlation between the concentration of a contaminant in the tissues of a certain species and the concentrations in the water and/or particulate material is good, the species can be used directly for chemical monitoring of water quality. When biomagnification occurs, the organisms offers the advantage of enabling detection of contaminants which might otherwise require sophisticated chemical analyses. Alternatively, biological tissues can be bulked together to give a greater total concentration for measurement by less sensitive techniques. At present, however, there are few species that occur world-wide which are known to be satisfactory for chemical analysis of water quality, although many species are useful at the local level (or possibly nationally in smaller countries).

Unless experimental observations are made on the toxicity of the contaminants to the organisms used for chemical monitoring, the results of tissue analysis cannot be used to assign any environmental significance to the measured levels, although they may indicate trends. However, tissue analysis provides useful information on the bioavailability, mobility and fate of contaminants in the aquatic environment. Organisms collected from their natural habitat also provide a time-integrated measure of ambient concentrations of a contaminant, averaging out temporal fluctuations.

Collecting organisms from the environment for chemical analysis is known as "passive biomonitoring". Deliberate exposure of organisms ("active biomonitoring") to chemicals or possibly to polluted environments can also provide useful information, particularly when chemical analysis of water samples is difficult due to extremely low concentrations. This technique has been found to be particularly useful for some organic chemicals which are bioaccumulated and which, in addition, may induce other effects such as histological changes (Philips, 1980; Lillelund *et al.*, 1987).

5.8.1 Criteria for organism selection

An aquatic organism must fulfil the following criteria before it can be used directly for contaminant monitoring:

- It should accumulate the contaminant of interest at the concentrations present in the environment without lethal toxic effects.
- There should be a simple correlation between the concentration of contaminant in the organism and the average contaminant concentration in the water body.
- Preferably, the organism should bioaccumulate the contaminant to a concentration high enough to enable direct analysis of the tissues.
- It should be abundant in, and representative of, the water body.
- It should be easy to sample and survive for long enough in the laboratory to enable studies of contaminant uptake to be performed.
- All the organisms used for comparison between sampling sites in a water body must show the same correlation between their contaminant concentrations and those in the surrounding water at all locations studied, under all conditions.

Suitable organisms are typically immobile, or at least unlikely to travel far from one place, so that they indicate the contaminant levels of the area in which they are collected rather than an area from which they may have migrated. Both plants and animals have been found suitable in many studies of metal, organic chemical and radioisotope pollution (Philips, 1980; Burton, 1986; Whitton, 1988; Samiullah, 1990). Fish are widely used as they are at the top of the food chain (and hence may bioaccumulate high concentrations from their food sources). As fish are often used as a human food source the results are directly relevant to human health (see section 5.11.1). Some aquatic plants, particularly mosses, have been shown to have a good correlation between tissue concentrations of some metals and the concentrations occurring in the surrounding water (Figure 5.6).

5.8.2 Necessary precautions when using biota for chemical analysis

There are many natural processes which affect the metabolism of organisms and their accumulation of contaminants. As a result, the age and the related size and weight of each individual used are important. Some organisms may accumulate contaminants throughout their life or only throughout their growth period. Consequently, in some species, older organisms may be expected to have higher concentrations of contaminants. Therefore, all organisms collected for monitoring should be of a comparable size or a similar age (see example in section 5.11.4). In some invertebrates the gonads may form a substantial proportion of total body weight and this proportion may vary between males and females. When measuring whole body contaminant concentrations it is, therefore, important that the sex of the organisms should be taken into consideration. The physiological condition of an organism may also affect bioaccumulation. In preparation for reproduction, or as a result of factors like starvation, body reserves such as fats or starches may be redistributed or used up. This is particularly important when contaminants are accumulated in certain tissue types only (e.g. organic chemicals in tissues with a high lipid content). Thus, the fitness and reproductive cycles of an organism, which are in turn related to the season of the year, must be considered.

Some metals are required for normal metabolic functions, for example as constituents of enzymes. Particular organisms, therefore, have a normal body concentration of those metals even when they have not accumulated any metal from contaminated water. This information may exist in the published literature for the species of interest and can provide guidelines for the expected baseline or "uncontaminated" levels within the monitoring species.

Before embarking on the use of organisms for biological monitoring additional information should be sought, or experimental analyses done, to determine whether the species of interest accumulate contaminants in certain tissues of the body more than others. For example, some metals such as Cd, Hg and Pb and organic chemicals may be accumulated mainly in organs, such

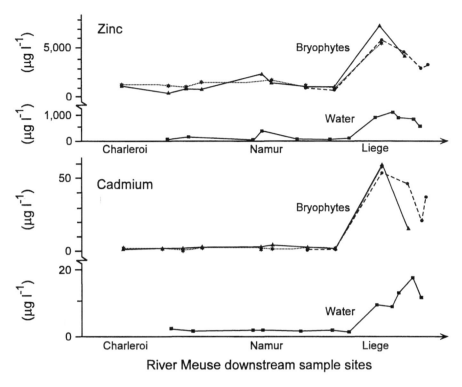

Figure 5.6 Concentrations of cadmium and zinc in water samples and mosses from the River Meuse, Belgium (After Burton, 1986, based on data from Empain, 1976)

as the liver, kidney and hepatopancreas, or in the exoskeleton of inverte-brates. Table 5.12 shows the accumulation of various organic chemicals in the organ and muscle tissues of the Rainbow Trout (*Salmo gairdnerii*). Per-sistent organic chemicals are frequently associated with tissues with a high lipid content. Therefore, it may be more economic, and effective, to collect and analyse only these tissues.

Some organisms regulate, usually by excretion, particular elements when they are exposed to levels in excess of normal body or environmental con-centrations. As a result correlations between body concentrations and environmental concentrations which previously existed, break down and the organism ceases to be useful as an indicator of environmental concentrations. This can be tested within the laboratory if the relevant information cannot be found in the published literature.

Samples of biological tissues for chemical analysis should be handled with the same high degree of cleanliness as for water samples, in order to avoid

Table 5.12   Bioaccumulation of micropollutants in Rainbow Trout (*Salmo gairdnerii*) exposed for six months to water from the River Rhine and to unchlorinated drinking water

Compound	Analytical limit	Drinking water		Rhine water	
		Organs	Muscle	Organs	Muscle
		(μg kg^{-1} wet weight)			
Trichlormethane	1	3.3	3.1	8.6	3.7
Tetrachlormethane	1	9.5	12	42	27
1,1,1-Trichlorethane	1	<1	<1	<1	<1
1,2-Dichlorethane	5	<5	<5	16	8.0
Trichlorethane	1	6.2	<1	8.4	<1
1,2-Dichlorbenzene	1	<1	2.0	140	10
1,3-Dichlorbenzene	1	<1	<1	30	6
1,4-Dichlorbenzene	1	<1	<1	100	10
1,2,3-Trichlorbenzene	1	<1	<1	15	1.3
1,2,4-Trichlorbenzene	1	<1	<1	190	7.2
1,2,3,4-Tetrachlorbenzene	1	<1	<1	46	7.1
1,2,3,5-Tetrachlorbenzene	1	<1	<1	58	<1
1,2,4,5-Tetrachlorbenzene	1	<1	<1	3.1	<1
Pentachlorbenzene	1	<1	<1	10	5.7
Hexachlorbenzene	1	1.2	1.4	200	40
α-HCH	1	<1	<1	1.1	<1
γ-HCH	1	<1	<1	2.0	<1
4,4'-DDT	1	<1	<1	5.8	<1
PCB-28	1	<1	<1	26	3.6
PCB-52	1	1.2	<1	12	3.4
PCB-101	1	2.3	2.6	26	5.7
PCB-153	1	3.0	3.9	30	6.5
PCB-138	1	3.7	3.7	29	6.0
PCB-180	1	1.2	1.4	8	2.1
Octachlorstyrene	1	5.4	9.7	17	8.7

Source: Friedrich, 1989

possible contamination and controls or special reference samples should also be analysed with each batch of samples (see sections 2.7.1 and 3.8.2).

5.9   Histological and morphological methods

In addition to the various effects discussed above, the presence of pollutants may lead to chronic and/or sub-lethal effects on the "health" of an organism which may be manifest as morphological or histological changes. When observed in the field, e.g. as symptoms of disease such as body tumours, it is often difficult to establish whether the effect is a normal occurrence of a disease in the population or due to an environmental stress such as pollution. Although attempts can be made to produce the same morphological effects in the laboratory, it is difficult to simulate accurately the precise environmental conditions to which the organism is accustomed, and thereby ensure that the results of the laboratory tests are a true reflection of the field situation. For this reason, at present, such methods have not been widely developed for routine monitoring purposes, but are occasionally used for research or special surveys.

An example of a simple morphological indicator of stress in aquatic organisms is the study of fish gills. During routine biological assessments or fishing exercises, it is possible to examine the gills of the fish caught, to check for signs of damage or inflammation. It is important that the examinations are carried out under water as well as while the fish is removed from the water. It may also be necessary to use a magnifying glass (Beim, pers. comm.). Under normal conditions the gills of fish are clean, bright pink in colour (due to the numerous blood vessels) and with undamaged, equally-spaced lamellae. Changes in gill colour, swelling or a covering of slime can all indicate stress. Frequent occurrence of such symptoms should be investigated further, perhaps with histological techniques. Effective use of such observation techniques requires detailed knowledge of the aquatic biology of a water body and the trained and experienced participation of a biologist.

5.10   Biological sampling strategies and techniques

5.10.1   Sampling strategies

Statistical treatment and analysis of biological data depends on an adequate sampling regime and appropriate numbers of samples. In general, the same principles for site selection, sampling frequency and number of samples collected for monitoring of water or sediment apply to biological assessments (see section 2.4.2). However, because many biological communities have a clumped (not random) distribution, quantitative sampling may be difficult. It is important to understand the distribution of the species or communities to be sampled so that methods for sample collection and data analysis can be selected properly. Therefore, preliminary surveys may be essential to indicate the number of samples (for a given method) required for a particular aquatic habitat, to achieve a certain degree of precision and confidence. Statistical methods relating to the sampling of benthic invertebrate communities are described by Elliott (1977) and general information on environmental sampling and analysis is given in Gilbert (1987). A worked example of statistical analysis of benthic invertebrate data is given in Chapter 10.

Samples of biota required for methods based on biotic indices, community analysis or species counts must often be taken from sites with specific hydrological features or substrate type. For example, samples for the Trent Biotic Index (see section 5.4.1) should be taken from shallow "riffle" sections of rivers. General guidelines for sampling biota are given by the International Standards Organization (ISO) in ISO Standard 7828 (ISO, 1985) and 9391 (ISO, 1991) and in the German standard DIN 38410 T.1 (1987). Full ecological assessment requires more than merely sampling the aquatic biota. Field workers should be properly trained to collect all the environmental information needed to aid interpretation of the biological data. For further information on field techniques see Bartram and Ballance (1996).

5.10.2 Sampling techniques

Sampling for biological assessments of water quality frequently require special methods, apparatus and precautions. There are many sampling devices for collecting benthic or planktonic organisms, many of which are available commercially, and some of which can be manufactured fairly simply. Each has advantages and disadvantages with respect to individual field situations as indicated in Table 5.13. Further information can be obtained in Edmondson and Winberg (1971), Vollenweider (1974), Elliott and Tullett (1978) and Bartram and Ballance (1996). Guidance on the use of different samplers for macroinvertebrates has been provided by ISO (1991). It must also be recognised that biological assessment may involve more than one group of organisms within the aquatic habitat, and that each group may require different types of apparatus in order to obtain quantitative samples.

Water samples taken for testing toxicity or for use in physiological tests should be taken with as much care as for chemical analysis, i.e. without any contamination, and preserved cool (and possibly in the dark). Samples which may be used for any form of microbiological analysis must be collected with sterilised glass bottles and kept cool and dark. When sampling biota for chemical analysis (bioaccumulation studies), particular species (see section 5.8), or even particular parts of an organism, have to be collected. For aquatic plants the growing tips are usually suitable. When collecting fish, it is necessary to record the name of the species, its length and general state of health, all of which can be seen without detailed investigation. Biological samples for chemical analysis should be handled with extreme care to avoid contamination, particularly from contact with sampling apparatus, storage vessels, dissecting instruments and the chemicals used in the preparation for analysis.

Table 5.13 Comparison of sampling methods for aquatic organisms

Sampler/ sampling mechanism	Most suitable organisms	Most suitable habitats	Advantages	Disadvantages
Collection by hand	Macrophytes, attached or clinging organisms	River and lake margins, shallow waters, stony substrates	Cheap – no equipment necessary	Qualitative only. Some organisms lost during disturbance. Specific organisms only collected
Hand net on pole (c. 500 μm mesh)	Benthic invertebrates	Shallow river beds, lake shores	Cheap, simple	Semi-quantitative. Mobile organisms may avoid net
Plankton net	Phytoplankton (c. 60 μm mesh), zooplankton (c. 150–300 μm mesh)	Open waters, mainly lakes	Cheap and simple. High density of organisms per sample. Large volume or integrated samples possible	Qualitative only (unless calibrated with a flow meter). Selective according to mesh size. Some damage to organisms possible
Bottle samplers (e.g. Friedinger, Van Dorn, Ruttner)	Phytoplankton, zooplankton (inc. protozoa), micro-organisms	Open waters, groundwaters	Quantitative. Enables samples to be collected from discrete depths. No damage to organisms	Expensive unless manufactured "in house". Low density of organisms per sample. Small total volume sampled
Water pump	Phytoplankton, zooplankton (inc. protozoa), micro-organisms	Open waters, groundwaters	Quantitative if calibrated. Rapid collection of large volume samples. Integrated depth sampling possible	Expensive and may need power supply. Sample may need filtration or centrifugation to concentrate organisms. Some damage to organisms possible

Continued

Table 5.13 Continued

Sampler/ sampling mechanism	Most suitable organisms	Most suitable habitats	Advantages	Disadvantages
Grab (e.g. Ekman, Peterson, Van Veen)	Benthic invertebrates living in, or on, the sediment. Macrophytes and associated attached organisms	Sandy or silty sediments, weed zones	Quantitative sample. Minimum disturbance to sample	Expensive. Requires winch for lowering and raising
Dredge-type	Mainly surface living benthic invertebrates	Bottom sediments of lakes and rivers	Semi-quantitative or qualitative analysis depending on sampler	Expensive. Mobile organisms avoid sampler. Natural spatial orientation of organisms disturbed
Corer (e.g. Jenkins)	Micro-organisms and benthic inverts living in sediment	Fine sediments, usually in lakes	Discrete, quantitative samples	Expensive unless made "in-house". Small quantity of sample
Artificial substrates (e.g. glass slides plastic baskets)	Epiphytic algae, attached invertebrate species, benthic invertebrates	Open waters of rivers and lakes, weed zones, bottom substrates	Semi-quantitative compared to other methods for similar groups of organisms. Minimum disturbance to community on removal of sampler. Cheap	"Unnatural habitat", therefore, not truly representative of natural communities. Positioning in water body important for successful use
Poisons (e.g. rotenone)	Fish	Small ponds or river stretches	Total collection of fish species in area sampled	Destructive technique
Fish net/trap	Fish	Open waters, river stretches, lakes	Cheap. Non-destructive	Selective. Qualitative unless mark recapture techniques used
Electro-fishing	Fish	Rivers and lake shores	Semi-quantitative. Non-destructive	Selective technique according to current used and fish size. Expensive. Safety risk to operators if not carried out carefully

Use and application of artificial substrates
When comparing biological data from more than one site it is important that the physical environment, especially the substrate and current velocity, is similar for all sites considered. Such problems can sometimes be overcome by the use of artificial substrates placed in the water body (e.g. synthetic foam blocks, glass slides, plastic baskets). These are also useful for situations where it is impossible to carry out biological sampling of the aquatic biota, for example when water levels are too high. The use of artificial substrates has also been incorporated into the design of some biological monitoring methods to enable the easy collection of microalgae, particularly periphytic species, which can then be used for physiological tests and measurement of chlorophyll or contaminants.

A technique particularly suitable for collecting macrozoobenthos, consists of baskets containing natural stones, glass balls or other inert particles of about 2 cm diameter or more. The basket can be suspended in the water by means of a fixed line and anchor. In the event of a pollution accident it is possible to lift the basket to the surface to observe whether any toxic effects on the biota have occurred. A similar method may be used for algae and other microphytes (Sládecková, 1962; Rott, 1991). Plastic plates, about 20 cm to 40 cm square, have been used successfully, either as single plates or as a sandwich of two plates a centimetre or more apart. The main problem associated with artificial substrates is placing them such that they are not subjected to vandalism, theft or destruction during floods.

5.11   Selection of biological methods: case studies

5.11.1   Impact assessment
The appropriate use of methods for biological assessment depends on the nature and severity of the anticipated impact. Many ecological methods work well in relation to organic matter inputs such as sewage discharges or effluents from food processing (e.g. dairies) or pulp and paper mills. They can be used to assess changes with time, survey areas affected by organic effluents and give a general indication of the ecological state of a water body. An example of the biological assessment of rivers and their improvement in water quality following the introduction of sewage treatment has been given in section 6.7.3 and Figures 6.34 and 6.35.

Toxic pollution can be assessed by means of toxicity tests or physiological and biochemical methods, including bioassay techniques. Toxicity tests do not indicate the overall effects on the aquatic ecosystem but, when undertaken regularly, can monitor possible changes in the environment (for better

Table 5.14   Mercury concentrations in muscle tissues of two different fish species from lakes and impounded reservoirs in Finland

Lake/reservoir	Roach (*Rutilus rutilus*)		Pike (*Esox lucius*)	
	Total Hg	Methyl Hg	Total Hg	Methyl Hg
	(μg g^{-1} fresh weight)			
Lake Pihlajavesi	0.33	0.31	0.92	0.87
Lake Seinäjärvi	0.29	0.29	0.60	0.58
Kalajärvi Res.	0.77	0.72	1.80	1.69
Kyrkösjärvi Res.	0.78	0.74	1.19	1.14
Porttipahta Res.	0.35	0.34	0.70	0.67

Source: After Samiullah, 1990 (based on data from Surma-Aha *et al.*, 1986)

or worse) and also act as an early warning mechanism for more severe effects. Sensitive physiological and biochemical methods are also particularly useful for detecting increased stress on the aquatic environment in advance of obviously noticeable effects. However, they are too expensive and sophisticated for routine assessments.

The impact of effluents containing metals or persistent organic chemicals can also be studied from the bioaccumulation of the contaminants in organisms living, or deliberately placed, near to the discharge. This form of assessment is particularly important where fish or shellfish are gathered as a source of human food and the regular monitoring of bioaccumulation can provide a measure of risk from toxic effluents to human populations. As a result of the potential human dietary intake of mono-methylmercury from contaminated fish in Finland, organic and inorganic mercury was measured in the food chain of two natural lakes and three reservoirs. Both inorganic and organic mercury concentrations were higher in the Pike (*Esox lucius*) than the Roach (*Rutilus rutilus*) because the Pike is a top predator feeding on other fish (Table 5.14). The higher mercury concentrations observed in the fish from the reservoirs was attributed to enhanced methylation or accumulation due to high concentrations of humic materials (Samiullah, 1990).

As with chemical monitoring, impact monitoring using biological methods should be applied upstream and downstream of the effluent expected to change the water quality (Figure 5.7). If this is not possible then a "control" site must be found in a similar water body unaffected by effluent discharges or other impacts. Methods based on artificial substrates can be particularly useful for sites with restricted access or no suitable habitats, or substrates, for the collection of test organisms. For example, mussels can be enclosed in

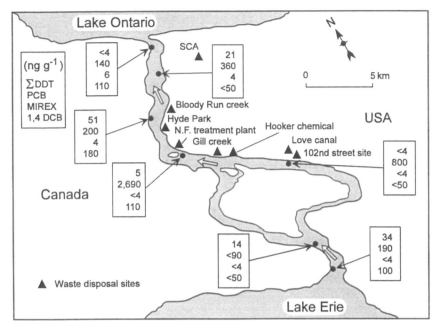

Figure 5.7   Use of biota (Spottail Shiners) for detecting the spatial distribution of organic contaminants in the Niagara river (After Kuntz, 1984)

suspended cages and later collected and analysed for bioaccumulation of metals or organic chemicals as illustrated in Figure 5.8.

5.11.2   Spatial surveys for contaminants and potential toxic effects

Acute toxicity, especially in fish, is usually so dramatic that it can be recognised by the appearance of dead fish floating in a water body. However, over large areas, it can be useful to carry out a spatial survey using biological methods, especially where a point source contains harmful substances which may be below the acute toxicity threshold but which may be bioaccumulated in certain organisms. These programmes can be very important for those people who live downstream of the effluent as well as acting as a protection mechanism for the aquatic environment (see Figure 5.7). Although any suitable aquatic organisms can be used, fish are recommended. Fish, especially predatory fish, are high up the food chain and have a long life span. The adults can provide a picture of the distribution of contaminants in the catchment area through their accumulation in body tissues. Their eggs and early life stages are also very sensitive to any environmental impact. This type of study gives information about the spatial distribution of harmful substances quickly and economically, and in a relatively short period of time.

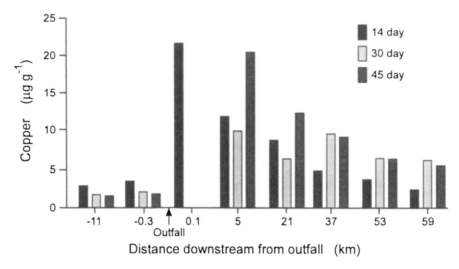

Figure 5.8 Accumulation of copper in the tissues of mussels *(Quadrula quadrula)* suspended in cages in the Muskingum river, USA (After Foster and Bates, 1978)

An example of a spatial survey for variations of pollutant concentrations with geographic region, which is also being used to detect trends over time, is the USA National Contaminant Biomonitoring Program (NCPB) (Samiullah, 1990). Fish are sampled, on alternate years, at 112 sites throughout the USA which are representative of all of the major river basins. Three composite samples, of no less than five fish each, are analysed from each site. Two samples are of bottom-dwelling fish species and the third is a predator fish. An example of the geographical variation observed in concentrations of total DDT in the fish tissue is given in Figure 5.9. Further results are available in Jacknow *et al.* (1986).

5.11.3   Early warning monitoring

Early warning monitoring can be used routinely in areas where there is a high risk of accidental pollution from many different sources, such as in the River Rhine (see Chapter 6), or it may be used to indicate changes in an existing situation, such as a regular discharge of toxic effluent. Regular sampling of biota to study bioaccumulation of toxic substances can also be useful for detecting any critical build up of toxins in the environment. Managerial action can be taken when levels are measured which may lead to toxic effects in the biota. For detecting acute toxicity a continuous "dynamic" test, e.g. using water routed through a by-pass, is particularly useful (see section 5.7.3). Standardised or special bioassays based on collected water samples have the

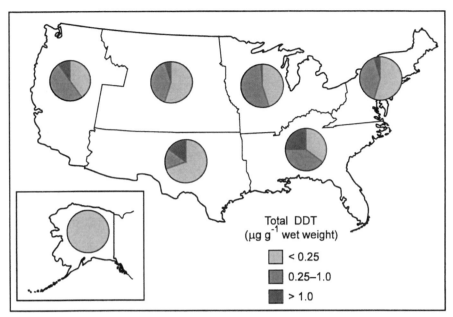

Figure 5.9 Variations in total DDT concentrations in selected freshwater fish species for different regions monitored as part of the U.S. National Contaminant Biomonitoring Program (NCBP) during 1980–1981 (After Samiullah, 1990, based on data from Jacknow *et al.*, 1986)

disadvantage that they are representative of the moment of sampling only. Therefore, it is necessary to take many samples in a short time to be sure that any critical concentration is actually sampled. This form of early warning monitoring is mainly required for the control of drinking water intakes.

5.11.4   Trends in biological concentrations and biological effects

Trends may be detected by repeating the same type of sampling and analysis at regular intervals for a long time. Biotic indices and chemical analysis of tissues of the same species collected regularly from the same site are particularly suitable methods for trend detection. However, particular care is required when interpreting the results of ecological methods since repeated analysis of the species composition from a given place may illustrate changes other than those directly caused by changes in water quality. For example, changes may be related to natural fluctuations in river discharge and exposure to light caused by climatic fluctuations. In addition, unless great care is taken to select organisms at certain times of year and at certain stages of the life cycle (see section 5.8.2), concentrations of contaminants in biota may reflect seasonal or annual dilution effects (i.e. wet and dry periods) of a fixed

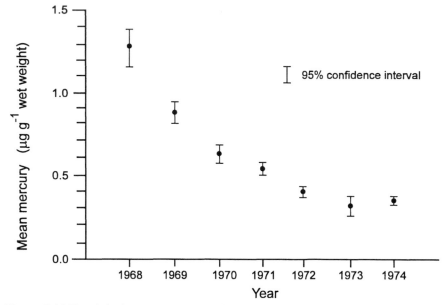

Figure 5.10 Trends in the mean concentration of mercury in axial muscle tissues of Roach (*Rutilus rutilus*) from the Mörrumsan river, Sweden after the removal of a phenyl mercury input to the river. Vertical bars represent 95 % confidence intervals (After Samiullah, 1990, based on data from Johnels *et al.*, 1979)

contaminant input. Therefore, the interpretation of effects on biota must be carried out in relation to all other aspects of the water body under investigation.

In order to study the recovery of mercury contamination in the River Mörrumsan (southern Sweden) following the removal of a phenyl mercury effluent from a paper mill, Roach (*Rutilus rutilus*) weighing 125 g were collected each year from 1968 to 1974. Mercury concentrations were measured in the axial muscle tissue and a clear decline was observed during the study of the recovery period (Figure 5.10).

5.11.5   Impact of accidental pollution

Determining the distribution of a contaminant in a water body after an accidental pollution incident, and thereby assessing the extent of the ecological damage, can be costly and time consuming, especially if chemical methods are used. However, bioassay procedures and simple toxicity tests can minimise the amount of chemical analysis required (see section 5.7.1). Applying such tests at the same sites and at regular intervals, following accidental pollution can indicate the recovery of the water quality to a condition where it can once again support aquatic life.

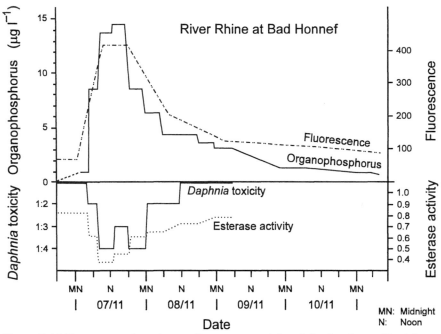

Figure 5.11 The recovery in water quality over several days following the accidental pollution of the River Rhine with pesticides by the "Sandoz" chemical plant at Basel, Switzerland in 1986, as indicated by two biological assessment methods

A combination of biological methods, which are simple and cheap to perform, was used to assess the recovery of the water in the River Rhine downstream of an accidental input of pesticides from the "Sandoz" chemical company in 1986. Figure 5.11 illustrates the recovery of the river water quality as indicated by the decrease in toxicity of the river water to *Daphnia* and a recovery in esterase (an enzyme) activity as the concentration of the contaminants in the river gradually declined.

5.12   Conclusions and recommendations

The importance of biological methods using plants and animals from all taxonomic levels has been demonstrated for all kinds of water quality monitoring and assessment programmes. The use of biota provides an integrated measure of all impacts on a water body as well as a possible historical interpretation of past pollution levels. Organisms can be used to provide early warning mechanisms of possible environmental damage and the impacts of effluents can also be tested and predicted before their discharge. Environmental degradation can be assessed by studying the communities of organisms present in the water body, thus giving an indication of effects on the aquatic ecosystem.

There is much international interest in standardising biological methods for use in major international monitoring programmes. Some bioassay and toxicity testing methods have already been accepted for world-wide use. The simplicity and low-cost of many methods enables monitoring to be carried out in many situations where the financial resources cannot support the sophisticated equipment required for chemical analysis of water quality. Basic methods, many of which have been described briefly in this chapter, can be adapted to local situations, climate, taxonomic expertise and the availability of trained personnel. Biological methods, therefore, are suitable for all three levels of monitoring: simple, intermediate and advanced (see Chapter 2), and their use is strongly recommended in conjunction with chemical and hydrological monitoring and assessment programmes.

5.13   References

Abel, P.D. 1989 *Water Pollution Biology*. Ellis Horwood Limited, Chichester.

Alabaster, J.S. [Ed.] 1977 *Biological Monitoring of Inland Fisheries*. Applied Science Publishers Ltd., London, 226 pp.

Ames, B.N., McCann, J. and Yamasaki, E. 1975 Methods for detecting carcinogens and mutagens with the Salmonella/mammalian microsome mutagenicity test. *Mutata. Res.*, **31**, 347-364.

Andersen, M.M., Riget, F.F. and Sparholt, H. 1984 A modification of the Trent Index for use in Denmark. *Water Research*, **18**, 141-151.

APHA 1989 *Standard Methods for the Examination of Water and Wastewater*. 17th edition, American Public Health Association, Washington DC., 1268 pp.

Armitage, P.D., Moss, D., Wright, J.F. and Furse, M.T. 1983 The performance of a new biological water quality score system based on macroinvertebrates over a wide range of unpolluted running water sites. *Water Res.* **17**(3), 333-347.

Backhaus, D. 1968a Ökologische Untersuchungen an den Aufwuchsalgen der obersten Donau und ihrer Quellflüsse, I. Voruntersuchungen. (Ecological investigations upon algae of the upper Danube and its spring streams, Part I. Pre-investigations). *Arch. Hydrobiol.*, **30**, 364-399.

Backhaus, D. 1968b Ökologische Untersuchungen an Aufwuchsalgen der obersten Donau und ihrer Quellflüsse, III. Die Algenverteilung und ihre Beziehung zur Milieuofferte. (Ecological investigations upon algae of the upper Danube and its spring streams, Part III. Distribution of algae and their relationship to the environment). *Arch. Hydrobiol.*, **34**, 130-149.

Balloch, D., Davies, C.E. and Jones, F.H. 1976 Biological assessment of water quality in three British rivers: the North Esk (Scotland), the Ivel (England) and the Taf (Wales). *Wat. Pollut Control*, **75**, 92-114.

Bartram, J. and Ballance, R. [Eds] 1996 *Water Quality Monitoring: A Practical*

Guide to the Design of Freshwater Quality Studies and Monitoring Programmes. Chapman & Hall, London.

Battarbee, R.W. and Charles, D.F. 1986 Diatom-based pH reconstruction studies of acid lakes in Europe and North America: a synthesis. *Water Air Soil Pollut.*, **30**, 347-354.

Beim, A.M. 1986a Effects of chemical pollutants on aquatic organisms. In: A.A. Kasparov and I.V. Sanotsky [Eds] *Toxicometry of Environmental Chemical Pollutants.* Moskow, 287-292.

Beim, A.M. 1986b Biological testing of industrial effluent. In: *Problems of Aquatic Toxicology, Biotesting and Water Quality Management.* EPA/600/9-86/024, U.S. Environmental Protection Agency, Athens, Georgia, 122-135.

Beim, A.M. 1986c Bioassays to protect water reservoirs against pollution by waste water (for Lake Baikal). In: UNEP/WHO *Hygienic Criteria of Drinking Water Quality.* Centre for International Projects GKNT, Moscow, 191-198.

Besch, W.K. and Juhnke, I. 1971 Un nouvel appareil d'étude toxicologique utilisant des carpillons. *Ann. de Limnologie*, **7**, 1-6.

Bhatti, N., Streets, D.G. and Foell, W.K. 1989 Acid rain in Asia. Paper presented at the "Workshop on Acid Rain in Asia", Bangkok, 13-17 November, 1989.

Breitig, G. and Von Tümpling 1982 *Ausgewählte Methoden der Wasseruntersuchung Bd. II. Biologische, mikrobiologische und toxikologische Methoden.* Fischer Jena.

Bringmann, G. and Kuehn, R. 1980 Comparison of the toxicity thresholds of water pollutants to Bacteria, Algae, and Protozoa in cell multiplication inhibition test. *Water Research*, **14**, 231-241.

Burton, M.A.S. 1986 *Biological Monitoring of Environmental Contaminants (Plants).* MARC Report No. 32, Monitoring and Assessment Research Centre, King's College London, University of London, 247 pp.

Cairns, J., Douglas, W.A., Busey, F. and Chaney, M.D. 1968 The sequential comparison index - a simplified method for non-biologists to estimate relative differences in biological diversity in stream pollution studies. *J. Wat. Pollut. Control. Fed.*, **40**, 1607-1613.

CEC (Commission of European Communities) 1978 Council Directive of 18 July 1978 on the quality of fresh waters needing protection or improvement in order to support fish life. (78/659/EEC), *Official Journal*, **L/222**, 1-10.

Chandler, J.R. 1970 A biological approach to water quality management. *Water Pollut. Control*, **69**, 415-422.

Coste, M. 1978 *Sur l'Utilisation des Diatomées Benthiques pour l'Appréciation de la Qualité Biologique des Eaux Courantes.* Thesis. Faculté des Sciences et des Techniques de l'Université de Franche-Comté, 147 pp.

Coste, M., Bosca, C., Dauta, A. 1991 Use of algae for monitoring rivers in

France. In: B.A. Witton, E. Rott and G. Friedrich [Eds] *Use of Algae for Monitoring Rivers.* Proceedings of an International Symposium, Düsseldorf, 26-28 May 1991, Published by Institute of Botany, Innsbruck University, Austria, 75-88.

De Kruijf, H.A., De Zwart, D., Ray, P.K. and Viswanathan, P.N. [Eds] 1988 *Manual on Aquatic Ecotoxicology.* Kluwer Academic Publishers, Dortrecht, 332 pp.

DIN 38410 T1 1987 German Standard Methods for Examination of Water, Wastewater and Sludge. *Biologisch-ökologische Gewasseruntersuchung Allgemeine Hinweise, Planung und Durchführung.* Berlin.

DIN 38410 T2 1990 German Standard Methods for Examination of Water, Wastewater and Sludge. *Biologisch-ökologische Gewasseruntersuchung Bestimmung des Saprobienindex.* Berlin.

DIN 38412 T.13 1983 German Standard Methods for Examination of Water, Wastewater and Sludge. *L 13: Bestimmung von Sauerstoffproducktion und Sauerstoffverbrauch im Gewässer mit der Hell-Dunkelflaschen — Methode SPG, SVG.* Berlin.

DIN 38412 T.14 1983 German Standard Methods for Examination of Water, Wastewater and Sludge. *L 14: Bestimmung der Sauerstoffproducktion mit der Hell-Dunkelflaschen — Methode unter Laborbedingungen, SPL.* Berlin.

Edmondson, W.T. and Winberg, G.G. 1971 *A Manual on Methods for the Assessment of Secondary Productivity in Fresh Waters.* IBP Handbook No. 17, Blackwell Scientific Publication, Oxford, 368 pp.

EIFAC (European Inland Fisheries Advisory Commission) 1975 *Report on fish toxicity testing procedures.* EIFAC Technical Paper No. 24, Food and Agriculture Organization, Rome.

Elliott, J.M. 1977 *Some Methods for the Statistical Analysis of Samples of Benthic Invertebrates.* 2nd edition, Freshwater Biological Association Scientific Publication No. 25, Freshwater Biological Association, Ambleside, UK, 156 pp.

Elliott, J.M. and Tullett, P.A. 1978 *A Bibliography of Samplers for Benthic Invertebrates.* Freshwater Biological Association Occasional Publication No. 4, Freshwater Biological Association, Ambleside, UK, 61 pp.

Empain, A. 1976 Les bryophytes aquatiques utilisés comme traceurs de la contaminations en métaux lourdes des eaux douces. *Mém. Soc. R. Bot. Belg.*, 7, 141-156.

Fjellheim, A. and Raddum, G.G. 1990 Acid precipitation: Biological monitoring of streams and lakes. *Sci. Total Environ.*, 96, 57-66.

Fjerdingstad, E. 1964 Pollution of streams estimated by benthal phytomicroorganisms I.E. saprobic system based on communities of organisms and ecological factors. *Int. Rev. Hydrobiol.*, 49, 63-131.

Fjerdingstad, E. 1965 Taxonomy and saprobic valency of benthic phytomicroorganisms. *Int. Rev. Hydrobiol.*, **50**, 475-604.

Fjerdingstad, E. 1971 Microbial criteria of environment qualities. *Ann. Rev. Microbiol.*, **25**, 563-582.

Foster, R.B. and Bates, J.M. 1978 Use of mussels to monitor point source industrial discharges. *Environ. Sci. Technol.*, **12**, 958-962.

Friedrich, G. 1973 Ökologische Untersuchungen an einem thermisch anomalen Fliessgewässer (Erft/Niederrhein). (Ecological investigations in a thermically abnormal river). *Schriftenreihe der Landesanstalt für Gewässerkunde und Gewässerschutz NW*, **33**, 125 pp.

Friedrich, G. 1989 The River Rhine. In: W. Lampert and K.-O. Rothhaupt [Eds.] *Limnology in the Federal Republic of Germany.* International Association for Theoretical and Applied Limnology, Plön, 18-24.

Friedrich, G. 1990 Eine Revision des Saprobiensystems. *Zeitschrift für Wasser und Abwasserforschung*, **23**, 141-152.

Friedrich, G. 1992 Objectives and opportunities for biological assessment techniques. In: P.J. Newman, M.A. Piavaux and R.A. Sweeting [Eds] *River Water Quality. Ecological Assessment and Control.* EUR 14606 EN-FR, Commission of the European Communities, Luxembourg, 195-215.

Friedrich, G., Araujo, P.R.P and Cruz, A.A.S. 1990 Proposta de desinvolvimento de uma sistema biologico-limnologico para analise e avaliacao de aguas correntes no estado do Rio de Janeiro, Brasil. *Acta Limnol. Brasil,* **III**, 993-1,000.

Friedrich, G. and Viehweg, M. 1984 Recent developments of the phytoplankton and its activity in the Lower Rhine. *Verh. Internat. Verein. Limnol.*, **22**, 2029-2035.

Gardeniers, J.J.P. and Tolkamp, H.H. 1976 Hydrobiologische kartering, waardering enschade aan de beekfauna in Achterhoekse beken. In: Th. v.d. Nes [Ed.] *Modelonderzoek '71-'74.* Comm. Best. Waterhuish. Gld., 26-29, 106-114, 294-296.

Ghetti, P.F. and Bonazzi, G. 1977 A comparison between various criteria for the interpretation of biological data in the analysis of the quality of running water. *Water Research*, **11**, 819-831.

Gilbert, R.O. 1987 *Statistical Methods for Environmental Pollution Monitoring.* Van Nostrand Reinhold Company, New York, 320 pp.

Hawkes, H.A. 1977 Biological classification of rivers: conceptual basis and ecological validity. In: J.S. Alabaster [Ed.] *Biological Monitoring of Inland Fisheries,* Applicd Science Publishers Ltd, London, 55-67.

Hellawell, J.M. 1977 Biological surveillance and water quality monitoring. In: J.S. Alabaster [Ed.] *Biological Monitoring of Inland Fisheries,* Applied Science Publishers Ltd, London, 69-88.

Hellawell, J.M. 1978 *Biological Surveillance of Rivers*. Water Research Centre, Medmenham, UK.

Hellawell, J.M. 1986 *Biological Indicators of Freshwater Pollution and Environmental Management*. Pollution Monitoring Series, K. Mellanby [Ed.], Elsevier Applied Sciences Publishers, London, 546 pp.

Hutchinson, G.E. 1967 *A Treatise on Limnology. Volume II Introduction to Lake Biology and the Limnoplankton*. John Wiley, New York, 1115 pp.

Hynes, H.B.N. 1960 *The Biology of Polluted Waters*. Liverpool University Press, Liverpool, 202 pp.

Hynes, H.B.N. 1970 *The Ecology of Running Waters*. Liverpool University Press, Liverpool, 555 pp.

ISO 1985 *International Standard 7828. Water Quality Methods of Biological Sampling - Guidance on Handnet Sampling of Aquatic Benthic Macro-invertebrates*. International Standards Organization.

ISO 1991 *Draft International Standard 9391. Water Quality Sampling in Deep Waters for Macro-invertebrates. Guidance on the Use of Colonisation, Qualitative and Quantitative Samplers*. International Standards Organization.

ISO-BMWP 1979 *Assessment of the Biological Quality of Rivers by a Macroinvertebrate Score*. ISO/TC147/SC5/WG6/N5, International Standards Organization, 18 pp.

Jaccard, P. 1908 Nouvelles recherches sur la distribution florale. *Bull. Soc. Vaud. Sci. Nat.*, **XLIV**, 163, 223-269.

Jacknow, J.J., Ludke, L. and Coon, N.C. 1986 Monitoring fish and wildlife for environmental contaminants: The National Contaminant Biomonitoring Program. US Fish Wildl. Serv., *Fish Wildl. Leafl.*, **4**, 15 pp.

Johnels, A., Tyler, G. and Westermark, T. 1979 A history of mercury levels in Swedish fauna. *Ambio*, **8**, 160-168.

Juhnke, I. and Besch W.K. 1971 Eine neue Testmethode zur Früherkennung akut toxischer Inhaltsstoffe im Wasser. *Gewässer und Abwässer*, **50/51**, S. 107-114.

Knie, J. 1978 Der Dynamische Daphnientest - ein automatischer Biomonitor zur überwachung von Gewässern und Abwässern. *Wasser und Boden*, **12**, S. 310-312.

Knöpp, H. 1968 Stoffwechseldynamische Untersuchungsverfahren für die biologische Wasseranalyse. (Metabolism - dynamics tests for biological water analysis). *Int. Revue ges. Hydrobiol. Hydrogr.*, **53** (3), 409-441.

Kolkwitz, R. and Marsson, M. 1902 Grundsätze für die biologische Beurteilung des Wassers nach seiner Flora und Fauna. (Principles for the biological assessment of water bodies according to their flora and fauna). *Kl. Mitt. d. Kgl. Prüfungsanstalt f. Wasserversorgung und Abwässerbeseitigung* **1**.

Kolkwitz, R. and Marsson, M. 1908 Ökologie der pflanzlichen Saprobien.

(Ecology of the plant saprobien). *Ber. d. Bot. Ges.*, **26**, 505-519.

Kolkwitz, R. and Marsson, M. 1909 Ökologie der tierischen Saprobien. (Ecology of the animal saprobien). *Int. Rev. ges. Hydrobiol. Hydrogr.*, **2**, 126-152.

Kotelevtsev, S.V., Stvolinskij, S.L. and Beim, A.M. 1986 *Ecologico-toxicological Analysis on Base of Biological Membranes.* Moscow University, Moscow, 106 pp, [In Russian].

Kozhova, O.M. and Beim, A.M. 1993 *Ecological Monitoring of Lake Baikal.* Ecology Publish., Moscow, 352 pp, [In Russian].

Kuntz, K.W. 1984 *Toxic Contaminants in the Niagara River, 1975-1982.* Tech. Bull. No. 134, Water Quality Branch, Inland Waters Directorate, Ontario Region, Burlington, Ontario.

Lang-Bertalot, H. 1979 Pollution tolerance of diatoms as a criterion for water quality estimation. *Nova Hedwigia*, **64**, 285-304.

LAWA 1976 *Die Gewässergütekarte der Bundesrepublik Deutschland.* Länderarbeitsgemeinschaft Wasser, Mainz, 16 pp.

Liebmann, H. 1962 *Handbuch der Frischwasser- und Abwasser-Biologie.* (Handbook of the biology of fresh water and waste water). Verlag R. Oldenbourg, München.

Levi, J., Henriet, C., Coutant, J.P., Lucas, M. and Leger, G. 1989 Monitoring acute toxicity in rivers with the help of the Microtox test. *Water Supply*, 7, 25-31.

Lillelund, K., de Haar, U., Elster, H.-J., Karbe, L., Schwoerbel, J. and Simonis, W. 1987 *Bioakkumulation in Nahrungsketten.* VCH, Weinheim, 327 pp.

Maron, D.M. and Ames, B.N. 1983 Revised methods for the *Salmonella* mutagenicity test. *Mutat. Res.*, **113**, 173-215.

Mason, C.F. 1981 *Biology of Freshwater Pollution.* Longman, Harlow, 250 pp.

Meybeck, M., Chapman, D. and Helmer, R. [Eds] 1989 *Global Freshwater Quality: A First Assessment.* Blackwell Reference, Oxford, 306 pp.

Moss, B. 1980 *Ecology of Fresh Waters.* Blackwell Scientific Publications, Oxford, 332 pp.

Mount, D.J. and Norberg, T.J. 1984 A seven-day life-cycle Cladoceran toxicity test. *Environ. Toxicol. Chem.* **3**, 425-434.

National Committee of the USSR for Hydrometeorology, Hydrochemical Institute 1987 *Methods for bioindication and biotesting inland waters. Volume 1.* Leningrad, 183 pp, [In Russian].

National Committee of the USSR for Hydrometeorology, Hydrochemical Institute 1989 *Methods for bioindication and biotesting inland waters. Volume 2.* Leningrad, 275 pp, [In Russian].

National Rivers Authority, 1994 *The Quality of Rivers and Canals in England and Wales (1990 to 1992) as Assessed by a New General Quality Assessment Scheme.* Water Quality Series No. 19, HMSO, London, 40 pp plus map.

Newman, P.J. 1988 *Classification of Surface Water Quality. Review of Schemes Used in EC Member States.* London, 189 pp.

Newman, P.J., Piavaux, M.A. and Sweeting, R.A. [Eds] 1992 *River Water Quality. Ecological Assessment and Control.* EUR 14606 EN-FR, Commission of the European Communities, Luxembourg.

NRC 1986 *Acid Deposition: Long-term Trends.* National Research Council, National Academy Press, Washington DC, 509 pp.

Nusch, E.A. 1980 Comparison of different methods for chlorophyll and pigments determination. *Arch. Hydrobiol.,* **14**, 14-36.

Nusch, E.A. 1989 Limnological investigation in the Westphalian Reservoirs and Ruhr River impoundments. In: W. Lampert and K.O. Rothhaupt [Eds] *Limnology in the Federal Republic of Germany,* Proceedings of the 24th Congress of the International Association of Theoretical and Applied Limnology, Munich, August 13-19, 1989, Plön, 53-57.

Nusch, E.A. and Koppe, P. 1981 Temporal and spatial distribution of phytoplankton as detected by *in vivo* and *in situ* fluorometry. *Verh. Internat. Limnol.,* **21**, 756-762.

Obst, U. and Holzapfel-Pschorn, A. 1988 *Enzymtische Tests für die Wasseranalytik,* Verlag Oldenbourg, Munich.

OECD 1983 *Guideline for testing of chemicals No. 471 "Genetic toxicology: Salmonella typhimurium, Reverse Mutation Assay",* Organisation for Economic Co-operation and Development, Paris.

OECD 1987 *The Use of Biological Tests for Water Pollution Assessment and Control.* Environment Monographs No. 11, Organisation for Economic Co-operation and Development, Paris, 70 pp.

Pantle, R. and Buck, H. 1955 Die biologische Überwachung der Gewässer und die Darstellung der Ergebnisse. (Biological monitoring of water bodies and the presentation of results). *Gas und Wasserfach,* **96**, 604.

Patrick, R. and Hohn, M.H. 1956 The diatometer - a method for indicating the conditions of aquatic life. *Proc. Am. Petrol. Inst., Sect. 3,* **36**, 332-338.

Pauw, N. and Vanhooven, G. 1983 Method for biological quality assessment of watercourses in Belgium. *Hydrobiologia,* **100**, 153-168.

Philips, D.J.H. 1980 *Quantitative Aquatic Biological Indicators.* Applied Science Publishers Ltd., London, 488 pp.

Raddum, G.G., Fjellheim, A. and Hesthagen, T. 1988 Monitoring of acidification through the use of aquatic organisms. *Verh. Int. Verein. Limnol.,* **23**, 2291-2297.

Ravera, O. 1978 *Biological Aspects of Freshwater Pollution.* Commission of the European Communities. Pergamon Press, Oxford.

Renberg, I. and Hellberg, T. 1982 The pH history of lakes in Southwestern

Sweden, as calculated from the sub-fossil diatom flora of the sediments. *Ambio*, **11**(1), 30-33.

Rodhe, H. and Herrera, R. [Eds] 1988 *Acidification in Tropical Countries*. SCOPE Report No. 36, John Wiley and Sons, Chichester, 405 pp.

Rott, E. 1991 Methodological aspects and perspectives in the use of periphyton for monitoring and protecting rivers. In: B. Whitton, E. Rott, and G. Friedrich [Eds] *Use of Algae for Monitoring Rivers*. Proceedings of an International Symposium, Düsseldorf, 26-28 May 1991, Published by Institute of Botany, Innsbruck University, Austria, 9-16.

Round, F.E. 1991 Use of diatoms for monitoring rivers. In: B. Whitton, E. Rott, and G. Friedrich [Eds] *Use of Algae for Monitoring Rivers*. Proceedings of an International Symposium, Düsseldorf, 26-28 May 1991, Published by Institute of Botany, Innsbruck University, Austria, 25-32.

Samiullah, Y. 1990 *Biological Monitoring of Environmental Contaminants: Animals*. MARC Report No. 37, GEMS-Monitoring and Assessment Research Centre, King's College London, University of London, 767 pp.

Schmitz, P., Krebs, F. and Irmer, U. 1994 Development, testing and implementation of automated biotests for the monitoring of the River Rhine, demonstrated by bacteria and algae tests. *Wat. Sci. Tech.*, **29**(3), 215-221.

Sládecek, V. 1973 System of water quality from the biological point of view. *Arch. Hydrobiol. Beih. Ergebn, Limnol.*, **7**(I-IV), 1-218.

Sládecková, A. 1962 Limnological investigation methods for the periphyton (Aufwuchs) community. *Bot. Review*, **28**(2), 286-350.

Smol, J.P., Battarbee, R.W., Davis, R.B. and Meriläinen, J. [Eds] 1986 *Diatoms and Lake Acidity*. Junk Publishers, Dordrecht.

Spynu, E.I. and Beim, A.M. 1986 Ecotoxicological assessment methods of chemicals and their use in environmental hygiene. In: UNEP/IRPTC *Toxicometry of Environmental Chemical Pollutants*. Centre for International Projects GKNT, Moscow, 283-300.

Surma-Aha, K., Paasivirta, J., Rekolainen, S. and Verta, M. 1986 Organic and inorganic mercury in the food chain of some lakes and reservoirs in Finland. *Chemosphere*, **15**, 353-372.

Tolkamp, H.H. 1985 Using several indices for biological assessment of water quality in running water. *Verh. Internat. Verein. Limnol.*, **22**, 2281-2286.

Tuffery, G. and Verneaux, J. 1967 Méthode de détermination de la qualité biologique des eaux courantes. Exploitation codifée des inventaires de la faune de fond. *Trav. Sect. P. et P.*, Cerafer, Paris, 23 pp.

UNESCO/WHO 1978 *Water Quality Surveys: A Guide for the Collection and Interpretation of Water Quality Data*. UNESCO Studies and Reports in

Hydrology, 23, United Nations Educational, Scientific and Cultural Organization, Paris/World Health Organization, Geneva, 350 pp.

US EPA 1978 *The Selenastrum capricornutum PRINTZ Algal Assay Bottle Test.* US EPA-600/9-78-018, U.S. Environmental Protection Agency.

US EPA 1985 *Short-term Methods for Estimating the Chronic Toxicity of Effluents and Receiving Water to Freshwater Organisms.* EPA/600/4-85/014, U.S. Environmental Protection Agency.

Vollenweider R.A. [Ed.] 1974 *A Manual on Methods for Measuring Primary Production in Aquatic Environments.* 2nd Edition, IBP Handbook No. 12, Blackwell Scientific Publications, Oxford, 225 pp.

Washington, H.G. 1984 Diversity, biotic and similarity indices: a review with special relevance to aquatic ecosystems. *Water Research*, **18**, 653-694.

Watanabe, T., Asai, K. and Houki, A. 1988 Numerical index of water quality using diatom assemblages. In: M. Yasuno and B.A. Whitton [Eds] *Biological Monitoring of Environmental Pollution*, Tokai University Press, Tokyo, 179-182.

Wetzel, R.G. 1975 *Limnology.* Saunders, Philadelphia, 743 pp.

Whitton, B.A. [Ed.] 1975 *River Ecology.* Studies in Ecology, Volume 2, Blackwell Scientific Publications, Oxford, 725 pp.

Whitton, B.A. 1988 Use of plants to monitor heavy metals in rivers. In: M. Yasuno and B.A. Whitton [Eds] *Biomonitoring of Environmental Pollution*, Tokai University Press, Tokyo, 159-162.

Whitton, B., Rott, E. and Friedrich G. [Eds] 1991 *Use of Algae for Monitoring Rivers.* Proceedings of an International Symposium, Düsseldorf, 26-28 May 1991, Published by Institute of Botany, Innsbruck University, Austria, 193 pp.

WHO 1984 *Guidelines for Drinking-Water Quality. Volume 2. Health Criteria and Other Supporting Information.* World Health Organization, Geneva, 335 pp.

WHO 1992 *GEMS/Water Operational Guide.* Third edition. World Health Organization, Geneva.

Woodiwiss, F.S. 1964 The biological system of stream classification used by the Trent River Board. *Chem. Ind.*, **11**, 443-447.

Wright, J.F., Furse, M.T. and Armitage, P.D. 1993 RIVPACS — a technique for evaluating the biological quality of rivers in the UK. *Eur. Water Pollut. Control*, **3**(4), 15-25.

Yasuno M. and Whitton, B.A 1988 [Eds] *Biological Monitoring of Environmental Pollution.* Tokai University Press, Tokyo.

Zelinka, M. and Marvan, P. 1961 Zur Präzisierung der biologischen Klassifikation der Reinheit fliessender Gewässer. (Making a precise biological classification of the quality of running waters). *Arch. Hydrobiol.*, **57**, 389-407.

Chapter 6*

RIVERS

6.1 Introduction

Rivers are the most important freshwater resource for man. Social, economic and political development has, in the past, been largely related to the availability and distribution of fresh waters contained in riverine systems. Major river water uses can be summarised as follows:

- sources of drinking water supply,
- irrigation of agricultural lands,
- industrial and municipal water supplies,
- industrial and municipal waste disposal,
- navigation,
- fishing, boating and body-contact recreation,
- aesthetic value.

A simple evaluation of surface waters available for regional, national or trans-boundary use can be based on the total river water discharge. The Colorado River, USA is an example where extraction of water has virtually depleted the final discharge to the ocean. The flow has been used almost completely by negotiated extraction and distribution to nearby states. Any increase in extraction and use would require diversion of a similar water quantity to guarantee the minimum flow required to meet all the water demands of the region.

Upstream use of water must only be undertaken in such a way that it does not affect water quantity, or water quality, for downstream users. Use of river water is, therefore, the subject of major political negotiations at all levels. Consequently, river water managers require high quality scientific information on the quantity and quality of the waters under their control. Provision of this information requires a network of river monitoring stations in order:

- to establish short- and long-term fluctuations in water quantity in relation to basin characteristics and climate,
- to determine the water quality criteria required to optimise and maintain water uses, and
- to determine seasonal, short- and long-term trends in water quantity and quality in relation to demographic changes, water use changes and management interventions for the purpose of water quality protection.

This chapter was prepared by M. Meybeck, G. Friedrich, R. Thomas and D. Chapman.

As with all freshwater systems, river quality data must be interpreted within the context of a basic understanding of the fluvial and river basin processes which control the underlying characteristics of the river system. Similarly, the design of the monitoring network, selection of sampling methods and variables to be measured must be based on an understanding of fluvial processes as well as the requirements for water use.

6.2   Hydrological characteristics

6.2.1   River classification

Rivers are complex systems of flowing waters draining specific land surfaces which are defined as river basins or watersheds. The characteristics of the river, or rivers, within the total basin system are related to a number of features. These features include the size, form and geological characteristics of the basin and the climatic conditions which determine the quantities of water to be drained by the river network.

Rivers can be classified according to the type of flow regime and magnitude of discharge (see below for further details). The flow regime may be subject to considerable modification by natural impoundments, lakes, dams, or water storage (see Chapter 8). Flow characteristics may also be changed by canalisation, or requirements for water uses, such as withdrawal for irrigation or other water supply needs, or by changes in flood characteristics due to modifications of the soil infiltration as a result of agriculture and urbanisation.

The classification of rivers according to their discharge is generally more satisfactory but has not, to date, been completely defined and accepted. However, there are certain specified discharge rates which are widely used to characterise river discharges and their annual variations. These include the average peak discharges, the monthly or annual average discharge and the average low discharge. A size classification based on discharge, drainage area and river width is given in Table 6.1. The distinctions are arbitrary and no indication of the annual variability in discharge is given. River discharge, particularly in arid and sub-tropical regions, may range from zero in the dry season to high discharge rates in large rivers during the rainy season. Very large rivers may also traverse many climatic zones and can have less variability than might be expected for the climatic conditions at the final point of discharge, such as the Mississippi and Nile rivers.

Rivers drain watersheds of varying dimensions. As indicated in Table 6.1, this area is directly related to the river discharge and width. Efficient drainage is achieved by means of a dendritic network of streams and rivers. As

Table 6.1   Classification of rivers based on discharge characteristics and the drainage area and river width

River size	Average discharge $(m^3\,s^{-1})$	Drainage area (km^2)	River width (m)	Stream order[1]
Very large rivers	> 10,000	> 10^6	> 1,500	> 10
Large rivers	1,000–10,000	100,000–10^6	800–1,500	7 to 11
Rivers	100–1,000	10,000–100,000	200–800	6 to 9
Small rivers	10–100	1,000–10,000	40–200	4 to 7
Streams	1–10	100–1,000	8–40	3 to 6
Small streams	0.1–1.0	10–100	1–8	2 to 5
Brooks	< 0.1	< 10	< 1	1 to 3

[1] Depending on local conditions

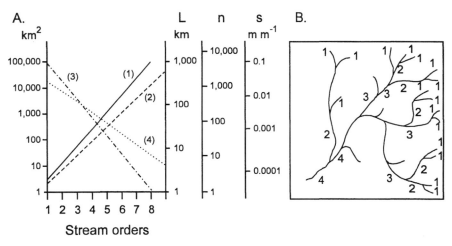

Figure 6.1   A. The relationship between stream orders and hydrological characteristics using a hypothetical example for a stream of order 8: (1) Watershed area (A); (2) Length of river stretch (L); (3) Number of tributaries (n); (4) Slope (m m^{-1}); **B.** Stream order distribution within a watershed

these increase in size from small to large, and then to the main river channel, the "order" in which they appear is a function of the watershed size. An example of an "ordered" network and the relationship of stream orders to other river characteristics is given in Figure 6.1.

River systems represent the dynamic flow of drainage water, which is the final product of surface run-off, infiltration to groundwater and groundwater discharge. The general relationships between these and the nomenclature for a river transect are summarised in Figure 6.2.

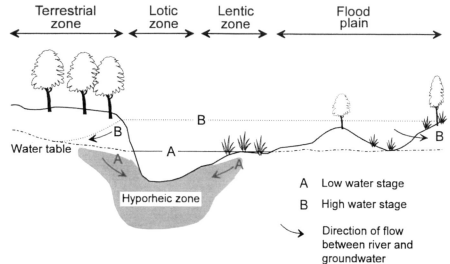

Figure 6.2 Generalised cross-section of a river showing the relationship between physical features and the low and high water stages

6.2.2   Velocity and discharge

Hydrological characteristics are determined by velocity and discharge. The velocity (sometimes referred to as flow) of the river water is the rate of water movement given as m s^{-1} or cm s^{-1}. The discharge (m^3 s^{-1}) is determined from the velocity multiplied by the cross-sectional area of a river. Cross-sectional area fluctuates with the change in water level or river stage. Similarly, a direct relationship exists between water level and velocity and measurement of level can be transformed directly into velocity. A velocity calibration can then be derived. With known cross-sectional area, the discharge Q can be derived by measurement of the water level. The level can be obtained using an appropriate gauge placed on the bank, or on another suitable structure in the river. It can often be measured automatically giving a continuous record (hydrograph) of the changing levels or direct discharge of the river. Further details of measurement of velocity, level and discharge are available in WMO (1980) and Herschy (1978). The discharge of a river is the single most important measurement that can be made because:

- it provides a direct measure of water quantity and hence the availability of water for specific uses,
- it allows for the calculation of loads of specific water quality variables,
- it characterises the origins of many water quality variables by the relationship between concentrations and discharge (see section 6.3.3), and
- it provides the basis for understanding river basin processes and is essential for interpreting and understanding water quality.

River regime

The size and geological formation of a watershed determines the river discharge regime. The discharge and its annual, as well as long-term, fluctuations are primarily influenced by the characteristics of the drainage basin. Climatic, meteorological, topographical and hydrological factors play a major role in the generation of river discharge.

Small watersheds usually result in low median discharges with extremely large ratios of peak and low discharge. In temperate humid climates, the annual variations between minimum and maximum discharges may reach two orders of magnitude. Larger watersheds produce more uniform discharges. Large rivers with a relatively uniform discharge regime during the year show a rather constant ratio of average peak and low discharge. Large seasonal variations in the discharge can be equalised and transformed into rather uniform discharges by the presence of reservoirs, storage dams or natural lakes along the river course. Examples include some of the alpine rivers which are impounded for hydroelectric power generation. The Rhine river, which is more than 1,000 km in length, has maintained a low-to-peak discharge ratio of 1:15 in the lower reaches, constantly over the last 20 years. This is because different parts of the catchment area contribute high discharge at different times of the year.

The principal factor causing large fluctuations in discharge is climate, which determines the distribution of rainfall over the year. The variability and resulting non-uniformity of discharge is moderate in temperate humid climates, but extreme for rivers in savannah areas and in certain sub-tropical regions. The composition and structure of the sub-soil are also important factors. Large differences can be observed between porous rocks, clays, marshy soils and fissured rocks. Such geological conditions of the drainage basin might cause variations in the discharge rates by a factor of two and in a few cases even more.

Vegetation also exerts an influence on the generation of river discharge because it largely determines the quantity of surface run-off. Fluctuations in discharge can be dampened by vegetation cover. In areas with little or no vegetation, rainfall results in immediate surface run-off.

For problems of water quality management, such as the disposal of wastewaters into rivers, low discharge conditions are used as the basis for the design of treatment facilities and control of the maximum permissible effluent disposal to rivers. At any higher river discharge rate, the ecological effects of a polluted effluent in a river are less harmful. Drought conditions can be critical for rivers serving as a water source for urban water supply.

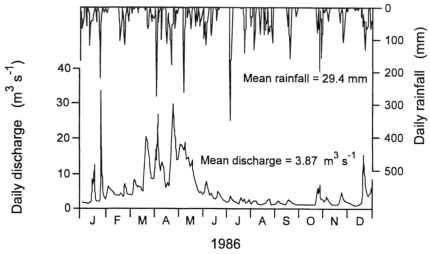

Figure 6.3 Comparison of daily discharge and daily rainfall for the Venoge river in 1986, measured at Ecublens-les Bois, Switzerland (see also Figures 6.28 and 6.29) (After Zhang Li, 1988)

Discharge characteristics

Most rivers are characterised by a condition called base flow or base discharge. This is the minimum amount of water moving within the individual river system, and in most cases is controlled by groundwater discharge. Figure 6.3 shows the characteristics of a storm controlled river basin by comparing daily discharge and daily rainfall in 1986 for the Venoge river draining the Jura mountains into Lake Geneva. Considerable increases in discharge Q above the base level occur during storm events. The increase in discharge is not only a function of the storm intensity but is also related to the duration of rainfall, soil water saturation, etc.

Discharge characteristics of rivers are greatly modified by the nature of the watershed. Changes in the infiltration rate to the groundwater system change the water run-off characteristics of the watershed. Loss of forest cover and the proportion of exposed bedrock to deep soils and sediments, all have profound effects on discharge characteristics. These changes are shown in Figure 6.4 which illustrates two river basins of the same dimensions but different infiltration rates, following a storm event of the same duration and intensity. With high infiltration within the watershed (Figure 6.4A) the peak flow or flood stage has a much lower discharge than in the basin of low infiltration (Figure 6.4B). In addition, the hydrograph in Figure 6.4A shows an increased duration of the flood stage indicating that the run-off from the watershed extends over a longer time scale. When a watershed is changed from type A to type B (e.g. by urbanisation) the river channel characteristics,

A.   B.

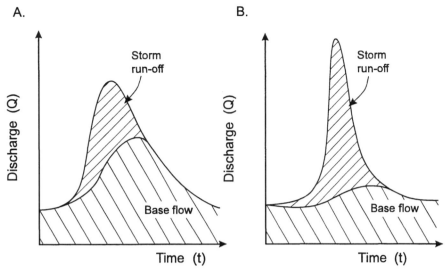

Figure 6.4 Theoretical hydrographs of a rain storm event of the same intensity in two basins of equal size but with different infiltration rates: **A.** High infiltration rate e.g. sandy soil, forest area; **B.** Low infiltration rate e.g. base crystalline rock, urbanised basin. The increase in base flow is different in each case

which were created under high infiltration rates, are no longer adequate to accommodate the high and rapid flood discharges. Under these conditions flooding becomes a major urban problem which may lead to overloading of the storm water collection facilities and sewage treatment systems. As a result, these discharges to rivers and coastal areas may contain higher than usual levels of organic matter, faecal bacteria and toxic substances. The resultant increases in biochemical oxygen demand (BOD) (see section 3.5.3) can lead to fish kills, and the high concentrations of faecal bacteria may restrict water use, including bathing in coastal waters. The construction of storm water ponds or reservoirs allows the collection and slow release of storm waters to the main river channel or coastal waters (see Chapter 8). Further details are available in UNESCO (1988).

Discharge may also be controlled by factors other than storm conditions. In all cold latitudes and mountainous regions the effects of freezing and thawing in glaciers are particularly important. Figure 6.5 shows the mean and maximum monthly discharges of the Alpine Rhône river, Switzerland at various locations from upstream (Figure 6.5A) to the river mouth at Lake Geneva (Figure 6.5D). Low discharge occurs during glacial freezing in the winter and high discharge occurs during the summer ice melt between May and October. The variations between winter low discharge and summer high discharge are less pronounced downstream. This is due to the increasing influence of storm

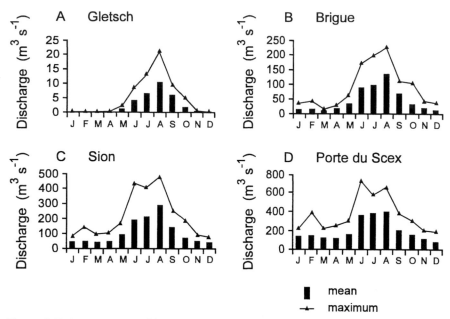

Figure 6.5 Long-term monthly means and maximum discharges at four stations on the Alpine Rhône river, Switzerland which is influenced by glacial melt (Modified from Burrus *et al.*, 1990)

related run-off and snow melt in the middle altitudes on the lower reaches of the river, and to less variation in discharge as a function of increasing basin area.

An additional feature which can be inferred from the hydrographs of the Alpine Rhône is the relationship of discharge to basin area. The larger the basin, the greater the damping effect on the variability in the discharge. This dependence is illustrated in Figure 6.6 by three theoretical hydrographs for river basins of different dimensions but with the same water regime.

6.2.3 Fluctuations in suspended solids

The concentration of total suspended solids (TSS) in rivers increases as a function of flow. Particles are derived by sheet, bank and gully erosion in the watershed and by the resuspension of particles deposited in the river bed. Rates of erosion are associated with climate, particularly the amount and intensity of rainfall, and can be modified by vegetative cover. Deforestation, or increased intensive agriculture results in large increases in erosion.

Although a general increase can usually be observed in suspended sediment concentration with increasing water discharge, it can be affected by a number of river basin processes. The relationship between discharge and

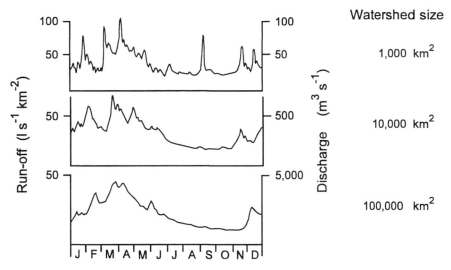

Figure 6.6 Theoretical annual variability in water discharge and run-off in river basins of various sizes but with the same water regime

suspended sediment for the River Exe, UK was discussed in section 4.3.3 and illustrated in Figure 4.4. Depending on the overall characteristics of the watershed, the peak in suspended sediment may, or may not, occur at the same time as the peak discharge.

6.2.4   In-stream velocities
In order to design effectively a water sampling programme, and to interpret river water monitoring results, some knowledge of the in-stream flows and velocity gradients is necessary. Within an even river channel, laminar flow occurs as indicated in Figure 6.7. Maximum velocities occur in the centre of the channel but are reduced to zero at the bank by frictional forces exerted by the shallow bank zone and the bank itself. The velocity gradient thus tends to force any influent waters from a tributary, industrial or municipal point source to the side of the river which they entered. A tributary entering a channel as illustrated in Figure 6.7 remains on the right hand side of the channel whilst laminar flow is maintained. Bends in the river, rapids or a waterfall induce mixing (bends by overturn and rapids and waterfalls by turbulent mixing). Concentration gradients, in the circumstances illustrated in Figure 6.7, follow the patterns indicated in the three cross-sections. Laminar flow, and the concentration gradients observed in the river sections, are normally maintained for less than one kilometre where perfect mixing occurs in small tributaries and turbulent rivers. Occasionally they are maintained for many hundreds of kilometres, as in the River Amazon downstream of the Rio

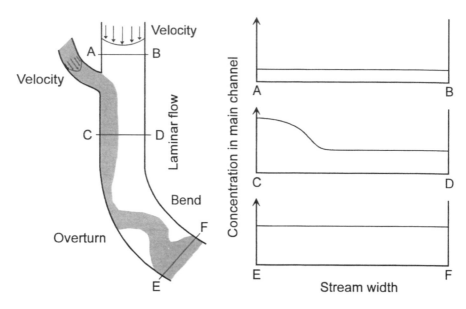

Figure 6.7 The pattern of mixing and associated variations in concentrations occurring upstream and downstream of a confluence with a river tributary

Negro confluence. The implications for sampling cross-sections of rivers are discussed in section 6.6.4 and further details are available in UNESCO (1982).

Other examples of the velocity characteristics of rivers are given in Figure 6.8. The cross-sectional velocities are shown for a uniform cross-section (as in Figure 6.8A) or variations in cross-section (such as in Figure 6.8B) where major changes in depth occur across the profile. Figure 6.8C indicates the vertical velocity profile where bottom friction results in deceleration of the bottom water. The energy released by the deceleration is transferred into the movement of coarser sediment particles along the bed of the river. Figure 6.8D illustrates the overturn associated with a river bend. The cross-section (Figure 6.8E) shows the direction of circulation, but as forward motion is also maintained the resultant motion is roughly helical.

6.3   Chemical characteristics

At a given river station water quality depends on many factors, including: (i) the proportion of surface run-off and groundwater, (ii) reactions within the river system governed by internal processes, (iii) the mixing of water from tributaries of different quality (in the case of heterogeneous river basins), and (iv) inputs of pollutants.

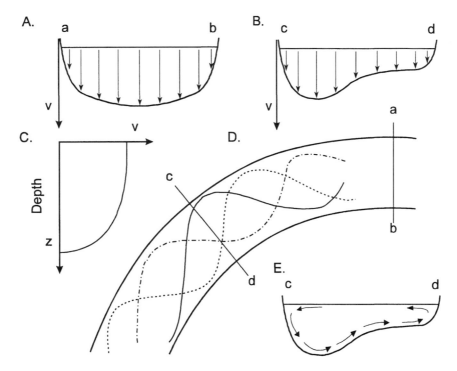

Figure 6.8 River flow and velocity (v) characteristics: **A** and **B.** Variations in velocity with depth at the points of cross-section indicated in D; **C.** Vertical velocity profile in the river; **D** and **E.** Mixing and direction of water circulation associated with a river bend

6.3.1   Origins of elements carried by rivers

In the absence of any human impact the concentrations, relative proportions, and rates of transport of dissolved substances in rivers are highly variable from one place to another, depending on their sources, pathways and interactions with particulates. Particulate matter composition is discussed in detail in Chapter 4. Figure 6.9 shows the main sources of elements to rivers:

- *Chemical weathering* of surficial rocks (Figure 6.9, sources 1, 2 and 3). The most abundant rock types are shales (33.1 per cent of continental outcrops), granite and gneisses (20.8 per cent), sandstone (15.8 per cent), carbonate rocks (15.9 per cent) and basalts (4.1 per cent). Although rarely found at the earth surface (1.3 per cent), the evaporitic rocks, gypsum and rock salt, may greatly influence surface waters due to their very high solubility. Most chemical weathering reactions derive from the attack of minerals, mostly aluminosilicates, by carbonic acid ($H_2O + CO_2$). This leads to the formation of major cations (Ca^{2+}, Mg^{2+}, Na^+, K^+) and of

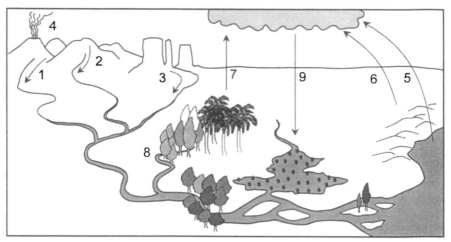

Figure 6.9 Natural sources of elements to rivers. 1, 2 and 3: Chemical weathering of surficial rocks; 4: Volcanic fallout; 5: Recycled oceanic aerosols; 6: Continental aeolian erosion; 7: Decay of vegetation; 8: Leaching of organic soils; 9: Atmospheric inputs

dissolved silica (SiO_2) and bicarbonates (HCO_3^-). The less soluble trace elements (e.g. Fe, Al, Ti) remain in residual soil minerals (i.e. oxides and clay minerals). Weathering of carbonate rocks (about 15 per cent of rock outcrops at the earth's surface) results in high concentrations of HCO_3^-, only half of which originates from carbonate minerals (the other half comes from atmospheric and soil CO_2). The sulphate anion is occasionally dominant when high proportions of pyrite (FeS) or gypsum ($CaSO_4.2H_2O$) occur. Chloride anion dominance is very seldom found in surface waters due to the scarcity of the NaCl mineral, but it may be important in coastal regions where it originates entirely from sea salt aerosols.

- *Atmospheric inputs* of natural origin (Figure 6.9, source 9). The amount of recycled oceanic aerosols (source 5: rich in Na^+, Cl^-, Mg^{2+}, SO_4^{2-}) falling on continents generally decreases exponentially from the coast inland, where the products of continental aeolian erosion (source 6: dust rich in Ca^{2+}, HCO_3^-, SO_4^{2-}), of vegetation decay (source 7: rich in N species) and volcanic fall-out (source 4: HCl, H_2SO_4) may be dominant.
- *Leaching of organic soils* (Figure 6.9, source 8). This process generates nitrogen (NH_4^+, NO_3^-) and dissolved organic matter (dissolved organic carbon and nitrogen [DOC and DON]) in surface waters.

Anthropogenic activities can enhance natural processes, such as erosion and soil leaching, increase inputs of natural compounds such as mineral salts and inorganic fertilisers to the river system, and add synthetic compounds which are mostly organic and not found in nature, such as solvents, pesticides, aromatic hydrocarbons, etc. The additional materials arising from increases in natural processes follow the same pathways, and behave in the same way, as compounds arising from soil leaching, such as fertilisers and pesticides. However, most urban pollutants enter rivers as point-sources, usually as treated or untreated sewage effluents.

6.3.2 Natural concentrations in rivers

In any region not yet affected by human activity, the variability in natural water quality depends on the combination of the following environmental factors (Meybeck and Helmer, 1989):

- the occurrence of highly soluble or easily weathered minerals of which the order of weathering is halite > gypsum > calcite > dolomite > pyrite > olivine,
- the distance to the coastline,
- the precipitation/river run-off ratio, and
- the occurrence of peat bogs, wetlands and marshes which release large quantities of dissolved organic matter.

Other factors include the ambient temperature, thickness of weathered rocks, organic soil cover, etc.

Examples of the concentrations of ions and nutrients occurring in pristine waters are presented in Table 6.2. The geographic variability of natural running waters is surprising and as a result no "world average quality" can be used as a reference to check if a given river is polluted or not. Careful investigation of pristine water quality from elsewhere in the watershed should be made for comparison.

The natural geographic variation of selected dissolved constituents in rivers is given in Table 6.3 for pristine streams and major world rivers (highly polluted rivers in Europe and North America are not included). Within a given region of 10^5 to 10^6 km^2, the distribution of natural ionic contents between tributaries (areas 10^3–10^4 km^2) may extend over several orders of magnitude, as for the Lower Amazon or the Mackenzie river tributaries (Figure 6.10). Internal processes which depend on the physical, chemical and biological conditions occurring in rivers, can also affect all components of water quality. Some major processes are listed in Table 6.4.

Dissolved trace element contents are very difficult to analyse correctly since samples are easily contaminated and analytical detection limits are sometimes higher than natural levels. The values given in Table 6.5 are the latest estimates found in the scientific literature for uncontaminated waters, resulting from utmost care in sampling, water treatment, and analysis. In routine surveys where adequate precautions are often not fully applied, the

Table 6.2 Geographic variability of dissolved major elements in pristine waters

	Electrical conductivity (µS cm⁻¹)	pH	Sum of cations (µeq l⁻¹)	SiO_2 (mg l⁻¹)	Ca^{2+} (mg l⁻¹)	Mg^{2+} (mg l⁻¹)	Na^+ (mg l⁻¹)	K^+ (mg l⁻¹)	Cl^- (mg l⁻¹)	SO_4^{2-} (mg l⁻¹)	HCO_3^- (mg l⁻¹)	Notes
A. Pristine streams draining most common rock types (corrected for oceanic aerosols)												
Granite	35	6.6	166	9.0	0.78	0.38	2.0	0.3	0	1.5	7.8	1
Gneiss	35	6.6	207	7.8	1.2	0.69	1.8	0.4	0	2.7	8.3	1
Volcanic	50	7.2	435	12.0	3.1	2.0	2.4	0.55	0	0.5	25.9	1
Sandstone	60	6.8	223	9.0	1.8	0.75	1.2	0.82	0	4.5	7.6	1
Shale			770	9.0	8.1	2.9	2.4	0.78	0.7	6.9	35.4	1
Carbonate	400	7.9	3,247	6.0	51	7.8	0.8	0.51	0	4.1	195	1
B. Pristine streams draining rare rock types or in a rare geological formation												
Amazonian clear waters	5.7	5.1	111	1.9	tr	0.13	1.2	1.4	0.7			2
Amazonian black waters	29.1	3.7	212	0.6	tr	0.02			1.2			2
Coal shale			40,700	7.0	87	121	600	10.2	14.8	1,400	652	2
Salt rock		8.0	312,000	1.2	607	68	6,300	7.7	9,400	1,330	183	2
C. Rivers influenced by evapotranspiration												
Horocallo R., Ethiopia	2,230	9.2	21,800	79	2.2	1.45	480	2.5	195	65	975	2
D. Rivers influenced by oceanic aerosols												
Clisson R., France	227	6.2		14.6	6.4	4.8	22.0	2.6	40.0	5.8	32.9	3

tr   Trace amount

1   Averages from a survey of 250 pristine streams in France (Meybeck, 1986) and from 75 sites world-wide, corrected for oceanic cyclic salts (Meybeck, 1987)

2   Compiled from various sources by Meybeck and Helmer, 1989

3   Analyses not corrected for cyclic salt. 60% of salts originate from oceanic atmospheric inputs. River located in SW France, on sands (Meybeck, 1986).

Source: After Meybeck and Helmer, 1989

Table 6.3 Natural ranges of dissolved constituents in rivers

	Streams (1–100 km²)				Rivers (100,000 km²)				Global average MCNC	
	Minimum		Maximum		Minimum		Maximum			
	(µeq l⁻¹)	(mg l⁻¹)	(µeq l⁻¹)	(mg l⁻¹)	(µeq l⁻¹)	(mg l⁻¹)	(µeq l⁻¹)	(mg l⁻¹)	(µeq l⁻¹)	(mg l⁻¹)
SiO_2[5] (µmole l⁻¹)	10	0.6	830	50	40	2.4	330	20	180	10.8
Ca^{2+}	3	0.06	10,500	210	100	2.0	2,500	50	400	8.0
Mg^{2+}	4	0.05	6,600	80	70	0.85	1,000	12.1	200	2.4
Na^+	2.6	0.06	15,000	350	55	8	1,100	25.3	160	3.7
K^+	3	0.1	160	6.3	13	0.5	100	4.0	27	1.0
Cl^-	2.5	0.09	15,000	530	17	0.6	700	25	110	3.9
SO_4^{2-}	2.9	0.14	15,000	720	45	2.2	1,200	58	100	4.8
HCO_3^-	0	0	5,750	350	165	10	2,800	170	500	30.5
Sum of cations	45		20,000		340		4,000		800	
pH	4.7		8.5		6.2		8.2			
TSS		3		15,000		10		1,700		150
DOC		0.5		40		2.5		8.5		4.2
POC		0.5		75						3.0
POC %				20						
TOC		1.5		25						2.0
$N-NH_4^+$						0.005		0.04		0.015
$N-NO_3^-$						0.05		0.2		0.10
N organic						0.05		1.0		0.26
$P-PO_4^{3-}$						0.002		0.025		0.010

Streams: Distribution based on 75 unpolluted monolithological watersheds from all countries in which the rock type proportion is close to the estimated global proportion of Meybeck (1987), particularly for the most soluble rocks; oceanic cyclic salts have been grossly subtracted.
Rivers: These figures are derived from the discharge-weighted distribution of constituents in 60 major rivers (basic data in Meybeck, 1979) without any correction of oceanic cyclic salts.
Minimum and maximum values correspond to 2 % and 98 % of the distribution except for nutrients which represent 10 % and 90 %.
MCNC (most common natural concentrations) corresponding to the median value obtained for the same 60 major rivers as above.

TSS Total suspended solids
DOC Dissolved organic carbon
POC Particulate organic carbon
TOC Total organic carbon
POC % is the percentage of organic carbon in the TSS
TSS, DOC, POC and nutrients data are mostly from Meybeck (1982)
Source: After Meybeck and Helmer, 1989

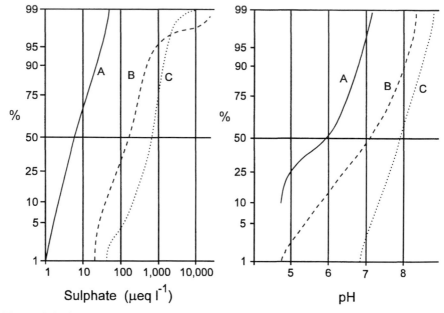

Figure 6.10 Cumulative distribution of sulphate and pH in unpolluted rivers and streams: **A.** Lower Amazon tributaries; **B.** Unpolluted French streams; **C.** Mackenzie river tributaries (Based on Reeder *et al.*, 1972; Stallard, 1980; Meybeck, 1986)

concentrations of dissolved trace elements may sometimes exceed (by one or even two orders of magnitude) the values reported in Table 6.5. Further information on the natural variations of river water quality can be found in Hem (1989) and in Berner and Berner (1987).

6.3.3   Variations of water quality with river discharge

Water quality variability depends on the hydrological regime of the river, i.e. the water discharge variability, the number of floods per year and their importance etc. (see section 6.2.2). During flood periods, water quality usually shows marked variations due to the different origins of the water: surface run-off, sub-surface run-off (i.e. water circulation within the soil layer), and groundwater discharge. Surface run-off is generally highly turbid and carries large amounts of total suspended solids, including particulate organic carbon (POC). Sub-surface run-off leaches dissolved organic carbon and nutrients (N and P) from soils, whereas groundwaters provide most of the elements resulting from rock weathering (SiO_2, Ca^{2+}, Mg^{2+}, Na^+, K^+).

The atmosphere is the source of most Cl^- and SO_4^{2-}, together with some Na^+, particularly in basins where there is no evaporitic rock or the mineral pyrite. Bicarbonate (HCO_3^-), the most common form of inorganic carbon

Table 6.4 Some major internal processes regulating the concentrations of selected water quality variables in rivers

	Water turbulence	Evaporation	Adsorption on sediments[1]	Primary production[2]	Oxidation of organic matter	
					In the water column[2]	In anoxic sediments[2]
pH		Increase		Increase	Decrease	Decrease
Electrical conductivity		Increase				
Calcium, bicarbonate	Precipitation[3]	Precipitation		Precipitation[4]		
Sodium, chloride potassium, calcium sulphate		Increase				
Nutrients	Volatilisation of NH_3		Decrease	Decrease (uptake)	Increase (release)	Decrease (denitrification) or increase (ammonification)
Dissolved O_2	Increase[5]			Increase	Decrease	Decrease
Dissolved organic carbon	Decrease (foam formation and oxidation)		Decrease	Increase	Decrease	
Dissolved metals			Decrease			Increase (desorption)
Organic micropollutants	Volatilisation		Decrease			

[1] During increased total suspended solids such as during floods
[2] In natural river systems these processes are of minor importance in river channels, except in highly eutrophic rivers; these processes can exert major influences on water quality in lakes, reservoirs and impoundments
[3] De-gassing of karstic waters
[4] Due to pH increase
[5] Re-aeration

found between pH 6 and 8.2, is derived partly from the dissolution of carbonate-bearing minerals and partly from CO_2 in the soil. When the river basin does not bear any carbonate rocks, the HCO_3^- content is largely derived from soil CO_2.

Changes in discharge, when compared to the simultaneous changes in concentrations of various substances, are of great value in indicating the major sources of substances. This is illustrated by the curves of concentration versus discharge in Figure 6.11A. Such curves may represent time scales varying from a single storm event to several years duration. Curve (1) shows a general decrease in concentration with discharge which implies increasing dilution of a substance introduced at a constant rate (e.g. major cations, possibly SiO_2, particularly when concentrations are high). This situation is also

Table 6.5   World average values of trace elements carried in solution by major unpolluted rivers

	Al	As	B	Cd	Cr	Co	Cu	F	Fe	Mn
Dissolved[1] concentration ($\mu g\ l^{-1}$)	40	1.0	30	0.001	0.1	0.1	1.4	100	50	10
% of total in solution[2]	0.13	25	40	2.5	0.25	2.5		20	0	2

	Mo	Ni	Pb	Sb	Sr	V	Zn
Dissolved[1] concentration ($\mu g\ l^{-1}$)	0.8	0.4	0.04		100		0.2
% of total in solution[2]	30	4	0.5	45	68	1.4	0.2

[1]   From Schiller and Boyle, 1985, 1987; Meybeck, 1988
[2]   Modified from Meybeck and Helmer, 1989

characteristic of point source discharges such as municipal sewage and many industrial point sources (see the example of PO_4^{3-} in Figure 6.12). Curve (2) shows a limited increase in concentration generally linked to the flushing of soil constituents (e.g. organic matter, nitrogen species) during run-off. Curve (3) is basically the same as curve (2) but a fall off in concentration occurs at very high discharges indicating dilution of the soil run-off waters. Curve (4) shows an exponential increase in concentration which occurs with TSS and with all substances bound to particulate matter. The curve represents the increase in particulate matter due to sheet erosion and bed remobilisation. Substances bound to such particulates include phosphorus, metals and organic compounds, predominantly pesticides and herbicides. Curve (5) is the hysteresis loop observed as time is introduced as an additional parameter to the sediment discharge relationship shown in curve (4). Such patterns can be seen for TSS, DOC and sometimes NO_3^-. The peak in sediment concentration occurs at X (advanced) before the occurrence of the peak discharges at Z. Curve (6) indicates a water source to the river with a constant, or near constant, concentration (e.g. Cl^- in rainfall, groundwater in karstic regions or an outlet from a lake).

The concentration discharge relationships (curves 1–6) indicated in Figure 6.11A are also illustrated as changes in concentration and discharge with time during a single storm event in Figures 6.11B and 6.11C. Simple relationships have been chosen to demonstrate their diagnostic capability in determining major sources of pollutants. Curve (1) in Figure 6.11A and B illustrates a simultaneous dilution effect and curves (2) and (4) illustrate

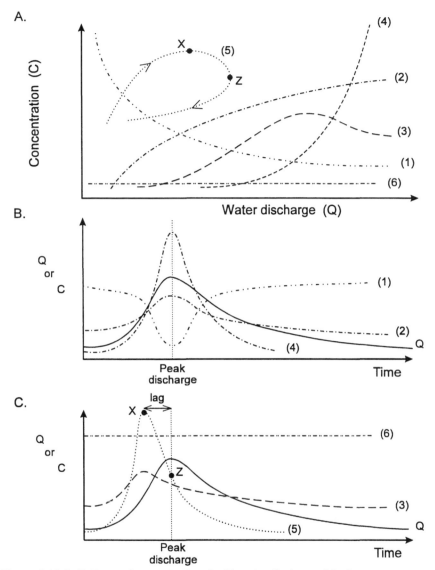

Figure 6.11 A. Patterns of concentration *C* with water discharge *Q* in rivers;
B. and **C.** The same patterns illustrated as synchronous (B.) and asynchronous (C.)
variations in concentration and discharge over time during a single storm event. Points
X and Z represent the maximum concentration and peak discharge respectively. For
further details see text. (Modified from Meybeck *et al.*, 1989)

simultaneous increases in concentration with discharge. Curves (3) and (5)
in Figure 6.11A and C illustrate moderate and rapid increases in concentration prior to the peak discharge. A pollutant is, however, usually derived from
a number of sources, resulting in a more complicated situation. This is

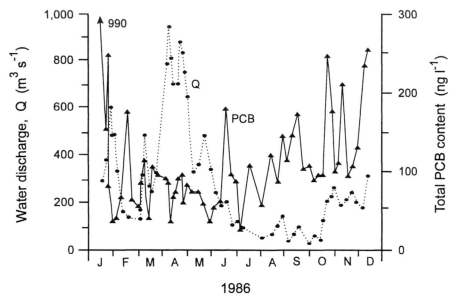

Figure 6.12 Water discharge and total PCB concentration in unfiltered water samples from the River Seine in Paris during 1986 (Based on Chevreuil *et al.*, 1988)

illustrated for total PCBs (polychlorinated biphenyls) in the River Seine, France (Figure 6.12). Three general relationships can be observed between PCB concentration and discharge:

- a high coincidence with high concentrations (PCBs occurring with a high discharge in January and November),
- a moderate relationship where peak concentrations of PCBs relate to only moderate peaks in discharge (as in early January and June), and
- maximum discharge associated with low PCB concentrations in March, April and May.

This indicates two major sources of PCB to this river system: an upstream source in which the PCB became adsorbed by sediment particles, and a more localised source providing soluble PCB to the river system. The sediment related source is active in January and some sediment exhaustion probably occurs since the soluble source predominates during the high discharges in March, April and May. Sediment with adsorbed PCB is re-accumulated in the river bed during summer slow flow and is reactivated by resuspension, thereby appearing as the major source in September and November.

Major sewage treatment plants release a constant flux of pollutants throughout the year which is diluted by the receiving river water. Figures 6.13A and 6.13B show the variation in orthophosphate with water discharge downstream of the Achéres sewage outfall on the River Seine. Most of the

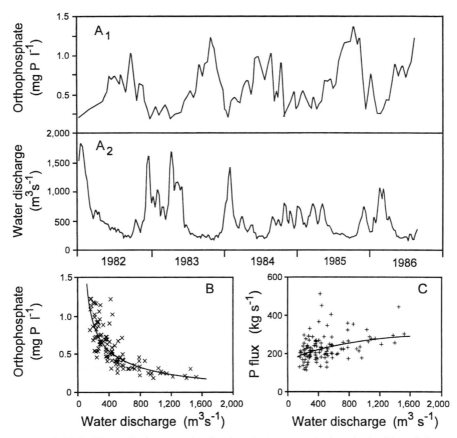

Figure 6.13 A. Water discharge and orthophosphate concentrations in the River Seine from 1982 to 1986 downstream of the Achéres sewage outfall; **B** and **C**. The orthophosphate shows a nearly perfect dilution with a nearly constant flux (Data courtesy of Service de la Navigation de la Seine, Rouen)

wastewaters of Paris are discharged at Achéres after primary or secondary treatment. The orthophosphate pattern shows a near-perfect dilution (Figure 6.13A, B) with a near-constant flux (Figure 6.13C). The weekly sampling frequency could be changed to a bimonthly, or even monthly, frequency if these relationships are taken into account.

6.3.4   Temporal variations in water quality: trends and fluxes

The temporal distribution of concentrations can be determined either with a low sampling frequency over a long time period or with a high frequency over a minimum period of one year (i.e. one hydrological cycle). When the sampling station represents a simple aquatic system, the data obtained are from a single statistical population which, in most cases, has either a normal

(Gaussian) or log-normal distribution. This can be checked on special probability paper where the cumulative frequency curve is plotted (see Chapter 10, and Figure 6.10 for the use of such plots in spatial distributions). If a river is permanently polluted, the data distribution is also normal or log-normal but shows a general shift towards higher concentrations for most compounds or towards lower concentrations, as for dissolved O_2. In the case of intermittent pollution (e.g. seasonal anthropogenic activities, periodic releases of pollutants, accidents), the statistical distribution of measurement data tends to show a distinct change which characterises two different populations. However, the data should be broken down into seasonal aggregations because marked seasonal variations can occur in water quality variables which do not reflect intermittent pollution. An example of statistical distributions is given in Figure 6.14 for the Laita river in Brittany. Such distributions of data (see also Chapter 10) are also useful for checking the proportion of time that water is fit for various uses defined by water quality standards or criteria.

Long-term trends in water quality cannot usually be determined for less than ten years of monitoring. The longest records (over 100 years) in Europe are for the River Thames in the UK and the River Seine in France. As the errors in annual average concentrations in rivers are usually at least ± 20 per cent, only considerable changes in water quality can be shown over short periods, such as the reduction of TSS downstream of a major reservoir following dam closure. Comparisons of time distributions at 10 to 20 year intervals are useful for determining whether the water quality at a river station has changed. A description of methods for trend detection is given in Chapter 10. Although very tedious to establish, trends are essential for assessing the efficiency of clean-up measures in a river basin, or for determining the necessary actions to combat pollution.

Measurement of fluxes of chemical compounds (e.g. nutrients, micropollutants, mineral salts, organic matter) is sometimes a key objective in river assessments at international boundaries, in lake tributaries, or upstream of the estuarine zone. Theoretically, fluxes (Φ) (in mass per unit time, usually t a^{-1}) are derived from the continuous measurements of both water discharge Q (m^3 s^{-1}) and concentrations C (mass per volume, usually mg l^{-1}) between time t_1 and t_2:

$$\Phi = \int_{t_1}^{t_2} C(t)Q(t)\delta t$$

However, continuous determination of concentration is rarely done and it is usually necessary to rely on discrete water quality information obtained at fixed sampling periods. The water discharge, however, may be recorded continuously with a river stage recorder (limnigraph) and determined with an

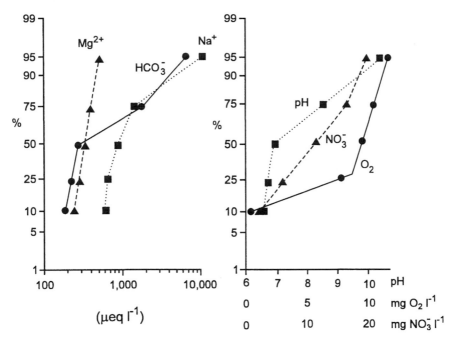

Figure 6.14 Statistical time distributions of Mg^{2+}, Na^+, HCO_3^-, pH, NO_3^- and O_2 in the Laita river, France. Results from monthly measurements over five years. (Data from Ministère de l'Environnement, Paris, 1976–1980 survey)

appropriate calibration curve for the sampling station. The actual concentrations occurring during the time between two chemical analyses must be extrapolated. For the fluxes of particulate pollutants (see Chapter 4), this can be done in two ways: by the constant concentration hypothesis or by the constant flux hypothesis:

- *Constant concentration:* The concentration C_i measured at time t_i is supposedly constant during a time interval δt_j around the time of sampling. The flux Ψ_j of dissolved material discharged during this interval is: $\Psi_j = Q_j C_j$ where Q_j is the average water discharge during the interval δt_j. The total flux for the whole period is then: $\Phi = \Sigma \Psi_j$. This assumption is particularly valid for compounds which increase in concentration during river floods (see Figure 6.11, curves 2, 3, 4 and 5), although the estimated flux is less than the real value.
- *Constant flux:* The flux, $\Psi_i = Q_i C_i$, measured at the sampling time t_i, is supposedly constant during the time interval δt_j. The total flux for the whole period is then: $\Phi = \Sigma \Psi_i$. This assumption is more valid for compounds which decrease in concentration during river floods (see

Figure 6.11, curves 1 and 6), although the estimated flux is greater than the real value. Other methods have been proposed by Thomas (1986) and Walling and Webb (1988).

6.4   Biological characteristics

6.4.1   Factors affecting biological communities in running waters

Flowing waters are complex ecosystems consisting of different habitats (biotopes) and biotic communities (biocoenoses). The physical structure of the ecosystem may be broadly divided into: the water body and stream bed (aquatic zone), the water exchange zone (lentic zone and flood plain) and the environment influenced by the water (terrestrial zone) (see Figure 6.2). For the purposes of water quality assessment the aquatic zone is the most important. The three zones are characterised by specific hydrological features which directly, or indirectly, govern the biological communities that thrive there. The characteristics of the habitats vary from the head-waters to the sites of eventual discharge to receiving waters. Consequently, the biological communities also vary, not only from site to site, but along the length of the river.

For successful colonisation of the flowing water mass, living organisms have to adopt a variety of basic, life strategies, principally:
- to exhibit growth patterns and survival techniques which can withstand the relatively short retention times,
- to exploit "refuge" spaces or boundary layers, or
- to have the ability to swim against the prevailing currents.

For small organisms unable to swim against the current, adaptations include a flattened or spindle-shaped body, as well as adhesive devices and the occupation of the spaces with little or no water flow. In this way the organisms are not carried away by the current and can benefit from the flowing water which provides continuous, rapid exchange of oxygen and nutrients.

In considering a flowing water ecosystem, two important aspects must be taken into account:
- Continuous water flow allows any input, such as an effluent, to have an effect locally as well as along the downstream course of the river.
- As river water is usually retained within the watercourse for relatively short periods (days to a few weeks) before being replaced from other sources, time dependent processes, such as growth or degradation, have only a limited time period within which they can show their effects.

A number of physico-chemical characteristics are of particular importance in determining the biological nature of river systems through their modification of suitable habitats. These characteristics are: flow rate, erosion and deposition, substrate nature, light, temperature and oxygen.

Flow rate

The velocity of water within a river has direct and indirect effects on the biota. It supports or carries organisms, determines the physical structure of the stream bed, and has considerable influence on surface exchange of gases. The roughness of the river bottom, as well as the flow pattern arising from it, are important for the formation of habitats in which organisms can survive (Figure 6.15). In most rivers, discharges vary seasonally, imposing seasonal changes on biological communities.

As flowing water allows turbulent support of a wide variety of types and sizes of particulate materials, the penetration of light into the water is often highly dependent on the stream velocity characteristics. This then has a direct bearing on the amount, type and distribution of the photosynthetic organisms which can colonise the various habitats.

Erosion and deposition

Rivers are subject to continuous change through erosion and deposition. In normal conditions, this can lead to the displacement of the stream bed and the channel line. Such effects may also be artificially magnified by human activities in the terrestrial zone (river bank modification for flood prevention) or by canalisation. Erosion which is intensified by human activity leads to the loss of habitats and a reduction in biological communities in the affected reaches of the river.

Occasional movement and displacement of deposited sediments and rocks are normal processes in flowing waters, and have little impact on biological communities. Permanent displacement on a large scale (particularly in areas of high erosion) frequently occurs in fast flowing rivers, especially in tropical countries, and tends to prevent colonisation by organisms.

Substrate

The substrates available for colonisation by biota vary considerably in rivers, such as solid rock, stones, gravel, sand or sludge. Roots, dead wood, as well as submerged spermatophytes, mosses, filamentous algae, reeds and floating leaf plants also form natural substrates. Artificial structures, such as concrete walls, wood and metal sheet piling, may also provide suitable, but limited, habitats. Fine-grained substrates (mud and sand) are preferentially colonised by diatoms, blue-green algae and higher plants, as well as by certain species

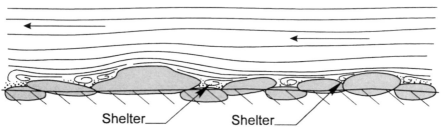

Figure 6.15 Water flow over a stony stream bed; the pockets of still water behind stones provide suitable habitats for colonisation by aquatic organisms

of worm, insects and other animals. Rocky bottomed streams do not usually support the growth of many macrophyte species but the surfaces of the rocks may provide suitable habitats for attached algae, such as certain diatoms.

Mobile animals generally prefer the sides of substrates which are sheltered from light and the water current (Figure 6.15). Consequently, the spaces between and beneath stones are particularly suitable habitats, as are moss layers and the standing crop of water plants. Sheltered spaces with slow flow conditions offer favourable living conditions to many invertebrates and also provide spawning grounds for fish. The pore spaces below the stream bed (hyporheic interstitial — see Figure 6.2) are also biologically important habitats, providing refuges for the early development stages of invertebrates and fish and a suitable medium for the self-purification processes carried out by micro-organisms.

Light
Light is required for photosynthesis by all river primary producers, i.e. the algae and macrophytes. Sub-surface light is usually exponentially absorbed in its downwards passage through the water column. The depth of the euphotic zone (i.e. the zone with sufficient light to support photosynthetic activity) in rivers is highly dependent on the water colour and the amount of suspended sediment present. Frequently, river and stream water which is very clear is fairly shallow allowing sufficient light penetration to support benthic algae or attached macrophytes. In the lower reaches of a river where the water is deeper, or more turbid, the euphotic zone may be wholly contained within the main water mass. Thus, except for the littoral shallows, plant colonisation of the sediments is not possible.

Temperature
Water temperature influences the rate of physiological processes of organisms, such as the microbial respiration which is responsible for much of the

self-purification that occurs in water bodies. Higher temperatures support faster growth rates and enable some biota to attain significant populations. Under natural conditions the temperature of running water varies between 0 °C and 30 °C. Higher temperatures (> 40 °C) usually only occur in volcanic waters and hot springs. In running water, the temperature normally increases gradually from the source of the river to its mouth. Cooling waters discharged to rivers, e.g. from industrial activities or from power generation, can lead to higher than normal water temperatures. These increased temperatures cause problems for sensitive organisms due to the increased oxygen demand (lowering oxygen saturation) and increased level of toxicity of harmful substances. They are sometimes also responsible for fish kills.

Oxygen
Oxygen is one of the most important factors for water quality and the associated aquatic life. Oxygen deficiency, even if it occurs only occasionally and for short periods, leads to a rapid decrease in the number of aquatic animals present, particularly the clean water species which depend on high oxygen levels, as well as most fish. In slow-flowing or impounded rivers, the effects of eutrophication (nutrient enrichment) can lead to deoxygenation of the sediments and possible remobilisation of nutrients and toxic trace elements, particularly from the sediments.

6.4.2 Pelagic communities
The pelagic communities are those swimming or floating organisms associated with the free water in the aquatic zone of the river, e.g. plankton and fish.

The phytoplankton of rivers and streams are only able to attain obvious populations when their growth rates are such that sufficient population doubling times can be attained within the retention periods of the watercourses. Increased light, higher temperature and lower turbidity, together with reduced velocity which provides longer watercourse retention times, tend to promote greater phytoplankton growth rates. River management practices such as abstraction or weir controls can lead to increased phytoplankton growth rates. Phytoplankton are particularly sparse, or absent altogether, in small streams and more free-flowing rivers, particularly where there is little or no natural, or artificial, input of nutrients. In many tropical and subtropical rivers, phytoplankton communities do not reach high densities due to the very high levels of turbidity caused by suspended solids arising from land-based erosion processes. Submerged vegetation is, therefore, also rare.

Zooplankton are small, to microscopic, animals which feed on primary producers or their products, or on other zooplankton. The juvenile stages of larger zooplankton (> 1 mm) may last for several days and, therefore, substantial populations can only be produced in rivers with very low velocities and warm water. However, smaller zooplankton such as rotifers can attain quite large populations.

Fish have been able to exploit all physically accessible river habitats. Their eggs often adhere to stones or weeds, or may be deliberately placed in specially constructed refuges. Consequently, highly specific physical conditions are necessary in a river for successful fish breeding. Fish communities are, therefore, sensitive to modification of the river regime (velocity, erosion, etc.) as well as to the input of toxic substances.

Migratory fish which return from the sea to spawning grounds in upstream stretches of rivers and their tributaries can be prevented from reaching their spawning grounds by physical barriers or chemical barriers along their migratory route. Physical barriers consist of weirs, locks and dams and chemical barriers are stretches of highly toxic or anoxic water in the river. The removal, or bypassing, of barriers to fish migration is fundamental to the restoration of water quality to a level which is acceptable by naturalists, fishermen, and the general public. Examples of schemes to restore self-sustaining populations of migratory salmonid fish can be found from several countries (e.g. Canada, UK, Japan) as well as for international rivers, such as the Rhine (IKSR, 1987). Such schemes provide visible evidence to the general public that river water quality has been improved.

6.4.3   Benthic and attached communities

The benthic communities of rivers and streams consist of those organisms which grow in, on, or otherwise in association with various bottom substrates. These communities are frequently used to assess changes in river water quality (see Chapter 5). As benthic organisms have limited mobility, their presence or absence is most likely to be associated with changes in their habitat or environmental conditions. The zoobenthos are primarily the invertebrate animals which live on, or are associated with, the river bed. They exhibit a very wide diversity of form, tolerance to environmental conditions and adaptation to survival in the different habitats of a river. A large proportion of the particulate organic carbon which enters a river, either as allochthonous or as autochthonous material, is processed by the zoobenthos.

Periphyton are microscopic plants which usually occur in quite thin layers on stones, rocks, sand and mud (epipelic algae). Epiphytic algae grow attached to the surfaces of higher plant stems, branches and leaves. These organisms are not easily swept away and their local nutrient environment is continuously replenished. In turbid conditions, caused by suspended solids or a phytoplankton bloom, algae on or near the bottom may not receive adequate light for photosynthesis and growth.

Macrophytes are plants which grow attached to, or rooted in, the substrate. The macrophyte species composition and abundance of established streams and rivers is usually fairly stable, but can show seasonal and annual differences. Some remain wholly submerged, whereas others produce surface-floating or truly aerial components. Since macrophytes require certain conditions of light penetration and nutrient availability, changes in the species composition and abundance can also be indicative of changes in water quality, especially eutrophication, or the physical characteristics of the river bed. Macrophytes can provide important refuges for small invertebrates, fish eggs and fry.

6.4.4 River zonation

Characteristic zones may be recognised in rivers and streams according to aspects of the habitats or biotic communities present, and the biological processes which occur along the length of the water course (Hawkes, 1975).

The aquatic zone of a river system is normally permanently submerged, and the associated communities are unable to withstand desiccation. The lentic zone and flood plain of a stream are the areas between the mean low water zone, e.g. the zone where reeds grow, and the mean high water limit. As a result, this area is subject to frequent, recurring fluctuations in water level. In large rivers the lentic zone may be very large and many metres wide, but in smaller rivers and streams it can be rather fragmented as the banks tend to be steep. As a result of increased erosion caused by human activity, particularly in deforested tropical areas, rivers and streams may become so deep that they cannot develop an active lentic zone.

Above the lentic zone and flood plain, at the mean high water level is the terrestrial zone. The lower limit of this zone is often indicated by the visible growth of small trees and bushes. The terrestrial zone is usually considered as part of the alluvial valley, at least from the limit to which the valley floods (the recent alluvial plain). There is a close inter-relationship between a stream and its valley, especially with old channels, backwaters, depressions and flood channels. In dry periods the water quality of these may be very different from the main river due to groundwater inputs, anoxic conditions, denitrification, etc. During floods, river water flows through these water bodies allowing the exchange of water, substrates and organisms. The soil water and nutrient budgets of the flood area are usually characterised by the river water, and the nature and shape of the soil surface is changed by erosion and siltation. This transitional zone (ecotone) between the river and the land is currently the subject of much scientific investigation. It is becoming increasingly appreciated that there is an ecological continuum from the aquatic zone to the flood plain and that this area provides the basis for self-purification within the ecosystem (Amoros and Petts, 1993).

In some flowing waters the natural secondary production (i.e. the biomass of aquatic invertebrates and fish) depends less on the primary production in the water itself than on the primary production of the surrounding terrestrial zone. This primary production arises from plants (leaves, roots, flowers and fruits) and also animals, e.g. insects, entering the water from the air, the river banks or flood zone, providing a significant source of nutrients for fish.

Habitat zonation

It is possible to recognise a number of zones based on a variety of habitats, from source to receiving waters. The main habitat classification is based on the erosion or deposition characteristics of the water stretches (Table 6.6). As discussed above, the interaction of erosion and deposition determines the dominant particle size within the habitat. The combinations of dominant particle size and associated organic material support typical assemblages of dominant primary, secondary (macroinvertebrate) and tertiary (fish) producers (Table 6.6). Within the macroinvertebrate communities, further characteristic groups can be identified according to their principal feeding mechanism (i.e. grazer, shredder, collector or predator).

Community zonation

In considering the biological communities which may be found in a river reach, fish have often been used to characterise zones. As fish are widespread, and have a specific ability to exploit many habitats, the species or genera that occur can be indicative of the water quality characteristics of the zone, particularly in association with other typical assemblages of organisms (Figure 6.16). The preferred habitats for fish may depend on such factors as suitable breeding sites (e.g. gravelly substrate, dense macrophyte growth or rapid flow), minimum oxygen concentrations or appropriate food supply. Such factors are usually typical of particular reaches of a river (see Table 6.6) and, therefore, habitat and community zonation are closely linked.

Ecological zonation

Zones can also be determined by the ratio of production (P) to degradation, as indicated by respiration (R) (i.e. the P:R ratio) together with the associated community structure of the macroinvertebrate populations (Figure 6.16). This can be a useful approach where fish are sparse or absent. The upper zone

Table 6.6 An example of river zonation based on the physical characteristics of riverine habitats and the major groups of organisms (northern temperate representatives) associated with each zone

Habitat type	Dominant particle size	Primary producers	Macroinvertebrates				Fish
			Grazers (scrapers)	Shredders (large particle detritivores)	Collectors (fine particle detritivores)	Predators	
Erosional	Coarse (> 16 mm) e.g. logs, branches, bark, leaves	Diatoms, mosses	Mollusca (Ancylidae) Ephemeroptera (Heptageniidae) Trichoptera (Glossosomatidae) Coleoptera (Psephenidae, Elminidae)	Plecoptera (Nemouridae, Pteronarcidae, Peltoperidae, Tipulidae)	Ephemeroptera (Heptageniidae, Baetidae, Siphlonuridae) Trichoptera (Hydropsychidae) Diptera (Simuliidae, Chironomidae, Orthocladiinae)	Plecoptera (Perlidae) Megaloptera (Corydalidae)	Cottidae (sculpins) Salmonidae (trout and char sp.)
Intermediate	Medium (> 1 mm < 16 mm) e.g. fragments of leaves, bark and twigs	Green algae, aquatic plants (e.g. *Heteranthera*, narrow-leaved *Potamogeton*)	Mollusca (Sphaeridae, Pleuroceridae, Planorbiidae)	Trichoptera (Limnephilidae) e.g. *Pycnopsyche*	Ephemeroptera (*Ephemera*)	Odonata (Corduligasteridae, Petalaridae) Plecoptera (Perlidae) Megaloptera (Sialidae)	Etheostominae (darters) Cyprinidae (*Rhinichthys*)
Depositional	Fine (< 1 mm) e.g. faeces, fragments of small terrestrial plants	Aquatic plants (e.g. *Elodea*)	Mollusca (Physidae, Unionidae) Trichoptera (Phryganeidae)	Trichoptera (Limnephilidae) e.g. *Platycentropus*	Oligochaeta Ephemeroptera (*Hexagenia*, Caenidae) Diptera (Chironomidae, Chironominae)	Odonata (Gomphidae, Agrionidae)	Cyprinidae (*Notropis*) Catostomidae (*Catostomus*)

Source: After Cummins, 1975

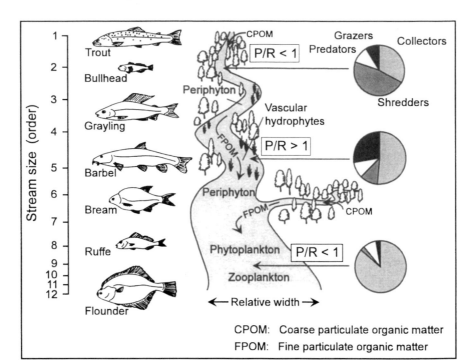

Figure 6.16 Example of a classification of north temperate rivers based on fish zones, together with the associated invertebrates (micro-organisms are present in all situations) and the ratios of production (P) to respiration (R) (After Vannote *et al.*, 1980)

is usually associated with a ratio: P/R < 1, and is dominated by organisms relying on filtering suspended matter or grazing on allochthonous material such as leaf detritus. The intermediate zone usually has a ratio: P/R > 1, and a dominance of organisms filtering suspended material and grazing on attached algae. The lower zone is characterised by a ratio: P/R < 1, and a dominance of organisms filtering suspended material.

6.5   Major water quality issues in rivers

6.5.1   Changes in physical characteristics
Temperature, turbidity and TSS in rivers can be greatly affected by human activities such as agriculture, deforestation and the use of water for cooling (see Table 6.7). However, in certain circumstances the same activities may have little effect in rivers. For example, tropical rivers may not be greatly affected by thermal wastes. In regions of high erosion (e.g. steep slopes, heavy rainfall, highly erodible rocks) where natural TSS already exceeds 1 g l^{-1} (see Chapter 4), intensive agriculture aggravates the natural erosion rates. The turbidity of many sub-arctic rivers, as well as some of those in the wet-tropics

Table 6.7   Major human activities affecting the physical characteristics of rivers

Activity	Temperature	Turbidity	Total suspended solids
Damming	$--$ to $++^1$	$--$	$---$
Cooling water discharge	$++$		
Domestic sewage discharge	$+$	$+$	$+$
Industrial wastewater discharge	$+$	$++$	$++$
Intensive agriculture	$+$	$+++$	$+++$
Navigation		$++$	$+$
Dredging		$++$	$++$

+ to + + + Slight to severe increase
– to – – – Slight to severe decrease
[1]   Depending on the depth of the water outlet with respect to the thermocline

(e.g. Zaire and Amazon basins), may be naturally very high due to the colour of dissolved humic substances.

6.5.2   Faecal contamination

Faecal contamination is still the primary water quality issue in rivers, especially in many developing countries where human and animal wastes are not yet adequately collected and treated. Although this applies to both rural and urban areas, the situation is probably more critical in fast-growing cities where the population growth rate still far exceeds the rate of development of wastewater collection and treatment facilities (Meybeck et al., 1989). As a result faecal coliform bacteria can be found in numbers of 10^6 per 100 ml for some rivers passing through cities such as Djakarta, New Delhi and others.

Where active collection and treatment of wastewater has been carried out since the late 1960s, counts of faecal coliform bacteria have stabilised or even fallen. Levels are now commonly between 100 and 10,000 coliforms per 100 ml.

6.5.3   Organic matter

The release into rivers of untreated domestic or industrial wastes high in organic matter results in a marked decline in oxygen concentration (sometimes resulting in anoxia) and a release of ammonia and nitrite downstream of the effluent input (Figure 6.17). The effects on the river are directly linked to the ratio of effluent load to river water discharge. The most obvious effect of organic matter along the length of the river is the "oxygen-sag curve" (Figure 6.17A) which can be observed from a few kilometres to 100 km downstream of the input. The eventual recovery in oxygen concentrations is enhanced by high water turbulence. Some industrial activities (e.g. pulp and paper production, palm oil extraction and sugar beet processing) may produce wastewaters with BOD_5 and COD values exceeding 1,000 mg l^{-1}. When

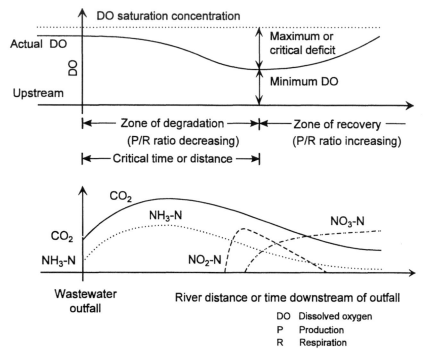

Figure 6.17 Typical changes in dissolved oxygen, nitrogen and carbon dioxide downstream of a wastewater input to a river (Based on Arceivala, 1981)

this wastewater is discharged into a river, oxygen can be completely depleted as in the Laita river (Figure 6.14).

When monitoring for the effects of organic matter pollution, stations should be located in the middle of the oxygen-sag curve (if the worst conditions are being studied) or at the beginning of the recovery zone, depending on the objectives of the programme. During preliminary surveys a complete longitudinal profile incorporating various hydrological features is necessary in order to choose the location of permanent monitoring stations.

6.5.4   River eutrophication

During the 1950s and 1960s, eutrophication (nutrient enrichment leading to increased primary production) was observed mostly in lakes and reservoirs. The increasing levels of phosphates and nitrates entering rivers, particularly in developed countries, have been largely responsible for eutrophication occurring in running waters since the 1970s. In small rivers (i.e. stream orders 3, 4 and 5, see section 6.2.1), eutrophication promotes macrophyte development, whereas in large rivers phytoplankton are usually more common than macrophytes. In such situations the chlorophyll levels may reach extremely

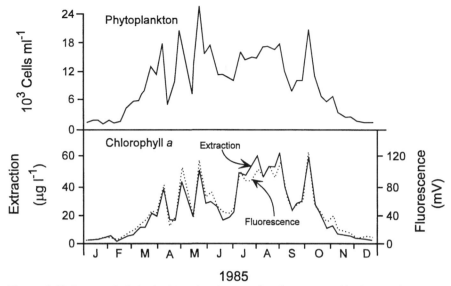

Figure 6.18 Seasonal phytoplankton development in a large eutrophic river — the Rhine (After Friedrich and Viehweg, 1987)

high values (up to 200 mg m^{-3}) as in the River Loire, France or the River Rhine in Germany (Figure 6.18). Eutrophication in river systems can also be caused by the construction of reservoirs and locks (used for navigation), both of which produce a marked decrease in flow velocities within the river (see Chapter 8). Table 6.8 summarises the effects of eutrophication in different types of running waters.

Eutrophication can result in marked variations in dissolved oxygen and pH in rivers during the day and night. During daylight, primary production (P) far exceeds the bacterial decomposition of algal detritus (R), and O$_2$ over-saturation may reach 200 per cent or more, with pH values in excess of 10 during the early afternoon. During the night, this pattern is reversed and O$_2$ levels may fall to 50 per cent saturation and the pH may fall below 8.5 (Figure 6.19). When such rivers also receive organic wastes, the diel (day and night) cycle still exists (Figure 6.20), but the average O$_2$ saturation is much lower and the peak O$_2$ level may not reach 100 per cent saturation. When respiration levels become greater than the primary production (i.e. R > P) in the downstream reaches of rivers, or in their estuaries, the O$_2$ concentration can decline dramatically. Occasionally this can result in total anoxia, as in some turbid estuaries during the summer period.

Diel variations in water quality cause major problems for monitoring and assessment of eutrophic rivers. Sampling at a fixed time of the day can lead

Table 6.8 The effects of eutrophication in running waters

Type of water	Eutrophication effects
Headwaters of streams in the shade	None
Headwaters and streams exposed to sun[1]	Promotion of macrophytes or periphyton growth including filamentous algae
Medium-sized rivers[2]	Promotion of periphyton and/or macrophyte growth
Major rivers[3]	Plankton growth, macrophyte growth
Locks on medium-sized rivers	Large increases in plankton populations and floating macrophyte growth

The construction of major dams which result in impoundments with water residence times greater than one week may lead to severe eutrophication if the nutrient loads are high enough to support algal growth. Monitoring and assessment of impoundments is discussed in Chapter 8.

[1] Average width < 1 m
[2] Average width > 1 m < 20 m; average depth < 2 m
[3] Average width > 20 m; average depth > 2 m

to a systematic bias in recorded O_2 and pH levels. Although chlorophyll and nutrients may also show some fluctuations in concentrations, these are generally within 20 per cent of the daily mean.

The changes in river water quality caused by eutrophication can be a major cause of stress to fish due to the release (at high pH) of gaseous NH_3, which is highly toxic to fish. In slow flowing, eutrophic rivers excessive phytoplankton growth can lead to problems for direct drinking water source intakes and any subsequent treatment processes. A rough estimate of the phytoplankton organic carbon (mg l^{-1}) can be obtained from a measurement of total pigments (i.e. chlorophyll a + phaeopigments (degraded pigments)). The minimum organic carbon present (mg l^{-1}) is approximately equal to 30 times the total pigments (mg l^{-1}), although this relationship has only been tested on western European rivers (Dessery *et al.*, 1984).

6.5.5 Salinisation

Increased mineral salts in rivers may arise from several sources: (i) release of mining wastewaters as in the Rhine basin (potash mines and salt mines) or as in the Vistula, Poland (salt mines), (ii) certain industrial wastewaters (as in Figure 6.14), and (iii) increased evaporation and evapotranspiration in the river basin (mainly in arid and sub-arid regions) resulting from reservoir construction, irrigation returns, etc.

Industrial and mining wastes result in increases in specific ions only, such as Cl^- and Na^+ from potash and salt mines, SO_4^{2-} from iron and coal mine wastes, Na^+ and CO_3^{2-} from some industrial wastes. However, evaporation affects all ions and as calcium carbonate reaches saturation levels, calcium

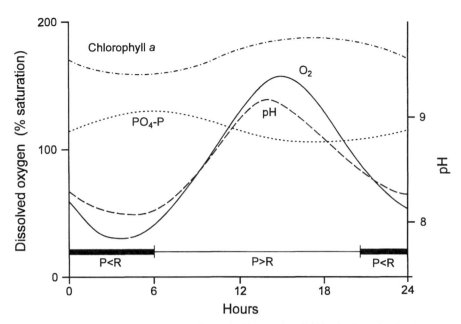

Figure 6.19 Theoretical variations in O₂ and pH associated with algal production in a eutrophic river (P production; R respiration)

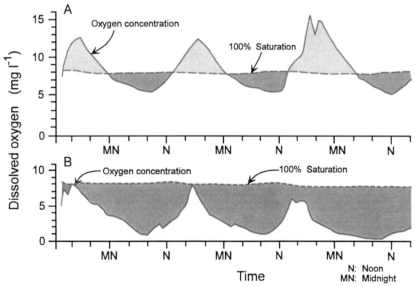

Figure 6.20 Modification of the diel oxygen cycle at two sites on the River Saar by the presence of organic matter pollution: **A.** Güdingen (unpolluted); **B**. Völklingen (polluted) (Based on Müller and Kirchesch, 1980)

sulphate-rich waters or sodium chloride-rich waters are produced. Ion content (principally Ca^{2+}, Na^+, Cl^-, SO_4^{2-}) can also be affected by other human activities such as domestic wastewater inputs, atmospheric pollution, the use of de-icing salts, fertiliser run-off, etc. However, the resulting increases in major ions are much less than those resulting from the three major causes of salinisation mentioned above. The changes in ionic contents and the ionic ratio of waters are very often linked to pH changes. Mine wastewaters are generally very acidic (pH ≤ 3), whereas industrial wastes may be basic (Figure 6.14) or acidic. Salinisation resulting from evaporation usually leads to more basic pH levels (see Table 6.2 where the pH reaches 9.2 in an arid region of Ethiopia).

As electrical conductivity is controlled by the major ion contents, continuous monitoring may be carried out using a conductivity meter linked to a recording device. Results should be checked occasionally for the relationship between the ion concentrations and the conductivity (see Figure 10.14).

Longitudinal profiles of chloride in rivers can help to determine the cause of salinisation such as for the Rhine (Figure 6.21) whereas long-term trends of concentration or fluxes assessed by carrying out measurements for over a decade or more can illustrate the results of legislative control of wastewater discharges and/or other remedial measures (Figure 6.22).

6.5.6 Acidification

Acidification can occur in running waters as a result of: (i) direct inputs of acidic wastewaters from mining or from specific industries, either as point sources (e.g. sewers) or diffuse sources (e.g. leaching of mine tailings), and (ii) indirect inputs through acidic atmospheric deposition, mainly as nitric and sulphuric acids resulting mostly from motor exhausts and fossil fuel combustion. In the latter case, acidification of surface waters may only take place if the buffering capacity of the river basin soil is very low. Low buffering capacity mainly occurs in areas of non-carbonate detrital rocks, such as sandstones, and of crystalline rocks such as granites and gneisses.

Point sources of acidic effluents to rivers may result in a substantial change in water quality downstream of the acid source. The extent of this kind of pollution is best monitored along longitudinal river profiles. Diffuse (i.e. atmospheric) acid inputs may occur over extended regions (up to 10^6 km^2), sometimes located far downwind of the pollutant sources (100–1,000 km), such as major cities, major smelters, refineries, coal-burning power plants, etc. In colder regions where snow melt has a significant hydrological influence, the accumulated acidic deposition in the snow

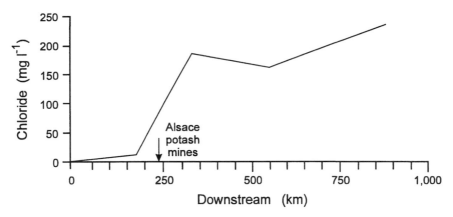

Figure 6.21 Chloride concentrations along the River Rhine showing the downstream influence of the Alsace potash mines, 1971 (Data from Commission Internationale pour la Protection du Rhin Contre la Pollution, Koblenz)

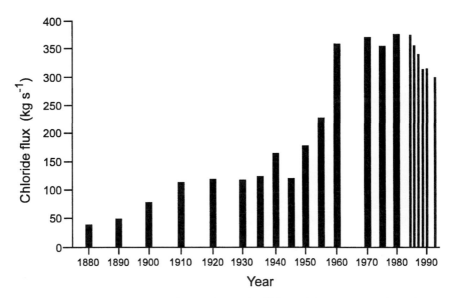

Figure 6.22 Long-term trends in chloride flux in the Rhine at Kleve-Bimmen (Data from Landesamt für Wasser und Abfall NRW, Germany)

may be released when it melts. This can cause a sudden acid pulse which may be more than one pH unit lower than normal (Jones and Bisson, 1984).

A particular problem associated with acidification is the solubilisation of some metals, particularly of Al^{3+}, when the pH falls below 4.5. The resultant increased metal concentrations can be toxic to fish (see also Chapter 7), and also render the water unsuitable for other uses.

The assessment of diffuse source acidification in rivers must start with a precise inventory of the sensitive area and of the potential atmospheric pollutant sources. The actual river monitoring programme must be combined with the monitoring of atmospheric deposition and should also include specific measurements, i.e. Al^{3+}, other selected metals, acid neutralising capacity (ANC) (see Chapters 3 and 7) and dissolved organic carbon. In the area affected by acidification, long-term hydrological trends and specific events, such as snow melt, should also be thoroughly investigated.

Pristine areas where no direct sources of pollution occur, such as headwater streams and lakes, are generally the best places to locate monitoring stations for acidification, although they may be difficult to reach. Ideally a comprehensive monitoring network should be based on a few streams sampled intensively throughout the year, as well as a great number of sites sampled only a few times a year to check the spatial distribution of acidification. Past records of acidification may be obtained through the study of diatom species distributions especially in sediment cores (see Chapters 4 and 7) and biological methods can be used to determine an index of acidification (see Chapter 5).

6.5.7   Trace elements

Trace element pollution results from various sources, mostly: (i) industrial wastewaters such as mercury from chlor-alkali plants, (ii) mining and smelter wastes, such as arsenic and cadmium, (iii) urban run-off, particularly lead, (iv) agricultural run-off (where copper is still used as a pesticide), (v) atmospheric deposition, and (vi) leaching from solid waste dumps.

In surface waters, at normal pH and redox conditions, most trace elements are readily adsorbed onto the particulate matter (see Table 4.1). Consequently, the actual dissolved element concentrations are very low. The monitoring of trace elements on a routine basis is very difficult (see Chapter 3). Ambient air is often highly contaminated with many pollutants, particularly lead. Therefore, as a result of contamination during sample handling, many of the existing routine measurements of dissolved trace elements give much higher values (10 to 100 times higher) than measurements made by specialised sampling and analysis programmes (Meybeck and Helmer, 1989). Unless large financial resources are available for materials and training, the routine monitoring of dissolved trace elements is not recommended. Instead, analysis should be made on particulate matter samples, either suspended or deposited (Chapter 4), or on biological samples (Chapter 5).

The deposition of sediments with adsorbed contaminants provides a useful mechanism for the rapid evaluation of the distribution and origins of

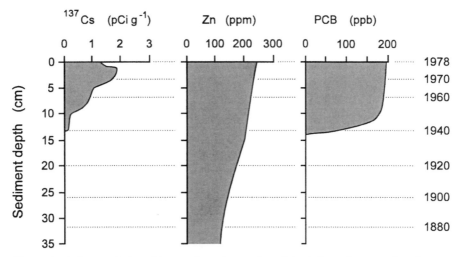

Figure 6.23 Accumulation of inorganic and organic pollutants in sediments taken from the mouth of the Old Rhine river (Based on Irion, 1982)

pollutants of low solubility in a river system. However, care must be taken to sample the fine bottom sediments which would have been recently accumulated and to take into account the variability of particle size in the interpretation of the data (see Chapter 4). Only rarely does long term accumulation allow the determination of the history of pollution in a river. An example from the River Rhine is given in Figure 6.23. Another example of the distribution of metals in river bed sediments is given in Figure 6.24 which shows the longitudinal profile of sediment contamination along the length of the Rio Paraiba Do Sul which runs through an industrialised regions of Brazil and which receives pollutants from three different states. This river is also the source of drinking water for the major urban centre of Rio de Janeiro.

6.5.8 Nitrate pollution in rivers

Nitrate concentrations in some rivers of western Europe are approaching the World Health Organization (WHO) drinking water guideline value of 50 mg l^{-1} NO$_3^-$ (Meybeck *et al.*, 1989). Urban wastewaters and some industrial wastes are major sources of nitrate and nitrite. However, in regions with intensive agriculture, the use of nitrogen fertilisers and discharge of wastewaters from the intensive indoor rearing of livestock can be the most significant sources. Heavy rain falling on exposed soil can cause substantial leaching of nitrate, some of which goes directly into rivers, but most of which percolates into the groundwater from where it may eventually reach the rivers if no natural denitrification occurs (see Chapter 9).

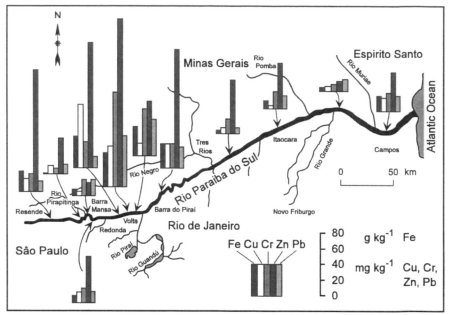

Figure 6.24 Metals in the sediments of the Rio Paraiba do Sul, an industrialised region of Brazil (After FEEMA, 1987)

Assessment of trends in nitrate concentrations should be undertaken on a long-term basis (i.e. frequent sampling for more than 10 years' duration). However, in rivers already affected by organic wastes, causing a reduction in dissolved O_2 concentrations, denitrification may occur in the river bed sediments releasing N_2 to the atmosphere. Specific studies to quantify this process involve nitrogen budgets in river stretches and/or studies of interstitial waters from the river bed. However, reduction of NO_3^- to N_2 does not usually result in a significant reduction of the NO_3^- load in rivers.

6.5.9   Organic micropollutants

Organic micropollutants (mostly synthetic chemicals manufactured artificially) are becoming a critical water quality issue in developed and developing countries. They enter rivers: (i) as point sources directly from sewers and effluent discharges (domestic, urban and industrial sources), (ii) as diffuse sources from the leaching of solid and liquid waste dumps or agricultural land run-off, or (iii) indirectly through long-range atmospheric transport and deposition. As an example, PCBs have been found in very remote sites of North America such as in the Isle Royale National Park in the middle of Lake Superior, where no human activity occurs. In some developing countries, agriculture is a major source of new chemical pollutants to

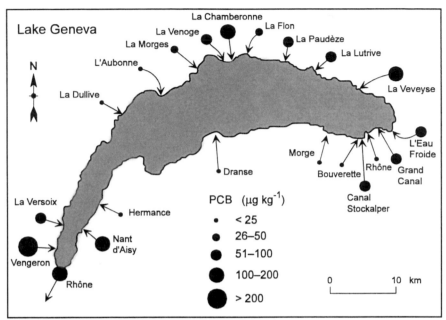

Figure 6.25 PCBs in the deposited sediments of the major tributaries of Lake Geneva (Based on Corvi *et al.*, 1986)

rivers, such as pesticides. The approach to monitoring these substances depends mostly on their properties, i.e. volatility, water solubility, solubility in lipids, photodegradation, biodegradation, bioaccumulation, etc. (see Table 2.4 for some general guidelines on the most appropriate media for the monitoring of various types of organic micropollutants). An example of the monitoring of PCBs is given in Figure 6.12 for the River Seine at Paris where PCBs were monitored bimonthly in connection with TSS and POC monitoring (see also Figure 4.12).

Another example of the use of deposited sediment is given in Figure 6.25 showing sediment-bound PCBs at the mouths of the rivers draining into Lake Geneva. Twenty-two rivers were analysed and the results are presented in five concentration class ranges. Of the three major rivers (Rhône, Dranse and Venoge), only the Venoge showed moderately high concentrations. The most contaminated sediments occurred at the mouths of the smaller tributaries. Care is needed to avoid misinterpreting the effect on the lake from the schematic data presented in Figure 6.25. The map indicates the rivers with the most polluted sediment (i.e. the highest concentrations of PCBs). However, the impact on the lake is determined by fluxes or loads, i.e. the total quantity of sediment delivered from any river multiplied by its pollutant concentration (see Chapter 7).

Table 6.9   Ecological impacts of direct or indirect modifications of the river bed

Activity/modification	Effects
Construction of locks	Enhancement of eutrophication Partial storage of fine sediments may result in anoxic interstitial waters
Damming	Enhancement of eutrophication (bottom anoxia, high organic matter in surface waters, etc.) Complete storage of sediments resulting in potential fish kills during sluice gate operation (high ammonia, BOD and TSS)
Dredging	Continuous high levels of TSS, and resultant silting of gravel spawning areas in downstream reaches Regressive erosion upstream of dredging areas may prevent fish migration
Felling of riparian woodland	Continuous high levels of TSS, and resultant silting of gravel spawning areas in downstream reaches Increased nitrate input from contaminated groundwaters
Flood plain reclamation and river bed channelisation	Loss of ecological diversity, including specific spawning areas Loss of biological habitats, especially for fish

BOD Biochemical oxygen demand
TSS Total suspended solids

6.5.10   Changes in river hydrology

Many human activities, directly or indirectly, lead to modification of the river and its valley which produce changes in the aquatic environment without major changes in the chemical characteristics of the river water (Table 6.9). Such changes can lead to loss of biological diversity and, therefore, biological monitoring techniques are most appropriate in these situations, supported by careful mapping of the changes in the river bed and banks.

Major modifications to river systems include changes to depth and width for navigation, flood control ponds, reservoirs for drinking water supply, damming for hydroelectric power generation, diversion for irrigation, and canalisation to prevent loss of flood plains of agricultural importance due to river meandering. All of these affect the hydrology and related uses of the river system. The dramatic changes in the Nile after construction of the High Aswan dam are examples of both predicted and unpredicted downstream effects (Meybeck *et al.*, 1989). Further discussion of the impoundment of rivers and the implications for assessment are given in Chapter 8.

6.6   Strategies for water quality assessment in river systems

Once polluting substances are introduced into a river, they are transported and transformed by physical, chemical, biological and biochemical processes. It is important to understand these various pathways in order to achieve the best sampling design and to determine the impact of the substance on the water system and the rates at which elimination may occur.

Sampling and analytical strategies for river assessments must be related to the present and future water uses. Two major concepts must be recognised in the design of assessment programmes which address water uses:

- Multiple use of river water may occur within any region of the river basin. Each use has different water quality requirements and user conflicts may occur. Ideally, water quality should meet the most stringent use requirements which, in virtually all cases, is the provision of good quality drinking water.
- There is always a responsibility for upstream users to ensure adequate water quality for the needs of downstream users.

Multiple use of a river system necessitates careful design of assessment programmes to ensure that the requirements of individual uses are accommodated in the monitoring strategy. Table 6.10 provides a structure for designing monitoring programmes to fit multiple water uses. Further details of the selection of specific variables are given in Tables 3.7 to 3.10. Many agencies may not be able to carry out all of the components listed in Table 6.10. Nevertheless, all the components for which facilities are available should be included and efforts should be made to upgrade monitoring capabilities until the total requirements of the scheme can be achieved routinely.

6.6.1   Physical transport and the use of bottom or suspended sediments

Physical transport is produced by riverine flow. Some pollutants may be carried in solution, in which case their transportation rates will be equivalent to the velocity of the river. There is also a tendency for dissolved pollutants to be diluted downstream, particularly if additional water derived from downstream tributaries is free of the same substance. Substances with low solubility are adsorbed by particles shortly after introduction into a river. Such substances are deposited in the river bed during low flow and are transported by resuspension at higher flows. However, most polluting substances occur in both adsorbed and soluble forms, with an equilibrium occurring between the solute and particle phases. The ratio between the two phases is subject to continuous change depending on the following conditions:

Table 6.10 Selection of sampling media, frequency and water quality variable groups in relation to assessments for major uses of, or impacts on, riverine systems

Purpose of sampling	Media	Sampling frequency	Filtered water < 0.45 µm	Total unfiltered sample	Particulate matter[1]	Bacteria	Major ions	Nutrients	Trace elements	Organics	Suspended sediment
ASSESSING WATER QUALITY FOR WATER USES											
Potable water	Water	C	x			x	x	x	x	x	x
Industrial water supply	Water	I/F	x	x			x		x	x	x
Irrigation	Water	R	x	x	x		x	x	x	x	
Contact recreation	Water	S,C		x		x					
Fisheries	Water, fish	R	x	x					x	x	x
ASSESSING POINT SOURCE POLLUTION											
Industrial wastewater disposal	Water, invertebrates, fish	R/F	x	x	x	x	x	x	x	x	x
Municipal wastewater disposal	Water, invertebrates, fish	R/F	x	x	x	x	x	x	x	x	
ASSESSING DIFFUSE POLLUTION											
Agricultural	Water, invertebrates, fish	S	x	x	x	x	x	x	x	x	x
Urban	Water, invertebrates, fish	S	x	x	x	x		x	x	x	x
Trophic status	Phytoplankton, macrophytes	R	x	x				x			

See also Chapter 3 for more detailed information on variable selection

C Ideally continuous or daily measurements (in practice frequency is usually related to the population served)

F Frequent, once per week to once per month

I Infrequent

R Regular, 2–12 times per year

S Seasonal, related to growing season, use periods etc.

[1] Fine deposited material or material collected by centrifugation

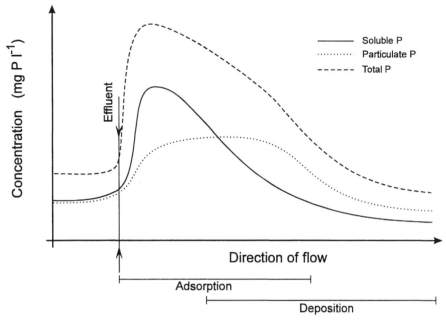

Figure 6.26 Schematic representation of the concentrations of soluble and particulate phosphorus downstream of a sewage treatment plant effluent discharging soluble phosphorus to a river

- The concentration in solution: if reduced, e.g. by biological activity, it will result in desorption from particles.
- The concentration of suspended solids: higher concentrations lead to increased particulate adsorption; lower concentrations result in less particulate adsorption.
- The composition of sediment particles including particle size, surface area and surface coatings in the provision of adsorption sites.

Phosphorus provides a good example of the physical and chemical transport of a pollutant. Figure 6.26 shows the theoretical concentration profiles downstream of a sewage treatment plant effluent discharging soluble phosphorus to a river. Soluble phosphorus shows a sudden increase in concentration downstream of the effluent. Particles start to adsorb soluble phosphorus rapidly to achieve an equilibrium condition. This is followed by a depositional phase where sediments high in phosphorus accumulate in the river bed. Once in the river bed, they can be subjected to resuspension and to slow desorption of phosphorus from the river bed back into solution. The particles remain in this dynamic condition until effectively removed by permanent deposition.

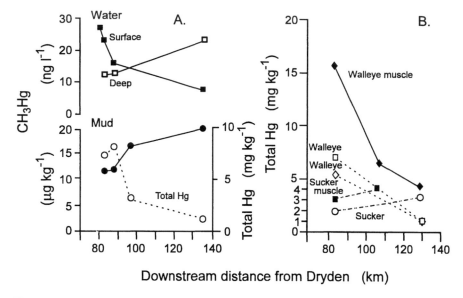

Figure 6.27 Total mercury in mud and fish and methyl-mercury in water and muds downstream of the city of Dryden in the English Wabegon river, Ontario, Canada (Modified from Jackson, 1980)

Physical and chemical transport have also been illustrated for mercury in the English Wabegon river system in North West Ontario, Canada by Jackson (1980). This work is summarised in Figure 6.27. Total mercury declined with sediment dilution with progressive distance from the source and was matched by a decline in methyl-mercury in the surface water due to elimination of particles. Methyl-mercury in the mud increased downstream and was paralleled by increasing concentration in the deeper waters due to release from the bottom sediment (Figure 6.27A).

6.6.2   Use of biological methods in rivers

The possible implications of anthropogenic influences on rivers have been mentioned above. These vary from activities in the terrestrial zone (such as agriculture and deforestation), to manipulation of the aquatic zone (such as canalisation, abstraction and weirs). However, the disposal of pollutants to watercourses also has an enormous impact on the biological communities. These effects may be direct, for example by toxicity or stimulation, or indirect, such as the results of changes in oxygen concentrations. A generalised scheme of different effects on biological communities downstream of a source of organic matter, has been discussed in Chapter 5 and illustrated in Figure 5.1. The number of different kinds of organisms decreases initially and then recovers. For toxic or inert pollution, population density decreases initially and recovers slowly, but following organic, non-toxic pollution it increases very rapidly after only a slight decline.

Most methods of biological assessment described in Chapter 5 can be applied to rivers. Some ecological methods were developed specifically for use in rivers, especially in relation to organic matter pollution, e.g. saprobic indices and biotic indices. However, these methods, which are based on presence or absence of indicator species, may need thorough testing and possible modification before use in a local river system where indigenous species may vary. Methods of ecological assessment, usually based on macroinvertebrate communities, are useful for determining the general condition of the water body or for long-term monitoring of the effects of point sources, such as sewage outfalls. An example of the use of biological assessment for classification of regional water quality is given in section 6.7.3.

Each ecological method (see Chapter 5) has specific requirements with respect to sampling the river, such as the type of sampler used and the precise conditions of flow and substrate (e.g. shallow, riffle section with stony bottom). For accurate determinations of complete benthic communities it is necessary to take sub-samples from all relevant habitats in the sampling site, e.g. stones, sand, mud, macrophytes, etc. The International Standards Organization (ISO) (ISO Standard 7828) has recommended procedures for biological surveys. Failure to adhere to appropriate procedures may prevent meaningful comparisons of the results between different rivers, or sites on the same river. However, in some situations rapid assessment may be possible in the field by examining live samples. For a more accurate calculation of a biological index samples must be kept cool and transported live to the laboratory, or they can be preserved with a suitable fixative solution. To obtain statistically accurate determinations of benthic populations for use in ecological methods it may be necessary to take a large number of samples on any single sampling occasion in order to obtain a representative population estimate (Elliott, 1977). Treatment of the data obtained during biological sampling is discussed in Chapter 10. When the number of species or individuals in samples is low, careful interpretation of the results by a trained hydrobiologist may be more meaningful than statistical treatment of the data. When there have been no major changes in a river system (such as modification of the hydrological regime or a new pollution source) a single survey of the benthic communities can represent the water quality over a long time span.

As planktonic communities are not usually well developed in river systems, methods based on samples of these organisms are not particularly suitable for biological assessment, except in laboratory-based bioassays or toxicity tests where samples of river water may be used with laboratory reared organisms (see Chapter 5). Some *in-situ* physiological methods, e.g. measurement of oxygen production or chlorophyll fluorescence (Figure 6.18), may be useful in lowland rivers where slow flow rates enable phyto- plankton communities to develop.

The macrophyte communities of some temperate rivers have been found to be useful for indicating degrees of organic pollution (Caffrey, 1987). Some macrophytes can be readily identified without removal, sampling or transport to the laboratory and, therefore, a rapid on-site assessment of water quality may be made. However, the degrees of organic pollution indicated may be quite rough and in some cases the variety of macrophyte species may be in- adequate to develop suitable indices.

The organisms of principal interest to man in rivers, as with lakes, are usu- ally the fish. In many areas they are exploited for food or recreation, and when large numbers die they are conspicuous, indicating an obvious deterio- ration in water quality. Many species are top predators in the food chain and are, therefore, susceptible to food chain biomagnification of toxic contami- nants (see Table 5.12). As a result they may bioaccumulate high levels of toxic compounds in their body tissues. In such situations the fish may present a health risk to man if consumed in sufficient quantity. For this reason fish are frequently used for biomonitoring the presence of contaminants in rivers, particularly in relation to industrial discharges. The study by Jackson (1980) (see Figure 6.27B) also provides an example of biological uptake of mercury, which also represents one of the major pathways followed by pollutants in river systems. In this example, methyl-mercury concentrations in pelagic fish followed surface water concentrations, whereas benthic organism concentra- tions followed bottom water concentrations. Biomagnification also occurred in this river system with increasing mercury concentrations up the food chain, from crayfish (crustacea) to suckers (fish) and to walleye (fish).

As fish are usually only associated with certain riverine habitats, and are very sensitive to changes in the physical, chemical and biological quality of their habitat, a survey of the species present gives a general indication of water quality. Samples of fish can be collected with nets or traps, or by elec- trical methods which temporarily stun the fish (see Table 5.13). Counting and identification can be carried out in the field. The sensitivity of fish species to changes in their environment also makes them useful for bioassay procedures and toxicity tests, or for use as "early warning" organisms in dynamic or con- tinuous bioassays (Boelens, 1987) (for further details see Chapter 5).

In certain circumstances, sampling rivers for the collection of indigenous fish or invertebrates, particularly for bioaccumulation studies or bioassays, may be difficult due to the physical nature of the river or its banks. In these situations artificial substrates, caged organisms or "dynamic" tests (see Chapter 5) are particularly useful. These methods allow comparable samples to be obtained from sites with different physical and hydrological conditions.

It must be stressed that the successful use of biological methods in water quality assessment requires the specialist knowledge of a trained hydrobiologist, together with adequately trained field personnel. A trained hydrobiologist can decide on the appropriate choice of bioindicators based on an understanding of the type of water quality information required, and on an interpretation of the biological, chemical, and physical nature of the river.

6.6.3 Sampling frequency

As stressed earlier, the measurement of discharge is an essential component of most sampling programmes. Without this, only qualitative surveys of the general condition of rivers can be obtained. For management of river water abstraction or water quality, a detailed background of information is necessary which includes discharge characteristics combined with an appropriate sampling strategy. Background information is particularly important with respect to diffuse sources of contaminants and for sediment associated variables which show exponential increases in concentrations with increasing discharge.

Figure 6.28 gives an example of the difficulty involved in adequately sampling a small river with a groundwater base flow and large, rapid responses of discharge to periodic rain storms and winter snow melt (see also Figure 6.3). The river (the Venoge, draining into Lake Geneva) was sampled monthly at its mouth throughout 1986 and 1987. The sampling site was adjacent to a permanent hydrographic station in order to obtain a continuous discharge record for the river. In addition to the monthly samples, this station was sampled intensively for four storm events using a centrifuge for the recovery of high flow sediment samples, and by an automatic water sampler for five storm events (see Figure 6.28). The automatic water samples were taken to aid an understanding of storm related river water quality processes.

Based only on monthly samples, the ability to ensure adequate sampling of all river stages is severely limited. This is illustrated in Figure 6.29 where the discharge frequency is shown as a curve representing the percentage of time at which specific discharges occurred throughout the two years of the records. The number of samples taken in relation to the discharge is shown on the two curves. At discharges below approximately 9 m^3 s^{-1} the water samples were collected in reasonable proportion to discharge. However, in 1986, no samples were taken at discharges greater than 9 m^3 s^{-1}. The discharges

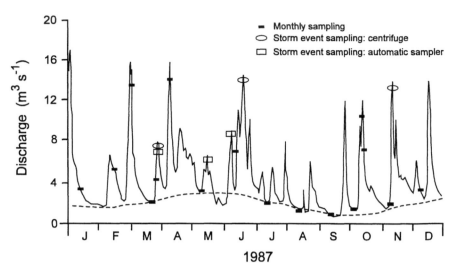

Figure 6.28 Daily discharge during 1987 at Ecublens-les Bois on the Venoge river. Monthly sampling intervals and the periods sampled for storm events are also indicated (After Zhang Li, 1988)

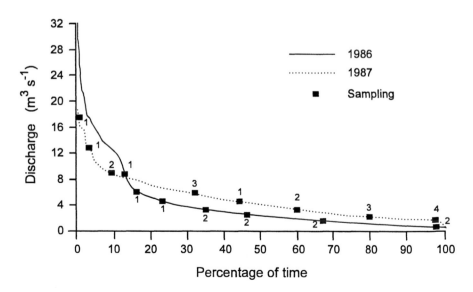

Figure 6.29 Annual cumulative discharge for 1986 and 1987 at Ecublens-les Bois on the Venoge river. The numbers correspond to the number of samples (see also Figures 6.3 and 6.28) (After Zhang Li, 1988)

greater than 9 $m^3 s^{-1}$ occurred up to 14 per cent of the time and included discharges reaching 32 $m^3 s^{-1}$. Thus, the sampling design totally missed significant high flow events. Therefore, high loads of sediment, and sediment related pollutants, were not represented in the data set and the importance of these variables was severely underestimated.

Fortunately, better representation was obtained in 1987, although only two samples were taken at discharges between 12 and 16 $m^3 s^{-1}$ which were not statistically representative. In the study of the Venoge these deficiencies were overcome by deliberate sampling during storm events. Such storm event sampling should be carried out wherever possible, particularly in small rivers with basin areas less than 1,000 km^2, where the random possibility of sampling high flow by constant frequency sampling is much reduced. In large rivers, with basins in excess of 100,000 km^2, constant frequency sampling of one month or less generally provides high quality, representative data without the need to sample specific storm events. Nevertheless, effort should be made to sample extreme events when they are adequately forecast. Extreme storm events are those which occur once in every 25 or 100 years and produce major river basin modifications.

Figure 6.30 shows the different strategies of sampling frequency in relation to different discharge measurement strategies and Table 6.11 indicates the relative cost and reliability of these monitoring strategies. Strategy A (discrete sampling without discharge data) is suitable for preliminary surveys to establish the requirements for more rigorous monitoring. Strategy B (discrete simultaneous measurement of concentration and discharge) permits calculation of instantaneous flux data and provides some information of the variability of concentration in relation to discharge. Strategy C is the most usual monitoring approach; discrete, usually equal interval, sampling for concentrations is carried out together with continuous discharge measurement. This allows a reasonable estimate of flux to be made by extrapolation of the water quality data between samples and calculation using the measured river discharge (see the example of the Venoge above). With this scheme, high discharge events are under-estimated and, therefore, the flux of sediment and related variables is also under-estimated.

In strategy D, sampling is controlled by the discharge with integrated samples taken during specified discharge rates. Continuous discharge data are available and, therefore, good average flux data may be determined. This strategy may be the optimum sampling regime that can be carried out reasonably without excessive sample numbers, although automatic sampling is necessary. A similar automated system is shown in strategy E, which is the same as strategy D except that the automatic sampler takes integrated

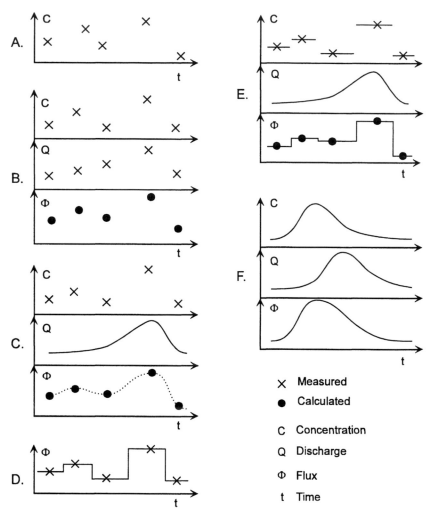

Figure 6.30 Sampling frequency required for water quality monitoring in rivers
A. Discrete sampling; no discharge data
B. Discrete sampling; discrete discharge data; discrete flux data
C. Discrete sampling; continuous discharge; extrapolated flux record
D. Water discharge integrated sampling; average flux data
E. Time integrated sampling; continuous discharge data; calculated average fluxes
F. Continuous records of concentration and discharge; continuous calculated fluxes

samples over a set time period chosen in relation to discharge. Good average flux data can be obtained, but short duration events may be under-represented. This kind of system, as with strategy D, can be set up for high frequency sampling during storm events as for the automated sampling from the Venoge (see Figure 6.28). The most advanced system of river sampling

Table 6.11 Relative costs and benefits of various monitoring strategies for chemicals in rivers (see also Figure 6.30)

Cost/benefit		Strategy A Hand sampling only	Strategy B Hand sampling plus discharge measurement	Strategy C Hand sampling plus discharge recording	Strategy D Discharge weighted automatic sampling	Strategy E Time weighted automatic sampling	Strategy F Continuous concentration and discharge recording
Field operational cost	Hand sampling	x	x to xx[1]	x to xx	x	x	x
	Discharge measurement[2]		x	x	x	x	x
Capital cost of apparatus	Discharge recorder			xx	xx	xx	xx
	Automatic sampler				xx	x	
	Concentration recorder						xxx
Reliability	Time variation	x	x to xx[3]	xx	x	x	xxx
	Flux determination		x	xx	xx[4]	xx[4]	xxx[4]

x   Low cost
xx Medium cost
xxx High cost

Strategies A to F are referred to in the text
[1] Depending on sampling frequency
[2] Establishment of gauge height-water discharge rating curve
[3] Linked to sampling frequency
[4] Reliability of flux determination when field samplers and/or concentration recorders are efficient 100 per cent of the time

is illustrated in strategy F. Continuous constituent measurement is carried out in parallel with continuous discharge recordings. This allows for continuous flux calculations and a very precisely calculated total flux for each variable measured. However, such a system is totally automated and requires the installation and maintenance of analytical probes (e.g. O_2, pH, conductivity, temperature, turbidity), automatic sampling apparatus and their associated electronic equipment. At present, few variables essential for water quality management and control can be measured on a truly continuous basis. The presence of toxic substances, however, may be monitored by continuous "dynamic" toxicity tests (see Chapter 5).

As illustrated in section 4.3.3, riverine fluxes of material are highly variable in time at a given sampling station. Generally, fluxes of TSS vary more than water discharge, while most major ion fluxes vary less than water discharge (due to decreasing concentrations with increasing discharge). Consequently, the optimum frequency for discrete sampling for flux determination is influenced by these relationships. The optimum sampling frequency is that frequency above which there is no significant gain in the accuracy of the flux determination with respect to other errors involved, such as analytical error and errors arising from the non-uniformity of the river section. Figure 6.31 illustrates the variation in the number of samples per month required for flux determination of TSS and major ions with the basin area of the river being sampled. For a given river basin the range of optimum sampling frequency is affected by the basin relief and climatic influences; steep, heterogeneous and dry basins need greater sampling frequencies than lowland, homogeneous and humid basins of the same size. In very small basins, more than four samples a day may be necessary, whereas for the Amazon and Zaire rivers, for example, a single sample a month may be sufficient.

6.6.4   Spatial distribution of samples

The spatial distribution of water quality stations within a river basin must be chosen in relation to the assessment objectives. Certain objectives, such as checking compliance with water quality guidelines for potable supply or other specific uses, require samples for determination of concentrations. For protecting water quality further downstream, e.g. in lakes, estuaries or the sea, it is necessary to know the loads or fluxes of river components.

The location of sampling stations should coincide with, or at least be near to, discharge measurement stations. In addition, water quality stations should be placed immediately upstream and downstream of major confluences and water use regions (e.g. urban centres, agricultural areas including irrigation zones, impoundments and major industrial complexes). Such locations are

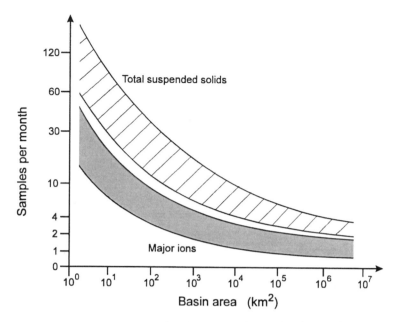

Figure 6.31 Optimum range of sampling frequencies for riverine fluxes of total suspended solids and major ions at the high water stage of rivers with increasing basin areas

illustrated schematically in Figure 4.1. Many rivers cross national, state or municipal boundaries and the responsible monitoring authorities should ensure that stations are also located at the boundaries of river input and river output for their particular regions. Such stations provide information on the suitability of water for use within the region, and determine whether water of inferior quality is being exported to downstream users.

A single station at the mouth of a river, at which water quality variables are measured, may be adequate for flux calculations. Such a station must be located far enough upstream of the junction with the receiving water body to prevent changes in level or discharge due to tidal or water level changes within the receiving water body. At the same time there must be no significant pollution inputs between the monitoring station and the receiving water body.

Very large rivers can be several kilometres wide at their downstream reaches and, therefore, sampling should be carried out through several vertical profiles across the river at each station. Vertical profile samples are normally mixed to give depth integrated measurements. Small and medium sized rivers may be sampled at one location provided that the river is well mixed and homogeneous with respect to water quality.

Table 6.12 Relative homogeneity in the concentration of suspended sediment and its composition (as inferred from the apatite phosphorus) across the River Rhône for three stations and three sampling episodes at Porte du Scex (samples taken at 0.2 m depth)

Date	Station	Suspended sediment (mg l^{-1})	Apatite phosphorus ($\mu g\ g^{-1}$)
6 Oct 1981	Right bank	58	460
	Centre	60	485
	Left bank	53	360
20 Oct 1981	Right bank	73	410
	Centre	79	430
	Left bank	77	460
4 Nov 1981	Right bank	36	410
	Centre	28	355
	Left bank	36	375

Source: After Burrus *et al.*, 1989

   With the exception of small rivers, the location of any sampling site needs to be evaluated for its suitability to represent that section of the river. This is normally done by means of a detailed study of a cross-section of the river. Ideally, the cross-section should be downstream of a physical structure resulting in mixing of the water, e.g. rapids, waterfalls or river bends producing a helical mixing pattern. The cross-section has to be checked for the efficiency of the mixing process by analysing a minimum of three locations within the cross-section: the left and right banks and the centre of the river. At each location a minimum of three depth samples should be taken: one below the surface, one in mid-column and one above the river bed (above the traction or carpet load). If these samples are homogeneous, then the single sampling site can be located at any point in the cross-section and can be assumed to be representative of the river water quality. An example of site evaluation has been described for the mouth of the Upper Rhône river on Lake Geneva by Burrus *et al.* (1989). By measuring particulate apatite phosphorus in sediments they were able to show considerable homogeneity of sediment composition throughout the cross-section of the river (Table 6.12).

   Laminar flow may confine a water mass to one river bank downstream of a major effluent (see Figure 6.7). When sampling is undertaken specifically to monitor the effects of the effluent, the sample site can be located downstream, on the same side of the river as the effluent. Sites further downstream, after the point of mixing or overturn, need to be chosen as described above. In large rivers a well-mixed site may be several kilometres below the source of the effluent.

6.6.5   Sampling methods

Most methods of river sampling are based on a bottle collection or water pump system. Difficulty can be experienced in collecting water samples from very wide rivers. It may be necessary to use a boat if samples cannot be taken from a bridge or to use a specially constructed cable system. When homogeneity of water quality has been established for the cross-section of the river, it may be possible to take certain samples from the river bank at fairly deep locations. However, the velocity of the water may make it difficult to submerge a water sampling device adequately. This may be overcome by using hydrodynamically designed samplers (streamlined and fish-like in shape) which have a depressor fin and tail to help maintain the correct orientation in relation to the flow of the water.

Details of field techniques for collecting samples are described in Bartram and Ballance (1996) and many kinds of water sampling devices are described in operational guides (e.g. WHO, 1992) or manufacturers' catalogues. Most devices operate by filling the sample container at a fixed depth or allowing operation such that a depth integrated sample can be obtained. Water pump intakes can also be set at fixed depths. Alternatively, a depth integrated sample can be obtained by hauling the pump inlet through fixed depths and allowing pumping to continue for a fixed time interval at each depth. Pumps can be used in this way to collect a series of depth integrated samples across a river, thereby giving a fully integrated, large volume sample for the river cross-section.

There are many automated systems available for sampling and analysing river water. The instrument used for the study of the Venoge mentioned above (section 6.6.3), could be powered externally or by internal batteries. It contained 24 one-litre bottles which were filled by a pumping system which could be set to operate at selected time intervals. This system allows the collection of 24 individual samples, or for each bottle to be filled for several small time intervals to provide 24 integrated one-litre samples. Alternatively, sampling may be triggered by a water level sensing device and samples may be collected for pre-selected time intervals during high discharge events. As such automated sampling instruments may be small, portable and versatile, they may be used for synoptic sampling in river basins or to monitor the changes in a flood wave as it progresses through the watershed. Similar systems with more numerous water bottles can also be obtained for permanent installation in association with discharge measurement stations.

Truly continuous water quality monitoring can only be carried out with submersible probes. Although these are only available for selected variables, e.g. O_2, pH, temperature, conductivity and turbidity, they provide very useful information. For security purposes the probes may be immersed in a flow of water pumped through a by-pass system housed within the gauging station. Alternatively, the probes may be suspended in the river from an anchored flotation system or overhead structure. The data can be continuously recorded using a pen and paper plotter or microcomputer. Highly sophisticated systems allow the digital data to be sent electronically to a central recording location or via satellite transmission to a distant location.

Sediment sampling is usually carried out by filtering water samples collected by the methods mentioned above. Filtration is normally done with 0.45 μm pore diameter filters, but these are subject to pore clogging. Such samples are suitable for determining total suspended solids but the final sample size is usually too small for comprehensive chemical analyses, e.g. of several trace elements. To overcome this problem, continuous flow centrifuges have been adapted for field use. Such instruments can process 6 litres per minute of water, with a recovery efficiency of approximately 98 per cent of particles greater than 0.45 μm and 90–95 per cent of particles greater than 0.25 μm. Many tens of grams of material can be obtained, even at low suspended sediment concentrations, by processing very large volumes of sample. Santiago *et al.* (1990) were able to obtain sufficient material for analysis of particle size, organic and inorganic carbon, nitrogen, forms of phosphorus, trace elements and organic pollutants, even from samples with a suspended solids concentration of 6 mg l^{-1}.

6.6.6   Laboratory techniques

Analytical techniques for water and sediment are described in Bartram and Ballance (1996) and various other specialised manuals. However, some general information is available in Chapters 3 and 4, including guidance on sample preservation methods.

An example of a river assessment programme from Germany, indicating the stages undertaken within the field and laboratory, is given in Figure 6.32. This system is based on comprehensive chemical analysis of water samples and excludes suspended sediment analysis. It is advisable to construct a flow diagram similar to Figure 6.32 for any monitoring programme prior to commencing the field work. This will ensure that an adequate quantity of sample is collected for all subsequent analyses.

6.6.7   Co-ordination of various types of monitoring

In the development of environmental quality assessment networks, multi-purpose monitoring is often the first type to be established. However, this

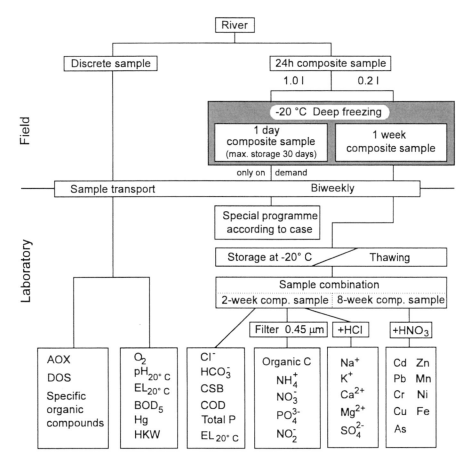

Figure 6.32 Sampling and analytical programme for river monitoring in Baden-Württemberg, Germany, 1986 (Based on Schmitz, 1989)

may eventually be developed into more specialised or sophisticated activities, such as basic surveys, trend monitoring or operational surveillance. An example of such a development is given in Table 6.13 from the Nordrhine-Westfalen region in Germany. However, it is more common for each monitoring programme to be designed by a different group of experts, each involved in different aspects of water quality management. Occasionally, the field and laboratory specialists are not involved in the preparation and planning of monitoring activities. Consequently, similar sampling and analysis can frequently be undertaken by two or more agencies with interests in water quality (e.g. agencies for agriculture, housing and development, fisheries and natural resources, environment or health) and sometimes within the same agency. In addition, monitoring may be repeated by non-governmental

Table 6.13 Example of a water quality monitoring system: Nordrhine-Westfalen, Germany

	Operations	Frequency
Nordrhine-Westfalen	Area: 34,000 km^2 Total length of rivers: 60,000 km Population density: 500 km^2 Rhine river stretch: 200 km (from km 639 to km 863)	
BASIC MONITORING		
Number of points	Approximately 3,000	
Programme (frequency)	Analysis of benthic organisms	Once a year
	Calculation of saprobic index	Once a year
	Analysis of basic physical-chemical variables: conductivity, temperature, pH, BOD, NH$_4$, Cl	Once a year
Data treatment	Classification into a seven stage water quality system. Water-quality report	Every year
	Water-quality map	Every 5 years
IMPACT MONITORING		
Number of points	Approximately 150 in selected catchment areas	
Programme (frequency)	Analysis of benthic organisms	Once a year
	Intensive chemical analysis: variables adapted to the special problems of the water course: basic variables, trace elements, organic micropollutants, etc.	13 times a year
Data treatment	Report as basis for management and control measures	
TREND MONITORING		
Number of points	Approximately 50 at selected sites	
Programme (frequency)	Analysis of benthic organisms	Once a year
	Intensive chemical analysis	Many times a year
Data treatment	Report on trends included into Federal trend monitoring programme	

BOD   Biochemical oxygen demand

organisations, such as private water treatment and supply companies. It is, therefore, highly recommended that:

- all information on water quality assessment is centralised by a co-ordinating authority for each main river basin,
- basic data should generally be freely accessible from programmes financed by public means,
- common data banks should be established for river basins, at either national or inter-state level, and
- all monitoring activities having similar objectives should be co-ordinated and harmonised to ensure maximum inter-comparability and proper inter-pretation (e.g. eutrophication monitoring and assessment, major trend stations, micropollutant monitoring in sediments).

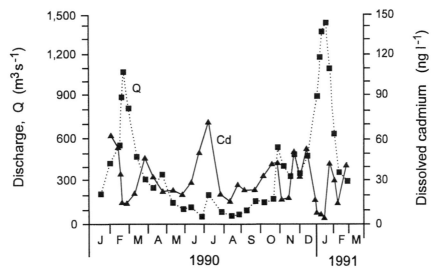

Figure 6.33 Dissolved cadmium and water discharge at the mouth of the River Seine at Poses during 1990 (After Cossa and Ficht, 1991)

6.7   Approaches to river monitoring and assessment: case studies

River assessments have increased greatly in number and sophistication over the last 30 years. The traditional practice of multi-purpose monitoring (see Chapter 2) is not always adequate for specific water quality issues. Many recent water quality assessment activities have combined the analyses of water, sediments, and biota.

6.7.1   Preliminary surveys

Before establishing the appropriate surveys of trace elements in the River Seine, France a very detailed survey was carried out at the river mouth station situated at Poses. All the necessary precautions were taken to avoid contamination during sampling, pretreatment and analysis (D. Cossa and M. Ficht, 1991 pers. comm.). As a result, the dissolved Cd concentrations were found to be more than ten times lower than previously reported in routine analyses. The variability was established on a weekly basis, pointing out a relative dilution during the high water stage and an unexpected peak in the summer (Figure 6.33). Consequently, weekly sampling was eventually chosen as appropriate for dissolved Cd since concentrations were much lower than water quality criteria and about 90 per cent of the total Cd discharged by the Seine was associated with particulate matter (D. Cossa, pers. comm.). Future monitoring should probably focus on suspended matter which can be analysed more easily and less frequently (e.g. bimonthly).

6.7.2 Multi-purpose monitoring

Multi-purpose monitoring should provide answers to many diverse questions concerning water quality at a fixed point on a river, or concerning the river water quality of a given region. Typical questions are:

- Is the water fit for drinking?
- Is the water fit for other major uses?
- Is the quality of the aquatic environment adequate to support the growth of the expected aquatic biota?
- What are the time variations in water quality?
- What are the long-term trends in water quality?
- What is the flux of pollutants (or nutrients) in the river?
- Where are the major pollutant sources (diffuse and point sources)?
- What is the regional distribution of water quality?
- Are pollution control measures adequate?

Each of these questions actually requires a specific monitoring activity and, therefore, multi-purpose monitoring is usually a compromise based on average station densities, average sampling frequencies and a restricted number of variables (depending on the financial resources of the monitoring agency). If the station density is mostly controlled by the technical and financial means available, the frequency and number of variables should be chosen by scientists in relation to the assessment objectives. Table 6.14 proposes a basis for the development of a multi-purpose river monitoring scheme in relation to the three stages of monitoring complexity discussed in Chapter 2.

6.7.3 Basic surveys

Basic surveys are used mainly when detailed spatial distribution of water quality is needed, perhaps as a starting point against which future changes can be measured. The accent is put on the station density rather than on sampling frequency. This enables the drawing of water quality maps or detailed river profiles identifying the various changes in quality. This type of water quality assessment requires a preliminary survey (inventory) of all potential sources of pollution in order to locate monitoring stations appropriately. Typical basic surveys combine all three media: biota, water and sediments.

The inter-comparability between stations for a given survey, and between surveys from one year to another, should always be carefully checked. As a result these surveys should not be carried out at periods when the aquatic system is very variable, such as during high water discharge, or at the beginning of biological cycles (Hines *et al.*, 1976). In the northern hemisphere, the end of the summer period is a suitable sampling time provided that there are no

Table 6.14 Development of a multi-purpose river quality monitoring programme for three different levels of monitoring complexity

Number of stations	Sampling frequency (per annum)	Water analysis	Sediment analysis	Biological surveys	Required resources
SIMPLE MONITORING					
10	6	$^{\circ}$C, pH, Cond., O$_2$, TSS, major ions, NO$_3^-$, visual observation			Small sampling team, general analytical chemistry laboratory[1]
INTERMEDIATE MONITORING					
100	6 to 12	as above, plus PO$_4^{3-}$, NH$_4^+$, NO$_2^-$, BOD, COD	Trace elements	Biological indices	Specialised chemistry laboratory, team of hydrobiologists
ADVANCED MONITORING					
100 to 1,000	> 12	as above, plus soluble organic pollutants, DOC, POC, chlorophyll, some trace elements	as above, plus organic micro-pollutants	as above, plus chemical analysis of target organisms	Major centralised laboratory plus regional teams, national research institute

Intermediate and advanced monitoring needs to be combined with hydrological surveys, especially continuous records of water discharge	COD Chemical oxygen demand DOC Dissolved organic carbon POC Particulate organic carbon
TSS Total suspended solids BOD Biochemical oxygen demand	[1] As required for agriculture or health departments

rain storms. The density of sampling stations in a given river basin can often be ten times higher than the number of regular or multi-purpose monitoring stations, but they may be sampled only once a year, or even less if no major environmental change is expected.

Water analyses for basic surveys usually include all general variables, i.e. nutrients, BOD, COD and major ions (see Chapter 3). The occurrence of micropollutants can be checked using deposited sediments from the river bed. A detailed visual inspection should also be made and biological samples collected (usually of benthic organisms) for eventual identification and counting in the laboratory. The results can be used to determine appropriate biological indices (see Chapter 5). In order to achieve maximum inter-comparability of results from these surveys, it is highly recommended that a limited number of field and laboratory personnel be used.

The final output of basic surveys is often presented in the form of water quality maps based on chemical data or biological assessment methods or longitudinal profiles of the river course. Figures 6.34 and 6.35 show the improvements, using colour coding, in water quality over 20 years in Nordrhine-Westfalen, Germany based on the biological classification described in Table 6.15. During these 20 years measures to improve water quality included the installation and maintenance of sewage collection and treatment facilities.

6.7.4 Operational surveillance

Operational surveillance provides local and regional planning and environmental management authorities with data for such purposes as checking compliance with guidelines and legislated standards and the efficiency of water and environmental protection measures. Such monitoring may be characterised by a high station density in a restricted area. For checking compliance with guidelines for a particular water use, continuous records of pH, oxygen and other variables (e.g. ammonia) amenable to automatic sampling are frequently used at the point of water intake. In some cases the same results can be used to set-up or validate water quality models of pollutant dispersion, oxygen balance, etc.

6.7.5 Trend and flux monitoring

Trend and flux monitoring are both characterised by a high sampling frequency. Trend monitoring may consider the time-averaged concentrations over the whole year, or the average concentrations at special sampling periods only, such as at low water when the dilution of pollutants from a point source is minimum, during summer periods when eutrophication effects are greatest, or during floods to check the maximum levels of total suspended solids and associated pollutants. For determination of trends (see also Chapter 10) it is important that samples are collected at (or data averaged for) equally spaced time intervals (e.g. annually, monthly). Before such monitoring programmes begin, preliminary surveys are conducted to determine the time-variability of the concentrations of interest in order to help optimise sampling regimes. The sampling frequency is much higher for compounds with highly variable concentrations, such as NH_4^+, than for more stable ones, such as Ca^{2+}. Where appropriate sampling frequencies cannot be attained, the annual average concentration may not be very precise and may prevent the establishment of trends, particularly if the rate of change over the years is lower than the precision of the values. For example, a 10 per cent annual increase in ammonia will not be apparent for many years if the precision of the annual average ammonia concentration is ± 30 per cent. An example of a long-term trend is given in Figure 6.22.

Table 6.15 Definition of quality grades used to classify running waters in Nordrhine-Westfalen, Germany

Quality grade I: *unpolluted to very mildly degraded*[1]
Sections of water bodies with pure, almost always oxygen saturated and nutrient-poor water; low bacteria content; moderately densely colonised, mainly by algae, mosses, flatworms and insect larvae; spawning ground for salmonid fish.

Quality grade I–II: *mildly degraded*[1]
Sections of water bodies with slight inputs of organic or inorganic nutrients but without oxygen depletion; densely colonised mostly by diverse species but dominated by salmonid fish.

Quality grade II: *moderately degraded*[1]
Sections of water bodies with moderate pollution but a good oxygen supply; a very great variety and density of individual species of algae, snails, crayfish and insect larvae; considerable stands of macrophytic plants; high yields of fish.

Quality grade II–III: *critically degraded*[1]
Sections of water bodies with inputs of organic, oxygen consuming substances capable of producing critical oxygen depletion; fish kills possible during short periods of oxygen deficiency; declining numbers of macro-organisms; certain species tend to produce massive populations; algae frequently cover large areas; usually high-yields of fish.

Quality grade III: *heavily polluted*
Sections of water bodies with heavy organic, oxygen depleting pollution; usually low oxygen content; localised deposits of anoxic sediment; filamentous sewage bacteria and colonies of non-motile ciliated protozoa predominate over the growths of algae and higher plants; occasional mass development of a few micro-organisms which are not sensitive to oxygen deficiency such as sponges, leeches and water lice; low fish yields; periodic fish kills occur.

Quality grade III–IV: *very heavily polluted*
Sections of water bodies with substantially restricted living conditions resulting from very severe pollution with organic, oxygen depleting substances, often combined with toxic effects; occasional total oxygen depletion; turbidity from suspended sewage; extensive anoxic sediment deposits, densely colonised by red blood-worm larvae or sediment, tube-dwelling worms; a decline in filamentous sewage bacteria; fish generally not present, unless only locally.

Quality grade IV: *excessively polluted*
Sections of water bodies with excessive pollution by organic, oxygen depleting sewage; processes of putrefaction predominate; prolonged periods of very low oxygen concentrations or total deoxygenation; mainly colonised by bacteria, flagellates and mobile ciliates; no fish stocks; loss of biological life in the presence of severe toxic inputs.

[1] Degraded refers to a slight deterioration in water quality which does not usually affect normal aquatic life and is considered acceptable, whereas polluted water causes the loss of most aquatic species and is unacceptable (see also section 1.2).

For flux monitoring, the sampling effort should be concentrated on the high flow periods because the maximum transport of any dissolved or particulate compounds occurs during maximum water discharge (see also section 6.6.3). Even for a constant source of pollutants the dilution process

Table 6.16 Flux monitoring: hypothetical example of discharge and flux distribution for elements increasing (A) and decreasing (B) with water discharge (Q)

Gauge height duration		Water discharge		Concentration (by vol)		% of annual flux	
Days	% of year	Average Q^1	% annual volume	Element A^1	Element B^1	Element A^1	Element B^1
2	0.6	1,000	5.6	500	15	24.0	2.2
10	2.7	500	13.9	250	20	29.8	7.5
30	8.3	250	20.9	125	30	22.4	16.8
70	19.2	125	24.4	70	40	14.6	26.2
253	69.3	50	35.2	30	50	9.2	47.3
Totals 365	100.0		100.0			100.0	100.0

^{1} Arbitrary units

is never perfect. Preliminary surveys should be used to determine the period of time responsible for any given percentage of the flux. A complete theoretical example is treated in Table 6.16. Flux monitoring for element A (which could, for example, be the total suspended matter and associated pollutants) should be focused on the 42 days of high discharge which correspond to 76 per cent of the annual flux. However, monitoring for element B (characteristic of dilution of a point-source pollutant in a river) should be spread over the whole year although, due to the imperfect dilution characteristics, 53 per cent of the river flux occurs during 31 per cent of the time. In both cases, the optimal number of samples should be taken in proportion to the flux distribution. For example, when taking 50 samples a year, in the case of A, 27 of them should be taken during the 12 days of maximum river water discharge (these can correspond to one, or several major floods). This may present some operational problems. However, automatic sampling can help simplify field work, although the equipment requires regular maintenance, usually on a weekly basis (see section 6.6.5).

For trend monitoring, individual samples can be combined to form time-averaged composite samples or the individual measurements can be combined to give a mean value. For flux monitoring, composite samples must be weighted according to the water discharge at the time of collection of each individual sample.

Long-term trends in rivers can also be assessed by the study of sediments from flood plains or old river channels. Figure 6.23 shows the changes in pollutant concentrations in a sediment core taken near the mouth of the old Rhine river in 1978 and dated with a ^{137}Cs profile. The most marked increase in contaminants (organic micropollutants) within the core occurred after 1940.

6.7.6   Early warning surveillance and associated networks

Continuous early warning surveillance is mostly carried out at water intakes for major drinking water treatment plants. It must give instantaneous (or very rapidly obtained) information on the overall water quality, especially the presence of toxic substances. Continuous measurement of dissolved substances is highly expensive and usually sophisticated. Other approaches are based on the continuous exposure of very sensitive organisms to river water. These biological monitoring methods are often based on the behaviour or activity of organisms such as fish (e.g. trout) or large invertebrates such as mussels and water fleas (e.g. *Daphnia* spp.) (see section 5.7.3). A biological early warning system is being developed for the Rhine using a suitable biotest for each trophic level (Schmitz *et al.*, 1994).

Automatic, continuous, chemical analysis is relatively cheap and easy for certain variables, such as dissolved oxygen, temperature, electrical conductivity, optical turbidity and pH, but is more difficult for ammonia, dissolved organic carbon and some specific major ions. It is very expensive and complicated for trace elements and organic pollutants, in addition to which the detection limits are relatively high. These factors restrict the use of automatic, continuous analysis for trace elements and organic pollutants to early warning stations.

An early warning network may be established on large rivers linking key stations upstream and downstream of water intakes. This is particularly useful for river stretches affected by major industrial areas, nuclear power plants, etc. Such networks are activated in the event of accidents (e.g. chemical spills and burst storage tanks), decreased efficiency of effluent treatment plants or unusual meteorological events (such as thunderstorms causing increased urban run-off with high BOD and resultant oxygen depletion in the river). Early warning networks are useful, not only for protecting major drinking water intakes, but also for protecting commercial fisheries, irrigated fields and livestock watering sources.

An example of an early warning system exists in Germany for the River Rhine which is an international water body characterised by extensive industrial and urban development. In addition, the river is an international shipping waterway and supplies 20 million people with drinking water. Therefore, a well organised, international warning system has been installed to help prevent accidental pollution reaching critical water intakes, and it has now been working successfully for many years. The system consists of eight main warning centres. The warning message may follow either a step by step route along the river course from one warning centre to the next, or a general alert may be given to all downstream stations. Each warning message is first passed on by telephone and then in written form by telex, telefax or letter. The message includes the following information:

1. Name of person, agency, etc., giving the message.
2. Date and time of accident.
3. Place of accident (i.e. name of place, kilometre position along the river, location (right, left or centre) and any other details).
4. Type of substance, the chemical compound involved, etc.
5. Quantity of substance discharged to the river, extent of accident and pollution effects.
6. Type of pollution effect (e.g. fishkill, colour, smell).
7. Preliminary actions taken.

After collecting the initial information more detailed data have to be provided by specialists in order to assess the impact and plan remedial action. This information includes:

- Quality criteria of the substance in question (e.g. solubility, density and reaction with water, air or other compounds).
- Potential effects on water quality, toxicity to man, fish or the ecosystem, self-purification ability and the risk category in water.
- Hydrological conditions in the river (discharge in m^3 s^{-1} and current velocity in m s^{-1}).
- Measured and calculated concentrations at the place of the accident, and downstream at any important sites. The expected time of arrival of the contaminant at given places such as drinking water intakes. Any initial action taken to minimise the effects.

When an alarm has been given but the danger has passed, it is also necessary to give an all-clear signal. Finally, it must be emphasised that in order to be effective early warning surveillance should be backed up by technical provisions (such as equipment for aeration) and by legal powers to initiate and carry out necessary preventative action (such as stopping waste disposal activities).

6.7.7   Surveys for water quality modelling

Water quality modelling can be a valuable tool for water management since it can simulate the potential response of the aquatic system to such changes as the addition of organic pollution, the building of small hydro-electric power plants, the increase in nutrient levels or water abstraction rates and changes in sewage treatment operations (such as the addition of tertiary treatment). For major projects the cost, including the necessary surveys, of producing a model can be only a few per cent of the total cost of the new operations or activities to be introduced into the river basin. The use of generally available models (i.e. models not produced as part of a specific project) should be carried out with caution and, where possible, any general model should be verified with data obtained from the water body for which its use is being considered.

Most existing river models are for oxygen balance and are based on BOD measurements, although more recent models include the influence of phytoplankton, macrophyte growth and benthic respiration. Models based on bacterial respiration are now also being developed (Billen, 1990 pers. comm.). Obtaining the necessary data for construction or verification of the models may require additional surveys, together with data from operational surveillance and multi-purpose monitoring networks. It is important to verify models if they are to be used routinely in the management of water quality (e.g. Rickert, 1984).

Krenkel and Novotny (1980) have listed the categories of variables required for oxygen balance modelling as:

- hydrological variables (e.g. river discharge),
- hydraulic variables (flow velocity, geometry of river bed, turbulence, etc.),
- oxygen sinks (e.g. benthic oxygen demand, nitrification of ammonia),
- oxygen sources (e.g. re-aeration, atmospheric exchange, primary production), and
- temperature.

Downstream of sewage effluent discharges from treatment plants using biological, secondary processes, bacterial activity may also need to be incorporated into the models.

6.8 Summary and conclusions

Rivers have to support a wide variety of activities including water supply for various uses (drinking water and irrigation of agricultural land being amongst the most important). Progressive urbanisation and industrial development has also led to increasing use of rivers for waste disposal activities. The pollution arising from these and other sources, such as use of agricultural pesticides, has led to the increasing need for rigorous assessment of river water quality. The complexity and components of such assessment programmes are defined by the water uses and their water quality requirements, as well as a need to protect the aquatic environment from further degradation.

Rivers are dynamic systems which respond to the physical characteristics of the watershed, which in turn are controlled by the local and regional geological and climatic conditions. The size of the watershed controls the fluctuations in water level, velocity and discharge. Extreme or rapid fluctuations are dampened as watershed size increases. The flow characteristics of a river are important to the understanding of the water mixing processes in the river channel, i.e. in association with laminar flow, helical overturn and turbulent mixing in channels with rapids and waterfalls. A basic knowledge of these processes is necessary for the correct siting of sampling stations within the watershed.

Determination of river discharge is extremely important for the measurement of the flux of material carried by the river and transported to downstream receiving waters. The changing concentrations of chemical variables relative to the changing volume and velocity of river waters can provide useful diagnostic information on the origins of contaminants. In general, with increasing discharge point sources of contaminants are diluted whereas diffuse sources show increased concentrations. Sediment related variables fluctuate with suspended solids concentrations which in turn are related to discharge.

Rivers can be characterised by particular communities of organisms which are dependent on certain conditions of discharge and the physical, chemical and structural effects that it has on the river bed and water quality. Changes in the structure or quality of the river resulting from anthropogenic activities often produce specific changes in the biological communities which, once identified, can be used to monitor changes in the river environment.

The media (water, sediment or biota) which are selected for monitoring depend on the environmental properties of the chemical variables of interest and the objectives of the programme. Sediment associated variables such as phosphorus, trace elements and some organic pollutants may be best evaluated by collection of suspended sediment samples using filtration or centrifugation methods. Contaminants occurring at very low dissolved concentrations, but which are accumulated in biota, may be studied by the collection and analysis of biological material, particularly if they are also accumulated by food organisms for potential human consumption.

6.9   References

Amoros, C. and Petts, G.E. [Eds] 1993 *Hydrosystèmes fluviaux*. Masson, Paris, 320 pp.

Arceivala, S.J. 1981 *Wastewater Treatment and Disposal — Engineering and Ecology in Pollution Control*. Pollution Engineering and Technology No. 15., Marcel Dekker Inc., New York.

Bartram J. and Ballance, R. [Eds] 1996 *Water Quality Monitoring: A Practical Guide to the Design and Implementation of Freshwater Quality Studies and Monitoring Programmes*. Chapman & Hall, London.

Berner, E.A. and Berner, R.A. 1987 *The Global Water Cycle, Geochemistry and Environment*. Prentice Hall, Englewood Cliffs, Mich., 397 pp.

Billen, G. 1990 GMMA, Université Libre de Bruxelles. Personal communication.

Boelens, R.G. 1987 The use of fish in water pollution studies. In: D.H.D. Richardson [Ed.] *Biological Indicators of Pollution*. Royal Irish Academy, Dublin, 89-109.

Burrus, D., Thomas, R.L., Dominik, J. and Vernet, J.-P. 1989 Recovery and concentration of suspended sediment in the Upper Rhône River by continuous flow centrifugation. *Hydrological Processes*, **3**, 65-74.

Burrus, D., Thomas, R.L., Dominik, B., Vernet, J.-P. and Dominik, J. 1990 Characteristics of suspended sediment in the Upper Rhône River, Switzerland, including the particulate forms of phosphorus. *Hydrological Processes*, **4**, 85-98.

Caffrey, J.M. 1987 Macrophytes as biological indicators of organic pollution in Irish rivers. In: D.H.D. Richardson [Ed.] *Biological Indicators of Pollution*. Royal Irish Academy, Dublin, 77-87.

Chevreuil, M., Chesterikoff, A. and Létolle, R. 1988 Modalités du transport des PCB dans la rivière Seine (France). *Sciences de l'Eau*, **1**, 321-337.

Corvi, C., Majeaux, C. and Vogel, J. 1986 Les polychlorobiphényles et le DDE dans les sédiments superficiels du Léman et de ses effluents, campagne 1985. *Report of the International Commission for the Protection of the Waters of Lake Geneva*, Lausanne, 207-216.

Cossa, D., Institute Français de Recherches et d'Etudes Marines, Nantes, and Ficht, A., Service de la Navigation de la Seine, Rouen 1991 Personal communication.

Cummins, K.W. 1975 Macroinvertebrates. In: B.A. Whitton [Ed.] *River Ecology*. Studies in Ecology, Volume 2, Blackwell Scientific Publications, Oxford, 170-198.

Dessery, S., Dulac, C., Lawrenceau, J.M. and Meybeck, M. 1984 Evolution du carbone organique particulaire algal et détritique dans trois rivières du Bassin Parisien. *Arch. Hydrobiol.*, **100**(2), 235-260.

Elliott, J.M. 1977 *Some Methods for the Statistical Analysis of Samples of Benthic Invertebrates*. 2nd edition, Freshwater Biological Association Scientific Publication No. 25, Freshwater Biological Association, Ambleside, UK, 156 pp.

FEEMA 1987 *Qualidade das Aguas do Estado do Rio de Janiero, Brazil*. Fundaçao de Tecnologia de Saneamento Ambiental, Rio de Janiero.

Friedrich, G. and Viehweg, M. 1987 Measurement of chlorophyll fluorescence within Rhine monitoring - results and problems. *Arch. Hydrobiol.*, Suppl. 29, 117-122.

Hawkes, H.A. 1975 River zonation and classification. In: B.A. Whitton [Ed.] *River Ecology.* Studies in Ecology, Volume 2, Blackwell Scientific Publications, Oxford, 312-374.

Hem, J.D. 1989 *Study and Interpretation of the Chemical Characteristics of Natural Water.* U.S. Geological Survey Water Supply Paper No. 2254, 263 pp.

Herschy, R.W. [Ed.] 1978 *Hydrometry. Principles and Practices.* John Wiley and Sons, Chichester, 511 pp.

Hines, W.G., Rickert, D.A. and McKenzie, S.W. 1976 *Hydrologic Analysis and River Quality Data Programs.* US Geological Survey Circular 715-D, United States Geological Survey, Arlington, VA., 20 pp.

IKSR 1987 *Rhine Action Plan.* International Commission for the Protection of the Rhine Against Pollution, Koblenz, Germany.

Irion, G. 1982 Sedimentdatierung durch anthropogene Schwermetalle. *Natur und Museum*, **112**, 183-189.

Jackson, T. 1980 Mercury speciation and distribution in a polluted river-lake system as related to the problems of lake restoration. In: *Proceedings of the International Symposium for Inland Waters and Lake Restoration.* United States Environmental Protection Agency/Organisation for Economic Cooperation and Development, Portland, Maine, 93-101.

Jones, H.G. and Bisson, M. 1984 Physical and chemical evolution of snow packs in the Canadian Shield (winter 1979 - 1980). *Verh. Internat. Ver. Limnol.*, **22**, 1786-1792.

Krenkel, P.A. and Novotny, V. 1980 *Water Quality Management.* Academic Press, New York, 671 pp.

Meybeck, M. 1979 Concentrations des eaux fluviales en éléments majeurs et apports en solution aux oceans. *Rev. Géol. Dyn. Géogr. Phys.*, **21**(3), 215-246.

Meybeck, M. 1982 Carbon, nitrogen and phosphorus transport by world rivers. *Am. J. Sci.*, **282**, 401-450.

Meybeck, M. 1986 Composition chimique naturelle des ruisseaux non pollués en France. *Sci. Géol. Bull. Strasbourg*, **39**, 3-77.

Meybeck, M. 1987 Global chemical weathering of surficial rocks estimated from river dissolved loads. *Am. J. Sci.*, **287**, 401-428.

Meybeck, M. 1988 How to establish and use world budgets of riverine materials. In: A. Lerman and M. Meybeck [Eds] *Physical and Chemical Weathering in Geochemical Cycles.* Kluver, Dordrecht, 247-272.

Meybeck, M., Chapman, D. and Helmer, R. [Eds] 1989 *Global Freshwater Quality: A First Assessment.* Blackwell Reference, Oxford, 306 pp.

Meybeck, M. and Helmer, R. 1989 The quality of rivers: from pristine stage to global pollution. *Palaeogeogr., Palaeoclimatol., Palaeoecol. (Global Planet. Change Sect.)*, **75**, 283-309.

Müller, D. and Kirchesch, V. 1980 Der Einfluß von Primärproduktion und Nitrifikation auf den Stoffhaushalt des staugeregelten Flusses. *Münchener Beiträge zur Abwasser-, Fischerei- und Flußbiologie*, **32**, 279-298.

Reeder, S.W., Hitchon, B. and Levinson, A.A. 1972 Hydrogeochemistry of the surface waters of the Mackenzie river drainage basin, Canada. I. Factors controlling inorganic composition. *Geochim. Cosmochim. Acta*, **36**, 825-865.

Rickert, D.A. 1984 Use of dissolved oxygen modelling results in the management of river quality. *J. Wat. Pollut. Control Fed.*, **56**(1), 94-101.

Santiago, S., Thomas, R.L., Loizeau, J.-L., Favarger, P.-Y and Vernet, J.P. 1990 Further discussion on the intercomparison of the trace metal concentrations and particle size of fluvial sediment recovered from two centrifuge systems. *Hydrological Processes*, **4**, 283-287.

Schiller, A.M. and Boyle, E.A. 1985 Dissolved zinc in rivers. *Nature*, **317**, 49-52.

Schiller, A.M. and Boyle, E.A. 1987 Variability of dissolved trace metals in the Mississippi River. *Geochim. Cosmochim. Acta*, **51**, 3273-3277.

Schmitz, P., Krebs, F. and Irmer, U. 1994 Development, testing and implementation of automated biotests for the monitoring of the River Rhine, demonstrated by bacteria and algae tests. *Wat. Sci. Tech.* **29**(3), 215-221.

Schmitz, W. 1989 River monitoring networks in the Federal Republic of Germany. In: W. Lampert and K. Rothhaupt [Eds] *Limnology in the Federal Republic of Germany*. International Association for Theoretical and Applied Limnology, Plön, 143-145.

Stallard, R.F. 1980 *Major Element Geochemistry of the Amazon River System*. MIR-WHOI, Woods Hole, WHOI-80, **29**, 362 pp.

Thomas, R.B. 1986 Calibrating SALT: a sampling scheme to improve estimates of suspended sediment yield. In: D. Lerner, [Ed.] *Monitoring to Detect Changes in Water Quality Series*, IAHS Publ. **157**, International Association for Hydrological Sciences, Wallingford, 79-88.

UNESCO 1982 *Dispersion and Self-purification of Pollutants in Surface Water Systems*. Technical Papers in Hydrology 23, United Nations Educational, Scientific and Cultural Organization, Paris, 98 pp.

UNESCO 1988 *Hydrological Processes and Water Management in Urban Areas*. Proceedings of Conference, Duisburg, 1988, United Nations Educational, Scientific and Cultural Organization, Paris.

Vannote, R.L., Minshall, G.W., Cummins, K.W., Sedell, J.R. and Cushing, C.E. 1980 The river continuum concept. *Can. J. Fish. Aquat. Sci.*, **37**, 130-137.

Walling, D.E. and Webb, B.W. 1988 The reliability of rating curve estimates of suspended sediment yield: some further comments. In: *Sediments Budgets*,

IAHS Publ. **174**, International Association for Hydrological Sciences, Wallingford.

WHO 1992 *GEMS/WATER Operational Guide*. Third edition. World Health Organization, Geneva.

WMO 1980 *Manual on Stream Gauging*. Publication No. 519. World Meteorological Organization, Geneva.

Zhang Li 1988 *Evaluation of Phosphorus Loadings to Lake Geneva from the Venoge and other Rivers, Switzerland*. Thesis No. 2333, University of Geneva, 104 pp.

Chapter 7*

LAKES

7.1 Introduction

The sampling of lakes for the purpose of assessing water quality is a complex process, as is the interpretation of the data obtained. Consequently, these activities must be based on a relatively complete understanding of basic limnology. The strategies employed for sampling and data interpretation are also controlled by lake use, the issue or lake problem being addressed, and the availability of resources for undertaking an assessment programme.

A detailed discussion of limnology is beyond the scope of this book and the reader is referred to the many texts which cover the subject in detail such as Hutchinson (1957), Golterman (1975), and Wetzel (1975). This chapter presents the more detailed descriptions and examples of processes which are essential to provide sufficient understanding to aid the design of sampling programmes and sensible interpretation of monitoring data. Further examples of the assessment of water quality in lakes using particulate material or biological methods are given in Chapters 4 and 5.

A lake may be defined as an enclosed body of water (usually freshwater) totally surrounded by land and with no direct access to the sea. A lake may also be isolated, with no observable direct water input and, on occasions, no direct output. In many circumstances these isolated lakes are saline due to evaporation or groundwater inputs. Depending on its origin, a lake may occur anywhere within a river basin. A headwater lake has no single river input but is maintained by inflow from many small tributary streams, by direct surface rainfall and by groundwater inflow. Such lakes almost invariably have a single river output. Further downstream in river basins, lakes have a major input and one major output, with the water balance from input to output varying as a function of additional sources of water.

Lakes may occur in series, inter-connected by rivers, or as an expansion in water width along the course of a river. In some cases the distinction between a river and a lake may become vague and the only differences may relate to changes in the residence time of the water and to a change in water circulation within the system. In the downstream section of river basins, lakes (as noted above) are separated from the sea by the hydraulic gradient of the river, or estuarine system. The saline waters of the Dead Sea, the Caspian and Aral

---

* This chapter was prepared by R. Thomas, M. Meybeck and A. Beim

seas are, therefore, strictly lakes whereas the Black Sea, with a direct connection to the Mediterranean via the Sea of Marmara, is truly a sea.

Lakes are traditionally under-valued resources to human society. They provide a multitude of uses and are prime regions for human settlement and habitation. Uses include drinking and municipal water supply; industrial and cooling water supply; power generation; navigation; commercial and recreational fisheries; body contact recreation, boating, and other aesthetic recreational uses. In addition, lake water is used for agricultural irrigation, canalisation and for waste disposal. It has been commonly believed that large lakes have an infinite ability to absorb or dilute industrial and municipal waste, and it is largely as a result of human waste disposal practices that monitoring and assessment are proving to be necessary in many large lakes.

Good water quality in lakes is essential for maintaining recreation and fisheries and for the provision of municipal drinking water. These uses are clearly in conflict with the degradation of water induced by agricultural use and by industrial and municipal waste disposal practices. The management of lake water quality is usually directed to the resolution of these conflicts. Nowhere in the world has lake management been a totally successful activity. However, much progress has been made particularly with respect to controllable point source discharges of waste. The more pervasive impacts of diffuse sources of pollution within the watershed, and from the atmosphere, are less manageable and are still the subject of intensive investigations in many parts of the world.

7.2   Characteristics and typology

7.2.1 Origins of lakes

In geological terms lakes are ephemeral. They originate as a product of geological processes and terminate as a result of the loss of the ponding mechanism, by evaporation caused by changes in the hydrological balance, or by infilling caused by sedimentation. The mechanisms of origin are numerous and are reviewed by Hutchinson (1957), who differentiated 11 major lake types, sub-divided into 76 sub-types. A full discussion is beyond the scope of this chapter, but a summary of the 11 major types of lake origin is given below (Meybeck, 1995).

Glacial lakes: Lakes on or in ice, ponded by ice or occurring in ice-scraped rock basins. The latter origin (glacial scour lakes) contains the most lakes. Lakes formed by moraines of all types, and kettle lakes occurring in glacial drift also come under this category. Lakes of glacial origin are by far the most numerous, occurring in all mountain regions, in the sub-arctic regions and on

Pleistocene surfaces. All of the cold temperate, and many warm temperate, lakes of the world fall in this category (e.g. in Canada, Russia, Scandinavia, Patagonia and New Zealand).

Tectonic lakes: Lakes formed by large scale crustal movements separating water bodies from the sea, e.g. the Aral and Caspian Seas. Lakes formed in rift valleys by earth faulting, folding or tilting, such as the African Rift lakes and Lake Baikal, Russia. Lakes in this category may be exceptionally old. For example, the present day Lake Baikal originated 25 million years ago.

Fluvial lakes: Lakes created by river meanders in flood plains such as oxbow and levee lakes, and lakes formed by fluvial damming due to sediment deposition by tributaries, e.g. delta lakes and meres.

Shoreline lakes: Lakes cut off from the sea by the creation of spits caused by sediment accretion due to long-shore sediment movement, such as for the coastal lakes of Egypt.

Dammed lakes: Lakes created behind rock slides, mud flows and screes. These are lakes of short duration but are of considerable importance in mountainous regions.

Volcanic lakes: Lakes occurring in craters and calderas and which include dammed lakes resulting from volcanic activity. These are common in certain countries, such as Japan, Philippines, Indonesia, Cameroon and parts of Central America and Western Europe.

Solution lakes: Lakes occurring in cavities created by percolating water in water-soluble rocks such as limestone, gypsum or rock salt. They are normally called *Karst lakes* and are very common in the appropriate geological terrain. They tend to be considered as small, although there is some evidence that some large water bodies may have originated in this way (e.g. Lake Ohrid, Yugoslavia).

Excluding reservoirs, many other natural origins for lakes may be defined, ranging from lakes created by beaver dams to lakes in depressions created by meteorite impact.

7.2.2 Classification of lakes

As noted in the brief discussion above the first level of classification of lakes is defined by their origin. However, in the context of lake use and assessments such a classification is of little value. Two other systems of classification which are based upon processes within lakes, and which are used universally, provide the basis upon which assessment strategies and interpretation are based. These are the physical or thermal lake classification and the classification by trophic level.

7.2.3 Physical/thermal lake types

The uptake of heat from solar radiation by lake water, and the cooling by convection loss of heat, result in major physical or structural changes in the water column. The density of water changes markedly as a function of temperature, with the highest density in freshwater occurring at 4 °C. The highest density water mass usually occurs at the bottom of a lake and this may be overlain by colder (0–4 °C) or warmer (4–30 °C) waters present in the lake. A clear physical separation of the water masses of different density occurs and the lake is then described as being stratified. When surface waters cool or warm towards 4 °C, the density separation is either eliminated or reaches a level where wind can easily induce vertical circulation and mixing of the water masses producing a constant temperature throughout the water column. In this condition the lake is termed homothermal and the process is defined as vertical circulation, mixing, or overturn.

The nomenclature applied to a stratified lake is summarised in Figure 7.1 in which three strata are defined:

- the epilimnion or surface waters of constant temperature (usually warm) mixed throughout by wind and wave circulation,
- the deeper high density water or hypolimnion (this is usually much colder, although in tropical lakes the temperature difference between surface and bottom water may be only 2–3 °C), and
- a fairly sharp gradational zone between the two which is defined as the metalimnion.

The name metalimnion is not commonly used and the gradation is normally referred to as the thermocline. The thickness of the epilimnion may be quite substantial, and it is dependent on the lake surface area, solar radiation, air temperature and lateral circulation and movement of the surface water. Commonly, it extends to about 10 m depth but in large lakes it can extend up to 30 m depth. Stratification in very shallow lakes is generally rare since they have warm water mixing throughout their water column due to wind energy input. However, winter or cold water stratification can occur even in the most shallow lakes under the right climatic conditions.

The interpretation of a shallow lake has never been satisfactorily defined, although there is a relationship between lake depth and surface area which controls the maximum depth to which wind induced mixing will occur. Therefore, an acceptable definition of a shallow lake (for the purposes of this discussion) is one which will overturn and mix throughout its water column when subjected to an average wind velocity of 20 km h^{-1} for more than a six hour period. As a general rule, wind exposed lakes of 10 m depth or less are defined as shallow water lakes.

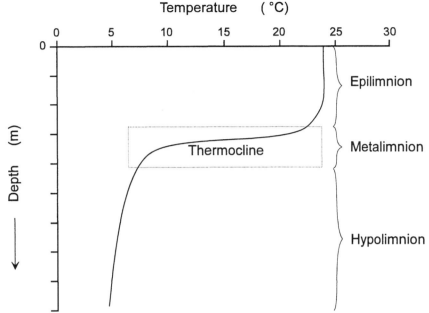

Figure 7.1 Typical temperature profile from a stratified lake in the temperate zone, showing the division of the water into three layers of different density

The thermal characteristics of lakes are a result of climatic conditions that provide a useful physical classification which is based upon the stratification and mixing characteristics of the water bodies. These characteristics are illustrated in Figure 7.2 with the lake types and terminologies defined as below. *Dimictic lakes* occur in the cool temperate latitudes. Overturn occurs twice a year, normally in the spring and autumn. Heating in the spring results in stratification with a warm water epilimnion during the summer. The autumn overturn results in homothermal conditions (at approximately 4 °C) which then cool to create a cold water inverse stratification during the winter months. Spring warming results in mixing and a re-establishment of the annual cycle. The stratification and mixing processes for a large dimictic lake are illustrated in Figure 7.3. This type of lake is the most common form of lake. Since the cool temperate latitudes encompass most of the world's industrial nations they have been subjected to the most intensive study and represent the greatest part of our limnological knowledge.

Cold monomictic lakes occur in cold areas and at high altitudes (sub-polar). The water temperature never exceeds 4 °C and they have a vertical temperature profile close to, or slightly below, 4 °C. They have winter stratification with a cold water epilimnion, often with ice cover for most of the year, and mixing occurs only once after ice melt.

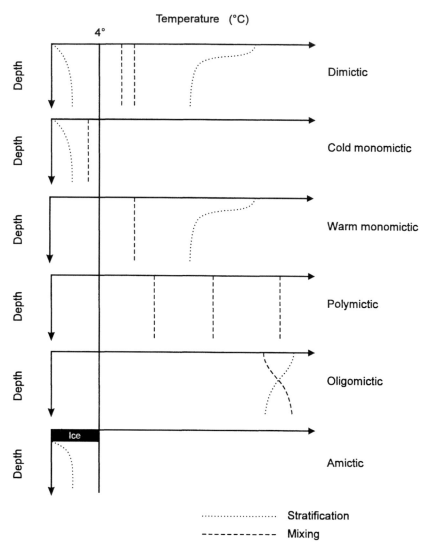

Figure 7.2 Lake thermal structure and classification based on mixing characteristics (After Häkanson and Jansson, 1983)

Warm monomictic lakes occur in temperate latitudes in subtropical mountains and in areas strongly influenced by oceanic climates. In the same way as their cold water counterparts, they mix only once during the year with temperatures that never fall below 4 °C.

Polymictic lakes occur in regions of low seasonal temperature variations, subject to rapidly alternating winds and often with large daily (diurnal) temperature variations. These lakes have frequent periods of circulation and

Temperature (°C)

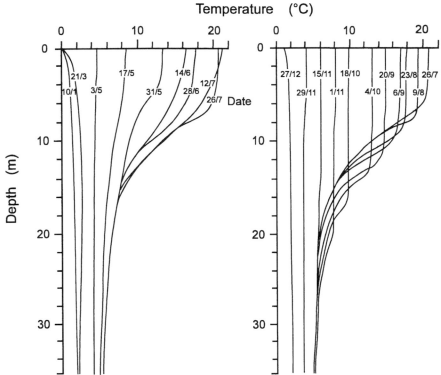

Figure 7.3 Annual thermal stratification and mixing in a dimictic lake (After Häkanson and Jansson, 1983)

mixing and may be subdivided into cold polymictic, which circulate at temperatures close to 4 °C, and warm polymictic which circulate at higher temperatures. As defined above, all shallow lakes fall within this category.

Oligomictic lakes occur in tropical regions and are characterised by rare, or irregular, mixing with water temperatures well above 4 °C.

Amictic lakes occur in the polar regions and at high altitudes. They are always frozen and never circulate or mix. Waters beneath the ice are generally at, or below, 4 °C depending on the amount of heat generated from the lake bed or by solar radiation through the ice. These lakes show an inverse cold water stratification.

In addition to the lake types noted above some additional characteristics and terminology need to be defined. Meromictic lakes are those which do not undergo complete mixing throughout the water column. Most deep lakes in tropical regions are meromictic, such as Lakes Tanganyika and Malawi. Complete mixing as described above characterises holomictic lakes. When lake stratification is due to density changes caused by salt concentrations

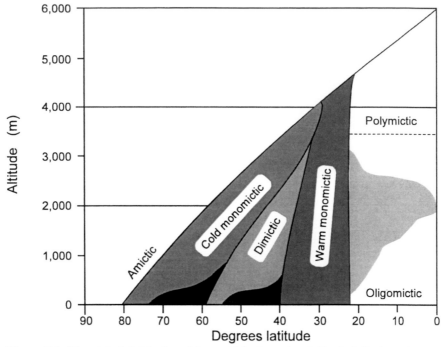

Figure 7.4   The global distribution of thermal lake types in relation to latitude and altitude (After Wetzel, 1975)

(normally the case in meromictic lakes), the gradient separating the upper layer from the denser layer is termed a chemocline or halocline, as distinct from temperature separation by a thermocline.

The physical or thermal classification of lakes, as described above, is largely climate controlled and is, therefore, related to latitude and altitude as summarised graphically by Wetzel (1975) (Figure 7.4).

7.2.4 Trophic status

The concept of trophic status as a system of lake classification was introduced by early limnologists such as Thienemann (1925, 1931) and Nauman (1932), and has been subject to continuous development up until the present time (Vollenweider, 1968; Pourriot and Meybeck, 1995). The process of eutrophication underlying this scheme is one of the most significant processes affecting lake management and is, therefore, described in more detail. The underlying concept is related to the internal generation of organic matter which is also known as autotrophic production. External inputs of organic matter from the watershed (allotrophy) produce dystrophic lakes rich in humic materials (see Figure 7.5). Such lakes may also be termed brown water

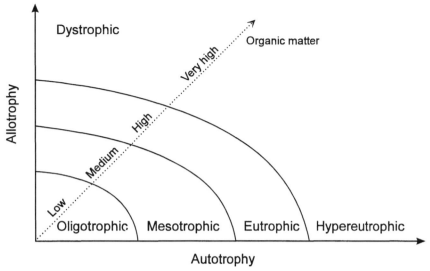

Figure 7.5 Trophic lake classification based on the primary source of carbon (Modified from Häkanson and Jansson, 1983)

or polyhumus lakes. In these lakes, most of the organic matter is derived from the surrounding watershed and internal carbon production is generally low.

The continuous process of eutrophication results from autotrophic production of internal organic matter by primary producers (i.e. the photosynthetic plants and algae) from the nutrients available within the lake. Nutrients are derived from external inputs to the lake or by internal recycling from the decay of organic matter and dissolution from bottom sediments. The process is illustrated schematically, in a simplified form, in Figure 7.6. Eutrophic lakes range from oligotrophic to hypereutrophic (Figure 7.5). In many shallow lakes, eutrophication may be manifest by macrophyte growth, rather than in the growth of phytoplankton. However, the efficient utilisation of the nutrients depends on the interplay of a number of factors which together define the growth conditions, and hence the resultant total biomass production at the primary producer level (see causes in Figure 7.6). The grazing of phytoplankton by zooplankton (secondary producers) and predation by fish (tertiary consumers) constitute the carbon transfer system of the lake. The efficiency of the system is dependent on two factors. Firstly, the quantity of biomass created at the primary producer level and, secondly, the species composition which determines the efficiency of grazing and the quantity and quality of the fish which terminate the internal food chain. With death, organisms at the primary, secondary and tertiary levels decay, resulting in the recycling of nutrients to the lake system.

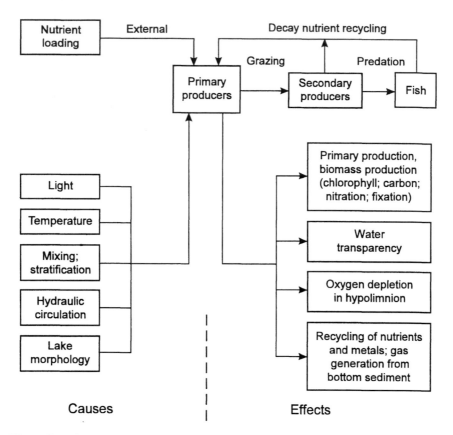

Figure 7.6 Causes and effects of eutrophication

The effects of eutrophication can be highly detrimental to lake water quality and severely limit the uses for which the water is suitable. Some effects are listed in Figure 7.6. However, it should be recognised that the detrimental effects result from the inefficient use of the phytoplankton biomass which, in turn, is derived from high nutrient availability. This is a result of changes in the dominance of algal species which are not consumed, or are ineffectively consumed, by zooplankton grazers.

The classification of lakes based on trophic level is shown in Figure 7.5. As mentioned above, these represent a continuous range of nutrient concentrations and associated biomass production. The names given to the classifications represent empirically defined intervals ranging from very low to very high productivity, which are defined below.

Oligotrophic lakes: Lakes of low primary productivity and low biomass associated with low concentrations of nutrients (N and P). In temperate regions the fish fauna is dominated by species such as lake trout and whitefish. These lakes tend to be saturated with O_2 throughout the water column.

Mesotrophic lakes: These lakes are less well defined than either oligotrophic or eutrophic lakes and are generally thought to be lakes in transition between the two conditions. In temperate regions the dominant fish may be whitefish and perch. Some depression in O_2 concentrations occurs in the hypolimnion during summer stratification.

Eutrophic lakes: Lakes which display high concentrations of nutrients and an associated high biomass production, usually with a low transparency. In temperate regions the fish communities are dominated by perch, roach and bream. Such lakes may also display many of the effects which begin to impair water use. Oxygen concentrations can get very low, often less than 1 mg l⁻¹ in the hypolimnion during summer stratification.

Hypereutrophic lakes: Lakes at the extreme end of the eutrophic range with exceedingly high nutrient concentrations and associated biomass production. In temperate regions the fish communities are dominated by roach and bream. The use of the water is severely impaired as is described below. Anoxia or complete loss of oxygen often occurs in the hypolimnion during summer stratification.

Dystrophic lakes: As defined previously, these are organic rich lakes (humic and fulvic acids) with organic materials derived by external inputs from the watershed.

Summary characteristics for each of these trophic lake types are given in Table 7.1.

7.2.5 Water balance

The water balance of a lake may be expressed simply as:

Input (major river + lake tributaries + precipitation) - Evaporation ± Groundwater = Output

Seasonal climate changes, particularly rainfall and solar heating, result in seasonal variations in the water balance producing a predictable seasonal variation in water level. Such fluctuations may affect water use, and many lakes have man-made control structures to dampen the water level fluctuations and optimise year-round water use. Longer-term changes in climate, such as dry and warm years as opposed to cool and wet years (particularly in the temperate zones), have major effects on the water levels of a lake. Considerable impairment of water use may occur as a result of these long-term cycles (e.g. navigation/dredging, coastal erosion, shifts in biological habitat, flooding and loss of wetlands).

The concept of water residence time (or turnover time) is particularly important. Residence time may be expressed theoretically as the lake water volume divided by the rate of total outflow. Residence times in normal lakes

Table 7.1 Nutrient levels, biomass and productivity of lakes at each trophic category

Trophic category	Mean total phosphorus (mg m^{-3})	Annual mean chlorophyll (mg m^{-3})	Chlorophyll maxima (mg m^{-3})	Annual mean Secchi disc transparency (m)	Secchi disc transparency minima (m)	Minimum oxygen (% sat^h)[1]	Dominant fish
Ultra-oligotrophic	4.0	1.0	2.5	12.0	6.0	< 90	Trout, Whitefish
Oligotrophic	10.0	2.5	8.0	6.0	3.0	< 80	Trout, Whitefish
Mesotrophic	10–35	2.5–8	8–25	6–3	3–1.5	40–89	Whitefish, Perch
Eutrophic	35–100	8–25	25–75	3–1.5	1.5–0.7	40–0	Perch, Roach
Hypereutrophic	100.0	25.0	75.0	1.5	0.7	10–0	Roach, Bream

[1] % saturation in bottom waters depending on mean depth

Sources: Häkanson, 1980; Häkanson and Jansson, 1983; Meybeck *et al.*, 1989

extend from a few days to many tens of years, or even a century or more. However, the theoretical residence time may be expected to occur only rarely, since it is based on the homogeneous mixing of lake waters. Depending on their thermal structure, deep water mixing and water circulation characteristics, most lakes may be characterised at any one time by different water masses which change from season to season, and possibly from year to year. Warm water in the summer epilimnion overflows the colder water hypolimnion and has a much shorter residence time than a mixed water mass. Conversely, deep waters with poor circulation have a much longer residence time than the theoretical value (Meybeck, 1995).

The concept of residence time is important because it provides some indication of the recovery, or self-purification rates, for individual water bodies. If a water body becomes polluted with a soluble toxic element, and the source of the pollutant is entirely eliminated, the residence time provides an indication of how long it will take to remove the polluted water and replace it with non-polluted water. For Lake Superior, elimination of a soluble toxic element would take 100 years, whereas for Lake Erie it would take only 2.4 years. However, this is an over-simplification since it does not take account of lake structure or the geochemistry of the element. Most shallow lakes (< 5 m deep) have a short residence time of one year or less.

7.3 Water quality issues

A water quality issue may be defined as a water quality problem or impairment which adversely affects the lake water to an extent which inhibits or prevents some beneficial water use. Since a water quality issue normally results from the deleterious effects of one or more human uses, major conflicts between users or uses may occur in lake systems subjected to multiple use. These kinds of water quality conflicts and associated human interactions bring considerable complexity to lake management. This often results in continuous arbitration between user-groups, and failure to take the necessary control actions to restore or maintain lake water quality. Since the problems of water quality relate to water use, water quality assessment strategies should be designed in these terms. The interpretation should be dependent on adequate knowledge of lake physics and structure, lake uses and associated water quality requirements, and the legislative powers or authorities which may be used to enforce and ensure compliance with standards. Monitoring data and the associated assessments are thus the basis upon which sound lake management should be conducted.

A number of issues affect lake water quality. These have mostly been identified and described in industrialised regions, such as the North American

Great Lakes, which have progressed from one issue to another in a sequence parallel to social and industrial development (Meybeck *et al.*, 1989). Meybeck *et al.* (1989) have highlighted that the developing nations of today are responsible for managing lakes which are being subjected to synchronous pollution from the simultaneous evolution of rural, urban and industrial development, as distinct from the historical progression which occurred in developed societies. This means that, in the developing world, multi-issue water quality problems must be faced with greater cost and complexity in assessment design, implementation and data interpretation. The current issues facing lake water quality are discussed below.

7.3.1 Eutrophication

There is much published literature available on the subject of eutrophication, both in specialised reports such as OECD (1982) and in the major limnological texts recommended in the introduction to this chapter. Numerous indices have been developed to measure the degree of eutrophication of water bodies. Many have been based on phytoplankton species composition, but are not recommended since they are complicated to undertake, difficult to interpret and are affected by local conditions. The most reliable methods are based on the classifications given in Table 7.1. Suitable alternatives for biological indicators are total particulate organic carbon (POC) and chlorophyll *a*, since they represent total biomass.

In its simplest expression, eutrophication is the biological response to excess nutrient inputs to a lake. The increase in biomass results in a number of effects which individually and collectively result in impaired water use. These effects are listed in Figure 7.6. Meybeck *et al.* (1989) highlight that eutrophication is a natural process which, in many surface waters, results in beneficial high biomass productivity with high fish yields. Accelerated, or human-induced, change in trophic status above the natural lake state is the common cause of the problems associated with eutrophication. Such human-induced changes may occur in any water body, including coastal marine waters, although the progression and effects of eutrophication are also mediated by climate. As a result warm tropical and sub-tropical lakes are more severely affected than their colder water counterparts.

High nutrient concentrations in a lake are derived from external inputs from the watershed. The final biomass attained is determined primarily by the pool of nutrients available for growth at the beginning of the growing season. The primary nutrients, such as nitrogen and phosphorus, are used until growth is complete and the exhaustion of the pool of either one of them places a final limit on the phytoplankton growth. By definition, the nutrient

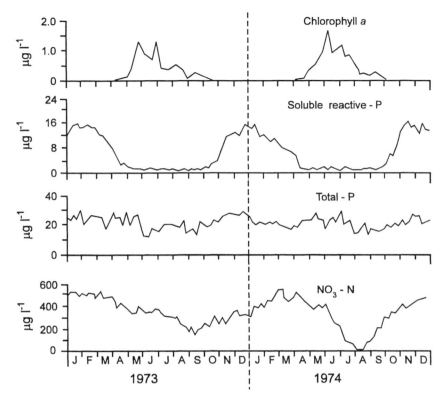

Figure 7.7 Seasonal cycling of nutrients and chlorophyll *a* in the surface waters of Lake Windermere, UK (After ILEC, 1987–1989)

which is exhausted is the limiting nutrient in any lake system. Meybeck *et al.* (1989) suggest that, in waters with a N/P ratio greater than 7 to 10, phosphorus will be limiting, whereas nitrogen will be limiting in lakes with a N/P ratio below 7.

Figure 7.7 shows the seasonal cycling of nutrient concentrations (soluble reactive P, total P, and NO_3-N) relative to primary production (as expressed by chlorophyll *a*) in the epilimnion of Lake Windermere, UK. Primary production occurs from April to late September, inversely to the depletion in soluble reactive P and NO_3-N. Soluble reactive P is reduced rapidly, virtually to zero, from a maximum concentration of about 16 μg l^{-1}, indicating that growth in this lake is limited by phosphorus.

Changes in transparency in lakes may be caused by increasing turbidity due to increasing concentrations of mineral material or to increasing plankton biomass. Increases in mineral matter are caused by:

- turbid in-flow in fluvial waters with high watershed erosion rates,

- resuspension of bottom sediment by wave action in shallow lakes, or shallow areas of lakes (when wave height and wavelength developed during storm events allow direct interaction with the bottom to take place), and
- shore line erosion due to wave impingement and to surface water gully erosion of unconsolidated shore line material.

High primary production is followed by death, settling, decay and deposition of the phytoplankton to the lake sediment. It has been calculated that in the decay process 1.5 to 1.8 g of oxygen is used per gram dry weight of mineralised organic matter. Oxygen is taken up from the water and, under highly eutrophic conditions, serious or complete depletion of hypolimnetic oxygen may occur with severe impacts on the lake system. Depletion or loss of oxygen in the hypolimnion of lakes used for drinking water supply presents a particularly serious problem. When combined with undesirable plankton species, it produces serious taste and odour problems (see Chapter 8 for further details). In addition, oxygen depletion and reducing conditions at the sediment–water interface result in the recycling of iron, manganese and related trace elements from the bottom sediments.

Figure 7.8 shows the seasonal change in dissolved oxygen throughout the water column of Lake Mendota, Minnesota together with the chlorophyll *a* values for the same period. Oxygen in the deeper waters progressively declines following the onset of primary production in April to a minimum of below 1 mg l^{-1} in the hypolimnion for July to October, during lake stratification. The system is re-oxygenated in late October following overturn and complete mixing of the lake water. In many lakes oxygen is completely used up and the hypolimnetic waters become anoxic. A typical example was the anoxia in the hypolimnion of the Central Basin of Lake Erie in the 1960s which led to the massive clean up and phosphorus control programme of the Lower Great Lakes of North America between 1972 and 1981. Lake Erie is a dimictic lake of the cold temperate zone similar to Lake Mendota.

Deep water anoxia may be more prevalent in highly productive tropical lakes. Lake Tanganyika is a warm, amictic lake with a stable thermocline and epilimnion over a hypolimnion with a constant temperature of 23.5 °C. Oxygen, which is saturated in the surface water, declines throughout the water column with the decay of settling organic detritus. Oxygen is totally depleted at about 100 m depth and H_2S becomes abundant, resulting in the deeper waters becoming uninhabitable for fish.

Depletion of oxygen in lake bottom waters and the onset of anoxia results in the re-mobilisation of phosphorus and other elements from lake sediments. This mechanism was described by Mortimer (1942) and has been the subject of many investigations. Project Hypo (Burns and Ross, 1971) followed the

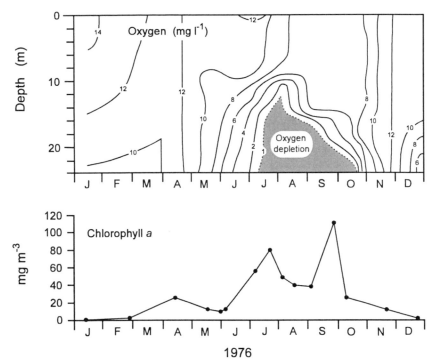

Figure 7.8 Oxygen concentrations and chlorophyll *a* in Lake Mendota, Minnesota during 1976. Depletion of oxygen in the hypolimnion in July, August and September is related to the period of high algal biomass (indicated by chlorophyll *a*) and is caused by biological degradation of the algae as they sink to the bottom (After ILEC, 1987–1989)

release of phosphorus during anoxia in the hypolimnion of Lake Erie. It was also noted that with decreasing redox potential, manganese was the first element to be released from bottom sediments to the hypolimnion, followed by the synchronous release of iron and phosphorus. The association of phosphorus with iron in bottom sediments has been well established (e.g. Williams *et al.*, 1976) and the reduction of Fe^{3+} to soluble Fe^{2+} results in the release of phosphorus. Other elements adsorbed by, or co-precipitated with, the oxides of iron and manganese are also released to the bottom waters under anoxic conditions. Significant quantities of hydrogen sulphide are not generally seen in freshwaters and were not seen in Lake Erie (Burns and Ross, 1971). Sulphate occurs in much lower concentrations in freshwaters than in marine waters and consequently, when reduced, it tends to precipitate as iron monosulphide. When all the sulphate has been reduced to sulphide, bacterial energy is derived by methane generation. Large quantities of this gas are generated under anoxic conditions in freshwaters. In anoxic water, nitrogen is commonly found as NH_4^+. As with most reduced forms (Mn^{2+}, Fe^{2+}, H_2S, CH_4), the presence of NH_4^+ may severely impair the use of the water, particularly as a drinking water source (due to odour, taste, precipitation of metals upon re-aeration, etc.). This loss can be one of the most detrimental effects of eutrophication.

Phosphorus released to the hypolimnion usually remains in the deep water mass during the period of oxygen depletion. At overturn when the water is re-aerated, the phosphorus is mixed throughout the water column, precipitating as it becomes adsorbed onto the insoluble Fe II hydroxides and oxides and fine particles in the water column. In this manner phosphorus regenerated from bottom sediments (internal loading or secondary cycling) is made available for the subsequent algal growth cycle.

7.3.2 Health related issues and organic wastes

When located in densely populated areas, lakes may become polluted by inadequately treated, or untreated, human and animal wastes. This causes eutrophication, makes the water unfit for human consumption and increases health risks to water users. In an overview of Lake Managua in Nicaragua, Central America, Swaine (1990 pers. comm.) indicated that 1,587.3 deaths per 100,000 inhabitants in Nicaragua are due to waterborne diseases which include typhoid and paratyphoid fever, amoebiasis, viral jaundice, dysentery and various parasitic diseases. He noted that the first sewers were built in Managua in 1922 and that by 1981 about 50 per cent of the 600,000 inhabitants were connected to the sewer system. As a result, in 1988 approximately 150,000 m^3 of untreated sewage were discharged to the lake. Coliform counts were routinely found in the order of 10^6 per 100 ml although no direct effect on human health had been established. However, with the high population centred in Managua there was a major effect on the health statistics of the country as a whole.

In many tropical lakes, the onset and progression of eutrophication results in increased growth of lake macrophytes, which in turn provide increased habitat for water fowl and snails which serve as secondary hosts for many human parasites such as *Schistosoma*.

In addition to bacterial and viral infection from poor or untreated sewage disposal, other major problems may occur. Wastes high in organic matter directly increase the chemical and biological oxygen demand in the receiving waters. This results in localised areas of oxygen depletion and the release of many trace elements by the reduction of iron and manganese (see above). In industrial societies, factories are often directly connected to the sewers and discharge metal and organic chemical pollutants to the sewage treatment system. In many cases, such effluents are highly toxic to aquatic organisms and should be subjected to analysis using bioassays (see Chapter 5). If found to be toxic, they must be analysed for trace elements and organic pollutants.

7.3.3 Contaminants

The major shift in management concern from nutrients to the issue of toxic chemicals in lake systems is largely due to the realisation that mercury contamination in Minimata Bay, Japan was responsible for major human health problems (D'Itri, 1971) together with the work on mercury transfer processes in freshwater carried out in Sweden (Jernalov, 1971). In parallel with the mercury problem, major concerns over the pesticide DDT have been investigated. The observation that DDT inhibited the hatching of lake trout eggs in Lake Michigan was a major factor in banning the use of DDT in North America in 1970 (Swaine, 1991 pers. comm.). Subsequent investigations of the toxicity and environmental cycling of metals and organic pollutants have resulted in the implementation of a variety of management strategies to alleviate real and perceived problems. Within this context, assessment of lake systems should provide the information upon which managerial actions can be taken. Such information includes:

- presence or absence of a metal or compound,
- concentrations of pollutants in water, sediment and biota,
- spatial distributions showing areas in which objectives or guidelines are exceeded (i.e. areas of non-compliance with regulatory requirements), and
- temporal changes in concentration resulting from changing socio-economic conditions and demography, or changes resulting from management intervention.

Sources of contaminants to lakes

The source of toxic pollutants to lakes is usually material derived from human activities. In many areas natural rock substrates may also result in high levels of toxic metals but these rarely cause human health problems since natural cycles of weathering, transport and deposition are slow processes. Human activity may be responsible for increasing the erosion rates (by mining and manufacturing use) of naturally occurring elements or by changing their chemical form, thereby allowing concentrations to reach levels which may be a hazard to aquatic ecosystems and to man. Organic compounds synthesised and manufactured industrially are not natural substances. Their occurrence in the environment is a direct manifestation of loss during manufacture, transportation and use of the compound.

Inputs, or general sources of contaminants to lakes may take a variety of pathways:

- Direct point sources, municipal and industrial effluent discharges.
- Diffuse agricultural sources: wash-off and soil erosion from agricultural lands carrying materials applied during agricultural land use, mainly herbicides and pesticides.
- Diffuse urban sources: wash-off from city streets, from horticultural and gardening activities in the sub-urban environment and from industrial sites and storage areas.
- Waste disposal: transfer of pollutants from solid and liquid industrial waste disposal sites and from municipal and household hazardous waste and refuse disposal sites.
- Riverine sources: inflow in solution, adsorbed onto particulate matter, or both. The cumulative input is the sum of contaminants from all of the rivers draining the watershed into a lake.
- Groundwater sources: groundwater systems polluted from point and diffuse sources (noted above) flowing into rivers, and directly into lake beds.
- Atmospheric sources: direct wet and dry atmospheric deposition to the lake surface amplified by the erosional recycling of atmospheric deposition on the drainage basin land surface. This latter process is defined as secondary cycling.

To determine loads and contaminant mass balances in any lake system, efforts must be made to estimate the contributions from each of the above noted sources. In many cases, (e.g. river inputs) detailed monitoring may be necessary to provide reasonable estimates of the total inputs.

Contaminant loads

The discussion on phosphorus above indicated that inputs have been controlled in order to achieve lake concentrations which, in turn, determined the level of productivity in the lake. A similar philosophy has been formulated in many nations in the management of toxic substances. Approaches have ranged from a regulated input level to total elimination of inputs to recipient water bodies.

Organisms react to the concentration and exposure time of a contaminant in the water body. The lake concentration, either in the water or the sediment, is a result of the contaminant load (mass per unit time) distributed in the lake, which is termed the loading (mass per unit volume [or area] per unit time). These differences can be confusing for water management policies which have established guidelines or objectives on the basis of concentrations rather than on an understanding of the input loads of the sources of contaminants.

The concept of inputs or loads with respect to river inflows to lakes, and the impact on local lake water, is illustrated below. However, in conditions where the range of the river discharge is greatly in excess of the range of contaminant concentrations, the following relationship does not hold true.

A. High river   × Low contaminant   = High   ⇒ Slow, whole
    water volume   concentration   load   lake deterioration

    e.g. $10^9 \, m^3 \, d^{-1} \times 10^{-3} \, g \, m^{-3} = 10^6 \, g \, d^{-1}$

B. Low river   × High contaminant   = Low   ⇒ Impairment close
    water volume   concentration   load   to river input

    e.g. $10^3 \, m^3 \, d^{-1} \times 10^{-2} \, g \, m^{-3} = 10 \, g \, d^{-1}$

In the case of A above there will be no effects on biota in the river since the concentration is low. However, the high lake load will result in the degradation of the receiving waters with long-term deleterious effects on the associated biota. In the second case (B), real effects will be observed on the biota in the river due to the high concentrations in a low water volume, but this in turn does not necessarily imply degradation of the whole lake which may adequately accommodate the low load delivered. Ordinarily, some form of control would be applied to the river input under scenario B, but would not apply under scenario A which actually represents a far worse condition for the general quality of the lake. The calculation of load can be illustrated for the Niagara River, using the same annual average river volume, as follows:

Average annual concentration in Niagara River	Annual load to Lake Ontario
$1 \, g \, m^{-3}$	26,380 t
$1 \, mg \, m^{-3}$	26.38 t
$1 \, \mu g \, m^{-3}$	26.38 kg
$1 \, ng \, m^{-3}$	26.38 g

It has been estimated by Swain (1990, pers. comm.) that for top predator fish in Lake Ontario to achieve a concentration of 25 ng g^{-1} TCDD (tetra chlorinated dibenzo dioxin) in their tissues (i.e. the guideline concentration for human consumption) the lake would require a load of 5 g a^{-1}.

The determination of load in a lake should result in the ability to produce a mass balance which provides an insight into the proportional contributions of different sources, and which allows a control strategy to be optimised. Examples of such mass balances are provided in Table 7.2 for lead in Lake Ontario. Despite some difficulty in estimating the direct deposition of lead to

Table 7.2   Lead mass balance for Lake Ontario, North America and proportional
loadings from major sources

	Inputs (t a^{-1})	% of total input	Deposited and outputs (t a^{-1})
Accumulating in sediment			725
Niagara River suspended solids	534	⎫	
All other rivers suspended solids	176	⎬ 69.6	
Solute — all rivers	195	⎭	
Shoreline erosion	50	3.8	
Dredged spoil	65	5.0	
Airborne input (Precip. chem.)	280	21.6	
Output suspended solids			547
Output solute			207
Total	*1,300*		*1,479*

Source: IJC, 1977

the lake surface by numerical models or direct measurement of land-based
rain chemistry stations, relatively good mass balances have been achieved.
Evaluation of the loads in the soluble and suspended solids phases display the
important role played by sediment in the cycling of contaminants in lakes
and reservoirs.

The major source of lead to Lake Ontario is fluvial, followed by atmospheric. The largest river load is from the Niagara River which drains from
Lake Erie. The output from Lake Erie to the Niagara River is estimated at
496 t, whereas the Niagara River load to Lake Ontario is 754 t, indicating a
direct contribution from other sources to the Niagara River of 258 t. Therefore, a reduction in load in Lake Erie will show a significant reduction to
Lake Ontario. Furthermore, much of the fluvial load to both lakes arises from
the recycling of lead deposited on the land surface from the atmosphere.

Pathways
Once in a lake system, pollutants follow a number of physical, chemical and
biological pathways which are dependent, to a large extent, on the chemical
characteristics of the element or compound. Elements and/or compounds
which are soluble in water (hydrophilic) are transported with the physical circulation of the water mass in the lake. In the simplest circumstances, such a
pollutant mixes throughout the lake water and is eliminated within the water
residence time of the lake, which may be hours or decades depending on the
lake. Pollution of a lake with a long turnover time is a potentially serious

event because remedial actions only result in a slow reduction or elimination of the contaminant.

The distribution of water soluble elements is actually more complex than presented above. The input of high loads or concentrations impacts directly on the lake margin. Dispersion into the lake is controlled by diffusion, water mass movements resulting from hydraulic flow and lake circulation as induced by wind direction and velocity. Another factor which has a direct effect on retention time and distribution of soluble elements is related to lake stratification. Soluble elements in warm river water entering a stratified lake mix rapidly as an overflow into the warm waters of the epilimnion. Since the epilimnion often represents a small fraction of the lake water volume, the mixing and the turnover time are accelerated. The elimination rate can then be considered as the epilimnion volume divided by the discharge during the period of stratification. Conversely, as river waters cool in the autumn, they may inter-stratify in the hypolimnion or directly under-flow to the deeper waters of the hypolimnion. In this case mixing tends to be less rapid and the elimination period extends in an unpredictable fashion (i.e. without some precise knowledge of deep water circulation in the lake). The processes described above can only be understood when detailed sampling of the water column at a number of lake sampling points is carried out. Additionally, the location of water withdrawal points may accelerate or retard removal of soluble compounds depending on the thermal structure of the water body (see Chapter 8 for more details).

Most toxic trace elements and many organic pollutants (e.g. PCBs, DDT) of low solubility (hydrophobic) or which are fat soluble (lipophilic) are predominantly adsorbed by particles of inorganic or biological origin. The distribution and elimination of these kinds of materials closely follow the processes of sediment sorting and deposition which, in turn, are the products of lake circulation, wave induced hydraulic energy and settling of particles of different sizes in waters of different temperatures. These processes are discussed in more detail in Chapter 4. Elimination is mainly by sedimentation as illustrated for Pb in Lake Ontario in Table 7.2.

The characteristics of a compound, therefore, not only determine pathways but also indicate in which lake compartment (water, sediment or organisms) the pollutants can most probably be found. It is also significant that, as a general rule, toxicity is inversely related to solubility (the greater the solubility the lower the toxicity). Many toxic elements and compounds are often adsorbed or scavenged by particles in the source rivers, or in the lake itself, and are wholly or partially removed from contact with most lake organisms by sedimentation processes. However, this general rule is an over-simplification since major exceptions, such as the methylation of mercury and incorporation of lipophilic compounds in the tissues of benthic organisms, may result in recirculation in the lake food web.

The final factor involved in the rate of elimination of an element from a lake is its persistence. Organic compounds chemically decay at a defined rate depending on the physico-chemical conditions of the receiving lake water. Compounds may break down completely or may be changed to a different form, e.g. DDT to DDD or DDE; aldrin to dieldrin, etc. In some cases the decay product or metabolite may be more toxic than the parent compound.

Bioaccumulation and biomagnification
The processes of bioaccumulation and biomagnification have been discussed in Chapter 5. These processes are extremely important in the distribution of toxic substances in freshwater ecosystems. Bioaccumulation and food chain amplification of concentrations of toxic compounds determine the exposure and consequent effects of these substances at each trophic level, including humans. Examples of food chain biomagnification of Hg and PCB in Lake Ontario are given in Figure 7.9. Not all substances are subject to biomagnification as is shown for zinc, lead and copper by organisms taken from the Lake Ontario food chain (Table 7.3).

Contaminant effects on lake biota
Current knowledge of the effects of contaminants on lake organisms is limited. High concentrations of pollutants such as mercury in fish and benthic organisms appear to have little effect upon the organisms themselves (Jackson, 1980). However, evidence that DDT and toxaphene have affected the hatching success of lake trout in Lake Michigan was instrumental in the banning of these compounds in North America (Swaine, 1991 pers. comm.). The emergence of mayflies (aquatic insects) in western Lake Erie ceased for a period from the 1950s to the late 1970s and is commonly believed to be due to the effects of persistent organochlorine compounds.

Evidence of the incidence of carcinoma (tumours) in some fish species in lakes affected by their proximity to urban centres has been presented by Black (1983). Black (1983) also determined that some of the tumours could be induced by sediment-bound PAHs (polychlorinated aromatic hydrocarbons). These findings have been supported by similar studies from estuarine and coastal marine environments (Malins *et al.*, 1987).

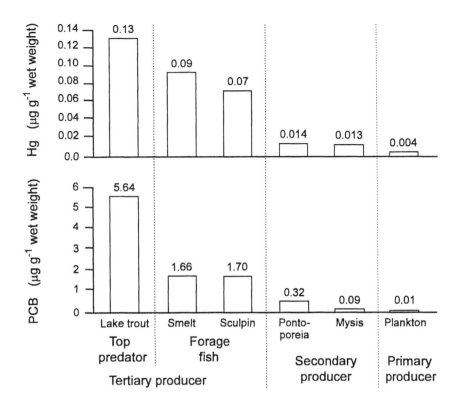

Figure 7.9 Biomagnification of mercury and PCB in the food chain of Lake Ontario (After Thomas *et al.*, 1988)

Table 7.3   Trace element concentrations in Lake Ontario biota

	Zinc	Lead	Copper
	\multicolumn{3}{c}{(μg g^{-1} wet weight)}		
Lake trout	10.35	0.61	1.30
Smelt	21.71	0.18	0.59
Sculpin	15.71	0.23	2.43
Pontoporeia	13.10	0.57	18.72
Mysis	13.66	0.46	0.90
Net plankton	15.16	1.07	0.92

Source: Thomas *et al.*, 1988

Effects of contaminants at lower trophic levels, particularly the phyto-plankton, have not been widely demonstrated. Work by Munawar and Munawar (1982) has shown that small species of phytoplankton, less than 20 μm in size, are more susceptible to contaminants then the more robust plankton greater than 20 μm in size. Changes in lake biota related to changes in contaminant loads are not well documented. It is possible that indigenous plankton populations with short life cycles adapt to the loads and concentrations to which they are constantly exposed.

The effects of lead on fish in lakes have been demonstrated by Hodson *et al.* (1983), who showed a physiological response in the synthesis of the enzyme ALA-D, as determined by blood analysis. Continuous exposure to elevated lead concentrations results in a darkening of the tail (BlackTail) and ultimately death.

Current techniques for assessing the effects of contaminants, both soluble and particulate, are discussed in Chapter 5. With respect to lakes, the best techniques are currently based on bioassay methods which will continue to be used until adequate research improves knowledge of the community and physiological responses to pollutant stress by the organisms naturally occurring in lakes.

7.3.4 Lake acidification

One of the major issues related to lakes in particular, and to freshwaters in general, is the progressive acidification associated with deposition of rain and particulates (wet and dry deposition) enriched in mineral acids. The problem is characteristic of lakes in specific regions of the world which satisfy two major critical conditions:

- the lakes must have soft water (i.e. low hardness, conductivity and dissolved salts), and
- the lakes must be subjected to "acid rain" or more precisely, to total deposition enriched in sulphate (H_2SO_4) and nitrogen oxides (NO_x) creating nitric acid.

All lakes have some acid buffering capacity due to the presence of dissolved salts from the watershed. However, in non-carbonate terrain, such as in areas of crystalline rocks or quartz sandstones, this buffering capacity is rapidly exhausted and free H_2 ions create a progressive acidification of the lake. This buffering capacity has been defined by Jeffries *et al.* (1986) who characterised lake sensitivity to acidification in terms of the acid neutralising capacity (ANC) as:

$$ANC = \Sigma \text{ base cations} - \Sigma \text{ strong acid anions}$$
$$([Ca]+[Mg]+[Na]+[K]) - ([SO_4]+[NO_3]+[Cl])$$

Jeffries *et al.* (1986) used an ANC scale to define lakes in Eastern Canada as follows:

ANC (μeq l^{-1})	Sensitivity
≤ 0	Acidified
0–40	Very sensitive
40–200	Sensitive
> 200	Insensitive

Lake sensitivity is a reflection of sub-surface geology and the associated soils (which may be easily mapped), whereas acidification is the result of acid deposition on lakes of differing sensitivity. These two factors have been well defined in Eastern Canada and are illustrated in the colour maps of Figures 7.10 and 7.11. The occurrence of high deposition (> 20 kg ha^{-1}) with highly sensitive terrain occurs as a band across the southern regions of the country from Central Ontario across Southern Quebec, almost to New Brunswick (Fraser, 1990 pers. comm.). Lake studies have indicated that the lakes in these areas have been subjected to progressive acidification and that any improvement will require substantial reduction in acid deposition from the 1982–1986 average levels shown in Figure 7.11. Summary statistics for pH and ANC for the lakes of Eastern Canada are given in Table 7.4.

The acidification of lake waters, together with the acid leaching of lake basin soils, results in the hydrolisation of the hydrated oxides of iron, manganese and aluminium with the resultant release of many trace elements which are toxic to aquatic organisms. Such release has been demonstrated in sediments from Norwegian lakes. An example is presented in Figure 7.12, in which depletion of zinc in the surface sediment occurred due to dissolution following lake acidification. The dissolution of aluminium has proved to be particularly dangerous for fish due to asphyxiation caused by the deposition of aluminium oxide on the gill filaments as it precipitates out of solution.

A correlation analysis of dissolved Al to H ion concentration in the lakes of Eastern Canada failed to show a good relationship. However, when the lakes were combined into a set of regional aggregates based on water hardness, geology and deposition, a clear relationship could be observed as is shown in Figure 7.13. Lake acidification is a major problem, limited geographically to those areas of the world where sensitive lakes occur downstream of major emissions of mineral acids. Localised problems exist in all industrial nations using large quantities of fossil fuels, but the major impacts currently occur in the Eastern USA, Canada, Central Europe and Scandinavia. Future problems may occur in China, Africa and other industrialising regions.

Table 7.4 pH and acid neutralising capacity (ANC) in lakes of Eastern Canada

| | pH | | | | | | Acid neutralising capacity | | | | | |
| | | | | Percentage | | | | | | Percentage | | |
	Sample (n)	Minimum pH	Median pH	≤ pH 5.0	≤ pH 5.5	≤ pH 6.0	Sample (n)	Minimum (μeq l^{-1})	Median (μeq l^{-1})	≤ 0 μeq l^{-1}	≤ 40 μeq l^{-1}	≤ 200 μeq l^{-1}
North West Ontario	1,080	5.03	7.10	0	0	4	1,078	−3	254	1	4	40
North East Ontario	1,820	4.00	6.85	6	10	20	1,805	−99	165	9	24	51
South Central Ontario	1,619	4.22	6.29	2	8	28	1,578	−66	63	3	32	83
Quebec	434	4.40	6.33	3	12	30	429	0	36	1	55	90
Labrador	198	4.84	6.36	1	2	20	182	2	46	0	43	95
New Brunswick	84	4.51	6.34	7	15	36	81	−22	38	11	51	96
Nova Scotia	232	4.20	5.20	39	63	82	198	−82	0	51	93	99
Newfoundland	270	4.84	6.17	3	14	44	176	−16	28	7	68	94

Source: Modified from Jeffries *et al.*, 1986

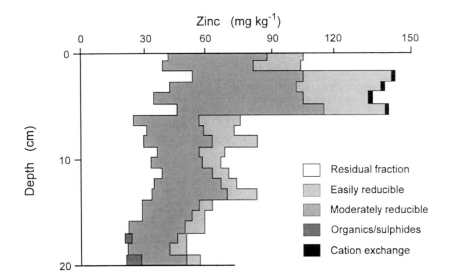

Figure 7.12 Depth profile of zinc in a sediment core taken from the acidified Lake Hovvatn, Norway showing dissolution of zinc from the surface layer (Modified from Golterman *et al.*, 1983)

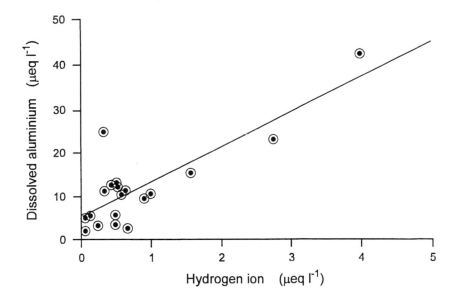

Figure 7.13 The relationship of dissolved aluminium to hydrogen ion concentration averaged for 19 geophysical regions in Eastern Canada (From Fraser, 1990 pers. comm.)

7.3.5 Salinisation

Salinisation is becoming a widespread water quality issue (Meybeck *et al.*, 1989) which often leads to the loss of usable water resources, including lake waters. In virtually all cases this is due to the mismanagement of the water for agricultural purposes. Natural salinisation occurs over long periods of time. In its simplest form it is caused by a change in the natural water balance of the lake, i.e. when the output is less than the input and the water balance is maintained by high levels of evaporation (Meybeck, 1995). This leads to a progressive increase in salt content, as can be seen in the Dead Sea, where salt concentrations reach 300 g l^{-1}.

The input of saline groundwater, which represents only a proportion of the total water balance, results in a meromictic condition for normal lakes. However, in shallow lakes, mixing may occur leading to the creation of lakes with brackish water or moderate salt composition. This condition often exists in coastal lakes, particularly when over-use of fresh groundwater allows marine salt water intrusion. Temporary saline lakes and saline groundwaters may also occur following marine flooding resulting from storm surges in flat, low-lying areas. With time, the normal water balance is restored by freshwater dilution, but flooding with saline water can always re-occur.

By far the most common form of lake salinisation is related to changes in the water balance and salt content caused by irrigation. Diversion of water from a lake to the land surface results in: i) loss of water during irrigation due to infiltration to groundwater and loss by evaporation and evapotranspiration, and ii) increased salt concentration in the water returning to the lake or river as a result of soil leaching processes. When combined, these produce a slow salinisation of the lake water. However, the process may be greatly accelerated if water losses during lake use fall below the water output rate of the lake. In this case lake levels also fall indicating that the lake water is being over-used. Control and diversion of the inlet waters to the Sea of Aral, Russia from the Amu Darya and Syr Darya rivers (for the purpose of large scale irrigation) have resulted in a disastrous lowering of the lake level. This has caused major disruption to other water uses, such as navigation, fisheries and water supply and has increased the salinity of the water due to the effects of uncompensated, high levels of evaporation.

7.3.6 Issue summary

The issues discussed above indicate that major impairment of water use can occur if efforts are not made to control or prevent the causes of the associated water degradation. A summary of the issues, causes and effects, and the associated requirements for the three levels of water quality assessment discussed in Chapter 1 are given in Table 7.5. To understand the degree to which a water body has been affected by any specific use or issue, some investigative monitoring and assessment is required. This assessment process, if undertaken in sufficient detail, provides the basis for the infrastructure required to design an appropriate control strategy. Once this strategy has been implemented, the assessment must be continued to ensure that the expected results, or improvements, in water quality are actually achieved. The assessment strategies and variables to be measured are discussed briefly in section 7.5 and in more detail in Chapters 2 and 3.

7.4 The application of sediment studies in lakes

The role of sediment studies in limnology has been a rapidly developing field over the last two decades. This importance is reflected in this book by the incorporation of a separate chapter dedicated to the monitoring and assessment of fluvial and lacustrine sediments (see Chapter 4). Therefore, the discussions included in this chapter relate only to special considerations for sediment studies within the limnological context. For more detail see Chapter 4.

Sediments interact with lake water and soluble constituents in such a manner that they give many unique insights into limnological processes. As a consequence, they provide important basic information on the geochemical origin, dispersion throughout the lake, and limnological fate of soluble and particulate constituents.

Sediment texture and mineralogy are closely inter-related. Finer particles of less than 64 μm consist mostly of clays, hydrated oxides of iron and manganese, and organic matter. These elements provide the geochemically active sites which allow for the uptake and release of chemical elements and compounds. Adsorption or desorption occur during transport and sorting, and are related to the solid/liquid concentration as a function of solubility, for the range of pH and redox conditions that occur in the transporting medium. A high solid contaminant concentration in equilibrium with river water, results in an initial desorption of the contaminant when entering the more dilute conditions of normal lake water. However, resuspension of fine bottom sediment occurs under storm conditions in a shallow lake, resulting in adsorption of elements from the water column and elimination under uniform settling conditions. Once settling has occurred, the contaminant (under normal pH and in oxygenated bottom waters) is effectively removed from the water column. Release back to the bottom water can occur under the following conditions:

- *Resuspension*: Physical disturbance of bed sediment and release of interstitial waters, including possible desorption of some contaminants to the water column.

Table 7.5   A summary of water quality issues, causes and effects in lakes together with the principal monitoring and assessment activities according to various levels of sophistication

Water quality issues	Causes	Water quality effects	Water quality assessment level[1]		
			I	(I) II	(I+II)+III
Eutrophication	Excess nutrients	Increased algal production Oxygen depletion in hypolimnion Release of Fe, Mn, NH_4 and metals in hypolim. Loss of biotic diversity at all trophic levels	Estimates of biomass Cell counting Chlorophyll *a* during stratification Transparency	Analysis of TP, SRP at frequent intervals Analysis of sources of P Vertical profiling + O_2 measur. Species composition	Full nutrient budget Vertical and spatial analysis Fe, Mn and other metal analyses
Health effects due to community waste	Human and animal organic waste	Bacterial and viral infection Other effects as for eutrophication	Simple test e.g. coliform	Bacterial counts Analysis of metals Toxicity bioassay	Test for viruses Analysis for organic pollutants
Acidification	Atmospheric deposition of sulphate and nitrogen oxides	Decrease in pH Increases in Al and heavy metals Loss of biota	Measure of pH	Measure of Σ cations and anions	Wet and dry atmospheric deposition Metal analyses (esp. Al)
Toxic pollution	Industrial waste disposal; Agricultural and municipal chemicals	Increased concentration of metals and organic pollutants in water sediment and biota Bioaccumulation and biomagnification	Simple bioassay e.g. *Daphnia magna*; Microtox	Analysis of water and sediment for heavy metals; full bioassay, biological analysis	Analysis of organic micro-pollutants. Full budget determination including atmospheric
Salinisation	Changes in water bal-ance; marine incursions; increased salt to soil leaching; mainly irrigation	Increasing salt concentrations	Conductivity Lake level measurement	Analysis of ionic content of water	Detailed hydrological balance; Salt budget inc. sources

P   Phosphorus
TP   Total phosphorus
SRP   Soluble reactive phosphorus
[1]   I, II, III Simple, intermediate and advanced level monitoring and assessment

- *Bioturbation*: The disturbance of sediment by benthic organisms leading to a re-distribution of the contaminants in the deeper sediment to the surface layers where they may be released to the water column (by the same processes as in resuspension).
- *Changes in pH and Eh*: These change the mineralogical composition of the sediment and release elements by solubilisation (e.g. release of adsorbed P and metals by reaction and solubilisation of iron ($Fe^{3+} \Rightarrow Fe^{2+}$) resulting from oxygen deficiency during bottom water anoxia).
- *Biological transformation*: Bacterial modification of trace elements such as Hg, As, Se and Pb by conversion to soluble or volatile organo-metallic complexes.
- *Changes in lake water concentration*: Reduction in water concentration by management control (e.g. phosphorus control) which changes the equilibrium concentrations and results in release from sediments, producing a new equilibrium condition.

Hydrophobic trace elements and compounds are adsorbed rapidly by fine sediment particles. Analysis of the fine particles throughout the lake and the observed concentration gradients identify the source of the element of interest. When the source is a river input, the same techniques applied to the river sediment allow the isolation and identification of the direct source of the pollutant. A fine-grain sediment pollutant concentration throughout the whole lake defines the transportation pathway within the lake system, and identifies the final "sink" regions.

The analysis of sediment at source (i.e. river inputs, coastal erosion, atmospheric deposition), together with estimates of deposition rates, allows for the computation of mass balances. Such mass balances have proved to be of considerable value in optimising control strategies by targeting the most amenable and cost effective sources to be controlled.

This summary of the use of sediment information in understanding water quality issues in lakes has demonstrated the need to include sediment measurements and sampling into lake monitoring and assessment programmes. However, the subject area is specialised and, ideally, studies should be undertaken by trained sedimentologists and geochemists. If this is not possible their advice and guidance should always be sought.

7.5   Assessment strategies

Appropriate monitoring strategies must be selected in relation to the objectives of the assessment programme (see Chapter 2). Sediment sampling has been discussed in Chapter 4 and is, therefore, not covered here. Water sampling is usually comparatively straightforward, although certain factors must always be taken into account. These factors include obtaining an adequate sample volume for all necessary analyses, cleanliness of samples and sample bottles, requirements for filtration of the sample and sample storage methods, etc. (see Chapter 3 and Bartram and Ballance (1996)). The measurement of phytoplankton chlorophyll, as an indication of algal biomass, is the most common routine biological monitoring strategy applied in lakes.

7.5.1 Sampling site location and frequency of sampling

The structure of the lake must be defined if a rational sampling design is to be used. Depth and temperature profiles must be obtained to establish the location of the thermocline (see later). The number of samples to be taken is related to lake size and morphology and to the water quality issue being addressed. Large lakes must be described, even in the simplest context, by more than one sample. For example, Lake Baikal in Russia requires a minimum of two sampling stations due to the occurrence of two sub-basins, and because the major fluvial input and output occur in the southern part of the lake. Consequently, the northern section of Lake Baikal has a significantly different physical and chemical structure. Lake Erie consists of three distinct basins and, therefore, a minimum of three stations is required for initial characterisation. As a general rule, samples should be taken from each section of a lake which can be regarded as a homogeneous water mass. A small lake with a single water mass may be described by a single sampling station.

If only one sample is taken, it should be located at the deepest part of the lake where oxygen deficits are likely to be greatest. A number of samples (3–15) should be taken throughout the water column, and there should be at least three sampling occasions per year. For temperate lakes, samples must be taken prior to spring stratification, late in the summer stratification and after the autumn overturn. In a large lake, such as Lake Ontario, many water masses may be identified, each of which can be represented by a single sampling station (El Shaarawi and Kwiatkowski, 1977). To identify the homogeneous water masses in Lake Ontario, statistical analyses were carried out on water quality surveys repeated at about one hundred stations. This kind of intensive water quality survey cannot be conducted without access to considerable resources (i.e. finance, man-power, equipment) and it is, therefore, necessary to use a logical sampling design based on purpose and limnological knowledge. Theoretical examples are described below.

7.5.2 Examples of sampling programme design

For the purposes of this guidebook sampling design is best described in relation to a number of precise water quality issues. Control and resolution of these issues require a clear understanding of the sources of materials and pollutants to a lake. Sources of pollution in water bodies are discussed in Chapter 1 and summarised in Table 1.4. An indication of the complexity of monitoring and assessment with respect to lakes is given (according to the three levels of complexity discussed in Chapter 1) in Table 7.5. Variables which should be measured for major uses of the water and major sources of pollution are suggested in Tables 3.7 to 3.10.

Eutrophication

Assessment of trends in eutrophication usually includes the monitoring of nutrients, major chemical ions and chlorophyll. A single station only is required in small lakes (Figure 7.14A), whereas a single station within each homogeneous sector is best for large lakes (Figure 7.14A). Samples should be taken frequently, ideally once a week but not less than monthly. These enable the understanding of seasonal changes and the computation of mean values (by season or by year). Samples should be taken using strings of bottles in relation to the physical structure of the lake (Figure 7.14B). For phytoplankton samples and associated chemical parameters, depth integrated samples of the epilimnion may be adequate. The results obtained can be used to illustrate the spatial and temporal variability of concentrations in lakes and to make comparisons of depth profiles over a season.

Spatial and temporal variations in silica (a major nutrient for diatoms) in Lake Geneva are shown in Figure 7.15. In February silica shows vertical homogeneity in the upper 250 m and a positive increase near the sediments due to a partial lake overturn. In May silica concentrations decrease in the top 5 m due to uptake by diatoms during a spring bloom. By July the silica is nearly exhausted in the epilimnion, but concentrations start to recover in October with decreased diatom activity. Concentrations become homogeneous throughout the water column in December due to the complete lake overturn. An alternative way of presenting similar seasonal depth data is by iso-contour lines as in Figures 7.8 and 7.19.

Input–output budgets

Assessment of the influence of a lake on the throughput of materials in a river–lake system may be accomplished by frequent sampling and analysis, close to the input and at the output of the lake (Figure 7.14A). Samples should be depth integrated and calculated as mass input per year and compared to mass output over the same time scale. The addition of open lake samples provides sufficient information to calculate material mass balances for the lake. It must be recognised, however, that such mass balances must take account

A. Sample site location

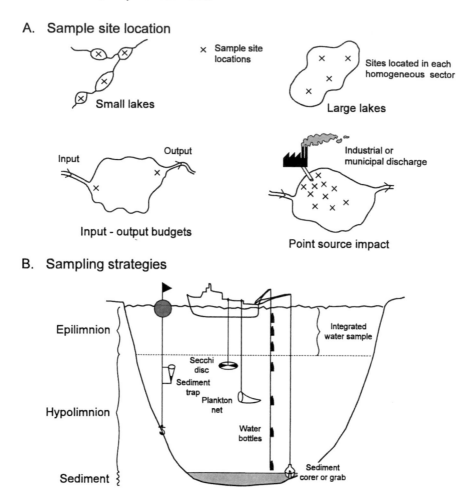

Figure 7.14 Theoretical examples of sample site location (**A**) and sampling strategies (**B**) in lakes

of all sources, including the atmosphere, and the significant role in recycling played by the sediments. Input–output measurements for nutrients and polluting substances are particularly important for verifying calculated predictions based on mathematical models.

Impacts of point sources

The localised effects of a discharge can be assessed by the random sampling of a scatter of points around the zone of influence of a point source, such as an industrial effluent or municipal sewage discharge pipe (Figure 7.14A). The effects of industrial outfalls should be determined by sampling and

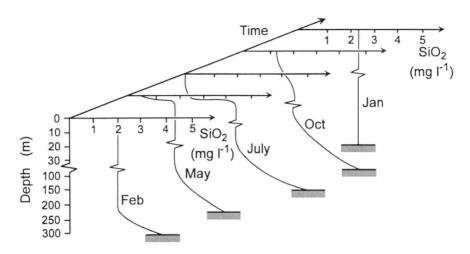

Figure 7.15 Seasonal variation in the vertical profile of dissolved silica in Lake Geneva. Note the break in the depth scale. (Data from the International Commission for the Protection of the Waters of Lake Geneva (CIPEL, 1984) and M. Meybeck, Université Pierre et Marie Curie, Paris (unpublished data))

analysing for the materials likely to be discharged from the industrial process, whereas the effects of sewage outfalls should be studied by measurements of bacterial populations, nutrients and toxic pollutants.

Comprehensive source assessment in lake basins
Comprehensive source assessment involves monitoring to determine the proportional contribution of different sources of specific materials. Frequent analysis of point and non-point (diffuse) sources is required to determine loads, and any changes in loads resulting from management actions. All rivers, as well as point and diffuse sources of industrial, urban, agricultural and natural materials (including the atmosphere), must be analysed to determine the complete impact of all human activities on water quality. Background conditions must also be established so that the anthropogenic contribution can be determined. Samples should be analysed at least monthly, but preferably with greater frequency for bacteria, nutrients and toxic pollutants. When combined with detailed, open lake monitoring and sediment analysis this approach is the most advanced and sophisticated with respect to lacustrine systems.

7.6   Approaches to lake assessment: case studies
Water quality assessments are conducted to define the conditions of the lake, usually with respect to its uses. Nutrients, trophic status and toxic substances

must generally be evaluated and a management strategy designed to maintain, or enhance, the use of the lake. Once implemented, assessments must be continued to ensure that goals are being achieved and water uses maintained. Some examples of this kind of assessment are given below.

There are many examples of the assessment of lakes in the published literature. However, few good examples exist from the developing world and, as is so often the case, most studies have been carried out in the western, industrial nations, particularly Scandinavia, North America and Western Europe. The bias towards wealthy nations provides good examples in the northern temperate zone and little detailed assessment information from elsewhere. The type of assessment undertaken is related to the resources of the nation carrying out the study. Consequently, examples of sophisticated biological monitoring for trace pollutants are usually restricted to the industrial nations. For these reasons, the following examples are taken primarily from the industrialised regions of the world, but it is hoped that they will encourage the development of increasingly sophisticated lake assessments in the developing regions.

7.6.1 General basic surveys

From the discussions above, it is clear that some primary information is required before a fairly detailed understanding of lake processes can be obtained. Two major lake characteristics must be established, bathymetry and thermal structure. Lake bathymetry is required:
- to check the occurrence of several sub-basins and to compute lake volumes for residence time calculations,
- as a basis for sediment mapping and sediment and water sampling and,
- as a basis for predicting the occurrence of specific organisms that may be required for future analysis.

Bathymetry is most conveniently obtained using a recording echo-sounder along straight lines run at a constant speed between known shore stations. Large lakes require the use of electronic positioning equipment and in these situations charting bathymetry is, therefore, expensive to perform. Small lakes may be surveyed using a measured, weighted line and the positions of each measurement determined from two or more angles to known targets, or land marks, on the shore. The use of a sextant is the traditional method of establishing position for each sounding. An example of lake bathymetry for Lake Vättern, Sweden is given in Figure 7.16.

The thermal structure of lakes has been discussed in detail in section 7.2.3. An understanding of lake processes cannot be achieved without a relatively detailed evaluation of the thermal structure. This can be done simply by rapid

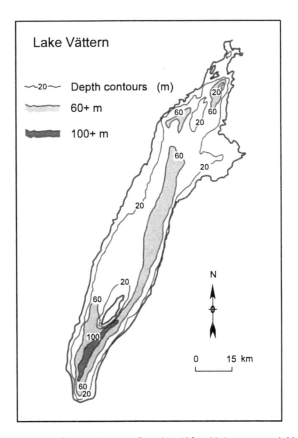

Figure 7.16 Bathymetry of Lake Vättern, Sweden (After Häkanson and Ahl, 1976)

measurement of the temperature of water samples taken throughout the water column, either by a single bottle lowered to different depths, by several bottles hung in series at fixed depths on a single cable, or by electronic depth and temperature profiling. Water bottles are closed by trigger weights (messengers) once the line has been set, and in some cases the bottles are equipped with reversing thermometers for *in situ* temperature measurement. Depending on the lake size, survey requirement and the availability of resources, temperature profiles may be taken at many stations or at a single point. An example of many profiles taken during a single survey cruise in Lake Ontario is given in Figure 7.17. However, in order to reconstruct the thermal characteristics of Lake Ontario such profiles must be taken many times during the year. An example of a single station thermal profile taken many times in order to determine seasonal thermal structure is given in Figure 7.3.

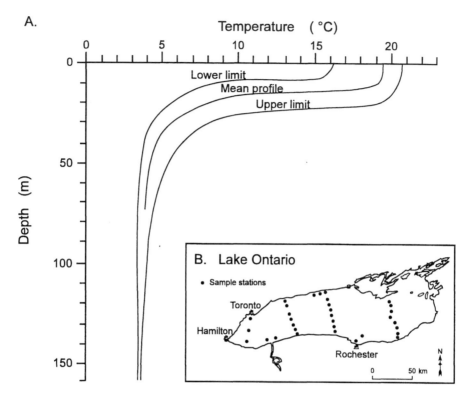

Figure 7.17 A. The summer stratification in Lake Ontario based on measurements made at 32 stations during a single cruise; **B.** The location of the measurement stations (After Dobson, 1984)

7.6.2 Assessment of trends

The simplest form of assessment is to establish the trend in concentrations of a nutrient, particularly phosphorus, during a period of holomixis when the lake is fully mixed and isothermal. Measurements to achieve this may be made at a single station, or more appropriately at a number of stations. Samples are normally taken in early spring before the onset of stratification. An example of this type of monitoring is given for Lake Constance from the early 1950s to 1988 in Figure 7.18. The trends are quite clear: phosphorus concentrations rise rapidly until the late 1970s when, following the large scale control of sewage effluents, a rapid decline can be observed.

   The assessment of long-term trends is related directly to lake management strategies whereas short-term, or seasonal, trend monitoring is required to understand the processes occurring within the lake and to provide more detailed information on a variety of indicators of lake condition. An example of seasonal trends is given for Lake Windermere, UK which was sampled on a

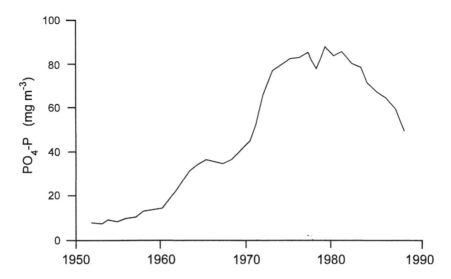

Figure 7.18 Long-term trends in phosphorus (PO4-P) in Lake Constance, Switzerland 1952–1988; measurements made at holomixis (After Geller and Güde, 1989)

weekly basis (Figure 7.7). The trends in chlorophyll *a* indicate the algal biomass which results in a summer depletion of soluble reactive phosphorus. The depletion of soluble reactive P occurs during stratification and the regeneration occurs during the winter when the water is mixed. No clear trends can be observed in total phosphorus concentrations, although a minor reduction occurs in summer resulting from soluble reactive P depletion. This suggests that the bulk of the total P is non-bioavailable and probably occurs bound with apatite and mineralised organic matter. The trend in NO_3-N is more interesting and indicates summer decreases probably related to algal production. In the summer (August) of 1974, nitrate was depleted almost to the point where it could have become limiting for algal growth. The seasonal trends portrayed for Lake Windermere are relatively simple because they represent trends in the surface waters.

A more complete understanding of nutrient and element cycling may be obtained through seasonal trend analysis of samples taken at many depths at one location. When the concentrations are contoured, as shown in Figure 7.19, they show a high degree of resolution in the seasonal development of the variables measured. This example from the Heiligensee, Germany summarises the seasonal cycling of O_2, H_2S, soluble reactive P and NH_4. Oxygen is distributed throughout the water column in December at concentrations of 10 mg l^{-1}. During the winter stratification caused by the January to March ice cover, some depletion of oxygen occurs in the hypolimnion, with

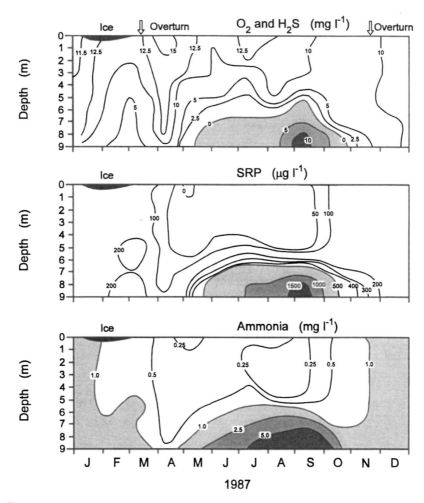

Figure 7.19 Isopleths summarising the seasonal cycling of oxygen and hydrogen sulphide, soluble reactive phosphorus (SRP) and ammonia in the Heilingensee, Germany in 1987 (After Adrian *et al.*, 1989)

re-oxygenation occurring almost to the bottom of the lake in April during the spring overturn. Summer stratification from May to November is characterised by progressive declines in oxygen concentrations in the deep water. Complete anoxia occurs in August and September, during which time major release of H_2S occurs from the bottom sediment. The cycling of soluble reactive P and ammonia conform to the seasonal cycle of oxygen, with the major release of soluble reactive P and NH_4 from the bottom sediments occurring at the time of oxygen depletion. Complete vertical mixing occurs during the autumn overturn in December, producing constant concentrations throughout the water column during the winter. This lake is hypereutrophic.

7.6.3 Spatial surveys of trace constituents

The distribution of contaminants in a lake can be determined by spatial surveys. These surveys can highlight areas in the lake which might be susceptible to possible effects of the contamination The distribution of the metals zinc, lead and copper in Lake Ontario is given in Figure 7.20. Analyses were performed on unfiltered water from samples which were grouped into zones of relative homogeneity. The metals Zn and Pb were grouped into three zones and Cu into four. The groups are numbered sequentially from the highest mean concentration in zone 1 to the lowest in zones 3 and 4. Mean values for each element are given in the table included in Figure 7.20. The distribution of the zones shows an inshore to offshore decline in the mean values of the three elements which probably relates to a decrease in particle concentrations. Zone one occurs in the extreme west end of the lake for all metals and extends to the north west shore for Zn and Pb, and also occurs on the south shore (just east of the central part of the lake) for Pb. The occurrence of zone 1 on the western shore is due to the steel manufacturing complex at Hamilton. The Pb on the southern shore is related to the presence of the city of Rochester, New York, USA and the north shore occurrence of high mean concentrations of Pb is adjacent to the city of Toronto. Zone 1, therefore, can be ascribed to anthropogenic sources of these elements, whereas the other zones reflect the mixing processes involved in the dispersion of these elements throughout the lake. Similar distributions for organic pollutants in the water of Lake Ontario have also been produced (Thomas *et al.*, 1988).

7.6.4 Biological monitoring

With the increasing concern over the potential impacts of toxic substances on the health of human populations, biological monitoring (see Chapter 5) has become an extremely important component of lake assessment programmes.

Examples of the analysis of toxic compounds in the tissues of organisms are available from the Great Lakes, as shown in Figure 7.9. These illustrate the biomagnification of pollutant concentrations within the food chain of Lake Ontario (for more details on bioaccumulation and biomagnification see section 5.8).

Organisms from a number of trophic levels in lakes may be used to assess the spatial and temporal variation of specific pollutants. Concentrations of contaminants in fish are normally used for these studies. An example of

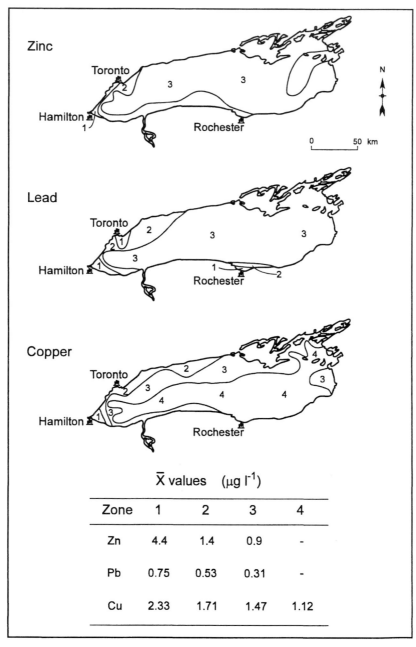

Figure 7.20 Distribution of total Zn, Pb and Cu in the waters of Lake Ontario together with the mean values for each zone. Zone 1 is the most polluted and zone 4 the least polluted (After Thomas *et al.*, 1988)

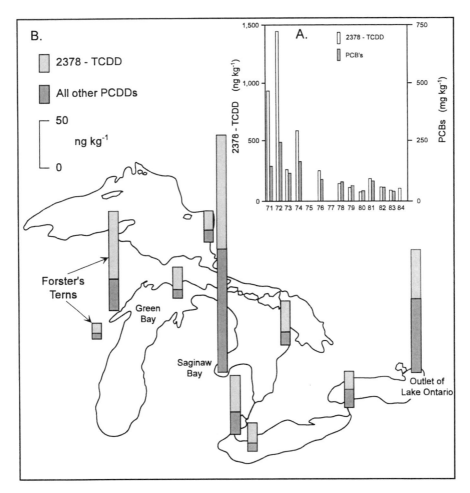

Figure 7.21 A. Trends in residues of chlorinated dibenzo-p-dioxins in eggs of Herring Gulls from the Great Lakes between 1971 and 1984. **B.** Variations in concentrations of chlorinated dibenzo-p-dioxins in the eggs of Herring Gulls and Forster's Terns from sites around the Great Lakes in 1983 (After Meybeck *et al.*, 1989)

probably the most successful biological monitoring programme undertaken in North America is given in Figure 7.21A. In this programme, a number of organic pollutants were analysed each year in the eggs of the Herring gull taken from bird colonies scattered throughout the Great Lakes. The Herring gull feeds on fish from the Great Lakes. The analysis of Herring gull eggs from Lake Ontario for PCB and 2378-TCDD (polychlorinated dibenzo-p-dioxins, particularly the "2378" tetra chlorinated form "2378-TCDD") showed a clear downward trend between 1971 and 1984. Figure 7.21B summarises the results for 1983 for PCBs and 2378-TCDD. These compounds occur throughout the lakes but particularly high concentrations have been found at Green Bay, Lake Michigan, Saginaw Bay in Lake Huron and at the outlet of Lake Ontario to the St. Lawrence river.

7.6.5 Sediment monitoring

Many examples of the use of sediment analyses to determine spatial and temporal distributions of various chemicals have been provided in Chapter 4. The spatial and temporal distribution of mercury in lake sediments has been combined into a single illustration by Häkanson and Jansson (1983) for Lake Ekoln in Sweden (Figure 7.22). The Hg distribution patterns in the lake indicated the major source as the River Fyris. Vertical profiles at the three core sites showed that Hg concentrations started to increase towards the end of the last century, around 1890, and have been rising up to the present time. The core from Graneberg Bay showed wide fluctuations in concentration since 1930 over the 35 cm of sediment thickness. However, these oscillations have been damped out in the other two cores. The source of Hg, as represented by the cores, was relatively constant over 40 years but remained at sufficiently high concentrations to give a continuous increase in concentrations further away from the source.

7.7   Summary and conclusions

Lakes are an essential source of freshwater for human populations. They provide water for a multitude of uses ranging from recreation and fisheries through to power generation, industry and waste disposal. As a consequence of the latter two uses most lakes have suffered water quality degradation to some extent or other. Degradation in the form of bacteriological pollution (health related), eutrophication, gradual increases in toxic pollutants and salinisation have all been observed. Even remote lakes are currently suffering degradation in water quality due to atmospheric deposition of mineral acids, nutrients and toxic chemicals.

The control of lake water quality is based on sound management practice (Jorgensen and Vollenweider, 1989) in relation to the required water uses and a reasonable, preferably detailed, knowledge of the limnological characteristics and processes of the water body of interest. The basic information required to characterise the ambient condition of a lake is based upon surveys of the various components of the lake system. These surveys provide the scientific basis for the development of the lake in order to meet the most specific requirements, and sensitive uses, to which it is subjected. Once this has been established a management protocol can be determined and implemented.

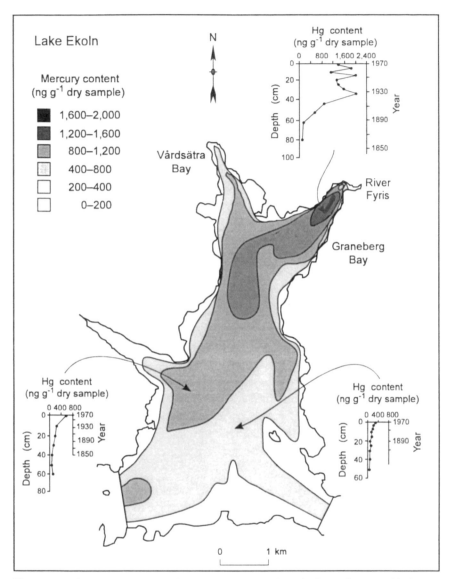

Figure 7.22 Spatial distribution of mercury concentrations in the sediments of Lake Ekoln, Sweden together with the temporal trends indicated by three sediment cores (After Häkanson and Jansson, 1983)

An assessment programme must be put in place right from the time of initial implementation of a specific lake management programme, in order to ensure that the lake water quality is adequate for the requirement of the management plan. Examples of such requirements include: increasing pH in an acidified lake; low bacterial counts to ensure potable water and contact recreation; appropriate phosphorus concentrations to maintain the desired levels of lake productivity; inputs of toxic substances which are low enough to ensure the concentrations in fish do not exceed a required guideline level; and preservation of all of the multitude of water uses required by human populations.

By undertaking properly designed lake assessment programmes, those agencies responsible for water management in the newly industrialising regions of the world will benefit from the lessons learned from an abundance of past errors in water quality management in industrialised countries.

7.8   References

Adrian, R., Dohle, W., Franke, C., Müller, B. and Stimpfig, A. 1989 Heiligensee - Studies on the ecology and interdependance of zooplankton species. In: W. Lampert and K. Rothhaupt [Eds] *Limnology in the Federal Republic of Germany.* International Association for Theoretical and Applied Limnology, Plön, 156-160.

Bartram, J. and Ballance, R. [Eds] 1996 *Water Quality Monitoring. A Practical Guide to the Design and Implementation of Fresh Water Quality Studies and Monitoring Programmes.* Chapman & Hall, London.

Black, J.J. 1983 Field and laboratory studies of environmental carcinogenesis in Niagara River fish. *J. Great Lakes Res.*, **9**, 326-334.

Burns, N.M. and Ross, C. 1971 Nutrient relationships in a stratified lake. In: *Proceedings of the 14th Conference on Great Lakes Research, Annual Meeting of the International Association for Great Lakes Research*, Toronto, 1970, 749-760.

CIPEL 1984 *Annual Report of the Franco-Swiss International Commission for the Protection of the Waters of Lake Léman*, Lausanne.

D'Itri, F. 1971 *The Environmental Mercury Problem.* Technical Report No. **12**, Institute of Water Research, Michigan State University, Lansing.

Dobson, H.F.H. 1984 *Lake Ontario Water Quality Atlas.* Environment Canada, Inland Waters Directorate (Ottawa), Scientific Series No. **139**, Canada Centre for Inland Waters, Burlington, Ontario, 59 pp.

El Shaarawi, A.A. and Kwiatkowski, R.E. 1977 A model to describe the inherent spatial and temporal variability of parameters in Lake Ontario, 1974. *J. Great Lakes Res.*, **3**, 177-183.

Fraser, A.S. 1990 Personal Communication, Rivers Research Branch, National Water Research Institute, Canada Centre for Inland Waters, Box 5050, Burlington, Ontario, Canada.

Geller, W. and Güde, H. 1989 Lake Constance-the largest lake. In: W. Lampert and K. Rothhaupt [Eds] *Limnology in the Federal Republic of Germany*. International Association for Theoretical and Applied Limnology, Plön, 9-17.

Golterman, H.L. 1975 *Physical Limnology*. Elsevier, Amsterdam, 489 pp.

Golterman, H.L., Sly, P.G. and Thomas, R.L. 1983 *Study of the Relationship Between Water Quality and Sediment Transport*. United Nations Educational, Scientific and Cultural Organization, Paris, 231 pp.

Häkanson, L. 1980 An ecological risk index for aquatic pollution control — a sedimentological approach. *Water Res.*, **14**, 957-1101.

Häkanson, L. and Ahl, T. 1976 V*attern-recenta Sediment ach Sedimentkemi*. National Swedish Protection Board, SNV PM 740, Uppsala, 167 pp.

Häkanson, L. and Jansson, M. 1983 *Principles of Lake Sedimentology*. Springer Verlag, Heidelberg, 316 pp.

Hodson, P.V., Whittle, D.M., Wong, P.T.S., Borgman, V., Thomas, R.L. Chav, Y.K., Nriagu, J.O. and Hallet, D.J. 1983 Lead contamination of the Great Lakes and its potential effects on aquatic biota. *Environ. Sci, Technol.*, **16**, 335-369.

Hutchinson, G.E. 1957 *A Treatise on Limnology. I. Geography, Physics and Chemistry*. John Wiley and Sons, New York, 1015 pp.

IJC 1977 *Pollution From Land Use Activities Reference Group (PLUARG). Annual Report to the International Joint Commission*, 1976, Windsor, Ontario.

ILEC 1987-1989 *Data Book of World Lake Environments - A Survey of the State of World Lakes*. International Lake Environment Committee, Otsu, Japan.

Jackson, T. 1980 Mercury speciation and distribution in a polluted river-lake system as related to the problems of lake restoration. In: *Proceedings of the International Symposium for Inland Waters and Lake Restoration (US EPA/OECD)*, September 1980, Portland, Maine, 93-101.

Jeffries, D.S., Wales, D.L., Kels, J.R.M. and Linthurst, R.A. 1986 Regional characteristics of lakes in North America. Part I Eastern Canada. *Water Air Soil Pollut.*, **31**, 555-567.

Jernalov, A. 1971 Release of methylmercury from sediments with layers containing inorganic mercury at different depths. *Limnol. Oceanogr.*, **15**, 958-960.

Jorgensen, S.E. and Vollenweider, R.A. [Eds] 1989 *Guidelines of Lake Management. Volume 1 Principles of Lake Management*. International Lake Environment Committee Foundation, Otsu, Japan and United Nations Environment Programme, Nairobi, 199 pp.

Malins, D.C., McCain, B.B., Brown, D.W., Varanasi, V., Krahn, M.M., Myers M.S. and Chan, S-L. 1987 Sediment associated contaminants and liver diseases in bottom-dwelling fish. *Hydrobiologia*, **149**, 67-74.

Meybeck, M. 1995 Le lacs et leur bassin versant. In: R. Pourriot and M. Meybeck [Eds] *Limnologie Générale*. Masson, Paris, 6-60.

Meybeck, M., Chapman, D. and Helmer, R. [Eds] 1989 *Global Freshwater Quality. A First Assessment*. Blackwell Reference, Oxford, 306 pp.

Mortimer, C.H. 1942 The exchange of dissolved substances between mud and water in lakes. *J. Ecol.*, **30**, 147-201.

Munawar, M. and Munawar, I.F. 1982 Phycological studies in Lakes Ontario, Erie, Huron and Superior. *Can. J. Bot.*, **60**, 1837-1858.

Nauman, E. 1932 *Limnologische Terminologie. Hanb. Biol. Arbert method.* Abt. IX, Teil 8, Urban and Schwarzenberg, Berlin, 776 pp.

OECD 1982 *Eutrophication of Waters. Monitoring, Assessment and Control.* Organisation for Economic Co-operation and Development, Paris, 154 pp.

Pourriot, R. and Meybeck, M. [Eds] 1995 *Limnologie Générale*. Masson, Paris, 956 pp.

Swaine, W.R. 1990, 1991 Ecologic Ltd, Ann Arbor, Michigan, personal communication.

Thienemann, A. 1925 Die binnengewasser Nittelewopas Eine limnologische Einfuhrung. *Binnengegewasses*, **1**, 1-225.

Thienemann, A. 1931 Der Produktionsbegriff in der Biologie. *Arch. Hydrobiol.*, **22**, 616-622.

Thomas, R.L., Williams, D.J., Whittle, M.D., Gannon, J.E. and Hartig, J.H. 1988 Contaminants in Lake Ontario - A case study. In: *Proceedings of the Large Lakes of the World Conference, Mackinaw, Michigan, 1987*, **3**, 328-387.

Vollenweider, R.A. 1968 *Scientific Fundamentals of the Eutrophication of Lakes and Flowing Waters with Special Reference to Nitrogen and Phosphorus as Factors in Eutrophication*. Technical Report DA5/SCI/68.27, Organisation for Economic Co-operation and Development, Paris, 250 pp.

Wetzel, R.G. 1975 *Limnology*. Saunders, Philadelphia, 743 pp.

Williams, J.D.H., Jaquet, J-M. and Thomas, R.L. 1976 Forms of phosphorus in the surficial sediments of Lake Erie. *J. Fish. Res. Board Can.*, **33**, 385-403.

Chapter 8*

RESERVOIRS

8.1 Introduction

Reservoirs are those water bodies formed or modified by human activity for specific purposes, in order to provide a reliable and controllable resource. Their main uses include:

- drinking and municipal water supply,
- industrial and cooling water supply,
- power generation,
- agricultural irrigation,
- river regulation and flood control,
- commercial and recreational fisheries,
- body contact recreation, boating, and other aesthetic recreational uses,
- navigation,
- canalisation, and
- waste disposal (in some situations).

Reservoirs are usually found in areas of water scarcity or excess, or where there are agricultural or technological reasons to have a controlled water facility. Where water is scarce, for example, reservoirs are mainly used to conserve available water for use during those periods in which it is most needed for irrigation or drinking water supply. When excess water may be the problem, then a reservoir can be used for flood control to prevent downstream areas from being inundated during periods of upstream rainfall or snow-melt. Particular activities such as power generation, fish-farming, paddy-field management or general wet-land formation, for example, are also met by constructing reservoirs. By implication, they are also water bodies which are potentially subject to significant human control, in addition to any other impact. Reservoirs are, nonetheless, a considerable, frequently undervalued, water resource: approximately 25 per cent of all waters flowing to the oceans have previously been impounded in reservoirs (UNEP, 1991).

Reservoirs range in size from pond-like to large lakes, but in relation to natural lakes the range of reservoir types and morphological variation is generally much greater. For example, the most regular, and the most irregular, water bodies are likely to be reservoirs. This variability in reservoirs, allied to management intervention, ensures that their water quality and process

This chapter was prepared by J. Thornton, A. Steel and W. Rast

behaviour is even more variable than may be characterised as limnologically normal. As reservoirs are so variable, it can often be misleading to make any general statements about them without significant qualification as to their type.

Reservoirs do, nevertheless, share a number of attributes with natural lakes and some are even riverine in their overall nature. Generally, all reservoirs are subject to water quality requirements in relation to a variety of human uses. The variation in design and operation of control structures in reservoirs can provide greater flexibility and potential for human intervention than in natural lakes (and, therefore, considerable scope for management and control) with the objective of achieving a desired water quality. However, the nature of the intervention or control can complicate the development and operation of water quality monitoring programmes, as well as the interpretation of resultant data (especially as the nature of these controls may vary over time, altering the responses of the system). For reservoirs, therefore, the assessment process must take full account of the direct management influences on the water body.

8.1.1 Definitions

Unfortunately, descriptive nomenclature of artificially created water bodies is frequently confusing. The term "dam" is often applied to both the physical structure retaining the water, and the water so retained. For the purposes of this chapter, dam will be used solely to describe the physical structure, and the term reservoir will be used to denote the artificially created water body. A reservoir is therefore:

- a water body contained by embankments or a dam, and subsequently managed in response to specific community needs; or
- any natural waters modified or managed to provide water for developing human activities and demands.

Reservoirs formed by a dam across the course of a river, with subsequent inundation of the upstream land surface are often called impoundments. Water bodies not constructed within the course of the river and formed by partially or completely enclosed water-proof banks (and usually filled by diverted river flows or pipes) are often referred to as off-river, or bunded, reservoirs. Reservoirs created by dams or weirs serially along a river course form a cascade (Figure 8.1).

Impoundments and off-river reservoirs, therefore, form the two main artificially created water body types in which differing amounts of human control are possible, and which are usually very different in their morphology. Impoundments tend to be larger, more sinuous and dendritic than off-river reservoirs, and the main control of their contained volume of water

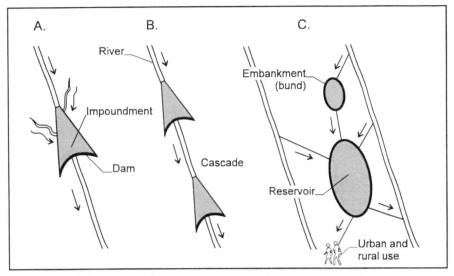

Figure 8.1 Various types of reservoirs: **A**. Reservoir created by damming a river; **B**. A cascade formed from a series of river impoundments (the upstream reservoir may be known as a pre-impoundment); **C**. Bunded or embanked reservoirs with controlled inflows and outflows to and from one or more rivers.

can usually only be exercised at an outlet (in other reservoirs both input and output are controllable). Off-river reservoirs are often virtually completely isolated from the local terrain by enclosing and water-proofed embankments, whereas impoundments are subject to considerable interaction with the flood plain within which they were formed.

The wide variation in size and application of reservoirs results in them forming an intermediate type of water body between rivers and natural lakes (Figure 8.2; see also section 1.1), although the simple implications of this may often be subject to considerable behavioural distortion because of artificial management control effects.

8.2 Construction and uses

Much of the variation in construction of reservoirs is the result of the original purposes for which they were constructed, although their principal uses can change over time. Other elements of variation in form are introduced as a result of regional geology and the availability of construction materials. For example, where clay deposits are plentiful, dam or embankment construction may be based on earthen structures with impervious clay cores, while in other areas dams may be constructed solely of massed concrete.

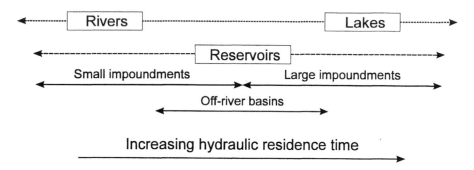

Figure 8.2 The intermediate position of reservoirs between rivers and natural lakes is determined by their water residence time and degree of riverine influence (Based on Kimmel and Groeger, 1984)

8.2.1 Typology

Figure 8.3 shows some generic reservoir forms. It is important to realise that such different forms, for similar volumes stored, can cause major differences in the physico-chemical and biological nature of the stored waters (see section 8.3).

Impoundments

Impoundments are formed from a variety of types of dam (Linsley and Franzini, 1972): earth-fill, gravity, arch, buttress and other, minor types. Earth-fill dams (e.g. Figure 8.3A) are the most common; about 85 per cent of dams in the 15 m to 60 m height range are of this type of construction (van der Leeden *et al.*, 1990). Arched dams are most commonly used in situations demanding extremely high walls, and account for about 40 to 50 per cent of the very large dams exceeding 150 m in height (Figure 8.3B). Small structures, of 3 m to 6 m in height, created to divert or impound water flow for various purposes can also be constructed by, for example, bolted timber structures. This form of dam typically has a life of only about 10 to 40 years. Novel technologies may also be used, such as inflatable dam structures which can be cheap to install and maintain, and may be collapsed in order to scour the dam bed.

Dammed rivers form the largest reservoirs; up to 150 km^3 in the case of the Aswan High Dam. Usually, their maximum depth is much greater than their mean depth (often nearly 3:1) and as a result a considerable proportion of their volume is derived from the quite shallow (i.e. littoral) areas which were created when the former riverine flood plain was permanently flooded. Impoundments tend to have higher watershed area to water surface area ratios

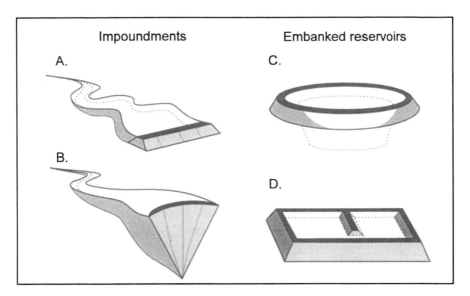

Figure 8.3 Examples of the principal reservoir configurations: **A.** shallow, 'U' shaped; **B.** deep, 'V' shaped; **C.** deep, regular; **D.** shallow, regular

than similar lakes. They also have a tendency to a sinuous form and a convoluted shoreline as a result of the artificial inundation of a terrain which would not otherwise contain a similar sized natural lake. This complex form and the tendency to have substantial shallows and deep sumps leads to horizontal and lateral heterogeneity in their physico-chemical and biological water quality variables (Figure 8.4) (see also section 8.3 and Figures 8.14 and 8.18). The differences mentioned above are highlighted by the comparison of reservoirs and natural lakes in North America given in Table 8.1.

Other reservoirs
Off-river reservoirs frequently have to be wholly constructed. This normally entails forming an embankment about the border of the reservoir site. Embankments are most often either concrete or clay-cored earth structures. Earth embankments are normally protected by stone or concrete on the water side. Further walling (baffles) may be placed within the reservoir site in order to cellularise the basin and control flow patterns. The reservoir volume is usually formed by excavation of the site, together with elevation of the embankment. As embankment slopes are normally in the order of 5:1, a high embankment would have a considerable ground-level footprint, and a compromise on embankment height is generally necessary in order to maximise the volume storage on any particular site.

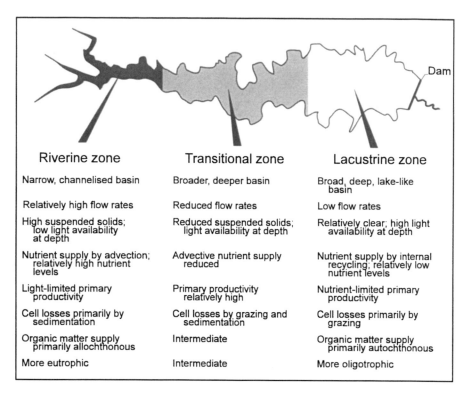

Riverine zone	Transitional zone	Lacustrine zone
Narrow, channelised basin	Broader, deeper basin	Broad, deep, lake-like basin
Relatively high flow rates	Reduced flow rates	Low flow rates
High suspended solids; low light availability at depth	Reduced suspended solids; light availability at depth	Relatively clear; high light availability at depth
Nutrient supply by advection; relatively high nutrient levels	Advective nutrient supply reduced	Nutrient supply by internal recycling; relatively low nutrient levels
Light-limited primary productivity	Primary productivity relatively high	Nutrient-limited primary productivity
Cell losses primarily by sedimentation	Cell losses by grazing and sedimentation	Cell losses primarily by grazing
Organic matter supply primarily allochthonous	Intermediate	Organic matter supply primarily autochthonous
More eutrophic	Intermediate	More oligotrophic

Figure 8.4 Longitudinal zonation of water quality conditions in reservoirs with complex shapes formed by damming rivers (Modified from Kimmel and Groeger, 1984)

Embanked reservoirs are generally very simple in shape (from near-circular to rectangular) and virtually uniform in depth (Figure 8.3C, D). Generally, mean depths are greater than 80 per cent of the maximum depths and the volumes contained are less than 100×10^6 m^3 (0.1 km^3). Their tendency to great regularity in shape is indicated by shoreline development ratios (D_L) of between 1 and 2, compared with values of 5 to 9 for impoundments (see Table 8.1 for an explanation of shoreline development). This regularity tends to reduce horizontal variability in physical, chemical and biological variables, particularly when associated with relatively short retention times. Apart from any management impacts, these morphological features influence the ecological behaviour of such reservoirs, and so must be taken into account fully during the planning of monitoring and assessment programmes (see sections 8.3 and 8.5).

Table 8.1 A comparison of geometric mean values for some selected variables from natural lakes and reservoirs in North America

Variable	Natural lakes	Reservoirs	Probability that mean values are equal
Number of water bodies (n)	107	309	
Morphometric variables			
Surface area (km^2)	5.6	34.5	0.0001
Watershed area (km^2)	222.0	3,228.0	0.0001
Watershed : surface area ratio	33.1	93.1	0.0001
Maximum depth (m)	10.7	19.8	0.0001
Mean depth (m)	4.5	6.9	0.0001
Shoreline development ratio (D_L)[1]	2.9 (n = 179)	9.0 (n = 34)	0.001
Areal water load (m a^{-1})	6.5	19.0	0.0001
Water residence time (years)	0.74	0.37	0.0001
Water quality variables			
Total phosphorus (mg m^{-3})	54.0	39.0	0.02
Chlorophyll *a* (mg m^{-3})	14.0	8.9	0.0001
Areal phosphorus load (g m^{-2} a^{-1})	0.87	1.7	0.0001
Areal nitrogen load (g m^{-2} a^{-1})	18.0	28.0	0.0001

[1] Shoreline development, D_L, is a crude measure of surface complexity and is defined as the ratio of a water body's actual shoreline length to the circumference of a circle having the same area as that of the water body. D_L is thus some measure of how nearly circular (≡ regular) is the plan form of a water body and typical values are:

Shape	D_L
Circle	1
5:1 Rectangle or Ellipse	≈1.5
10:1 Triangle	≈2.5
Natural Lakes	2–5
Impoundments	3–9

Sources: Hutchinson, 1957; Leidy and Jenkins, 1977; Thornton *et al.*, 1982; Ryding and Rast, 1989

8.2.2 Uses

Reservoirs are generally used for water storage along a river course, for off-river water storage, or for inter-basin water transfer schemes. The latter usually require a variety of storage and transfer structures. An understanding of the natural and artificial influences common to these various uses helps in the interpretation or prediction of the physical, chemical and biological behaviour of the reservoir.

Impoundments and cascades

Impoundments formed by dams are the most common form of large reservoirs (Veltrop, 1993) and can occur singly or in series on a water course as a cascade, where water passing through or released from one impounded section flows into a second, and so on. Examples include the Volga-Kama cascade in Russia (Znamenski, 1975) the Vlatava cascade in the Czech republic (Straskrabá and Straskrabová, 1975), the Tietê complex in Brazil (Tundisi, 1981) and the Highland Lakes of central Texas, USA (Rast, 1995). Depending on the topography of the land surrounding the river upstream of the dam wall, these impoundments can be dendritic to varying extents. In extreme cases, the impounded waters are riverine in appearance and with complex shorelines. In less extreme cases, the impoundment basins can mimic natural lakes, being more rounded and less riverine.

The creation of cascades can yield water quality benefits to the impounded sections downstream as a result of the retention of contaminants in the upstream impoundment. In such cases, the upstream impoundment has sometimes been referred to as a pre-impoundment. Many chemical and biological variables can be affected in lakes and reservoirs downstream of a major impoundment, as illustrated in Figure 8.5.

Off-river storage
Off-river storage is usually achieved by constructing reservoirs (with embankments) off the main course of the river, although usually relatively close to it. They may be operated singly or in series. These reservoirs are frequently used for water supply purposes, acting as a means of controlling and modifying water quality and quantity transferred from the river to the treatment plant. They may also be used as a local resource for activities such as fish-farming or agricultural purposes (particularly rice cultivation) (Schiemer, 1984). Off-river storage reservoirs are generally filled and emptied by pumping, although some are designed to capture flood flows which enter the reservoir by gravity via a diversion channel or similar mechanism.

Inter-basin transfers
Inter-basin transfer schemes are artificial hydrological systems which, because of their operation and potential impacts, require specific attention. Inter-basin transfer schemes often occur in water-poor regions, where water resources engineering efforts have been traditionally directed towards providing water from a variety of sources to centres of human settlement (Petitjean and Davies, 1988). This requires water to be drawn from watersheds outside of that in which the settlement was established, necessitating the transfer of water from one watershed to another. Transfers may be achieved in several ways, including pumping, but the most common involves creating a network of canals, balancing reservoirs and storages which make optimum use of gravity flows. While these integrated systems generally make use of natural water courses wherever possible to convey waters

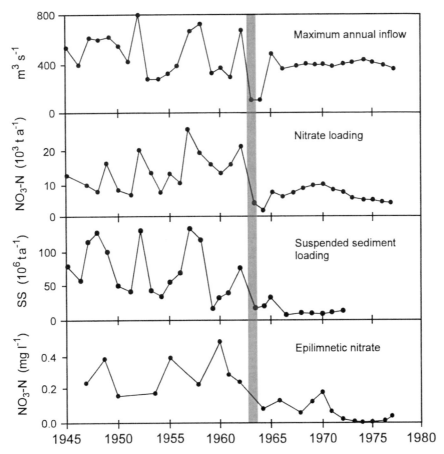

Figure 8.5 Effects of an upstream reservoir (Lake Powell) on some selected water quality characteristics of the downstream Lake Mead, USA. The shaded bar indicates the completion of the dam and the formation of the reservoir. (After Prentki and Paulson, 1983)

between catchments, they occasionally create balancing reservoirs that bear a greater resemblance to off-river storages than to the more traditional form of in-stream impoundment because the inflow and outflow are artificially created (i.e. the basin previously had no natural outlet and was internally drained).

Variants of such schemes include pumped-storage facilities where waters are alternately drained from, and returned to, reservoirs usually as a means of generating hydro-electricity during periods of peak demand (using power generated by other sources at other times to pump the water back up-gradient). This same technique has been used during drought periods in the

Vaal River, South Africa, to reverse the river flow and augment water availability in upstream water supply reservoirs (DWA, 1986). Each of these operating regimes presents specific challenges for the monitoring and characterisation of water quality.

8.3   Special characteristics of reservoirs

The form and operation of reservoirs can have such an influence on their water quality characteristics that it is usually necessary to adopt some modifications to the more common strategies for assessment of water quality used in natural lakes. Reservoir characteristics are, therefore, described in some detail in relation to their form and operation in the following sub-sections.

Reservoirs in densely populated or agricultural areas, and which receive little management control, have a tendency to be highly enriched although some also have a more rapid flushing regime than natural lakes which partly masks this enrichment effect (see section 8.4.1). There is, therefore, a potentially high dependency of productive capacity on management regime, source water qualities, and internal chemical and biological process rates.

Reservoirs show many of the same basic hydrodynamic, chemical and biological characteristics as the natural lakes described in Chapter 7. However, even when reservoirs are formed by the damming of a single river, creating water bodies which may look very much like natural lakes, the operating regime determined by the purpose for which the reservoirs were created (e.g. hydro-power generation, irrigation or domestic water supply) may significantly alter their physico-chemical character and biological responses. These responses may also be made more difficult to interpret as a result of the changing demands placed on such reservoirs over time (such as additional recreational uses). In most instances, the peculiar form of a reservoir, its mode of operation, and an unnatural location and shape may cause considerable, actual variation of the basic limnological behaviour (Straskrabá *et al.*, 1993).

Reservoirs formed by river impoundment undergo great changes in water quality during the early stages of their formation whilst a new ecological balance is becoming established. Balon and Coche (1974) were among the first to identify and document the sequence of eutrophication, recovery, and stabilisation following impoundment using data from Lake Kariba, Zimbabwe, the then-largest reservoir in the world (volume 160 km^3; area 5,100 km^2). Their work showed that the release of organically-bound elements from flooded vegetation, excreta and soils could result in an initial high level of biological production (sometimes called a trophic surge), suggesting a very different water quality than would actually be seen in the future. The duration

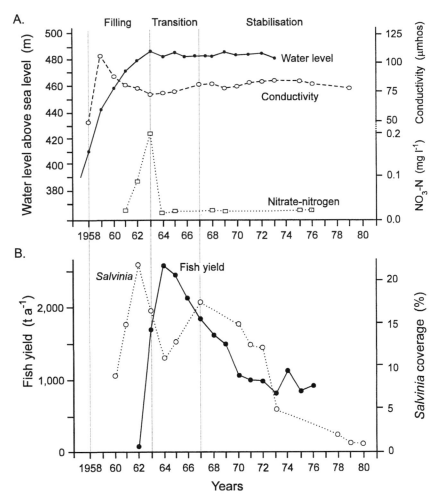

Figure 8.6 The changes occurring after the filling of Lake Kariba, Zimbabwe, showing the transition to stabilised conditions. **A**. Variations in water level, salinity and nitrate-nitrogen. **B**. Coverage by the aquatic plant *Salvinia molesta* and total yields from the inshore fishery on the Zimbabwean side of the reservoir. (Based on Marshall and Junor, 1981 and Marshall *et al.*, 1982)

of this increased productivity varies with the amounts of source material present within the flooded river basin, the nature of the soils and previous land uses, the elemental concentrations in the source waters and the degree of basin clearing prior to inundation (Figure 8.6). In these transient conditions, particularly careful assessment of the water quality is essential for the effective management of the reservoir. Some general effects of impounding running waters are summarised in Table 8.2.

Table 8.2   Changes in water quality characteristics arising in water bodies created by impoundment compared with the characteristics of a river prior to impoundment

Characteristics of river stretch	Not impounded	Minor impoundment	Major impoundment/ chain of impoundments
Flow velocity at mean low water discharge	20 cm s^{-1}	5–10 cm s^{-1}	5 cm s^{-1}
Residence time in the area of impoundment	na	1–5 days	> 5 days
Mean depth of water	1.5 m	2–5 m	5 m
Physical O_2 input per unit of volume	xxx	xx	x
Day/night fluctuations of O_2 concentration; tendency to O_2 supersaturation	x	xx	xxx
Self-purification of organic pollution[1]	x	xx	xxx
Ammonia oxidation	xxx	xx	xx
Fine sediment deposition at low discharges	—	x	xxx
Resuspension of fine sediment at high discharges	xxx	xx	x
O_2 depletion in sediment	—	x	xxx
Turbidity due to suspended sediment	xxx	xx	x
Secondary pollution, algal growth and turbidity due to algae	x	xx	xxx
Development of higher aquatic plants	x	xx	xxx

x to xxx	Minor to significant effect	na	Not applicable
—	Negligible or no effect	[1]	Resulting from biological activity

The following sub-sections give some examples of ways in which the water quality characteristics of reservoirs are influenced by reservoir type and operation to an extent that significantly different monitoring and assessment strategies may be required from those recommended for natural lakes.

8.3.1   Thermal characteristics
Most reservoirs experience the same thermal structure development as lakes. If sufficiently deep they become dimictic in temperate areas, or monomictic in polar and tropical areas, during their annual cycle. Many reservoirs are, however, polymictic due to their relatively shallow depths or the effects of

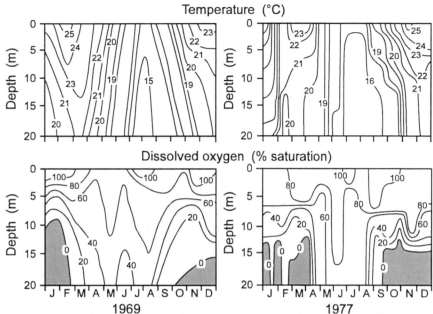

Figure 8.7 Variation in the seasonal temperature and dissolved oxygen profiles of Lake Chivero, Zimbabwe 1969 and 1977 (Modified from Thornton, 1982)

enhanced flow induced turbulence. In part, their tendency towards thermal stratification is determined by the geography of their basins. Reservoirs constructed on broad flood plains are generally less deep and more susceptible to multiple mixing that those constructed in confined river channels, canyons or ravines. A particular problem associated with thermal stratification is the potentially large variations in water quality that may be induced by internal waves (seiches) at fixed-level off-takes at the reservoir shores. These seiches can have a range sufficiently great to present epilimnetic water (warm, oxygenated and with a high algal density) and hypolimnetic water (cold, anoxic, concentrated Fe^{3+}, Mn^{3+}, H_2S, etc.) alternately to the outlet. This can create particular problems for water users requiring a constant water quality. Similar events may occur transiently under periods of sustained wind-stress.

In tropical reservoirs, as with lakes, the onset of stratification occurs over a smaller temperature range as a result of the higher ambient temperatures in these systems (Serruya and Pollingher, 1983). Figure 8.7 shows summarised data for oxygen and temperature for Lake Chivero, Zimbabwe (volume 250×10^6 m^3; area 26 km^2) (Thornton, 1982). The isopleths illustrate the monomictic condition typical of many warm water lakes, but also show the same reservoir undergoing multiple annual mixings in 1977 (a rare event in

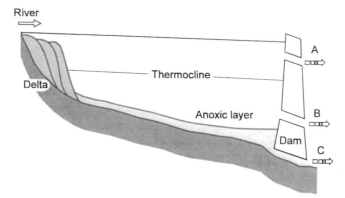

Figure 8.8 The quality of water withdrawn from a reservoir determined by the depth
and location of water withdrawal points (A, B, C) in relation to the thermocline:
A Oxygenated epilimnetic water; B Anoxic hypolimnetic water; C Anoxic,
sediment-laden water (After Cole and Hannan, 1990)

this reservoir). Reservoirs can exhibit greater intra-annual variability than
natural lakes and their annual variation can be affected by changes in operational protocols and water usage. These characteristics must be considered
when drawing possible conclusions about time-based trends in water quality
within reservoirs.

Impoundments can also exhibit a range of thermal patterns at varying times
of the year because of fluctuating water levels caused by their operating
regimes. They may be deep enough to stratify while at full supply level (or
maximum capacity) but shallow enough to mix repeatedly or constantly at
lower stage levels. This depth variation may also cause atypical behaviour,
such as in the hydroelectric dams in northern Sweden where winter drawdown (spring filling) prevents the ice covered waters from exhibiting the
amictic characteristics of their natural relatives. Similarly, few reservoirs
would be expected to be meromictic; although the solar ponds of the Sinai,
Israel, are possibly an exception to this generalisation.

The use of multi-level off-takes and particularly valves which scour water
from the bottom of the reservoir (Figure 8.8) can significantly modify the
thermal structure and chemical gradients of reservoirs by selectively remov-
ing either surface, mid-depth or bottom waters. Not only do the resulting
discharges modify the downstream flow regimes and have potential downstream impacts (Davies, 1979), but they can also selectively remove cold,
anaerobic waters (the hypolimnion) from the reservoir, leaving an apparently
unstratified water body behind. Similarly, the hydraulic influences of these
off-takes can create some unique "short-circuit" effects within the water
body which, when combined with the volumes of water generated by the

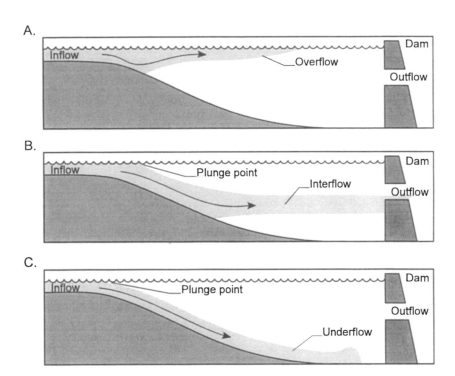

Figure 8.9 Density flows in reservoirs: **A.** Overflow of warm incoming waters. **B.** Interflow. **C.** Underflow (After Moore and Thornton, 1988)

large watershed to surface area ratio, encourage an underflow of denser water that can modify the impact of riverine inflows on the reservoir water quality (Figure 8.9).

In some reservoirs, facilities are specifically provided to modify substantially, or even to eliminate completely, thermal stratification. If thermal stratification is eliminated in a deep reservoir (e.g. mean depth 20 m), the sediments experience abnormal conditions of temperature and oxygenation within what would otherwise be the summer stagnation period. Thus, relative to a natural lake, a completely atypical thermal and hydrodynamic regime can be established. Stratification control facilities can also greatly enhance the natural turbulence in reservoirs, causing considerable modification of the vertical and horizontal distributions of chemical variables, together with biological organisms and their production dynamics. This type of control may be exercised by harnessing input energy when pumping water into the reservoir (e.g. by jetting the input water), or by a separate system of air- or water-pumping.

8.3.2 Chemical characteristics

Chemical events which occur in reservoirs are not unique compared with lakes, although their timing and intensity may be unusual. Inflow–outflow velocities and water body morphometry can also dramatically affect the within-lake characteristics of the reservoir itself (Figure 8.10). In reservoirs subject to thermal stratification (as in similar lakes), the suppression of vertical transport processes at the thermocline normally allows an oxygen gradient to occur which can cause anoxic conditions to develop in the hypolimnion. The rich sediments of most reservoirs would then almost certainly ensure the movement of copious quantities of iron and manganese back into the water column, with the possible eventual formation of large quantities of hydrogen sulphide. As mentioned before, these events can significantly affect the users of the reservoir if water is withdrawn from the hypolimnion. Hypolimnetic deoxygenation is more common in tropical and sub-tropical reservoirs because of the higher rates of decomposition occurring at the ambient temperatures of those reservoirs. De-oxygenation is a particular problem at the first inundation of tropical impoundments. This is due to the oxygen demand of the decaying, submerged carbonaceous materials exceeding the available oxygen supply until a more stable aquatic regime is established.

When assessing the distributions of chemical variables in reservoirs, and their associated process rates, it is always necessary to bear in mind the potential effects of substantial draw-offs. These can directly affect the depth distributions of water quality variables and result in quite unusual chemical and biological profiles. This is illustrated by metalimnetic oxygen minima created by selective withdrawal of waters at that depth (Figure 8.11).

In some reservoirs, denitrification can result in a substantial reduction in the inorganic nitrate-nitrogen concentrations compared with the concentrations of the source waters (see example in Figure 8.17). This is usually associated with a greater instability in the water column due to the enhanced internal turbulence in the reservoir, its relative shallowness, long water retention time (e.g. > 6 months) and reduced dissolved oxygen content of the water. If the dissolved oxygen at the sediment surface is maintained at about 10 per cent saturation combined with a slow circulation in the remainder of the water column, considerable quantities of nitrogen gas can be released from the sediment into the overlying water. In areas where source waters are contaminated by agricultural run-off, the combined effects of storage (acting as a buffer or barrier between source water and supply to consumers) and denitrification can provide enormous benefits to water supplies by reducing the exposure of consumers to surface water derived nitrates (see Figure 8.17). In

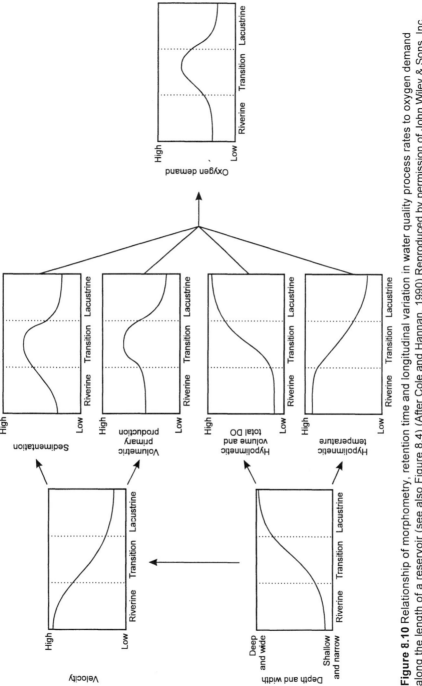

Figure 8.10 Relationship of morphometry, retention time and longitudinal variation in water quality process rates to oxygen demand along the length of a reservoir (see also Figure 8.4) (After Cole and Hannan, 1990) Reproduced by permission of John Wiley & Sons, Inc.

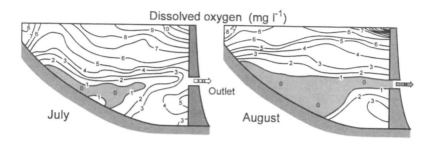

Figure 8.11 An example of modifications in the depth distribution of dissolved oxygen by currents caused by mid-depth water withdrawal (Based on Ebel and Koski, 1968)

deeper, more oxygenated reservoirs, denitrification is a less effective process particularly if the sediments have been exposed and dry out, as may occur in drought years (Whitehead, 1992).

One aim of thermal stratification control in reservoirs (or any other managed water body) is often to eliminate the chemical problems associated with stratification, rather than just to alter the thermal structure of the water mass. Complete destratification can ensure substantial dissolved oxygen concentrations from surface to sediments, as illustrated in Figure 8.12 for a deep reservoir , i.e. mean depth > 15 m (volume 19.6×10^6; area 1.3 km^2) (compare with the natural stratification illustrated in Figures 7.9, 7.22 and 8.7). Maintenance of an oxidised microzone at the sediment surface is also quite possible. This prevents the interchange of many noxious or undesirable substances between water and sediments, either directly or because they remain bound to the iron complexes formed in aerobic conditions. The sediment biota also experience warm, oxygenated conditions. Fully aerobic sediments are less effective at denitrification, although some nitrate loss can still occur by nitrification, especially if the sediments are rich in carbonaceous materials (O'Neill and Holding, 1975). An alternative approach to destratification is to re-aerate the hypolimnetic waters, by direct supply of oxygen without disrupting the thermal profile (Bernhardt, 1978). By this means, the cool hypolimnetic water is maintained as a water supply or fisheries resource even through the warm months of the year. Such management actions (usually in the pursuit of adequate drinking water quality) completely alter the chemical characteristics of a reservoir from its more natural conditions.

In addition to control of thermal stratification, a variety of other interventions such as pre-impoundments, chemical modification of water quality within smaller reservoirs (e.g. addition of a coagulant such as alum or an

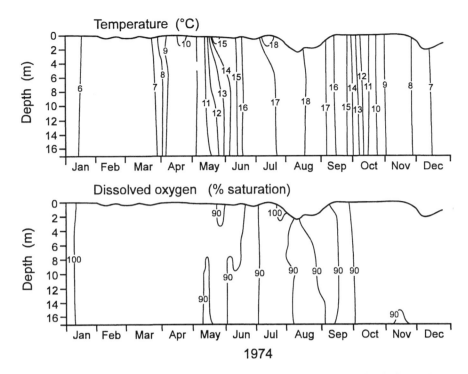

Figure 8.12 Uniformity in annual temperature and dissolved oxygen depth distributions in Queen Elizabeth II reservoir, UK achieved by jet-mixing in the water column. Compare with an unmanaged situation in a similar reservoir in Figure 8.18.

inhibitor such as copper sulphate) and sediment manipulation, can all significantly alter the chemical characteristics of a reservoir. Similarly, the transfer of water between basins can result in a reservoir having a very different ionic composition to that which would occur naturally in the receiving basin. Petitjean and Davies (1988) document such effects from a variety of interbasin transfer schemes throughout the world.

8.3.3 Biological characteristics

As with chemical properties, biological characteristics of reservoirs are fundamentally the same as for natural lakes, but the peculiar physico-chemical conditions which may occur in managed reservoirs can also result in biological production and ecological successions quite different in degree and timing from an equivalent natural lake. The biology of reservoirs is particularly influenced by:

- the effects of greater flow-through and turbulence,
- the nutrient loading they normally receive,

- the plankton populations of the inflows relative to those in the reservoir itself, and
- the management controls applied and their effects.

An understanding of the biological characteristics of reservoirs, in addition to their physical and chemical features, is important to their management. Biological assessment techniques are commonly included in reservoir assessments because some uses of reservoir water (e.g. drinking water supply, fisheries, recreation) are particularly subject to the effects of biological communities (especially algal populations) and their interactions (e.g. between fish and plankton groups).

Enhanced turbulence, such as produced by artificial destratification measures, can maintain phytoplankton in suspension when they would otherwise sink to the sediments, or can affect productivity by carrying them into deeper waters with insufficient light penetration to support photosynthetic production (Horn and Paul, 1984). The increased turbulence can also maintain greater amounts of organic material in suspension, sustaining filter-feeding zooplankton when food resources might otherwise be much lower and limiting to their growth (Steel, 1978a; Horn, 1981).

The extreme water level fluctuations of some reservoirs also pose biological problems for the establishment of rooting macrophytes and spawning areas for fish. When highly managed, the unusual vertical profiles of temperature and chemical variables present abnormal conditions for benthic organisms.

A major problem for reservoirs (as for lakes), and a driving force behind much of their management, is the biological outcome of eutrophication (see section 8.4.1), particularly as manifest in the enhanced phytoplankton growth. The resultant phytoplankton population densities cause problems for recreational users, for treatment processes in drinking water supply and directly to consumers. Users may be merely reluctant to use the water because of its taste and odour but, at worst, they may experience a toxic effect. Bunded reservoirs used for water supply are particularly prone to benthic mats of *Oscillatoria* (a filamentous blue-green alga) which imparts considerable earthy or musty tastes and odours to the water (van Breeman *et al.*, 1992). Many strains of cyanobacteria (blue-green algae) can produce an active intra-cellular toxin, particularly when the phytoplankton are senescent and decaying. Toxin production may occur naturally or as a result of management attempts to control large phytoplankton populations, for example, by the application of copper sulphate. The toxins released may then have a direct effect on water consumers (livestock and humans). Cyanobacterial toxins are becoming an increasing source of concern for water managers where eutrophic reservoirs are used for recreation, aquaculture or potable supplies (Lawton and Codd, 1991).

Warm water reservoirs (tropical/sub-tropical) may have a particular problem with floating plants, such as *Salvinia molesta* in Lake Kariba and *Eichhornia crassipes* in Lake Brokopondo, South America (Van der Heide, 1982). Such exceptional macrophyte growths can significantly reduce dissolved oxygen concentrations in the water as they decompose, interfering with fishing and restricting potable and industrial use.

Standard approaches to biological assessment and management of reservoirs are complicated by the extreme variety, location and uses of many reservoirs. As in natural lakes, trophic relations in reservoirs vary from simple food chains to highly complex food webs; the latter often being characteristic of the conditions in reservoirs created by flooding river valleys. In reservoirs with simple morphometry, the possible absence of littoral zones and macrophyte refuges for herbivorous zooplankton can increase the zooplankton's exposure to predation by fish. Large fish stocks can deplete herbivorous zooplankton populations to levels which enable phytoplankton populations to grow unrestrained by any grazing by zooplankton (i.e. to the limits of nutrient or energy availability). Factors which reduce fish stocks (and consequently their predation on zooplankton) can enable zooplankton populations to reach levels which can potentially deplete phytoplankton populations through grazing effects, even when phosphate (required for algal growth) is in excess. In reservoirs with a varied morphology (different shoreline habitats as well as sheltered and exposed and deep and shallow areas), many other inter-trophic effects may also take place because the niche diversity is sufficiently high to allow the coexistence of much more varied biological communities (Benndorf *et al.*, 1988). A particularly clear understanding of the interactions of such factors is necessary for a valid assessment of the biological nature of these types of water body.

Microbiology

The typical microbiological characteristics of reservoirs appear to be the same as for lakes (Straskrabová and Komárková, 1979). The main source of energy for bacterioplankton is decomposable organic material. With short retention times, this material is mainly derived from the inflows into the reservoir. As the retention time increases in larger reservoirs relative to the outflow rates, the main source of energy becomes the primary productivity within the reservoir. Seasonal changes in bacterial populations relate mostly to seasonal variations in source water conditions and the primary productivity in the reservoir.

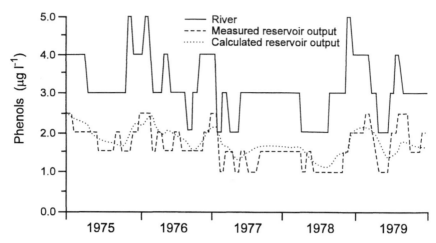

Figure 8.13 Calculated and measured phenol concentrations in water withdrawn from the Honderd en Dertig reservoir, The Netherlands, compared with the incoming river water. The reduction in phenol concentrations in the reservoir water is achieved by microbiological decomposition during water storage. (After Oskam, 1983)

Microbiological activity is important in the degradation of organic compounds of industrial origin which are frequently present in the source waters of lakes and reservoirs in highly populated and industrialised areas (Oskam, 1983). Together with dilution and the elimination of the source of the organic contaminants, microbial breakdown is a major factor in maintaining a satisfactory drinking water quality in some reservoirs, e.g. Honderd en Dertig, the Netherlands (volume 32 x 10⁶ m³; area 2.2 km²) (Figure 8.13). This process can be further enhanced by the serial operation of reservoirs, if available.

Many reservoir source waters are subject to gross viral and bacterial pollution because of effluent disposal and run-off. Reservoirs can form a barrier to the survival of pathogenic micro-organisms; water storage for 20 days or more normally reduces river water coliform concentrations to about 1 per cent of their initial value, due to the relatively alien reservoir temperature, exposure to ultra-violet radiation, and ionic conditions. This effect appears to be maintained even when reservoirs are subject to substantial mixing for thermal and chemical control purposes.

8.3.4 Trophic classification

Reservoirs exhibit a range of trophic states in a manner similar to the natural lakes described in Chapter 7. Individually, however, they can also do so over their length and can exhibit considerable spatial heterogeneity (Figures 8.4 and 8.15). Providing a trophic state description of a confined, riverine reservoir is very difficult and debate has taken place concerning the relevance of such classifications in such reservoirs (e.g. Lind, 1986). Their trophic states may range from nutrient rich (eutrophic) in their upper reaches to nutrient poor (oligotrophic) closer to the dam wall (see Figure 8.4). This tendency is enhanced when they consist of several basins with discrete morphologies and distinct characteristics, such as embayments or human settlements, influencing their responses to external stimuli (Kimmel and Groeger, 1984; Kennedy and Gaugush, 1988; Thornton *et al.*, 1990). Classifying whole reservoirs as oligotrophic on the basis of sampling stations located over the "deep hole" adjacent to the dam wall can, therefore, be misleading if more nutrient rich water is generally characteristic of the major part of the reservoir system.

While some investigators have attempted to create volume-weighted or area-weighted characterisations, it appears that most investigators continue to base trophic classifications on a single sampling site (Straskrabá *et al.*, 1993). Some justification for this can be derived from the fact that water withdrawals are more common at the dam end of reservoirs than elsewhere. In an attempt to provide an appropriate comparative framework for classification of impoundments, several workers have devoted considerable effort towards defining trophic concepts across climatic regions (Ryding and Rast, 1989; Thornton and Rast, 1993). Although trophic classifications can often be ascribed to lakes and reservoirs in the temperate zone on the basis of single indicators, such as hypolimnetic oxygen regime, such simple classifications are less applicable in other latitudes, requiring the more refined approaches described by these workers. At a minimum, it may be necessary to segment a reservoir into multiple regions of distinct water quality and trophic character as described below.

8.4 Water quality issues

Despite the complexity of environmental and social issues surrounding reservoirs, particularly large impoundments (Petr, 1978; UNESCO/UNEP 1990; Thornton *et al.*, 1992), large numbers presently exist and others continue to be built. In order to manage these systems in an environmentally-sound manner, and to minimise the occurrence of conditions that interfere with the beneficial uses of their waters, regular collection and analysis of data is required. In this respect, reservoirs are not different from natural lakes. However, reservoirs present specific challenges that are rarely present in natural lakes. Thus, whilst the water quality issues of concern in reservoirs parallel and overlap those of natural lakes, certain differences and specific conditions also exist. A summary of the major issues and assessment implications relevant to lakes and reservoirs is given in Table 7.5. Further information on assessment strategies and selection of appropriate monitoring variables is available in Chapters 2 and 3.

The following sub-sections highlight some issues of particular relevance to reservoirs and the assessment of their water quality. Some issues of major concern in lakes are less of a general problem in reservoirs. For example, reservoirs are less susceptible to acidification than are natural lakes. The influence of the source water can be so dominant in the overall chemistry of some reservoirs, as a result of high flow-through, that gradual water quality modifications are more a function of the source rivers and streams than of the characteristics of the reservoir itself.

8.4.1 Eutrophication
Reservoirs, like natural lakes, are affected by the process of eutrophication. Some reservoirs undergo the full process of gradual ageing that ultimately converts lakes to wetlands and to terrestrial biomes (Rast and Lee, 1978). Unlike natural lakes, however, most reservoirs have a design life (typically estimated to be greater than 30 years) which reflects the period over which the structure is capitalised as well as the on-going demand for the water that the reservoir must supply. Exceptions to this generalisation do exist, for example, in the eastern Cape Province of South Africa where several reservoirs have been constructed (primarily as salinisation and sediment control structures) with significantly shorter life expectancies (DWA, 1986). These reservoirs tend to exhibit the symptoms of eutrophication (reduction in depth, excessive plant growth, etc.) much more rapidly than natural lakes or reservoirs constructed elsewhere. Despite such exceptions, many reservoirs do mimic natural lakes in their transition from an aquatic environment to a terrestrial environment over time. That time period, however, is generally much shorter than the geological time scale associated with most natural lakes.

Reservoirs are also susceptible to cultural eutrophication or an increased rate of ageing caused by human settlement and activities in the watershed. Some impoundments may be more productive than their natural counterparts because of their generally larger watershed areas and terminal location. For any given unit area load of pollutant generated from a specific type of land usage, the larger watershed area results in a greater load being delivered to the water body. To some extent, however, this greater load is often moderated by a generally higher flushing rate (or shorter water residence time) than typical for a similarly-sized natural lake (e.g. Table 8.1). This higher flushing rate is principally related to the greater watershed area delivering the water load to the reservoir.

As described above, considerable artificial control of the normal responses associated with increasing eutrophication is possible in reservoirs. In some situations, thermal stratification and any associated chemical deterioration can be controlled. Appreciable depression of primary production can be achieved by artificially circulating phytoplankton into waters below the euphotic zone (zone of phytoplankton production), resulting in a decrease in population photosynthesis and production relative to respiration and sedimentation (Steel, 1975). Thus a reduction in the efficiency of primary production can take place even in the presence of what might otherwise be overwhelming quantities of phosphorus and nitrogen. Alternative forms of biomanipulation are also possible in some reservoirs by exploiting the effects of controlling fish stocks (see section 8.3.3); either directly or by controlling water level at critical spawning times.

Other measures to reduce nutrient concentrations within reservoirs include the use of flow by-passes, pre-impoundments, scour valves discharging nutrient-rich hypolimnetic (bottom) water, and modifications to the operating regime which can alter the effects of water and nutrient loads within the reservoir. The inlet to Delavan Lake, Wisconsin (volume $54.8 \times 10^6 \, m^3$; area $7.2 \, km^2$), for example, has been modified so that nutrient-rich high flows will be passed through the dam without entering the main body of the lake (University of Wisconsin-Madison, 1986). This example is a purposely designed short-circuit effect created by the proximity of the inlet and outlet and enhanced by the construction of a diversion structure to accentuate the flow of water towards and over the dam during flood events. In contrast, the modified inlet permits river waters to mingle gradually with the waters of the main lake basin during lower flows.

In many circumstances greater water quality control is often achieved by a combination of approaches to control eutrophication. For example, the operational water quality management of the Wahnbach impoundment in Germany (volume $40.9 \times 10^6 \, m^3$; area $1.99 \, km^2$) (LAWA, 1990), developed over a number of years, uses a combination of phosphorus elimination by treatment of inputs, pre-impoundment and hypolimnetic re-aeration to produce high quality drinking water (Figure 8.14). Many other procedures (such as sediment removal, flocculation and flow control) have also been shown to be potential modifiers of the response of a reservoir to eutrophication (Jorgensen and Vollenweider, 1989; Ryding and Rast, 1989).

Operational manipulations may not always result in beneficial effects, even within the basins of eutrophic reservoirs. Drawdown caused by the withdrawal of aerobic surface waters for drinking water or irrigation supply purposes can make a reservoir more susceptible to overturn. This can

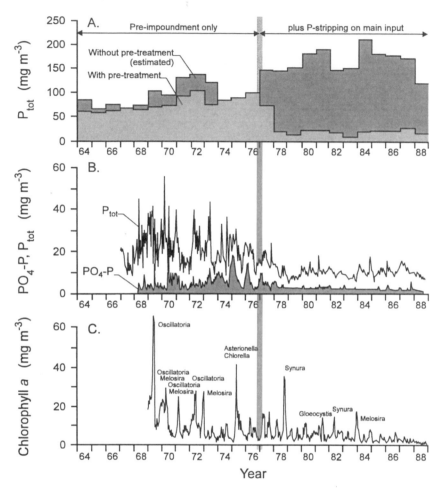

Figure 8.14 Monitoring to determine the effectiveness of measures taken to prevent the Wahnbach reservoir, Germany from becoming eutrophic. The measures consisted of a combination of pre-impoundment, phosphorus elimination in some source waters, and hypolimnetic aeration. **A.** Reductions achieved in the phosphorus inputs to the reservoir between 1964 and 1988 compared with the estimated concentrations if phosphorus reduction measures had not been applied. **B.** Total phosphorus and phosphate concentrations in the reservoir water over the same period. **C.** Associated chlorophyll a concentrations in the reservoir and the principal algal species present. (Modified from LAWA, 1990)

introduce nutrient-rich bottom waters to the euphotic zone, enhancing algal growth and creating an intensified oxygen demand throughout the water column. An extreme example of such an impact is reported by Robarts *et al.* (1982) from Hartbeespoort Dam, South Africa. Passing anaerobic waters through turbines or similar precision machinery can also cause severe damage to the equipment. The initial eutrophication of Lake Kariba (see section 8.3) resulted in the (unintentional) passage of anaerobic waters through the hydroelectric turbines, damaging them so badly that they had to be replaced (Balon and Coche, 1974). McKendrick (1982) described similar disruptive impacts from withdrawal of anaerobic water for drinking water treatment at Lake McIlwaine (now Lake Chivero), Zimbabwe. Neither of these latter cases resulted in any visible water quality change within the reservoir basins. Similarly, release of anoxic, hypolimnetic water from the eutrophic Iniscarra reservoir, Ireland, caused fish mortalities in the downstream waters of the River Lee (Bryan, 1964).

8.4.2 Public health impacts and contaminants
In addition to the range of potential health impacts arising from water-borne diseases and water-related disease vectors described in Chapter 7, the construction of a reservoir can provide an avenue for transmittal of diseases and their occurrence in areas where such diseases, and natural immunities, did not previously exist. This is especially true in the tropics, where the construction of reservoirs has been implicated in the epidemic spread of river blindness (onchocerciasis), bilharzia (schistosomiasis), and guinea worm disease (dracunculiasis). Volta Lake, Ghana, is an extreme and well-documented example of a reservoir where the spread of these diseases almost decimated the local and migrant populations attracted to the reservoir, necessitating an extensive and on-going international control programme (Thanh and Biswas, 1990). Similar, but less extreme, examples have been reported from other large-scale water resources development programmes in Africa, Asia and South America (Thanh and Biswas,1990; Bloom *et al.*, 1991).

Industries and urban settlements adjacent to, or in the drainage basin of, reservoirs can further compound human health problems through the release of contaminants, and through the discharge of wastewaters to these water bodies or their source waters (as in lakes; see Chapter 7). In many cases, the presence of a reservoir can attract development that could not otherwise be sustained due to a lack of process and drinking water. Contaminants of particular concern in reservoirs with respect to human and animal health are:
- synthetic organic compounds,
- nitrates,
- pathogenic bacteria,
- viruses, and
- cyanobacterial toxins.

Section 8.3 has already provided some detail of the particular impacts of these contaminants in reservoirs, and of how the conditions within a reservoir can affect them, as well as how management techniques may reduce their impacts on water quality and water use (see also section 8.6). Toxic metals may be a problem in some areas, although the normal processes of adsorption to particulate material and subsequent deposition in the more quiescent regions of a reservoir or pre-impoundment are generally the only effective means of containment in such cases (see also Chapter 7). Sediment oxygenation, as may be achieved in managed reservoirs (see section 8.3.2), is also a possible means of ensuring retention of contaminants within the basin.

8.4.3 Salinisation

Many reservoirs are situated in arid and semi-arid areas where surface water is naturally scarce and for these reservoirs the problem of salinisation (also referred to as salination or mineralisation) can be severe. Irrigation of these areas can lead to leaching of salts from the soils, and their transport in return flows to the reservoirs. This can be aggravated by the highly seasonal rainfall of these areas increasing the evaporative concentration of the ambient salinities in the water bodies during the dry season. In the reservoirs of the eastern Cape Province of South Africa, for example, salinities can exceed a median value of 1,000 mg l^{-1} measured as total dissolved solids (TDS) (DWA, 1986). The South African Department of Water Affairs (DWA, 1986) gives a range of median TDS concentrations, for 13 major reservoirs throughout the country, of between 137 mg l^{-1} and 955 mg l^{-1}. The extreme values of TDS measured in these reservoirs ranged from 99 mg l^{-1} to 2,220 mg l^{-1}. Salinities in this range limit the possible uses of the waters because all but a few salt-tolerant crops, and most industrial and consumptive uses, of the water are seriously affected by high salinities (see Chapter 3).

8.5   Sampling strategies

As with natural lakes, sampling strategies for reservoirs must be based upon the objectives of the assessment programme as set out in Chapter 2. The modifying influences of off-take locations, selective-level withdrawals, dendricity, and linear gradations of water quality along the main lengths of impoundments, are complicating factors that must be considered in developing an appropriate reservoir sampling regime. The principal purpose(s) for which the reservoir was created should also be considered. For multi-purpose reservoirs, the use requiring the highest water quality should generally be used as the basis for determining an appropriate quality objective, and for guiding water quality management programmes. Some possible monitoring strategies in relation to principal water uses are given in Table 8.3.

Table 8.3 Suggested monitoring strategies for some major uses of reservoir waters

Principal water use	Main sample site location	Sampling frequency[1]	Hydrodynamic considerations	Physical and chemical measurements	Biological methods
Potable water supply	At outlet to supply	Continuous, daily to weekly	Thermal stratification, short-circuit flows	Temperature, diss. oxygen, colour, turbidity, suspended solids, odour, pH, organic compounds, metals, nitrate	Coliforms, pathogens, phytoplankton species, chlorophyll a
Industrial water supply[2]	At outlet to supply	Continuous, daily to weekly	Thermal stratification	Temperature, pH, hardness, dissolved and suspended solids, major ions	Pathogens[2]
Power generation	Close to outlet	Daily to weekly	Thermal stratification, internal water currents	Conductivity, dissolved and suspended solids, major ions, dissolved oxygen	
Irrigation supply	Representative open water site(s) and/or outlet(s)	Weekly to monthly		pH, total dissolved solids, sodium, chloride, magnesium	Faecal coliforms
Fisheries and recreation	Representative open water site(s)[3]	Weekly to monthly	Thermal stratification	Suspended solids, dissolved oxygen, BOD, ammonia	Phytoplankton species, chlorophyll a, fish biomass
Aquaculture	At inlet and open water site(s)	Daily to weekly	Thermal stratification	Temperature, suspended solids, dissolved oxygen, ammonia, pesticides	Chlorophyll a
Flood control	Inflow(s) and outflow(s)	Monthly to annual		Suspended solids, turbidity	

[1] Depending on the variables and as set by statute
[2] Requirements vary according to industrial use, e.g., food processing should require pathogen monitoring
[3] The number and location of the sites depends on the complexity of the reservoir basin and its discrete water masses

Desirable water quality objectives may vary with the location of the off-take structures within the reservoir basin. For example, if the dendricity of the reservoir effectively isolates a bay in which an irrigation water off-take is located, a lower quality water may be accepted or tolerated in that bay than in the main body of the reservoir where a potable water off-take might be located. The depth distributions of water quality variables should be considered in a similar manner, as illustrated by the metalimnetic oxygen minimum created by selective withdrawal of waters at that depth (see Figure 8.11).

8.5.1   Selection of sampling site

As noted above, the assessment of reservoir water quality should reflect the uses of the water, as well as the conditions at the site where water is withdrawn. In impoundments this is often at a location close to the dam wall. Thus, the popular selection of the "deep hole" (generally near the dam wall) as the principal sampling site in such reservoirs is based on an intuitive application of this convention. It is also noteworthy that when intensive sampling has been carried out throughout a reservoir basin to generate a "true" mean for the more common trophic state variables such as nitrogen, phosphorus and chlorophyll concentrations, the volume-weighted mean values have often been statistically indistinguishable from those measured in the deep hole or main basin. This is probably due to the greater volume represented by the water at the deeper dam-end of the impoundment.

In regularly shaped, off-river reservoirs, it is commonly found that the water quality may be characterised by a few notional sub-basins, even if they do not physically exist. The number and extent of such volumes are greatly influenced by the operational regime and physical structure of the reservoir. For example, in a well-mixed, near-circular reservoir, the main water body (except for a small volume near the inlet) is usually sufficiently homogeneous to be well characterised by a single sampling site. The characteristic pattern of water masses within a given reservoir may generally be deduced by a preliminary survey (see Chapter 2) under representative conditions, from which an appropriate, general sampling strategy may be subsequently derived (Steel, 1975). Internal walls or similar internal structures can frequently give a good indication of the likely sub-basins for sampling purposes. Therefore, in many respects, sampling station selection for more regularly shaped reservoirs is similar to that detailed in Chapter 7 for natural lakes.

In physically more complex reservoirs, the degree of dendricity and sinuosity arising from river impoundment can result in actual, distinct differences in water quality over the length of an impounded water course (see Figure 8.15). In these instances, a single sampling site does not usually provide an adequate assessment of water quality for the entire impoundment. In very large impoundments several sampling sites, defined by basin morphology or geography, are more appropriate. In Lake Kariba, for example, five such basins were defined and used as the basis for water quality classification (Balon and Coche, 1974). In effect, for modelling and management purposes, Lake Kariba can be considered as a cascade of five small lakes, each discharging into its immediate neighbour in a downstream direction. Therefore, the "deep hole" sampling site approach should be considered only as an operational approach and it should not be used in situations where a scientific and statistically valid assessment of water quality is required.

As with natural lakes, the monitoring of inflows is generally carried out in addition to the selected site(s) described above as a means of providing an early warning of possible changes in water quality at the water withdrawal points or in the various basins.

8.5.2 Sampling for horizontal and vertical characterisation

The need to provide statistically valid descriptions of water quality in reservoirs poses a number of concerns, fundamentally due to the possible horizontal, vertical and temporal variations in water quality in larger reservoirs. This is particularly true in impoundments, where their size, bathymetry, sinuosity, complex shorelines and operation all tend to allow relatively local water quality conditions to be maintained (see section 8.3). Vertical stratifications of varying complexity are a further complicating factor. Designing a sampling programme to deal adequately with such variability requires specific recognition of these concerns (for details of statistical approaches see Appendix 10.1). A statistically valid mean result for the whole reservoir would be applicable, for example, to the siting of withdrawal points or recreational facilities, identification of water quality variables of concern and significant differences in these variables between sites, and development of valid functional relationships amongst reservoir water quality variables (a prerequisite for modelling reservoir water quality).

A practical example of site selection for horizontal and vertical characterisation of a reservoir is given by Thornton et al. (1982) and described briefly below. This example represents the advanced level of monitoring as defined in Chapter 2. The optimal location of the minimum number of sampling stations and depths required to characterise the water quality along the length of the reservoir were determined in DeGray Lake, USA (Figure 8.15). Preliminary data were obtained from a series of close interval, horizontal and vertical transects in order to distinguish statistical similarities and differences between the sampling sites.

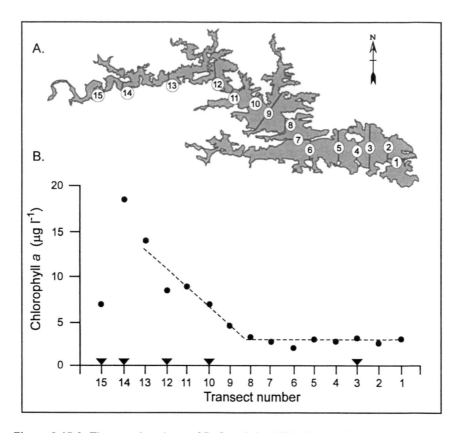

Figure 8.15 A. The complex shape of DeGray Lake, USA showing the sampling transects chosen for a detailed study. **B.** Chlorophyll *a* concentrations determined for each transect. Arrows indicate the sites which represent the overall water quality of the reservoir (see text for further details). (After Thornton *et al.*, 1982)

DeGray Lake was theoretically partitioned into 15 transects, averaging 5 stations per transect. The reservoir was sampled for total phosphorus, turbidity and chlorophyll *a* at 0, 2, 4, 6, and 10 m, and at 5 m intervals thereafter to the bottom, in July 1978, and in January and October 1979. Statistical analysis of the results showed that transects 1 through 5 had similar means for all variables over all dates. Thus, one station was sufficient to characterise the entire area represented by these five transects. The type of variation amongst the mean values of transects 6 through 13 suggested that two stations were necessary to characterise the water quality in that area. Transects 14 and 15 were either distinct or varied in a manner similar to transects 6–13 and, therefore, these two transects each needed a separate sampling station. Based on this analysis, a minimum of five sampling stations (on transects 3,

10, 12, 14 and 15) was required to characterise statistically DeGray Lake along its longitudinal water quality gradients (see Figure 8.15).

In addition to the number of sampling stations, it is also necessary to consider the number of samples needed from each sampling station to characterise the vertical variability in water quality. The number of samples required is influenced by the data variability and the desired precision of the estimate and can be obtained using statistical methods (see Appendix 10.1). Such approaches (e.g. Thornton *et al.,* 1982) reflect an advanced approach to monitoring and assessment.

The approach used by Thornton *et al.* (1982) and others (e.g. Gaugush, 1986) identify areas or segments within a reservoir which exhibit statistically distinct water quality. They do not, however, address how this information can, or should, be used to assess the overall water quality or trophic state of a reservoir (given the presence of longitudinal and vertical gradients in water quality, such as those identified by Kimmel and Groeger (1984)) (see Figure 8.4). This approach actually highlights the fact that the water quality or trophic status of highly dendritic and sinuous reservoirs, in particular, cannot be readily characterised by a single average value.

As a result of the complexity involved in completely assessing the water quality of reservoirs, the most logical approach to the management of such water bodies is to assess water quality with respect to the most sensitive use of the water at the point of withdrawal of that water. Thus, for water supply reservoirs this would be at the point of withdrawal, while recreational water bodies would require a more comprehensive assessment, possibly covering several segments in the reservoir (see Table 8.3). Although this site- and use-specific approach still does not provide specific guidance for the management of water quality along the course of the reservoir (or in the reservoir as a whole), it does provide an approach that is practical (and practicable) given the present state of knowledge and available techniques for management of water quality (Ryding and Rast, 1989).

8.5.3 Variable selection
The variables measured as part of any given sampling programme should reflect consideration of the uses to which the water is put as well as any known or anticipated impacts on the water quality. Collection of data pertinent to such usage, and the assessment of actual and potential water quality problems, should provide as full as possible an understanding of the sources, impacts and subsequent behaviour of pollutants, as well as aiding the identification, evaluation and outcome of appropriate and possible mitigation methods. Although the general selection of variables is described in Chapters

1, 3 and 7, operational procedures may also influence such selection in reservoirs because slight modifications in their operation can often resolve their water quality problems. For example, in Lake Chivero modification of the outlet structure to permit multi-level withdrawal, and its relocation to an up-wind position, resolved many of the algal problems previously experienced with the original off-take (McKendrick, 1982). Similarly, de Moor (1986) showed that small variations in the timing of releases from the Vaalharts Diversion Weir could obviate simuliid problems in the Vaal River, South Africa (the blackfly *Simulium* spp. is the vector of river blindness).

8.5.4 The use of management models

Reservoir management facilities often provide a considerable range of options which may be used to attain a water quality suited for the main purposes of the reservoir. In order to make the most effective use of the management options, it is essential that there is a clear understanding of the interplay between the effects of the management control measures and the natural dynamics of the reservoir system. This understanding is necessary to exploit effectively processes beneficial to a particular reservoir water use, and to ensure as far as possible that no unacceptable detriment is caused for other uses.

A useful tool for such informed control is a model of the reservoir system. Models range from empirical, statistical descriptions to analytical, mathematical models (see section 10.7). Either approach would normally imply the need for a specialised assessment, including surveys and data analyses appropriate to the task of constructing such models. Operational management may make day-to-day use of a combination of models. Regular operational surveillance may, therefore, also have to include sampling sites and variables required as input data for reservoir management models, even though such variables may not be specifically required for assessment of the water quality use or issue of concern (see Table 2.2).

Straskrabá and Gnauck (1985) give a wide-ranging account of modelling in freshwater systems. It is important to realise that there is not likely to be a single "best" model or model type for a given reservoir. Experience shows that overly complex mathematical models which attempt a complete description of reservoir dynamics are invariably not used, because of their extensive data demands, often complex nature and high relative costs of operation. However, operational management can be aided by quite simple models (Steel, 1978a,b). Figure 8.13 illustrates the prediction of a contaminant concentration in a reservoir under different operating conditions. Appropriate modelling, sometimes based on water bodies with similar characteristics, may be of greater benefit when considering facilities to be included at the construction phase or to be added at a later date, such as phosphorus elimination by pre-impoundment (Benndorf and Pütz, 1987a,b) or mixing the water within reservoirs with air (Goossens, 1979) or water-jets (Cooley and Harris, 1954).

8.6 Approaches to reservoir assessment

There are numerous examples of water quality assessments of reservoirs in the published literature and the majority of examples (in terms of number) are from industrialised nations of Europe and North America. Nevertheless, in relation to the total number of natural lakes and reservoir assessments in the developing world, there are more examples of reservoir assessments.

8.6.1 General basic surveys

The same types of bathymetric and thermal surveys undertaken in natural lakes are necessary to establish the physical environment of reservoirs. The methods given in Chapter 7 can also be applied to reservoirs, although some differences can be identified.

Generally, the bathymetry of reservoirs is easier to determine than that of natural lakes for the simple reason that reservoir basins were once dry land. The construction of reservoirs requires detailed design drawings and land surveys which then also identify the subsequent bathymetry in detail. This feature simplifies the process of water quality modelling prior to impoundment by allowing the assessor to make forecasts based on the predetermined full supply level contour.

In contrast to lakes, the thermal regime of reservoirs can be more variable (see section 8.3). Unnatural variability is added as a result of the terminal location of some reservoirs at the downstream end of a drainage basin, their tendency to develop density currents, their operating regime (which can result in modification of the thermal profile), and morphology (which tends to be less deep than natural lakes and, therefore, more likely to encourage periodic water mixing) (Ward and Stanford, 1979).

8.6.2 Operational surveillance

In reservoirs used mainly for drinking water supply, it is imperative that all management and monitoring should be appropriate for that main purpose. Of particular consequence are those substances identified as potentially harmful to consumers, and which may interfere with, or resist, standard water treatment technologies. The widespread use of pesticides in many areas of the world presents a particular problem in this respect. Some of the more persistent pesticides which enter water courses may eventually be carried into

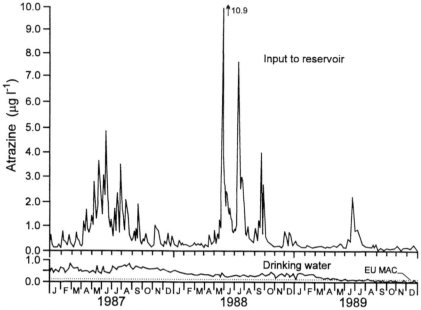

Figure 8.16 A comparison of fluctuations in the concentrations of the herbicide atrazine in the inlet water to the Hullern water supply reservoir, Germany, with the concentration in the drinking water supply derived from the reservoir. EU MAC: European Union Maximum Allowable Concentration (After LAWA, 1990)

reservoirs. Figure 8.16 shows the results of monitoring over a four year period the inlet of the Hullern reservoir in northern Germany (volume 11×10^6 m^3; area 1.5 km^2) for the widely used herbicide, atrazine. Large seasonal fluctuations in pesticide concentrations entering reservoirs can be dampened by water storage in the reservoir, thereby reducing possible problems at the subsequent water treatment stages. These effects are particularly useful if the reservoir water is to be used as a source for drinking water supplies. After initial difficulty in meeting the maximum admissible concentration (MAC) for atrazine as given by the European Union Drinking Water Directive (80/778/EEC), the final drinking water supply derived from the Hullern reservoir (Figure 8.16) eventually attained the MAC in 1989/90 (LAWA, 1990).

Nitrate monitoring forms a routine aspect of water quality assessments in reservoirs concerned with drinking water supply in Europe because of a European MAC for this variable of 11.3 mg l^{-1} NO$_3$-N (the new World Health Organization guideline value is given as 50 mg l^{-1} NO$_3^-$ [≡ 11.3 mg l^{-1} as NO$_3$-N] ; see Table 3.4). Nitrates are a potentially serious problem for drinking water supplies in areas of intensive agriculture, especially where

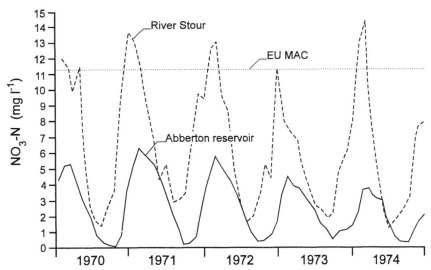

Figure 8.17 Annual variation in the nitrate-nitrogen concentrations in the River Stour, UK compared with the concentrations in the waters of the Abberton reservoir which is supplied by the river. De-nitrification in the reservoir helps to reduce the nitrate-nitrogen concentrations to below guideline levels for drinking water. (After Slack, 1977)

inorganic nitrogenous fertiliser is used heavily. For example, within Europe, the eastern UK is an area of high nutrient inputs to water bodies combined with relatively low rainfall. In this area, nitrates accumulated in soils are washed into water courses in large quantities during the rains of late autumn and early winter. These fluctuations in nitrate loads can cause considerable problems for water supplies drawn from affected water courses. Figure 8.17 shows the results of routine measurements of NO_3-N concentrations in the River Stour in eastern Britain, and the associated Abberton reservoir (volume 25.7×10^6 m^3; area 4.9 km^2). Without the reservoir and its associated de-nitrification processes, the responsible water company would require specialised treatment facilities to deal with this extreme seasonal problem. The reservoir, therefore, not only serves as a multi-purpose water body, but also ensures a suitable quality water for drinking water supply.

8.6.3 Eutrophication studies
Reservoirs have been subjected to much, long-term investigation of the effects of nutrient enrichment, principally to attempt to understand its effects in reservoirs and to identify possible ways of reducing any nutrient associated problems. Figure 8.18 illustrates a typical situation in the highly eutrophic King George VI reservoir (volume 20.2×10^6 m^3; area 1.4 km^2) in the southern UK prior to the intervention of any management regime. The

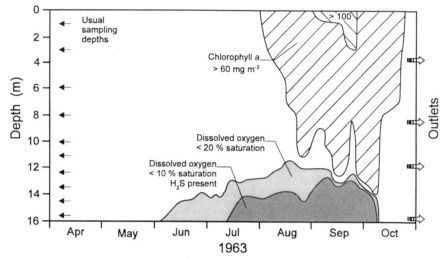

Figure 8.18 Depths distributions of chlorophyll *a* and dissolved oxygen concentrations in relation to the water withdrawal points in the highly eutrophic King George VI reservoir, UK, during the summer months, prior to the application of any management regime (compare with the managed reservoir in Figure 8.12) (After Steel, 1972)

routine sampling of the water column at fixed depths (indicated in Figure 8.18) for temperature and dissolved oxygen was combined with the results of a special survey of phytoplankton species and biomass. The results showed that the water body thermally stratified, resulting in deoxygenation of the hypolimnion and an eventual build-up of hydrogen sulphide. At the same time, a substantial population of the cyanobacterium *Aphanizomenon flos-aquae* developed in the epilimnion. The possible (shore-line) outlets for drinking water supply are also shown in Figure 8.18. From mid-June onwards (as summer water demands increased) it was difficult to extract water of a suitable quality for treatment by the available water treatment processes. When internal seiching was also considered, the ability to select a particular water layer of desired water quality was also compromised. Detailed assessments such as these led to the search for solutions to the problems of thermal stratification and algal production in nutrient rich reservoirs (see sections 8.3 and 8.4 for examples of management solutions).

8.6.4 Assessment of trends

The routine collection of information for operational surveillance in reservoirs often allows the associated long- and short-term trends in water quality variables to be identified without additional effort. However, changes in reservoir operational procedures or water use over time can influence long-term trends and/or their significance to reservoir water use. Inter-annual variation may be induced for operational reasons as well as being a function of climatic and other anthropogenic factors.

8.7 Summary and conclusions

Reservoirs are essential sources of freshwater for consumptive (e.g. drinking or process water) and non-consumptive (e.g. fisheries and recreational) human use. They have also contributed to industrial development by providing cheap sources of hydropower and hydroelectric power and many continue to provide hydroelectric power throughout the world. Although the development of reservoirs is usually economically beneficial to communities, in some geographic zones they can have negative impacts, such as encouraging the spread of water-borne diseases (e.g. bilharzia and river blindness). Some large impoundments may even require the displacement of the local populations while attracting new people to the reservoir shores.

Reservoirs are, essentially, managed water bodies and, therefore, there is a particular need for the managers to understand their physics, chemistry and biology. This knowledge is acquired through assessment of water quality data gathered during well-planned monitoring programmes. In turn, the management operations, themselves, can affect the water quality characteristics and behaviour of the reservoir. There is also a particular need for further evaluation of monitoring strategies in relation to operational changes over time.

The objectives of the assessment programme and the associated monitoring strategies will often be governed by the operational requirements for a specified water quality. Nevertheless, in many situations, the monitoring activities will be similar to those carried out in natural lakes, although the interpretation and assessment of the data can differ markedly as a result of the modifying influences of the operating regime of the reservoir. Awareness of the effects of the location and depth of water withdrawal points on the water quality characteristics and behaviour within the reservoir is essential to their management to achieve the optimum water quality for intended uses. In addition, in order that managers may make the maximum use of physical and water quality data gathered from reservoirs, it is necessary to determine and understand the effects of inflows and outflows, retention time, in-basin facilities and actions, and morphometry and location of the reservoir within its watershed.

Reservoirs, and their associated water quality, can benefit from the ability to site and to construct them, within the limitations of topography, to best operational advantage. The water quality of established reservoirs can be managed by altering operating regimes, changing water uses, and blending waters of varying quality. In water-poor climates, the development of elaborate inter-basin transfer schemes provides additional opportunities to ensure safe and reliable supplies of water throughout countries and regions. Such schemes, however, can increase the likelihood of the transfer of undesirable species and habitat destruction.

Unlike natural lakes, reservoirs create environmental impacts, as well as suffer from impacts due to human activity. By modifying the flow regime of natural water courses (in some cases completely diverting river flows from one watershed to another) reservoirs can alter species compositions both upstream and downstream, change thermal regimes, and also modify the chemical content of waters. In addition, changes in the quality and quantity of reservoir discharges, arising from changes in operational regime, into rivers and downstream lakes can affect the water quality of the receiving water body. The impact of these, possibly variable, water quality discharges should be taken into account in monitoring and assessment programmes for the receiving water bodies.

8.8   References

Balon, E.K. and Coche, A.G. 1974 *Lake Kariba: A Man-made Tropical Ecosystem in Central Africa*, Monographiae Biologicae **24**, Dr W. Junk, The Hague.

Benndorf, J. and Pütz, K. 1987a Control of eutrophication of lakes and reservoirs by means of pre-dams. I. Mode of operation and calculation of the nutrient elimination capacity. *Wat. Res.*, **21**(7), 829-838.

Benndorf, J. and Pütz, K. 1987b Control of eutrophication of lakes and reservoirs by means of pre-dams. II. Validation of the phosphate removal model and size optimization. *Wat. Res.*, **21**(7), 839-842.

Benndorf, J., Schultz, H., Benndorf, A., Unger, R., Penz, E., Kneschke, H., Kossatz, K., Dumke, R., Hornig, U., Kruspe, R., and Reichel, S. 1988 Food web manipulation by enhancement of piscivorous fish stocks: long-term effects in the hypertrophic Bautzen reservoir. *Limnologica*, **19**, 97-110.

Bernhardt, H. 1978 Die hypolimnische Belüftung der Wahnbachtalsperre. [The hypolimnetic aeration of the Wahnbach reservoir]. *GWF-Wasser/Abwasser*, **119**(4), 177-182 [In German].

Bloom, H., Dodson, J.J., Tjitrosomo, S.S., Umaly, R.C. and Sukimin, S. 1991 *Inland Aquatic Environmental Stress Monitoring*. Biotrop Special Publication **43**, Southeast Asia Regional Centre for Tropical Biology, Bogor, 262 pp.

Bryan, J.G. 1964 Physical control of water quality. *J. Brit. Watwks Assn*, 19 pp.

Cole, T.M. and Hannan, H.H. 1990 Dissolved oxygen dynamics. In: K.W. Thornton, B.L. Kimmel and F.F. Payne [Eds] *Reservoir Limnology: Ecological Perspectives*. Wiley, New York, 71-107.

Cooley, P. and Harris, S.L. 1954 The prevention of stratification in reservoirs. *J. Inst. Wat. Engrs.*, **8**, 517-537.

Davies, B.R. 1979 Stream regulation in Africa: A review. In: J.F. Ward and J.A. Stanford [Eds] *The Ecology of Regulated Streams*, Plenum Press, New York, 113-142.

de Moor, F.C. 1986 Invertebrates of the Lower Vaal River, with emphasis on the Simuliidae. In: B.R. Davies and K.F. Walker [Eds] *The Ecology of River Systems*. Monographiae Biologicae **60**, Junk, The Hague, 135-142.

DWA 1986 *Management of the Water Resources of the Republic of South Africa*. Department of Water Affairs, Government Printer, Pretoria.

Ebel, W.J. and Koski, C.H. 1968 Physical and chemical limnology of Brownlee Reservoir, 1962-1964. *Fish. Bull. Bureau Commercial Fisheries*, **67**, 295-335.

Gaugush, R.F. 1986 *Statistical Methods for Reservoir Water Quality Investigations*. U.S. Army Corps of Engineers Waterways Experiment Station Instruction Report E-86-2, Vicksburg, 214 pp.

Goossens, L.H.J. 1979 *Reservoir Destratification with Bubble Columns*. Delft University Press, Delft, 200 pp.

Horn, W. 1981 Phytoplankton losses due to zooplankton grazing in a drinking water reservoir. *Int. Revue ges. Hydrobiol.*, **66**(6), 787-810.

Horn, H. and Paul, L. 1984. Interactions between light situation, depth of mixing, and phytoplankton growth during the Spring period of full circulation. *Int. Revue ges. Hydrobiol.*, **69**(4), 507-519.

Hutchinson, G.E. 1957 *A Treatise on Limnology. Volume I, Geography, Physics and Chemistry*. Wiley, New York, 1,015 pp.

Jorgensen, S.E. and Vollenweider, R.A. 1989 Remedial techniques. In: S.E. Jorgensen and R.A Vollenweider [Eds] *Guidelines of Lake Management*. International Lake Environment Committee, Otsu, Shiga and United Nations Environment Programme, Nairobi, 99-114.

Kennedy, R.H. and Gaugush, R.F. 1988 Assessment of water quality in Corps of Engineers reservoirs. *Lake Reserv. Manage.*, **4**, 253-260.

Kimmel, B.C. and Groeger, A.W. 1984 Factors controlling primary production in lakes and reservoirs: A perspective. In: *Lake and Reservoir Management*. Report EPA-440/5-84-001, United States Environmental Protection Agency, Washington, D.C., 277-281.

Lawton, L.A. and Codd, G.A. 1991 Cyanobacterial (blue-green algal) toxins and their significance in UK and European waters. *J. Inst. Water Environ. Manag.*, **5**(4), 460-465.

LAWA 1990 *Limnologie und Bedeutung ausgewählter Talsperren in der Bundesrepublik Deutschland*. [Regional Water Study Group: Limnology and

Importance of Selected Impoundments in the German Federal Republic] Länderarbeitsgemeinschaft Wasser , Weisbaden, 280 pp [In German].

Leidy, G.R. and Jenkins, R.M. 1977 *The Development of Fishery Compartments and Population Rate Coefficients for Use in Reservoir and Ecosystem Modelling*. US Army Corps of Engineers Waterways Experiment Station Contract Report Y-77-1, Vicksburg, A1-A8.

Lind, O.T. 1986 The effect of non-algal turbidity on the relationship of Secchi depth to chlorophyll-a. *Hydrobiologia*, **140**, 27-35.

Linsley, R.K. and Franzini, J.B. 1972 *Water-Resources Engineering*. McGraw-Hill Book Company, New York, 690 pp.

Marshall, B.E. and Junor, F.J.R. 1981 The decline of *Salvinia molesta* on Lake Kariba. *Hydrobiologia*, **83**, 477-484.

Marshall, B.E., Junor, F.J.R. and Langerman, J.D. 1982 Fisheries and fish production on the Zimbabwean side of Lake Kariba. *Kariba Studies*, **10**, 175-231.

McKendrick, J. 1982 Water supply and sewage treatment in relation to water quality in Lake McIlwaine. In: J.A. Thornton [Ed] *Lake McIlwaine: The Eutrophication and Recovery of a Tropical African Man-made Lake*. Monographiae Biologicae **49**, Junk, The Hague, 202-217.

Moore, L. and Thornton, K.W. 1988 *Lake and Reservoir Restoration Guidance Manual*. United States Environmental Protection Agency EPA-440/5-88-002, Washington, D.C.

O'Neill, J.G. and Holding, A.J. 1975 The importance of nitrate reducing bacteria in lakes and reservoirs. In: *The Effects of Storage on Water Quality*. Water Research Centre, Medmenham, 91-115.

Oskam, G. 1983 Quality aspects of the Biesbosch reservoirs. *Aqua*, **6**, 1156, 497-504.

Petitjean, M.O.G. and Davies, B.R. 1988 Ecological impacts of inter-basin transfers—some case studies, research requirements and assessment procedures. *S. Afr. J. Sci.*, **84**, 819-828.

Petr, T. 1978 Tropical man-made lakes—their ecological impact. *Arch. Hydrobiol.*, **81**, 368-385

Prentki, R.T. and Paulson, L.J. 1983 Historical patterns of phytoplankton productivity of Lake Mead. In: *Aquatic Resources Management of the Colorado River Ecosystem*. Ann Arbor Science, Ann Arbor, 105-123.

Rast, W. 1995 *Salinity Model of the Highland Lakes of Central Texas, USA*. Water Resour. Invest. Report, US Geological Survey, Austin.

Rast, W. and Lee, G.F. 1978 *Summary Analysis of the North American (US Portion) OECD Eutrophication Project: Nutrient Loading — Lake Response Relationships and Trophic State Indices*. Report EPA-600/3-78-008, United States Environmental Protection Agency, Corvallis, 454 pp.

Robarts, R.D., Ashton, P.J., Thornton, J.A., Taussig, H.J. and Sephton, L.M. 1982 Overturn in a hypertrophic, warm, monomictic impoundment (Hartbeespoort Dam, South Africa). *Hydrobiologia*, **97**, 209-224.

Ryding, S.-O. and Rast, W. 1989 *The Control of Eutrophication of Lakes and Reservoirs*. Man and the Biosphere **1**, Parthenon Press, Carnforth, 314 pp.

Schiemer, F. 1984 *Limnology of Parakrama Samudra, Sri Lanka: A Case Study of an Ancient Man-made Lake in the Tropics*. Developments in Hydrobiology **12**, Dr W. Junk, The Hague, 236 pp.

Serruya, C. and Pollingher, U. 1983 *Lakes of the Warm Belt*. Cambridge University Press, Cambridge, 569 pp.

Slack, J.G. 1977 Nitrate levels in Essex river waters. *J. Inst. Wat. Eng. Sci.*, **31**, 43-51.

Steel, J.A. 1972 Limnological research in water supply systems. In: R.W. Edwards and D.J. Garrod [Eds] *Conservation and Productivity of Natural Waters*. Symp. Zool. Lond. **29**, 41-67.

Steel, J.A. 1975 The management of Thames Valley reservoirs. In: *The Effects of Storage on Water Quality*. Water Research Centre, Medmenham, 371-419.

Steel, J.A. 1978a The use of simple plankton models in the management of Thames valley reservoirs. *DVBW-Schriftenreihe Wasser*, **16**, 42-59.

Steel, J.A. 1978b Reservoir algal productivity. In: A. James [Ed.] *Mathematical Models in Water Pollution Control*. John Wiley & Sons, Chichester, 107-135.

Straskrabá, M. and Gnauck, A.H. 1985 *Freshwater Ecosystems Modelling and Simulation*. Developments in Environmental Modelling, **8**, Elsevier, Amsterdam, 309 pp.

Straskrabá, M. and Straskrabová, V. 1975 Management problems of Slapy reservoir, Bohemia, Czechoslovakia. In: *The Effects of Storage on Water Quality*. Water Research Centre, Medmenham, 449-483.

Straskrabá, M., Tundisi, J.G. and Duncan, A. 1993 *Comparative Reservoir Limnology and Water Quality Management*. Developments in Hydrobiology Volume DH 77, Kluwer Academic Publishers, Boston, 291 pp.

Straskrabová, V. and Komárková, J 1979 Seasonal changes in bacterioplankton in a reservoir related to algae: 1. Numbers and biomass. *Int. Revue ges. Hydrobiol.,*. **64**(3), 285-302.

Thanh, N.C. and Biswas, A.K. 1990 *Environmentally-sound Water Management*. Oxford University Press, Delhi, 276 pp.

Thornton, J.A. 1982 *Lake McIlwaine: The Eutrophication and Recovery of a Tropical African Man-made Lake*. Monographiae Biologicae **49**, Junk, The Hague, 251 pp.

Thornton, J.A. and Rast, W. 1993 A test of hypotheses relating to the comparative limnology and assessment of eutrophication in semi-arid man-made lakes. In: M. Straskrabá, J.G. Tundisi and A. Duncan [Eds] 1993 *Comparative Reservoir Limnology and Water Quality Management*. Developments in Hydrobiology **77**, Kluwer Academic Publishers, Boston, 1-24.

Thornton, J.A., Williams, W.D. and Ryding, S.-O. 1992 Emigration, economics and environmental pollution in southern Africa. In: R. Herrmann [Ed] *Managing Water Resources During Global Change*, AWRA Technical Publication Series TPS-92-4, American Water Resources Association, Bethesda, 609-617.

Thornton, K.W., Kimmel, B.L. and Payne, F.F. 1990 *Reservoir Limnology: Ecological Perspectives*. Wiley, New York, 246 pp.

Thornton, K.W., Kennedy, R.H., Magoun, A.D. and Saul, G.E. 1982 Reservoir water quality sampling design. *Water Resources Bull.*, **18**, 471-480.

Tundisi, J.G. 1981 Typology of reservoirs in southern Brazil. *Verh. int. Ver. Limnol.*, **21**, 1,031-1,039.

UNEP 1991 *Freshwater Pollution*. UNEP/GEMS Environment Library, No 6, United Nations Environment Programme, Nairobi, 36 pp.

UNESCO/UNEP 1990 *The Impact of Large Water Projects on the Environment*. Proceedings of an International Symposium, Paris, 1986, United Nations Educational, Scientific and Cultural Organisation, Paris, 570 pp.

University of Wisconsin-Madison 1986 *Delavan Lake: A Recovery and Management Study*. Water Resources Management Workshop, Institute for Environmental Studies, Madison, 332 pp.

van Breeman, L.W.C.A., Dits, J.S. and Ketelaars, H.A.M. 1992 Production and reduction of geosmin and 2-methylisoborneol during storage of river water in deep reservoirs. *Wat. Sci. Tech.*, **25**, 233-240.

Van der Heide, J. 1982 *Lake Brokopondo: Filling Phase Limnology of a Manmade Lake in the Humid Tropics*. PhD Dissertation, University of Amsterdam.

van der Leeden, F., Troise, F.L. and Todd, D.K. 1990 *The Water Encyclopedia*. Second Edition. Lewis Publishers, Boca Raton, 808 pp.

Veltrop, J.A. 1993 Importance of dams for water supply and hydropower. In: A.K. Biswas, M. Jellalli and G.E. Stout [Eds] *Water for Sustainable Development in the Twenty-first Century*. Oxford University Press, Delhi, 104-115.

Ward, J.F. and Stanford, J.A. [Eds] 1979 *The Ecology of Regulated Streams*. Plenum Press, New York.

Whitehead, P. 1992 Examples of recent models in environmental impact assessment. *J. Inst. Water Environ. Manag.*, **6**(4), 475-484.

Znamenski, V.A. 1975 The role of hydrological factors with respect to water quality in reservoirs. In: *The Effects of Storage on Water Quality*. Water Research Centre, Medmenham, 567-573.

Chapter 9*

GROUNDWATER

9.1   Introduction

Water from beneath the ground has been exploited for domestic use, livestock and irrigation since the earliest times. Although the precise nature of its occurrence was not necessarily understood, successful methods of bringing the water to the surface have been developed and groundwater use has grown consistently ever since. It is, however, common for the dominant role of groundwater in the freshwater part of the hydrological cycle to be overlooked. Groundwater is easily the most important component and con-stitutes about two thirds of the freshwater resources of the world and, if the polar ice caps and glaciers are not considered, groundwater accounts for nearly all usable freshwater (see Table 1.1). Even if consideration is further limited to only the most active and accessible groundwater bodies (estimated by Lvovitch (1972) at 4×10^6 km^3) then they constitute 95 per cent of total freshwater. Lakes, swamps, reservoirs and rivers account for 3.5 per cent and soil moisture accounts for only 1.5 per cent (Freeze and Cherry, 1979). The dominant role of groundwater resources is clear and their use and protection is, therefore, of fundamental importance to human life and economic activity.

It is easy for the importance of groundwater in water supplies to be under-estimated. It is customary to think of groundwater as being more important in arid or semi-arid areas and surface water as more important in humid areas. However, inventories of groundwater and surface water use reveal the world-wide importance of groundwater. The reasons for this include its convenient availability close to where water is required, its excellent natural quality (which is generally adequate for potable supplies with little or no treatment) and the relatively low capital cost of development. Development in stages, to keep pace with rising demand, is usually more easily achieved for ground-water than for surface water.

In the USA where groundwater is important in all climatic regions, it accounts for about 50 per cent of livestock and irrigation water use, and just under 40 per cent of public water supplies. In rural areas of the USA, 96 per cent of domestic water is supplied from groundwater (Todd, 1980). Some very large cities are totally dependent on groundwater. In Latin America, many of the continent's largest cities, Mexico City, Lima, Buenos Aires and Santiago, obtain a significant proportion of their municipal water supply

This chapter was prepared by J. Chilton

from groundwater. In the valley of Mexico City, over 1,000 deep wells supply 3,200 x 10^6 m^3 day^{-1}, which is about 95 per cent of the total supply to a population of nearly 20 million people (Foster *et al.*, 1987). In Europe also, groundwater has always played a major part in water supplies. The proportion of groundwater in drinking water supplies in some European countries in 1988 was (UNEP, 1989):

Denmark	98 %	Netherlands	67 %
Portugal	94 %	Luxembourg	66 %
Germany Fed. Rep.	89 %	Sweden	49 %
Italy	88 %	United Kingdom	35 %
Switzerland	75 %	Spain	20 %
Belgium	67 %	Norway	15 %

Many of the major cities of Europe are, therefore, dependent on groundwater. In Africa and Asia, most of the largest cities use surface water, but many millions of people in the rural areas are dependent on groundwater. For many millions more, particularly in sub-Saharan Africa, who do not as yet have any form of improved supply, untreated groundwater supplies from protected wells with handpumps are likely to be their best solution for many years to come.

Water is drawn from the ground for a variety of uses, principally community water supply, farming (both livestock and irrigated cultivation) and industrial processes. Unlike surface water, groundwater is rarely used *in situ* for non-consumptive purposes such as recreation and fisheries, except occasionally where it comes to the surface as springs. Consequently, groundwater quality assessment is invariably directed towards factors which may lessen the suitability of pumped groundwater with respect to its potability and use in agriculture and industry.

The overall goal of a groundwater quality assessment programme, as for surface water programmes, is to obtain a comprehensive picture of the spatial distribution of groundwater quality and of the changes in time that occur, either naturally, or under the influence of man (Wilkinson and Edworthy, 1981). The benefits of well designed and executed programmes are that timely water quality management, and/or pollution control measures, can be taken which are based on comprehensive and appropriate water quality information. Each specific assessment programme is designed to meet a specific objective, or several objectives, which are related in each case to relevant water quality issues and water uses (see Chapter 2).

Two principal features of groundwater bodies distinguish them from surface water bodies. Firstly, the relatively slow movement of water through the ground means that residence times in groundwaters are generally orders of magnitude longer than in surface waters (see Table 1.1). Once polluted, a groundwater body could remain so for decades, or even for hundreds of years, because the natural processes of through-flushing are so slow. Secondly, there is a considerable degree of physico-chemical and chemical interdependence between the water and the containing material. The word groundwater, without further qualification, is generally understood to mean all the water underground, occupying the voids within geological formations. It follows, therefore, that in dealing with groundwater, the properties of both the ground and the water are important, and there is considerable scope for water quality to be modified by interaction between the two, as described in section 9.3. The scope for such modification is in turn enhanced by the long residence times, which depend on the size and type of the groundwater body (Figure 1.2). To appreciate the particular difficulties of monitoring ground-water bodies, it is necessary first to identify and to discuss briefly those properties of ground and water that are relevant to the occurrence and move-ment of groundwater. This is done in the following sections. Only a brief summary is possible, but further information is available in Price (1985) which gives a general introduction to the subject. Comprehensive descrip-tions of hydrogeology are given by Freeze and Cherry (1979), Todd (1980) and Driscoll (1986).

9.2 Characteristics of groundwater bodies

9.2.1 Occurrence of groundwater

Groundwater occurs in many different geological formations. Nearly all rocks in the upper part of the Earth's crust, whatever their type, origin or age, possess openings called pores or voids. In unconsolidated, granular materials the voids are the spaces between the grains (Figure 9.1a), which may become reduced by compaction and cementation (Figure 9.1d). In consolidated rocks, the only voids may be the fractures or fissures, which are generally restricted but may be enlarged by solution (Figure 9.1e,f). The volume of water con-tained in the rock depends on the percentage of these openings or pores in a given volume of the rock, which is termed the porosity of the rock. More pore spaces result in higher porosity and more stored water. Typical porosity ranges for common geological materials are shown in Table 9.1.

Only a part of the water contained in the fully-saturated pores can be abstracted and used. Under the influence of gravity when, for example, the water level falls, part of the water drains from the pores and part remains held by surface tension and molecular effects. The ratio of the volume of water that will drain under gravity from an initially saturated rock mass to the total

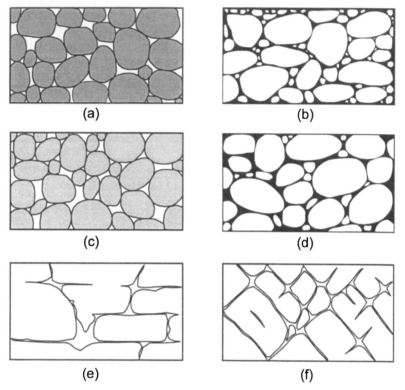

Figure 9.1 Rock texture and porosity of typical aquifer materials (Based on Todd, 1980)
a) Well sorted sedimentary deposit with high porosity
b) Poorly sorted sedimentary deposit with low porosity
c) Well sorted sedimentary deposit of porous pebbles, resulting in high overall porosity
d) Well sorted sedimentary deposit in which porosity has been reduced by cement deposited between grains
e) Consolidated rock rendered porous by solution
f) Consolidated rock rendered porous by fracturing

volume of that rock (including the enclosed water) is defined as the specific yield of the material, and is usually expressed as a percentage. Typical values are shown in Table 9.1.

Groundwater is not usually static but flows through the rock. The ease with which water can flow through a rock mass depends on a combination of the size of the pores and the degree to which they are inter-connected. This is defined as the permeability of the rock. Materials which permit water to pass through them easily are said to be permeable and those which permit water to pass only with difficulty, or not at all, are described as impermeable. A layer of rock that is sufficiently porous to store water and permeable enough to transmit water in quantities that can be economically exploited is called an

Table 9.1 Porosity and specific yield of geological materials

Material	Porosity (%)	Specific yield (%)
Unconsolidated sediments		
Gravel	25–35	15–30
Sand	25–45	10–30
Silt	35–50	5–10
Clay	45–55	1–5
Sand and gravel	20–30	10–20
Glacial till	20–30	5–15
Consolidated rocks		
Sandstone	5–30	3–15
Limestone and dolomite	1–20	0.5–10
Karst limestone	5–30	2–15
Shale	1–10	0.5–5
Vesicular basalt	10–40	5–15
Fractured basalt	5–30	2–10
Tuff	10–60	5–20
Fresh granite and gneiss	0.01–2	< 0.1
Weathered granite and gneiss	1–15	0.5–5

Sources: Freeze and Cherry, 1979; Todd, 1980; Driscoll, 1986

aquifer. Groundwater flow may take place through the spaces between the grains or through fissures (Figure 9.1), or by a combination of the two in, for example, a jointed sandstone or limestone. For any aquifer, distinguishing whether inter-granular or fissure flow predominates is fundamental to understanding the hydrogeology and to designing monitoring systems, particularly for point source pollution incidents.

When rain falls, some infiltrates into the soil. Part of this moisture is taken up by the roots of plants and some moves deeper under the influence of gravity. In the rock nearest to the ground surface, the empty spaces are partly filled with water and partly with air. This is known as the unsaturated or vadose zone (Figure 9.2), and can vary in depth from nothing to tens of metres. In the unsaturated zone, soil, air and water are in contact and may react with each other. Downward water movement in the unsaturated zone is slow, less than 10 m a^{-1} and often less than 1 m a^{-1}. Residence times in the unsaturated zone depend on its thickness and can vary from almost nothing to tens of years. At greater depths, all the empty spaces are completely filled with water and this is called the saturated zone. If a hole is dug or drilled down into the saturated zone, water will flow from the ground into the hole

Ground surface

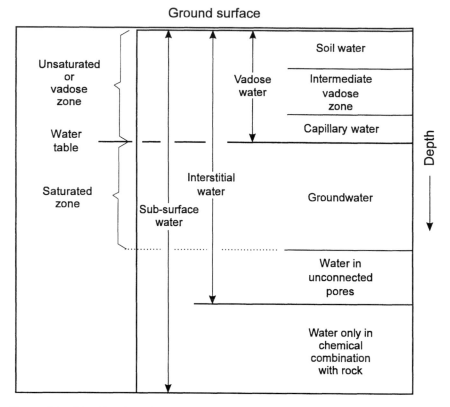

Figure 9.2 Classification of sub-surface water (Modified from Driscoll, 1986)

and settle at the depth below which all the pore spaces are filled with water. This level is the water table and forms the upper surface of the saturated zone, at which the fluid pressure in the pores is exactly atmospheric. Strictly speaking, the term groundwater refers only to the saturated zone below the water table. All water that occurs naturally beneath the Earth's surface, including saturated and unsaturated zones, is called sub-surface water (Figure 9.2).

A further important way of characterising aquifers is useful in any discussion of the development and protection of groundwater resources. An unconfined aquifer is one in which the upper limit of the zone of saturation, i.e. the water table, is at atmospheric pressure. At any depth below the water table the pressure is greater than atmospheric and at any point above the water table the pressure is less than atmospheric. In a confined aquifer the effective aquifer thickness extends between two impermeable layers. At any point in the aquifer, the water pressure is greater than atmospheric. If a well is drilled through the confining layer into the aquifer, water will rise up into

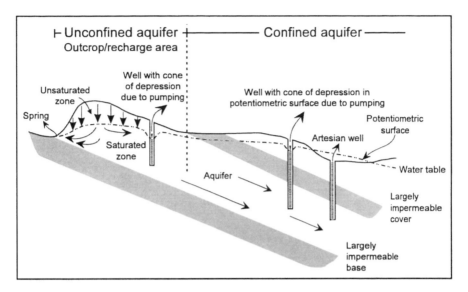

Figure 9.3 Schematic cross-section illustrating confined and unconfined aquifers

the well until the column of water in the well balances the pressure in the aquifer. An imaginary surface joining the water level in many wells in a confined aquifer is called the potentiometric surface. For a phreatic aquifer, that is the first unconfined aquifer to be formed from the land surface, the potentiometric surface corresponds to the water table found in wells or boreholes. The flow directions of groundwater are perpendicular to the isolines of the potentiometric surface. An example is given later in Figure 9.26.

If the pressure in a confined aquifer is such that the potentiometric surface comes above ground level, then the well will overflow and is said to be artesian. The two types of aquifer are shown in Figure 9.3. Clearly, a confined aquifer with thick overlying impermeable clays is likely to be much less vulnerable to pollution than an unconfined aquifer.

9.2.2 Groundwater flow

The flow of water through an aquifer is governed by Darcy's Law, which states that the rate of flow is directly proportional to the hydraulic gradient:

$$Q = -KiA$$

where Q is the rate of flow through unit area A under hydraulic gradient i. The hydraulic gradient dh/dl is the difference between the levels of the potentiometric surface at any two points divided by the horizontal distance between them. The parameter K is known as the hydraulic conductivity, and is a measure of the permeability of the material through which the water is

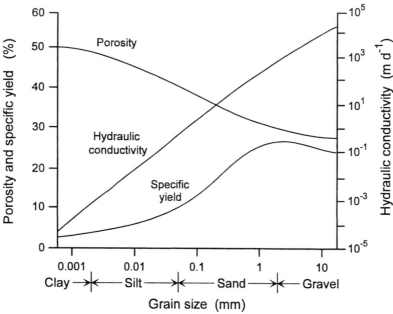

Figure 9.4 Porosity, specific yield and hydraulic conductivity of granular materials (Modified from Davis and De Wiest, 1966)

flowing. The similarity between Darcy's Law and the other important laws of physics governing the flow of both electricity and heat should be noted. For clean, granular materials, hydraulic conductivity increases with grain size (Figure 9.4). Typical ranges of hydraulic conductivity for the main geological materials are shown in Figure 9.5.

Darcy's Law can be written in several forms. By substituting V for the term Q/A, it can take the form:

$$V = -K\frac{dh}{dl}$$

in which the term V is commonly referred to as the specific discharge. The specific discharge does not represent the actual velocity of groundwater flow through the aquifer material. To determine specific discharge, the volumetric flux Q is divided by the full cross-sectional area, which includes both solids and voids. Clearly, flow can only take place through the pore spaces, and the actual velocity of groundwater flow can be calculated if the porosity is known. Most aquifer materials in which inter-granular flow predominates have porosities of 20 to 40 per cent, so the actual groundwater flow velocity is three to five times the specific discharge. However, this average velocity in the direction of groundwater flow does not represent the true velocity of water particles travelling through the pore spaces. These microscopic

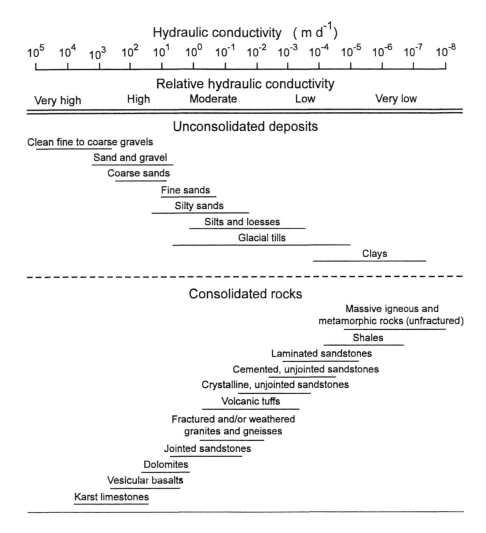

Figure 9.5 Range of hydraulic conductivity values for geological materials (Based on Driscoll, 1986 and Todd, 1980)

velocities are generally greater, as the inter-granular flow pathways are irregular and longer than the average, linear, macroscopic pathway (Figure 9.6). The true microscopic velocities are seldom of interest, and are in any case almost impossible to determine.

Darcy's Law provides a valid description of the flow of groundwater in most naturally occurring hydrogeological conditions. Thus, in general, it holds true for fractured rock aquifers as well as granular materials, although there are particular problems associated with the analysis of flow in fractured

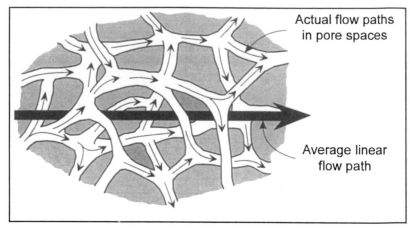

Actual flow paths
in pore spaces

Average linear
flow path

Figure 9.6 Average linear and microscopic flow paths (Based on Freeze and Cherry, 1979)

media which are outside the scope of the present discussion. Fissure flow conditions are characterised by low porosity and high permeability, and very high flow velocities may result, especially where a small number of fissures are enlarged by solution (Figure 9.1e). Extensive development of solution in limestone areas can result in karst terrain, characterised by solution channels, closed depressions, sink holes and caves, into which all traces of surface drainage may disappear. Extreme groundwater velocities of several kilometres per day have been observed in tracer experiments in such areas. These conditions can be very favourable for groundwater abstraction, but aquifers of this type are often highly vulnerable to all types of pollution and difficult to monitor.

Although most of the flow in a fractured rock aquifer is through the fissures, very slow groundwater movement may occur through the interconnected voids of the matrix. An aquifer of this type is characterised by high porosity and low permeability in the matrix with low porosity and high permeability in the fractures. The volume of groundwater stored in the matrix may be 20 to 100 times greater than the volume in the fractures. Many limestones, some sandstones and certain types of volcanic rocks exhibit these characteristics, and such aquifers can be said to have dual porosity and permeability. One of the most intensively used and well studied aquifers of this type is the chalk of north west Europe.

9.2.3 Groundwater flow systems
The origin of fresh groundwater is normally atmospheric precipitation of some kind, either by direct infiltration of rainfall or indirectly from rivers,

lakes or canals. Groundwater is, in turn, the origin of much stream-flow and an important flow component to lakes and oceans and is, therefore, an integral part of the hydrological cycle. Water bodies such as marshes are often transitional between groundwater and surface water (see Figure 1.1). At all places where surface and groundwater meet, the inter-connection between the two must be appreciated in relation to the design of water quality assessment programmes.

Within the context of the overall cycle, the groundwater flow system is a useful concept in describing both the physical occurrence and geochemical evolution of groundwater. A groundwater flow system is a three dimensional, closed system containing the flow paths from the point at which water enters an aquifer to the topographically lower point where it leaves. Infiltration of rainfall on high ground occurs in a recharge area in which the hydraulic head decreases with depth and net saturated flow is downwards away from the water table. After moving slowly through the aquifer down the hydraulic gradient, groundwater leaves the aquifer by springs, swamps and baseflow to streams or to the ocean in a discharge area. In a discharge area, the hydraulic head increases with depth and the net saturated flow is upwards towards the water table. In a recharge area, the water table often lies at considerable depth beneath a thick unsaturated zone. In a discharge area, the water table is usually at, or very near, the ground surface. Rivers, canals, lakes and reservoirs may either discharge to, or recharge from, groundwater and the relationship may change seasonally or over a longer time span.

Examples of groundwater flow regimes under humid and semi-arid climatic conditions are shown in Figure 9.7. In large, deep aquifers, groundwater may be moving slowly, at rates of a few metres per year, from recharge to discharge areas for hundreds or thousands of years. In small, shallow aquifers, recharge and discharge areas may be much closer or even adjacent to each other, and residence times can be restricted to a few months or years (Figure 9.7). In arid and semi-arid areas, groundwater discharge areas are often characterised by poor quality groundwater, particularly with high salinity. In these areas, groundwater discharge occurs from seepages or salt marshes with distinctive vegetation, known as salinas or playas, in which evapotranspiration at high rates, over a long time, has led to a build up of salinity.

Water movement in the unsaturated zone cannot be determined by the same methods as for the saturated zone because water will not freely enter void spaces unless the soil-water pressure is greater than atmospheric. Therefore, special methods must be used to measure water movement in the unsaturated zone, and the most common are tensiometers, psychrometers and neutron moisture logging. Groundwater flow in the unsaturated zone is

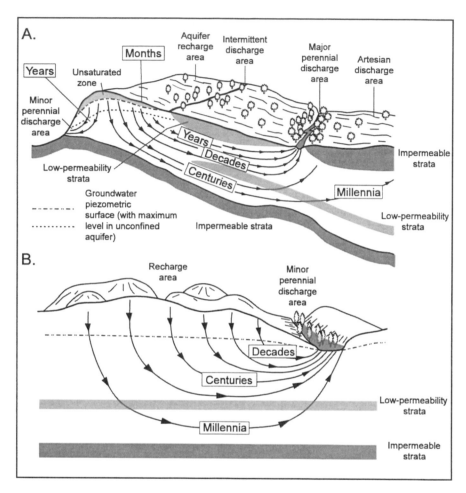

Figure 9.7   Groundwater flow systems **A.** Humid regions **B.** Arid regions. The residence periods indicated are typical order-of-magnitude values from time of recharge to point of discharge. (After Foster and Hirata, 1988)

dominantly vertical, and is related to the rate of infiltration, the storage capacity of the vadose zone and the depth to the water table. Downward movement in granular materials is generally a few metres per year or less, and residence times of tens of years are thus possible where the unsaturated zone is thick. In fissured or dual porosity aquifers exposed at the surface, rapid infiltration through the fissures can by-pass the unsaturated zone completely, dramatically reducing residence times and making the aquifer highly vulnerable to pollution.

9.2.4 Matrix of aquifer types

It is usual to classify all rock types as igneous, sedimentary or metamorphic, depending on their origins. Igneous rocks are formed by the cooling and solidifying of molten rock or magma, after it has been either forced (intruded) into other rocks below ground or extruded at the ground surface as a volcanic eruption. Sedimentary rocks are formed as a result of the deposition of particles (often resulting from the weathering and erosion of other rocks) usually under water on the sea bed, or in rivers, lakes and reservoirs. Metamorphic rocks are formed by the alteration of other rocks under the action of heat and pressure.

Most of the world's important regional aquifers are of sedimentary origin. Igneous and metamorphic rocks are generally far less important as sources of groundwater. As a result of their method of formation, igneous and metamorphic rocks often have such low initial porosity and permeability (Table 9.1, Figure 9.5) that they do not form important aquifers. In time, and particularly with elevation to the ground surface and prolonged exposure to weathering, joints may be formed and opened up so that sufficient porosity and permeability is created to allow the igneous and metamorphic rocks to be called aquifers. Locally these weathered rocks can be vital sources of water. In contrast, sediments usually start their geological lives with a high porosity and a permeability which is closely related to grain size (Figure 9.4). With time, they may be buried and compacted, and cement may form between the grains (Figure 9.1d); such consolidated or indurated sedimentary rocks may have low porosity and permeability and only become aquifers by virtue of their joints and fractures.

Table 9.2 represents an attempt to construct a matrix of major aquifer types. The aquifers in the upper part of the table are largely unconsolidated and relatively young in geological terms. They are, therefore, invariably found close to the surface and form aquifers at shallow depths, usually less than 100 to 200 m. Much older sediments which started life as unconsolidated materials are found at greater depths. There are, for example, ancient glacial deposits many millions of years old but these have been deeply buried and compacted and have lost so much of their original porosity that they can no longer be considered as aquifers. Unconsolidated materials in which inter-granular flow dominates are, therefore, restricted to relatively shallow depths (Table 9.2). Unconsolidated materials are widely distributed and very important as aquifers. They are characterised by generally good natural water quality because of the short to moderate residence times and regular recharge, at least in humid areas, but are often highly vulnerable to anthropogenic influences.

Table 9.2 Major aquifer types

| Geological environment | Principal aquifer features | | | | | Examples of groundwater bodies | | |
	Formations	Lithologies	Class	Dominant porosity	Groundwater flow regime	Small shallow	Large shallow	Large deep
Glacial deposits	Eskers, kames, terraces, fans, moraines, buried valleys	Sands, gravels mixed sands and gravels, boulders, clay lenses	SU	P	I	Canada, Northern USA, Denmark		
Fluvial deposits	Terraces, fans, buried valleys	Sands, silts, clay lenses	SU	P	I	World-wide		
	Alluvium	Sand, silts	SU	P	I	World-wide	Netherlands N Germany Indo-Gangetic Plain	
Deltaic deposits	Alluvium	Fine sands, silts	SU	P	I		Lower Indus Bangladesh Mekong	
Aeolian deposits	Dune sands	Sands	SU	P	I	Netherlands		
	Loess	Silts	SU	P	I	?	China	
Marine deposits	Limestones, dolomites	Oolites, marls	SC	P,S	F,I	Caribbean Islands	Florida	N Europe chalk Ogallala (USA)
	Karst limestones		SC	S	F	Caribbean Islands	Yucatan, Jaffna Yugoslavia	
	Sandstones	Cemented sand grains	SC	S	F,I			Nubian, Karoo Great Artesian Basin (Australia)

Continued

Table 9.2 Continued

Geological environment	Principal aquifer features					Examples of groundwater bodies		
	Formations	Lithologies	Class	Dominant porosity	Groundwater flow regime	Small shallow	Large shallow	Large deep
Volcanics	Ashes	Disaggregated fragments	IU	P	I			
	Lavas	Fine-grained crystalline	IC	S	F	Hawaii	Deccan basalts Columbia River Plateau (USA)	Karoo basalts
	Tuffs	Cemented grains	IC	S	I,F	Central America		
Igneous and metamorphic	Granites, schists, gneisses (fresh)	Crystalline	IMC	S	F			
	Granites, gneisses, schists (weathered)	Disaggregated crystalline	IMC	S	F,I		Africa, India, Sri Lanka, Brazil	

Rock class and degree of consolidation: S Sedimentary; I Igneous; M Metamorphic; U Unconsolidated; C Consolidated

Dominant porosity: P Primary; S Secondary

Groundwater flow regime: I Intergranular; F Fissure

Consolidated formations produce shallow and deep aquifers (Table 9.2). Sedimentary formations originally formed at shallow depths and having high inter-granular porosity may have been deeply buried, compacted, cemented and subsequently brought back to the ground surface by major earth movements. If secondary porosity has developed (Table 9.1), they then form shallow aquifers in which both fissure flow and inter-granular flow may be important (Table 9.2). Such aquifers may be highly productive and have generally good natural groundwater quality if there is regular recharge, but may be especially vulnerable to pollution because of the possibility of very rapid flow in fissures (Figure 9.5).

Similarly, igneous and metamorphic rocks having low original porosity when they were formed at great depths within the earth's crust, can form important (but usually relatively low yielding) shallow aquifers when brought to the earth's surface and exposed to weathering processes. At greater depths, without weathering most of the fissures are closed and insufficient secondary porosity or permeability develops for such rocks to be considered aquifers (Table 9.2).

Igneous volcanic rocks come to the ground surface as molten lava or as ejected fragmentary material. These fragments may be welded by the great heat and gasses associated with volcanic activity and form tuffs. Fine grained crystalline lavas have almost no primary porosity. Their usefulness as aquifers depends on features such as cooling joints, the blocky nature of some lava types, and the weathering of the top of an individual flow before the next one is extruded over it. Shallow volcanic aquifers have generally good natural water quality for the same reasons as other shallow aquifers, except that in areas of current or recent volcanic or hydrothermal activity, specific constituents, particularly fluoride may present problems.

Deep, consolidated sedimentary formations are characterised by slow groundwater movement, long residence times, ample opportunity for dissolution of minerals and, therefore, often poor natural water quality. As a result of their great depth, such formations are often confined beneath thick sequences of low permeability clays or shales, and are generally less vulnerable to anthropogenic influences.

9.2.5 Contaminant transport

In water quality investigations, the groundwater flow system is considered in terms of its ability to transport dissolved substances or solutes, which may be natural chemical constituents or contaminants. Solutes are transported by the bulk movement of the flowing groundwater, a process termed advection. However, when a small volume of solute, either a contaminant or an artificial

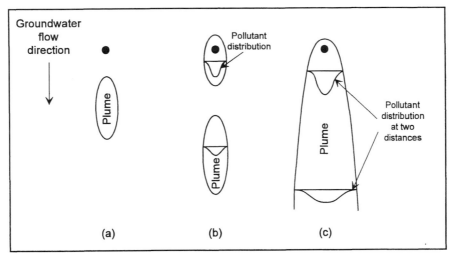

Figure 9.8 Plume of contaminant generated from a) a slug source or spill, b) an intermittent source, and c) a continuous source (Modified from Barcelona *et al.*, 1985)

tracer, is released into an aquifer it will spread out from the expected advective flow path. Instead it forms a plume of diluted solute, which broadens both along, and perpendicular to, the flow direction (Figure 9.8). Two processes can contribute to this phenomenon (Freeze and Cherry, 1979). The first is molecular diffusion in the direction of the concentration gradient due to the thermal-kinetic energy of the solute particles. This process is only important at low velocities. Much more important is the mechanical dispersion which arises from the tortuosity of the pore channels in a granular aquifer and of the fissures in a fractured aquifer, and from the different speeds of groundwater flow in channels or fissures of different widths.

In dual porosity aquifers, groundwater solute concentrations may be significantly different in the fissures than in the matrix. In either diffuse or point source pollution, the water moving in the fissures may have a concentration higher than the water in the matrix, known as the pore-water; hence diffusion from fissures to matrix will occur. In time, if the source of pollutant is removed, the solute concentration in the fissures may decline below that of the pore-water, and diffusion in the opposite direction may result. Groundwater sampling at this stage by conventional methods would sample the fissure water and perhaps miss an important reservoir of pollutant in the rock matrix.

If pollutants find their way into aquifers in the immiscible phase, transport will be governed by completely different factors from those which determine groundwater flow, notably the density and viscosity of the immiscible fluid (Lawrence and Foster, 1987). The aromatic hydrocarbons are less dense, and generally more viscous, than water. In the immiscible phase they tend to "float" at the water table (Figure 9.9), and subsequent lateral migration depends on the hydraulic gradient. In this position they can rise with rising water levels, and during subsequent recession, the hydrocarbon may be held by surface tension effects in the pore spaces of the unsaturated zone. In contrast, the chlorinated solvents in the immiscible phase have a considerably higher density and lower viscosity than water, a combination of properties that can result in rapid and deep penetration into aquifers (Schwille, 1981, 1988). These solvents may reach the base of an aquifer, where they could accumulate in depressions or migrate down slope, irrespective of the direction of groundwater flow.

An important factor controlling the persistence of solvents in groundwater is the extent to which the immiscible phase can displace water from a fine grained porous matrix of, for example, a fissured limestone aquifer. This depends on the surface tension properties of the immiscible fluid relative to water and to the minerals of the aquifer matrix (Lawrence and Foster, 1987). It seems likely that considerable excess pressure would be required to enable the immiscible phase of most organic fluids to penetrate into the matrix. Once in the aquifer, an immiscible body of dense solvent could act as a buried pollution source, perhaps for many years or decades. The rate of dissolution will depend on the solubility of the particular compound in water, the rate of groundwater flow and the degree of mixing that is permitted by the distribution of the solvent in relation to the local hydraulic structure and flow pathways. The complexity of behaviour by an immiscible contaminant presents considerable problems for the design of assessment programmes and for the interpretation of the monitoring data.

9.2.6 Chemical characteristics of groundwater

Since groundwater often occurs in association with geological materials containing soluble minerals, higher concentrations of dissolved salts are normally expected in groundwater relative to surface water. The type and concentration of salts depends on the geological environment and the source and movement of the water. A brief description of the principal natural chemical changes that occur in groundwater systems is given in section 9.3. The following paragraphs outline the main ways in which the natural characteristics of groundwater affect water quality for different uses.

A simple hydrochemical classification divides groundwaters into meteoric, connate and juvenile. Meteoric groundwater, easily the most important, is derived from rainfall and infiltration within the normal hydrological cycle (Figure 9.7) and is subjected to the type of hydrochemical evolution

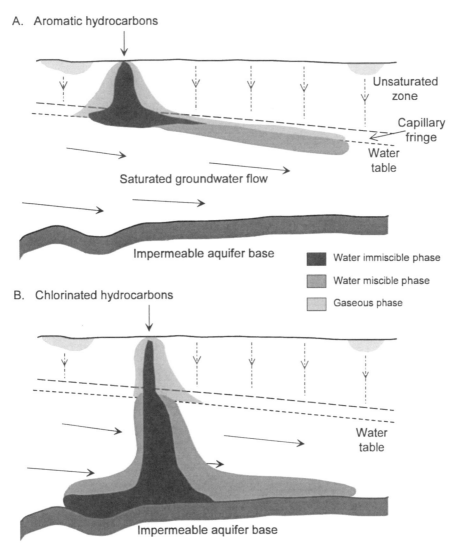

Figure 9.9 Generalised distribution of hydrocarbon phases down a groundwater gradient following a surface spillage (After Lawrence and Foster, 1987 in Meybeck *et al.*, 1989)

described in section 9.3. Groundwater originating as sea water which has been entrapped in the pores of marine sediments since their time of deposition is generally referred to as connate water. The term has usually been applied to saline water encountered at great depths in old sedimentary formations. It is now accepted that meteoric groundwater can eventually become equally saline, and that entrapped sea water can become modified and moved from its original place of entrapment. It is doubtful whether groundwater exists that meets the original definition of connate water, and the non-generic term formation water is preferred by many authors. Connate water is, perhaps, useful to describe groundwater that has been removed from atmospheric circulation for a significant period of geological time. Formation waters are not usually developed for water supplies because of their high salinity. However, they may become involved in the assessment of saline intrusions caused by the overpumping of overlying aquifers.

Juvenile groundwater describes the relatively small amounts of water which have not previously been involved in the circulating system of the hydrological cycle, but are derived from igneous processes within the earth. However, juvenile groundwater often cannot be distinguished geochemically from meteoric groundwater that has circulated to great depths and become involved in igneous processes. True juvenile waters unmixed with meteoric water are rare and of very localised extent, and are not normally associated with the development and assessment of fresh groundwater resources.

The natural chemical quality of groundwater is generally good, but elevated concentrations of a number of constituents can cause problems for water use, as described in Chapter 3. The relative abundance of the constituents dissolved in groundwater is given in Table 9.3, and a summary of the natural sources and range of concentrations of the principal constituents of groundwater is given in Table 9.4.

High iron levels in groundwater are widely reported from developing countries, where they are often an important water quality issue. Consumers may reject untreated groundwater from handpump supplies if it has a high iron concentration in favour of unprotected surface water sources with low iron levels but which may have gross bacteriological pollution. The situation is made worse in many areas by the corrosion of ferrous well linings and pump components. Obtaining representative samples for iron in groundwater presents particular difficulties because of the transformations brought about by the change in oxidation/reduction status which occurs on lifting the water from the aquifer to ground level.

9.2.7 Biological characteristics of groundwater

Groundwater quality can be influenced directly and indirectly by microbiological processes, which can transform both inorganic and organic constituents of groundwater. These biological transformations usually hasten geochemical processes (Chapelle, 1993). Single and multi-celled organisms have become adapted to using the dissolved material and suspended solids in the water and solid matter in the aquifer in their metabolism, and then

Table 9.3 Relative abundance of dissolved constituents in groundwater

Major constituents (1.0 to 1,000 mg l⁻¹)	Secondary constituents (0.01 to 10.0 mg l⁻¹)	Minor constituents (0.0001 to 0.1 mg l⁻¹)
Sodium	Iron	Arsenic
Calcium	Aluminium	Barium
Magnesium	Potassium	Bromide
Bicarbonate	Carbonate	Cadmium
Sulphate	Nitrate	Chromium
Chloride	Fluoride	Cobalt
Silica	Boron	Copper
	Selenium	Iodide
		Lead
		Lithium
		Manganese
		Nickel
		Phosphate
		Strontium
		Uranium
		Zinc

Source: After Todd, 1980

releasing the metabolic products back into the water (Matthess, 1982). There is practically no geological environment at, or near, the earth's surface where the pH and Eh conditions will not support some form of organic life (Chilton and West, 1992). In addition to groups tolerating extremes of pH and Eh, there are groups of microbes which prefer low temperatures (psychrophiles), others which prefer high temperature (thermophiles) (Ehrlich, 1990), and yet others which are tolerant of high pressures. However, the most biologically favourable environments generally occur in warm, humid conditions.

Micro-organisms do not affect the direction of reactions governed by the thermodynamic constraints of the system, but they do affect their rate. Sulphides, for example, can be oxidised without microbial help, but microbial processes can greatly speed up oxidation to the extent that, under optimum moisture and temperature conditions, they become dominant over physical and chemical factors.

All organic compounds can act as potential sources of energy for organisms. Most organisms require oxygen for respiration (aerobic respiration) and the breakdown of organic matter, but when oxygen concentrations are depleted some bacteria can use alternatives, such as nitrate, sulphate and carbon dioxide (anaerobic respiration). Organisms which can live in the presence of oxygen (or without it) are known as facultative anaerobes. In contrast, obligate anaerobes are organisms which do not like oxygen. The

Table 9.4   Sources and concentrations of natural groundwater components

Component	Natural sources	Concentration in natural water
Dissolved solids	Mineral constituents dissolved in water	Usually < 5,000 mg l^{-1}, but some brines contain as much as 300,000 mg l^{-1}
Nitrate	Atmosphere, legumes, plant debris, animal excrement	Usually < 10 mg l^{-1}
Sodium	Feldspars (albite), clay minerals, evaporites such as halite, NaCl, industrial wastes	Generally < 200 mg l^{-1}; about 10,000 mg l^{-1} in sea water; ~ 25,000 mg l^{-1} in brines
Potassium	Feldspars (orthoclase, microcline), feldspathoids, some micas, clay minerals	Usually < 10 mg l^{-1}, but up to 100 mg l^{-1} in hot springs and 25,000 mg l^{-1} in brines
Calcium	Amphiboles, feldspars, gypsum, pyroxenes, dolomite, aragonite, calcite, clay minerals	Usually < 100 mg l^{-1}, but brines may contain up to 75,000 mg l^{-1}
Magnesium	Amphiboles, olivine, pyroxenes, dolomite, magnesite, clay minerals	Usually < 50 mg l^{-1}; about 1,000 mg l^{-1} in ocean water; brines may have 57,000 mg l^{-1}
Carbonate	Limestone, dolomite	Usually < 10 mg l^{-1}, but can exceed 50 mg l^{-1} in water highly charged with sodium
Bicarbonate	Limestone, dolomite	Usually < 500 mg l^{-1}, but can exceed 1,000 mg l^{-1} in water highly charged with CO_2
Chloride	Sedimentary rock (evaporites), a little from igneous rocks	Usually < 10 mg l^{-1} in humid areas; up to 1,000 mg l^{-1} in more arid regions; approximately 19,300 mg l^{-1} in sea water and up to 200,000 mg l^{-1} in brines
Sulphate	Oxidation of sulphide ores, gypsum, anhydrite	Usually < 300 mg l^{-1}, except in wells influenced by acid mine drainage; up to 200,000 mg l^{-1} in some brines
Silica	Feldspars, ferromagnesian and clay minerals, amorphous silica, chert and opal	Ranges from 1–30 mg l^{-1} but as much as 100 mg l^{-1} can occur and concentrations may reach 4,000 mg l^{-1} in brines
Fluoride	Amphiboles (hornblende), apatite, fluorite, mica	Usually < 10 mg l^{-1}, but up to 1,600 mg l^{-1} in brines
Iron	Igneous rocks: amphiboles, ferromagnesian micas, FeS, FeS_2 and magnetite, Fe_3O_4. Sandstone rocks: oxides, carbonates, sulphides or iron clay minerals	Usually < 0.5 mg l^{-1} in fully aerated water; groundwater with pH < 8 can contain 10 mg l^{-1}; infrequently, 50 mg l^{-1} may be present
Manganese	Arises from soils and sediments. Metamorphic and sedimentary rocks and mica biotite and amphibole hornblende minerals contain large quantities of Mn	Usually < 0.2 mg l^{-1}; groundwater contains > 10 mg l^{-1}

Source: Modified from Todd, 1980

presence or absence of oxygen is, therefore, one of the most important factors affecting microbial activity, but not the only one. For an organism to grow and multiply, nutrients must be supplied in an appropriate mix, which satisfies carbon, energy, nitrogen and mineral requirements (Ehrlich, 1990).

Most micro-organisms grow on solid surfaces and, therefore, coat the grains of the soil or aquifer. They attach themselves with extra-cellular polysaccharides, forming a protective biofilm which can be very difficult to remove. Up to 95 per cent of the bacterial population may be attached in this way rather than being in the groundwater itself. However, transport in the flowing groundwater is also possible. The population density of micro-organisms depends on the supply of nutrients and removal of harmful metabolic products (Matthess, 1982). Thus, in general terms, higher rates of groundwater flow supply more nutrients and remove the metabolic products more readily. Microbe populations are largest in the nutrient-rich humic upper parts of the soil, and decline with decreasing nutrient supply and oxygen availability at greater depths. Many sub-surface microbes, however, prefer lower nutrient conditions. In the presence of energy sources, such as organic material, anaerobic microbial activity can take place far below the soil and has been observed at depths of hundreds and even thousands of metres. The depth to which such activity is possible is determined by the nutrient supply and, in addition, pH, Eh, salt content, groundwater temperature and the permeability of the aquifer.

Microbiological activity primarily affects compounds of nitrogen and sulphur, and some of the metals, principally iron and manganese. Sulphate reduction by obligate aerobes is one of the most important biological processes in groundwater. Nitrogen compounds are affected by both nitrifying and denitrifying bacteria. Reduction of nitrate by denitrifying bacteria occurs in the presence of organic material in anaerobic conditions, leading to the production of nitrite which is then broken down further to elemental nitrogen. The possibility of enhancing natural denitrification is currently receiving attention in relation to the problem of nitrate in groundwater. Under aerobic conditions, ammonia (which may be produced during the decomposition of organic matter) is oxidised to nitrite and nitrate. Likewise iron can be subjected to either reduction or oxidation, depending on the Eh and pH conditions of the groundwater. In favourable microbiological environments, massive growth of iron bacteria can cause clogging of well screens and loss of permeability of aquifer material close to wells, and may require special monitoring and remedial action.

Micro-organisms can break down complex organic materials dissolved in groundwater. Under anaerobic conditions, microbial breakdown proceeds either as a methane fermentation or by reduction of sulphate and nitrate (Matthess, 1982). Microbial decomposition has been demonstrated for a whole range of organic compounds, including fuel hydrocarbons, chlorinated solvents and pesticides. Under ideal conditions, all organic materials would eventually be converted to the simplest inorganic compounds. In practice, complete breakdown is never reached, and intermediate products of equal or even greater toxicity and persistence may be produced. Nevertheless, the possibility of using indigenous or introduced microbial populations and enhancing the nutritional status of their environment is currently an important area of research in the quest for effective techniques for remediation of contaminated aquifers (NRC, 1993).

A principal microbiological concern in groundwater is the health hazard posed by faecal contamination. Of the four types of pathogens contained in human excreta, only bacteria and viruses are likely to be small enough to be transmitted through the soil and aquifer matrix to groundwater bodies (Lewis *et al.*, 1982). The soil has long been recognised as a most effective defence against groundwater contamination by faecal organisms, and a number of processes (see section 9.3) combine to remove pathogens from infiltrating water on its way to the water table. Not all soils are equally effective in this respect. In addition, many human activities which can cause groundwater pollution involve the removal of the soil altogether (section 9.4). Bacteriological contamination of groundwater remains a major concern, especially where many dispersed, shallow dug wells or boreholes provide protected but untreated domestic water supplies.

In summary, microbiological processes may influence groundwater quality both positively and negatively. The former include reducing nitrate and sulphate contents of groundwater, and removal of organic pollutants. The latter include the production of hydrogen sulphide and soluble metals, production of gas and biofilm fouling of well screens and distribution pipes.

9.3   Water–soil–rock interactions

9.3.1 Natural hydrochemical evolution

In treating groundwater, the importance of the physical properties of both ground and water have been summarised and used in section 9.2 to assist in defining a matrix of groundwater bodies. Similarly, it follows that the chemical properties of both ground and water are important in determining the quality of groundwater. The natural quality of groundwater (as described briefly in section 9.2.6) is, therefore, controlled by the geochemistry of the

Table 9.5   Average composition of igneous and some sedimentary rocks

		Sedimentary rocks		
Element	Igneous rocks (ppm)	Sandstone (ppm)	Shale (ppm)	Carbonates (ppm)
Si	285,000	359,000	260,000	34
Al	79,500	32,100	80,100	8,970
Fe	42,200	18,600	38,800	8,190
Ca	36,200	22,400	22,500	272,000
Na	28,100	3,870	4,850	393
K	25,700	13,200	24,900	2,390
Mg	17,600	8,100	16,400	45,300
Ti	4,830	1,950	4,440	377
P	1,100	539	733	281
Mn	937	392	575	842
F	715	220	560	112
Ba	595	193	250	30
S	410	945	1,850	4,550
Sr	368	28	290	617
C	320	13,800	15,300	113,500
Cl	305	15	170	305
Cr	198	120	423	7.1
Cu	97	15	45	4.4
Ni	94	2.6	29	13
Zn	80	16	130	16
Co	23	0.33	8.1	0.12
Pb	16	14	80	16
Hg	0.33	0.057	0.27	0.046
Se	0.05	0.52	0.60	0.32

Source: Hem, 1989

lithosphere, the solid portion of the earth, and the hydrochemistry of the hydrosphere, the aqueous portion of the earth.

The average composition of igneous and sedimentary rocks is shown in Table 9.5. About 95 per cent of the earth's crust to a depth of 16 km is composed of igneous rocks (Hem, 1989). Ninety-eight per cent of the crust consists of the common alumino-silicates, made up of various combinations of the uppermost seven elements in the first column of Table 9.5, plus oxygen. Most usable groundwater occurs at depths of < 1 km, where sedimentary rocks predominate. Moreover, it is clear from section 9.2 that igneous rocks generally have low porosity and permeability and are much less important as aquifers on a global scale than sedimentary rocks.

Atmospheric precipitation infiltrating through the soil dissolves CO_2 produced by biological activity. The resulting solution of weak carbonic acid dissolves soluble minerals from the underlying rocks. A second process operating during passage through the soil is the consumption by soil organisms of some of the oxygen which was dissolved in the rainfall. These reactions occur in the soil and the top few metres of the underlying rock. In temperate and humid climates with significant recharge, groundwater moves continuously and relatively rapidly through the outcrop area of an aquifer (Figure 9.7); hence contact time with the rock matrix is relatively short. Readily soluble minerals will be removed, but insufficient contact time exists for less soluble minerals to be taken up. Groundwater in the outcrop areas of aquifers is likely to be low in overall mineralisation, with the natural constituents depending on the materials of which the rocks are made.

In igneous rocks, the restricted opportunity for reactions to take place is accentuated by the fact that groundwater storage and flow is predominantly in fissures, giving short residence times and low contact surface area. Groundwater in igneous rocks is, therefore, often exceptionally lightly mineralised, although characterised by high silica contents (Hem, 1989). Pure siliceous sands or sandstones without a soluble cement also contain groundwater with very low total dissolved solids (Matthess, 1982). In such aquifers, the dissolved constituents which are present come mainly from other sources, such as rainfall and dry deposition, especially sodium, chloride and sulphate which, in coastal regions, may exceed calcium, magnesium and bicarbonate. Sulphate may also be produced by the oxidation of metallic sulphides which are present in small amounts in many rock types. The presence of soluble cement may produce increased concentrations of the major ions. Groundwaters in carbonate rocks have pH values above 7, and mineral contents usually dominated by bicarbonate and calcium.

In many small and/or shallow aquifers the hydrochemistry does not evolve further. If, however, an aquifer dips below a confining layer, a sequence of hydrochemical processes occurs with progressive distance down gradient from the outcrop. These processes are clearly observed in the Lincolnshire Limestone of eastern England (Edmunds, 1973). This Middle Jurassic limestone outcrops in eastern England, and dips eastwards at less than one degree beneath confining strata consisting of clays, shales and marls (Figure 9.10). Dissolution of calcium and bicarbonate is active in the recharge area, such that Ca^{2+} is soon saturated with respect to calcite (Edmunds, 1973). As the water moves down dip, further modifications are at first limited. By observing the Eh of pumped samples, a sharp redox barrier was defined some 10–12 km east of the onset of confining conditions (Figure 9.10). An Eh of

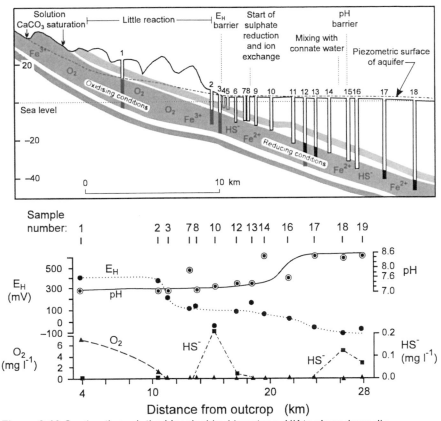

Figure 9.10 Section through the Lincolnshire Limestone, UK to show down dip oxidation and reduction processes (After Edmunds, 1973)

+ 440 mV on the oxidising side, buffered by the presence of dissolved oxygen fell sharply to + 150 mV, corresponding with the complete exhaustion of oxygen. The water then became steadily more reducing down dip, as demonstrated by the presence of sulphide species (Figure 9.10), detectable by smell at many of the pumping boreholes. Reducing conditions also led to an increase in solubility of iron and denitrification of nitrate to N_2.

Beyond the redox barrier, there was a distance of several kilometres before any further chemical changes could be observed. Calcium was buffered until about 14 km from the outcrop. Eastwards of this point, Na^+ in the groundwater gradually increased at the expense of Ca^{2+} (Figure 9.11). This process of ion exchange of sodium on the aquifer particles for calcium in the water tends towards equilibrium, and produces a natural softening of the water. Calcium concentrations were almost insignificant and the reaction was practically complete by about 20 km from the outcrop. Bicarbonate rose

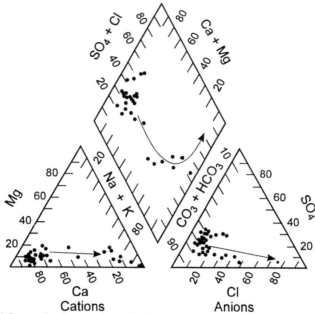

Figure 9.11 Down dip changes in major ions in the Lincolnshire Limestone, UK (After Downing and Williams, 1969)

gradually and the pH increased until buffering occurred at 8.3 (Figure 9.10). Sulphate concentration was constant in the oxidising water, probably reflecting the lack of additional solution of sulphate after initial infiltration. There was a sudden decrease in sulphate concentration 3 km beyond the redox barrier, resulting from sulphate reduction. Twenty-three kilometres away from the confining layer, the increase in Na^+ was greater than the equivalent concentration of Ca^{2+}. This was accompanied by an increase in chloride, and marks the point at which meteoric water moving down dip mixes with much older formation water, as shown in the triangular diagram used to illustrate groundwater analyses (Figure 9.11).

The observed hydrochemical changes down dip can thus be interpreted in terms of oxidation/reduction, ion exchange and mixing processes. Similar sequences have been described from the Atlantic Coastal Plain of the USA (Back, 1966) and the UK Chalk (Edmunds *et al.*, 1987), and probably occur widely in humid, temperate regions. In arid and semi-arid areas, evapotranspiration rates are much higher, and exceed rainfall for much of the year. Recharge is limited, and often occurs in very restricted areas. In the example in Figure 9.7, recharge occurs in a mountainous region of somewhat heavier rainfall and travels slowly through the aquifer, dissolving soluble salts as it goes. The long residence times and incomplete flushing of soluble minerals produce groundwaters which are generally of the sodium chloride type. In the discharge area, evapotranspiration is often too great to allow a perennial stream to develop. Salt is concentrated in soil and water by direct evaporation and a salt marsh or sabkha may be formed.

9.3.2 Reactions related to anthropogenic effects

Physico-chemical reactions between soil or rock and water are of considerable importance when evaluating or predicting the nature of anthropogenic impacts on groundwater quality. In this respect the unsaturated zone, and particularly the soil, deserves special attention since it represents the first and most important natural defence against groundwater pollution (Lewis *et al.*, 1982; Foster, 1985; Matthess *et al.*, 1985). This is as a result of its position between the land surface and the water table and because a number of processes of pollutant attenuation are more favoured by the environments of the soil and the unsaturated zone (Figure 9.12).

Water movement in the unsaturated zone is largely vertical and normally slow. The chemical condition (see section 9.3.1) is usually aerobic and frequently alkaline. Thus, as suggested by Foster and Hirata (1988), there is considerable potential for:

- interception, sorption and elimination of pathogenic bacteria and viruses,
- attenuation of trace elements and other inorganic compounds by precipitation, sorption or cation exchange, and
- sorption and biodegradation of many hydrocarbons and synthetic organic compounds.

The widths of the respective lines in Figure 9.12 indicate that most of these processes occur at their highest rates in the more biologically active soil zone, because of its higher content of clay minerals and organic matter and the much larger bacterial population. However, as noted in section 9.2.7, many activities which can cause groundwater pollution involve the complete removal, or by-passing, of the soil zone. The characteristics of the soil also influence the scope for nutrient and pesticide leaching from a given agricultural activity and whether acid aerial deposition is neutralised. The continuation of these processes at greater depths, albeit to a lesser degree, is more likely in sedimentary materials (in which inter-granular flow predominates) than in consolidated rocks (in which flow is largely restricted to fissures). Some of these processes are applicable to a range of possible constituents of percolating groundwater, while others are more restricted in their effects (Table 9.6).

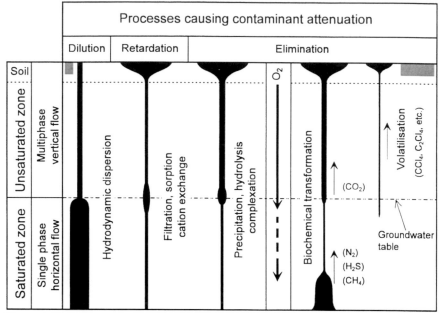

Figure 9.12 Processes causing contaminant attenuation in groundwater systems. The thickness of the corresponding line indicates the relative importance of the process in the soil, above, at and below the groundwater table. (After Foster and Hirata, 1988 and Gowler, 1983)

9.4   Groundwater quality issues

Groundwater quality is the sum of natural and anthropogenic influences. Multi-purpose monitoring may be directed towards a whole range of ground-water quality issues and embrace many variables. Other categories of monitoring may concentrate on a single quality issue such as agricultural pesticides, an industrial spill, saline intrusion, etc. Table 9.7 lists and categorises the human activities that may potentially pollute groundwater and identifies the main pollutants in each case. These have been linked to the principal uses to which groundwater is put and to three levels of industrial development, so that the most important current and future groundwater quality issues can be identified. A major subdivision into urban, industrial and agricultural is made, although it is clear that there is considerable overlap between the first two. Some of the activities generating serious pollution risks are common to highly industrialised, newly industrialising and low development countries (mainly agricultural based economies), but those presenting the most serious threats differ significantly (Table 9.7). The most important groundwater quality issues world-wide are described briefly in the following sections.

Table 9.6 Processes which may affect constituents of groundwater

Constituent	Physical		Geochemical						Biochemical	
	Dispersion	Filtration	Complexation	Ionic strength	Acid-base	Oxidation-reduction	Precipitation-solution	Adsorption-desorption	Decay, respiration	Cell synthesis
Cl⁻, Br⁻	xx									
NO_3^-	xx			x	x	xx			xx	xx
SO_4^{2-}	xx		x	x	xx	xx		x	x	
HCO_3^-	xx		x	xx	xx		xx		xx	
PO_4^{3-}	xx		xx	x	xx		xx	xx	xx	xx
Na^+	xx			x				xx		
K^+	xx			x				xx		
NH_4^+	xx		xx	x	xx	xx		xx	xx	xx
Ca^{2+}	xx		x	xx			x	xx		
Mg^{2+}	xx		x	xx			x	xx		
Fe^{2+}	xx		xx	xx	xx	xx	xx	xx		
Mn^{2+}	xx		xx	xx	xx	xx	xx	xx		
Fe^{3+} and Mn^{4+} oxyhydroxides	xx	xx			xx	xx	xx			
Trace elements	xx		xx	xx		xx	xx	xx		
Organic solutes	xx		xx	x	xx	xx	x	x	xx	xx
Micro-organisms	xx	xx				xx			xx	xx

xx Major control
x   Minor control

Source: UNESCO/WHO, 1978

Table 9.7 Principal activities potentially causing groundwater pollution

Activity	Principal characteristics of pollution					Stage of development[1]			Impact of water use		
	Distribution	Category	Main types of pollutant	Relative hydraulic surcharge	Soil zone by-passed	I	II	III	Drinking	Agricultural	Industrial
Urbanisation											
Unsewered sanitation	ur	P–D	pno	x	✓	xxxx	xx	x	xxxx		x
Land discharge of sewage	ur	P–D	nsop	x		x	x	x	xx	x	x
Stream discharge of sewage	ur	P–L	nop	xx	✓	x	x		xx	x	x
Sewage oxidation lagoons	u	P	opn	xx	✓	x	xx	x	xx		x
Sewer leakage	u	P–L	opn	x	✓			xx	x		x
Landfill, solid waste disposal	ur	P	osnh		✓	x	xx	xxx	x		x
Highway drainage soak-aways	ur	P–L	so	xx	✓	x	xx	xx	xx	x	x
Wellhead contamination	ur	P	pn		✓	xxx	x		xxx		x
Industrial development											
Process water/effluent lagoons	u	P	ohs	xx	✓	x	xx	xx	xx		x
Tank and pipeline leakage	u	P	oh		✓	x	xx	xxx	xx		xx
Accidental spillages	ur	P	oh	xx		x	xx	xxx	xxx		xx
Land discharge of effluent	u	P–D	ohs	x		x	xx	xx	x	x	x
Stream discharge of effluent	u	P–L	ohs	xx	✓	x	x	x	x	x	x
Landfill disposal residues and waste	ur	P	ohs		✓	x	xxx	xxx	xx		x
Well disposal of effluent	u	P	ohs	xx	✓		x	x	xx		x
Aerial fallout	ur	D	a				x	xx	x	x	x
Agricultural development											
Cultivation with:											
Agrochemicals	r	D	no			x	xx	xxx	xxx	x	x
Irrigation	r	D	sno	x		xx	xx	x	xxx	xxxx	x
Sludge and slurry	r	D	nos	x		x	x	x	xx	x	xx
Wastewater irrigation	r	D	nosp	x			xx	x	xx	xx	xx

Continued

Table 9.7 Continued

Activity	Principal characteristics of pollution					Stage of development[1]			Impact of water use		
	Distribution	Category	Main types of pollutant	Relative hydraulic surcharge	Soil zone by-passed	I	II	III	Drinking	Agricultural	Industrial
Livestock rearing/crop processing:											
Unlined effluent lagoons	r	P	pno	x	✓	x	x	xx	x	x	x
Land discharge of effluent	r	P–D	nsop	x	✓	x	x	xx	x	x	x
Stream discharge of effluent	r	P–L	onp	x	✓	x	x	xx	x	x	x
Mining development											
Mine drainage discharge	ru	P–L	sha	xx	✓	x	xx	xx	xx	x	x
Process water/sludge lagoons	ru	P	hsa	xx	✓	x	xx	xx	xx	x	x
Solid mine tailings	ru	P	hsa		✓	x	xx	xx	xx	x	x
Oilfield brine disposal	r	P	s	x	✓		x	x	xx	x	x
Hydraulic disturbance	ru	D	s		na		x	x	xx	x	x
Groundwater resource management											
Saline intrusion	ur	D–L	s	xx	na	x	x	xx	xxx	xxx	xx
Recovering water levels	u	D	so	na	na			x	x		x

Distribution: u Urban; r Rural

Category: P Point; D Diffuse; L Line

Types of pollutant: p Faecal pathogens; n Nutrients; o Organic micropollutants; h Heavy metals; s Salinity; a Acidification

x to xxxx Increasing importance or impact

na Not applicable

[1] Stages of development: I Low development; II Newly industrialising; III Highly industrialised

The differentiation between point and diffuse sources has been discussed in Chapter 1. Although there is overlap between the two (Table 9.7), as many small point sources may become a diffuse source, the distinction is of fundamental importance, especially in relation to prevention and control of groundwater pollution, and hence in the design of monitoring undertaken to evaluate the effectiveness of pollution control measures. It is apparent, for example, from Table 9.7 and from the brief descriptions below that groundwater quality issues may be global, national or regional, affect the whole of one or more aquifers, or be restricted to the immediate vicinity of a single contaminant source. Therefore, the scale and type of monitoring operations required depend on the issue, or combination of issues, and the size of groundwater body affected. These are described further in section 9.5 and illustrated by examples in section 9.6.

9.4.1 Unsewered domestic sanitation

The impetus of the International Water Supply and Sanitation Decade (from 1981 to 1990) produced considerable efforts in many countries to improve health by investment in water supply and sanitation programmes. These often comprise the provision of largely untreated rural and small-scale urban water supplies from groundwater and the construction of unsewered, on-site sanitation facilities using various types of latrines. Under certain hydrogeological conditions, unsewered sanitation can cause severe groundwater contamination by pathogenic micro-organisms and nitrate, which may largely negate the expected health benefits of such programmes. In some circumstances, therefore, these two low-cost technologies may be incompatible (Lewis et al., 1982; Foster et al., 1987).

Unsewered sanitation consists of the installation of either septic tanks or pit latrines of the ventilated, dry or pour-flush types. There are important differences between the two in relation to the risk of groundwater contamination (Foster et al., 1987). Septic tank soak-aways discharge at higher levels in the soil profile than pit latrines and conditions may be more favourable for pathogen elimination. Pit latrines are often deep excavations (to allow a long useful life) and the soil may be entirely removed. The hydraulic loading from septic tank soak-aways is likely to be less than for some of the pit latrine types. Septic tanks are lined and their solid effluent of high nitrogen content is periodically removed, whereas most pit latrines are unlined and the solid material remains in the ground. For these reasons, septic tanks are likely to pose a less serious threat to groundwater than pit latrines. If domestic wastewater is also discharged to unsewered sanitation, there is an added risk of groundwater contamination by the increasing range of organic compounds used in household products such as detergents and disinfectants.

The impact of unsewered sanitation is felt particularly in relation to drinking water (Table 9.7). Contamination of groundwater supplies by unsewered sanitation has been the proven vector of pathogens in numerous disease outbreaks. In unconsolidated deposits, filtration, adsorption and inactivation during migration through one metre or less of unsaturated, fine-grained strata normally reduces pathogen numbers to acceptable levels (Lewis *et al.*, 1982). Problems usually arise only where the water table is so shallow that on-site sanitation systems discharge directly into the saturated zone. The risk to groundwater may be enhanced by persistent organisms, and considerable uncertainty remains about the persistence in aquifers of some pathogens, especially viruses.

Whilst bacteriological contamination of shallow wells and boreholes in all types of geological formations is believed to be widespread, migration of pathogens through unconsolidated strata to deep water supply wells is unlikely. In these situations, bacteriological contamination is most probably direct and localised, reflecting poor well design and/or construction and sanitary completion rather than aquifer pollution. Thus, in many large cities, the impact of unsewered sanitation may not be significant because municipal water supplies are drawn from surface water and then treated, or drawn from relatively distant and deep, well-protected aquifers. Often the most serious problems arise in medium to smaller sized towns and in densely populated peri-urban and rural areas where local, shallower, and often untreated, groundwater sources are used. In these circumstances, direct pollution of the source at the wellhead by the users, by livestock and by wastewater may be a serious problem.

The nitrogen compounds in excreta do not represent such an immediate hazard to groundwater as pathogens, but can cause more widespread and persistent problems. Unsewered sanitation has been shown to cause increased nitrate concentrations in the underlying groundwater at many localities in, for example, South America, Africa and India. An example is shown in Figure 9.13. The impact is often demonstrated by rising nitrate concentrations in public supply wells, as in Greater Buenos Aires (Foster *et al.*, 1987) and Bermuda (Figure 9.14). It is possible to make semi-quantitative estimates of the concentration of persistent and mobile contaminants, such as nitrate and chloride, in groundwater recharge by using the following formula (Foster and Hirata, 1988):

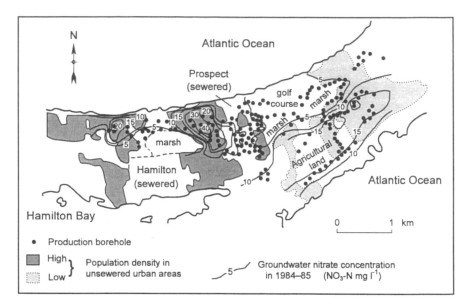

Figure 9.13 Correlation of high nitrate concentrations in groundwater with densely populated areas in Bermuda with unsewered sanitation (Redrawn from Thomson and Foster, 1986)

$$C = \frac{1,000 \times a \times A \times F}{0.365 \times A \times U + 10I}$$

where: C = the concentration (mg l⁻¹) of the contaminant in recharge

a = the unit weight of nitrogen or chloride in excreta (kg capita⁻¹ a⁻¹)

A = population density (persons ha⁻¹)

F = proportion of excreted nitrogen oxidised to nitrate and leached

U = non-consumptive portion of total water use (l capita⁻¹ day⁻¹)

I = natural rate of rainfall infiltration (mm a⁻¹)

Greatest uncertainty surrounds the proportion of the total nitrogen load that will be oxidised and leached into the groundwater recharge.

Groundwater pollution by unsewered sanitation is most likely to occur where soils are thin or absent, where fissures allow rapid movement and where the water table is shallow. A further example of groundwater pollution of this type is described by Lewis *et al.* (1980) in weathered metamorphic rocks in southern Africa. Groundwater quality assessment activities related to unsewered sanitation can generally be considered as surveillance (as defined in Chapter 2).

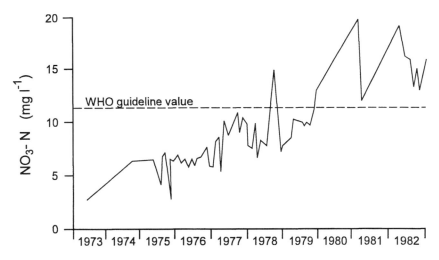

Figure 9.14 Composite trend of average nitrate concentrations in public supplies derived from the Bermuda central groundwater lens (After Thomson and Foster, 1986)

9.4.2 Disposal of liquid urban and industrial waste

An example of groundwater pollution from disposal of liquid waste is described by Everett (1980). Liquid wastes containing chromium and cadmium from metal plating processes were disposed of, virtually untreated, into lagoons recharging directly to a glacial sand aquifer in Long Island, New York. No records were kept of the types and quantities of the wastes, and disposal continued in this way for many years. Eventually chromium was detected in a water supply well, and extensive investigations were undertaken to define the lateral and vertical extent of the plume of polluted groundwater (Figures 9.15 and 9.16). The difference of behaviour between cadmium and chromium should be noted since their maximum concentrations do not coincide. Further examples of groundwater pollution of this type are described by Williams *et al.* (1984) and Foster *et al.* (1987).

Typical methods of wastewater disposal include infiltration ponds, spreading or spraying onto the ground surface and discharge to streams or dry stream beds which may provide a rapid pollution pathway to underlying, shallow aquifers. Current practices are reviewed by Foster *et al.* (1994). In some areas, deep soak-aways or abandoned wells are used for the disposal of liquid domestic, industrial or farming waste into aquifers. There are many thousands of such wells in the USA. Even if the intention is to dispose of the waste at depth, improper sealing or corrosion of well linings often produces

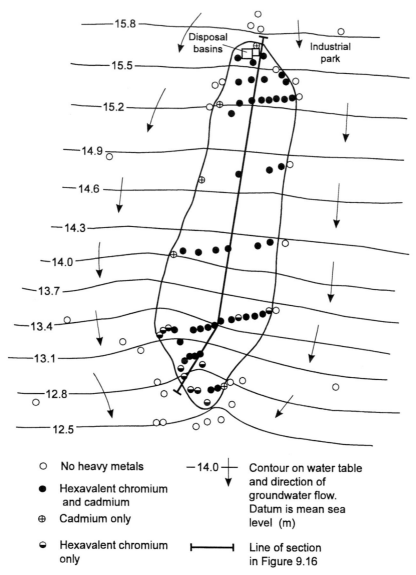

Figure 9.15 Monitoring wells, water table contours and the contaminant plume associated with disposal of liquid industrial waste, Long Island, New York (Modified from Everett, 1980)

leaks and subsequent pollution of the shallow groundwater which is used for water supplies. In recent years, attention has been given to the possibility of injecting treated municipal sewage into aquifers to enhance recharge or establish hydraulic barriers against saline intrusion. Well established examples

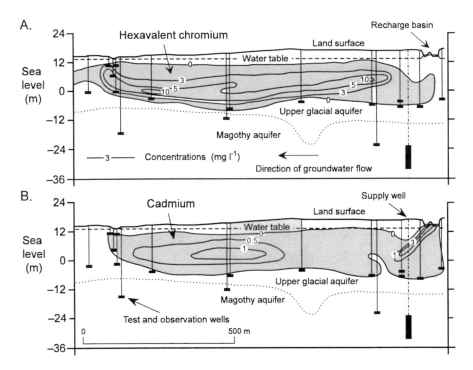

A.

Hexavalent chromium

Sea level (m)

B.

Cadmium

Sea level (m)

Supply well

Figure 9.16 Vertical distribution of hexavalent chromium and cadmium along the centre line of the plume resulting from disposal of liquid industrial waste, Long Island, New York (Modified from Perlmutter and Lieber, 1970 in Everett, 1980)

of disposal by this method are described for a humid temperate environment by Beard and Giles (1990), and for more arid climates by Bouwer (1991) and Idelovitch and Michail (1984).

In urban areas which do have mains sewerage systems, an economical method of partial treatment of sewage is wastewater stabilisation by retention in shallow oxidation lagoons before subsequent discharge into rivers or onto land or subsequent re-use for irrigation (see section 9.4.8). These lagoons are often unlined and, over suitably coarse-textured soils, may have high rates of seepage loss, especially immediately after construction and each time the lagoon is cleaned. This can have a considerable local impact on groundwater quality, particularly in relation to nitrogen and trace organic compounds and where groundwater is used for drinking water supplies (Geake *et al.*, 1986).

Sanitary sewers are intended to be watertight; if they were completely so, they would present no threat to groundwater. In practice, leakage is a common problem, especially from old sewers, and may be caused by defective pipes, poor workmanship, breakage by tree roots, settlement and rupturing from soil slippage or seismic activity. The problem could be greater but for the fact that the suspended solids in raw sewage can clog the cracks and the soil around the pipes and be, to some extent, self-sealing.

9.4.3 Disposal of solid domestic and industrial waste

The most common method of disposal of solid municipal waste is by deposition in landfills. A landfill can be any area of land used for the deposit of mainly solid wastes and they constitute important potential sources of groundwater pollution (Everett, 1980; Foster *et al.*, 1987). Sanitary landfills are usually planned, located, designed and constructed according to engineering specifications to minimise the impact on the environment, including groundwater quality. The engineering methods adopted include lining and capping, compaction of fill and control of surface water inflow. Of the 100,000 landfills in the USA, only about ten per cent could be described as sanitary (Everett, 1980), and the remainder are open dumps.

The principal threat to groundwater comes from the leachate generated from the fill material. The design of a sanitary landfill aims to minimise leachate development by sealing the fill from rainfall, run-off and adjacent groundwater. For significant leachate to be produced, a flow of water through the fill is required. Possible sources include precipitation, moisture in the refuse, and surface water or groundwater flowing into the landfill. Therefore, the volume of leachate generated can be estimated by a simple water balance (Foster and Hirata, 1988), and is greater in humid regions with high rainfall, plenty of surface water and shallow water tables, and is much less in arid regions.

The chemical composition depends on the nature and age of the landfill and the leaching rate. Most municipal solid wastes contain little hazardous material, but solid wastes of industrial origin may contain a much higher proportion of toxic constituents, such as metals and organic pollutants. The most serious threats to groundwater occur where there is uncontrolled tipping rather than controlled, sanitary landfill. There may be no record of the nature and quantity of materials disposed of, which could include hazardous industrial wastes, such as drums of toxic liquid effluent. In Europe and North America this situation existed until the mid-1970s, when uncontrolled tipping was replaced by managed, sanitary landfill facilities. Abandoned landfills can, therefore, represent a potential hazard to groundwater for years or decades.

9.4.4 Accidents and leaks

Groundwater pollution incidents from major industrial complexes are becoming more common and are often the subject of major, expensive investigations and clean-up activities. The causes include accidents during

		Probability of immiscible phase		
		Low	Moderate	High
Duration of discharge	Continuous	Leaking industrial sewers and lagoons	Infiltration from landfill waste disposal	Leaking underground storage tanks
	Intermittent	Courtyard drainage	Effluent soak-aways	
	Single pulse			Major spillages

Figure 9.17 The range of possibilities for groundwater contamination by industrial organic compounds (After Lawrence and Foster, 1991)

transportation, spillages due to operational failures and leaks due to corrosion or structural failure of pipes or tanks. The result is a point source of pollution which may be short and intense in the case of an accidental spillage or small but continuous in the case of an undiscovered leak. In either case, dangerous chemicals may be discharged into the environment and serious groundwater pollution can be caused. Petroleum and petroleum products are the most important because they are widely used. Fuel stations with buried tanks are universal and pollution of this type is not restricted to industrial, or even urban, areas. Moreover, accidents during road or rail transport may result in spillages of, for example, industrial solvents in rural areas. Very few steel underground tanks are protected from corrosion and it has been estimated that up to 25 per cent of fuel storage tanks in the USA leak (Canter *et al.*, 1987).

A major control over how a pollution plume develops and migrates in groundwater, and hence on how it should be monitored, is the mechanism by which the contaminant enters the sub-surface. An overnight leakage of several thousands of litres of solvent will produce a different plume to that caused by contaminated drainage regularly infiltrating from an industrial site. In the former situation, significant leaks of fuel oils, oil derivatives or chlorinated solvents are likely to result in the immiscible phase penetrating into the aquifer, whilst in the latter situation shallow contamination of the aqueous phase will result. An indication of the probability of the immiscible phase liquid being present in the groundwater is shown in Figure 9.17. Two examples of contrasting pollution incidents of this type are summarised by Lawrence and Foster (1991). Monitoring of the development of the plume has to take account of the behaviour of the light and dense immiscible phase contaminants described in section 9.2.5.

The cost of aquifer restoration measures and/or provision of alternative water supplies after major incidents of this type may run into many millions of US dollars. Much of this expenditure may depend on the information obtained from the groundwater quality assessment programme, the design and operation of which becomes especially critical if a legal case is involved.

9.4.5 Acid deposition

Of the total groundwater resources, that part which is in active circulation and potentially available for abstraction for whatever purpose is derived mainly from rainwater infiltration through the soil to underlying aquifers. The acidity of this groundwater depends on the inputs of acidity from internal (biological, soil and rock) and external (atmospheric and anthropogenic) sources, and on the ability of the soil to attenuate them (Kinniburgh and Edmunds, 1986). Some groundwater is, therefore, vulnerable to acid deposition, a collective term for a wide variety of processes by which acidic substances are transferred from the atmosphere to vegetation, land, or water surfaces. For convenience, acid deposition can be divided into wet and dry deposition.

The nature, origins and effects of acid deposition have been most studied in the highly industrialised areas of Europe and North America, where the problem of acid deposition is felt most acutely. In the UK, it is thought that about two thirds of the present acidity of the rain can be attributed to sulphur dioxide emissions from the burning of fossil fuels and one third to nitrous emissions from internal combustion engines. The average pH of rainfall over much of the UK is 4.2–4.5. Dry deposition of both emissions is also an important source of acidity.

The attenuation of acidity in the infiltrating rainwater depends on the time available for reactions with the soil and the rock. The residence time of water in the soil and the aquifer is, therefore, an important factor in determining the acidity of groundwater, together with the mineral composition of the aquifer material. Thus limestone aquifers with their high calcium carbonate content are well buffered and will not be adversely affected by acid deposition. Other non-carbonate aquifers may be protected by the neutralising of acid infiltration as it passes through overlying strata. The most vulnerable aquifers are shallow sands, sandstones and shales with relatively short residence times. The most serious consequences of acidification of groundwater are the increased mobilisation of trace elements, especially aluminium, in soils and aquifers, and the increased solubility of some metals in water distribution systems, both resulting from the lowering of the pH.

9.4.6 Cultivation with agrochemicals

Agricultural land-use and cultivation practices have been shown to exert major influences on groundwater quality. Under certain circumstances serious groundwater pollution can be caused by agricultural activities, the influence of which may be very important because of the large areas of aquifer affected. Of particular concern is the leaching of fertilisers and pesticides from regular, intensive cultivation, with or without irrigation, of cereal and horticultural crops. The changes in groundwater quality brought about by the clearing of natural vegetation and ploughing up of virgin land for new cultivation are also important. The impact of cultivation practices on groundwater quality is greatest, as are most anthropogenic effects, where relatively shallow, unconfined aquifers are used for potable supply in areas where there is no alternative.

The impact of modern agricultural practices on groundwater quality became fully apparent in some regions of industrialised countries during the latter part of the 1970s. Detailed scientific investigations demonstrated that high rates of leaching to groundwater of nitrate and other mobile ions occurred from many soil types under continuous cultivation sustained by large applications of inorganic fertilisers. In the USA, for example, fertiliser use doubled between 1950 and 1970 from 20 to 40 million tons, and the percentage of nitrogen in all fertilisers increased from 6 to 20 per cent. A similar pattern occurred in Europe (OECD, 1986) and is now also occurring in rapidly industrialising countries such as India, in response to growing population and food demands. In the developing world, annual consumption of nitrogen fertiliser has more than tripled since 1975 (Conway and Pretty, 1991).

Monitoring of groundwater for compliance with legally imposed allowable nitrate concentrations, and to observe trends, is based on discharge samples from production boreholes, i.e. the water that goes to the consumer. Where there is a thick unsaturated zone in the aquifer recharge area (Figure 9.3), much higher nitrate concentrations from the most recent fertiliser applications may be observed (Foster *et al.*, 1986). The slow movement in the unsaturated zone means that there is a time lag (sometimes 10 to 20 years in the chalk aquifers of Western Europe), dependent on the thickness of the unsaturated zone, before the full impact of agricultural intensification is felt in public supply sources. Similarly, changes in land use intended to reduce nitrate leaching will take time to have an effect on nitrate concentrations in

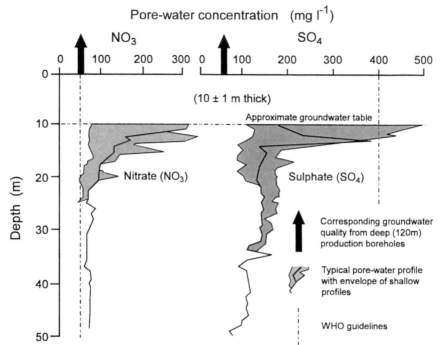

Figure 9.18 Saturated zone pore-water profiles from the Triassic Sandstone Aquifer in Yorkshire, UK. Vertical stratification of groundwater quality means that pump discharge samples do not fully reflect severity of pollution (After Parker and Foster, 1986)

groundwater supplies. Much of the detailed research on nitrate leaching through the unsaturated zone was based on vertical profiling by extracting matrix water from undisturbed drilling samples by centrifugation. Repeat profiling of this type has been employed to examine the development of nitrate profiles in the unsaturated zone (Geake and Foster, 1989). This approach, although expensive, is likely to be required in monitoring to evaluate the effects of land use changes employed in a groundwater protection programme.

Pore-water profiling has also been extensively employed in the saturated zone in the study of diffuse pollution, and has demonstrated considerable stratification of groundwater quality (Parker and Foster, 1986). Groundwater quality assessed on the basis of pumped samples from deep water supply boreholes may not reflect the severity of pollution already present in the upper part of the aquifer (Figure 9.18). Pumped samples from boreholes completed at different depths in the aquifer, and depth sampling in deep boreholes have also demonstrated quality stratification (Figure 9.19). By the time pollutants are recorded at significant concentrations in deep pumping

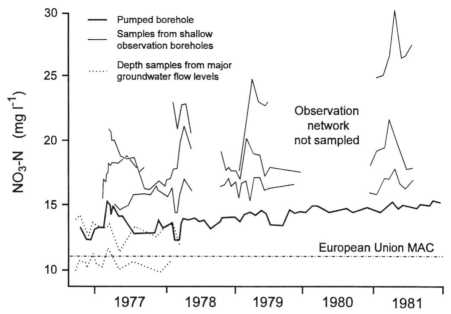

Figure 9.19 Groundwater quality stratification in the Chalk of Eastern England (After Parker and Foster, 1986)

boreholes, a large volume of the aquifer will already be polluted, and effective remedial action may be impossible.

The examples in Figures 9.18 and 9.19 are from consolidated aquifers in which fissure flow is important. Such aquifers are particularly prone to marked vertical variations in groundwater quality, but thick aquifers with inter-granular flow may also exhibit such quality stratification. Thus, in the USA, a high proportion of the rural population in agricultural areas obtain their domestic water supplies from shallow, private boreholes, which suffer the impact of nitrate pollution to a much greater extent than deeper, public supply boreholes (Figure 9.20).

Evidence of the impact of tropical agriculture on groundwater quality is now becoming available. The results of a recent study (Table 9.8) show a wide variation in nitrate leaching losses, resulting from differences in soil and crop types, fertiliser application rates and irrigation practices. Nitrogen leaching losses beneath paddy cultivation are likely to be low because the flooded, largely anaerobic soil conditions during the growing season restrict oxidation of nitrogen to nitrate. In tropical regions, groundwater is most vulnerable to leaching of nitrate where (Chilton et al., 1995):

- the soil and unsaturated zone are thin and permeable,
- several crops a year are grown,

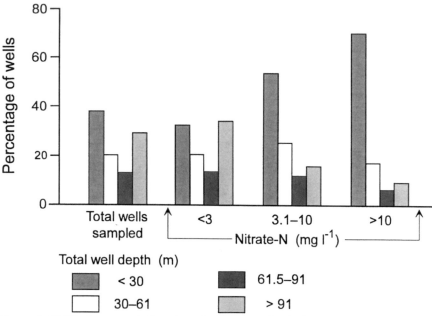

Figure 9.20 Nitrate concentrations in wells of various depths in the USA (After Madison and Brunett, 1985 in Meybeck *et al.*, 1989)

- fertiliser inputs are high, and
- excess irrigation leads to rapid leaching of nutrients beyond the root zone.

Much less attention has been given to the leaching of pesticides from agricultural land to the underlying groundwater, in spite of the dramatic increase in the use of pesticide compounds over the last 20 years. The use of some of the most toxic, particularly the organochlorine insecticides, is now banned or rigorously controlled. However, new compounds are continually being introduced. All are, to a greater or lesser degree, manufactured to be toxic and persistent. With improved analytical techniques, instances of pesticide occurrence in groundwater are increasingly being reported.

Many of the pesticide compounds which are widely used are unlikely to be leached significantly to groundwater from normal agricultural use as their physico-chemical properties, and methods and rates of application are not such as to promote leaching. In addition, many of the processes described in Figure 9.12 are very active in relation to pesticide compounds. There are, however, exceptions, such as the carbamate insecticides and herbicides of the carboxyacid, triazine and phenylurea groups which have been encountered in groundwater at levels which give cause for concern, in spite of the relatively rapid biodegradation rates quoted for these compounds. The published values for sorption and degradation of common pesticide compounds are for fertile,

Table 9.8 Summary of the results of nitrate investigations in tropical environments

Features	Sri Lanka	Barbados	India	Mexico
Aquifer type	Coastal dune sands	Coral limestone	Thin alluvium	Thick alluvium
Depth to water (m)	1–3	20–80	10	5–15
Travel time to water table or borehole screen	10 days	Days	1 year	25–40 years
Potential for dilution within aquifer	Low–moderate	Low	Low	High
Principal crops	Onions, chillies	Sugar cane	Rice	Wheat
Crops per year	2–3	1	2	1
Length of period of intensive cultivation (years)	10–20	> 30	20	> 30
Fertiliser type	Urea, triple super-phosphate	24-0-18 NPK	Urea	Urea, anhydrous ammonia
Application rates (kg N ha^{-1} a^{-1})	400–500	123	300	120–220
Leaching losses (kg N ha^{-1} a^{-1})	60–120	35–70	Small	?
Current groundwater nitrate concentrations (mg NO$_3$-N l^{-1})	10–30	5–9	2–5	2–5

Source: After Chilton *et al.*, 1995

organic, clay-type soils. Much less favourable values must be expected if pesticide compounds are leached into the unsaturated and saturated zones of aquifers. Figure 9.12 indicates that the attenuation processes are much reduced in effectiveness below the soil. Even a very small percentage (1–5 per cent) of the applied pesticide leaving the soil layer could be persistent and mobile enough in the groundwater system to present a problem in relation to the very low concentrations recommended in drinking water guidelines.

Both scientific investigation and routine monitoring of pesticides in groundwater present significant difficulties. A strategy needs to be developed to assist water utilities and regulatory authorities to protect groundwater from contamination by pesticides. It is not practical to monitor for all of the many pesticides in regular use. General screening is prohibitively expensive where potable supplies are based on a few high-discharge sources, and quite out of the question for large numbers of small, rural supplies. To be effective, the available resources for sampling and analysis should be focused by the following approach:

- Existing usage data, if available, or new surveys are required to select those compounds which are widely used.
- The pesticides and their metabolites most likely to be leached to groundwater can be identified from their physical and chemical properties.
- Toxicity information is required for the pesticides and their metabolites.

In this way, monitoring can be concentrated on the high-risk compounds, those which are widely used, mobile, toxic and persistent (Chilton *et al.*, 1994). Within such a strategy, monitoring resources should be concentrated on the most vulnerable aquifers. The hydrogeological environments which are likely to be most vulnerable to the leaching of pesticides are those with shallow water tables and coarse textured soils low in organic matter (Chilton *et al.*, 1995). These include coastal and island limestones, sands and some alluvial deposits. In many countries, the trend of growth in pesticide use closely follows that of inorganic fertilisers. Thus, in spite of the difficulties outlined above, monitoring for pesticides in drinking water is likely to become of increasing importance.

9.4.7 Salinity from irrigation

Increasing salinity resulting from the effects of irrigated agriculture is one of the oldest and most widespread forms of groundwater pollution (Meybeck *et al.*, 1989). Many, but not all, instances of waterlogging and salinisation are related to low irrigation efficiency and lack of proper drainage measures. Over-irrigation without adequate drainage can cause rises in groundwater level which result in soil and groundwater salinisation from direct phreatic evapotranspiration. The addition of further excess irrigation water to leach salts from the soil merely transfers the problem to the underlying groundwater. Additional contributions to increased salinity may come from the dissolution of salts from soil and aquifer material by the rising groundwater, and from the greatly increased infiltration from irrigated land leaching out the salts present in desert or semi-desert soils, at the time the land is first brought under irrigation.

The total area of irrigated cultivation in the world is about 270×10^6 ha (WRI, 1987), about 75 per cent of which is in developing countries. Up to half of this land may be affected by salinity to some extent, and productivity is seriously affected on seven per cent of the total (Meybeck *et al.*, 1989). Salinisation of fertile crop lands is occurring at a rate of $1-1.5 \times 10^6$ ha a^{-1} (Kovda, 1983), and irrigated land is going out of production at a rate of 30 to 50 per cent of the rate at which new land is being brought under irrigated cultivation. The fall in crop production and the consequent economic costs of this loss of agricultural land are enormous and difficult to quantify. The increase in salinity of groundwater also has a significant impact on drinking water and industrial uses (Table 9.7).

Preventing or alleviating the problem of groundwater salinity requires more efficient irrigation combined with effective drainage. Adequate drainage requires a general lowering of the water table to below 2–3 m from the ground. This can be achieved by open ditches, tile drains or pumping from wells. However, even if drainage measures are implemented there are often problems associated with the disposal of saline water, and irrigation return flows have a serious impact on surface water quality in many places (Meybeck et al., 1989). Assessment of river water quality has often provided the best indication of overall trends in salinity (Williams, 1987).

9.4.8 Use of urban wastewater for irrigation

Greatly improved urban water supplies with individual domestic connections have led to increasing water consumption and an increasing problem of wastewater disposal. The projected effluent production (Pescod and Alka, 1984) for some of the fastest growing major cities of the world (e.g. Mexico City and São Paulo) indicates that in these areas, water demands are escalating rapidly, and it is imperative that effluent re-use forms an integral part of the overall water resource management strategy. Agricultural re-use is consistent with the need to maintain and increase food production for rapidly expanding urban populations (Pescod, 1992). Effluent re-use is not a new concept, but controlled wastewater irrigation has not been extensively employed in developing countries because the installation of sewered sanitation has generally lagged behind improved provision of water reticulation systems. As more large-scale, urban, water-borne sewerage systems are installed, it is inevitable that effluent re-use for irrigation will become more important.

When effluent is to be re-used for irrigation, both public health and agronomic effects must be considered when assessing the possible impact on underlying groundwater. Public health considerations centre on the presence of pathogenic organisms in the effluent (Pescod and Alka, 1984) which may be transported with the infiltrating irrigation water. However, municipal wastewater may also contain hazardous trace elements, organic compounds and nutrients (Foster et al., 1994). As in the case of unsewered sanitation, the likelihood of pathogens reaching the water table depends largely on their survival times and retention or adsorption in the soil. Research in the Lima area (Geake et al., 1986) has demonstrated that under the soil conditions and irrigation practices at that site, elimination of faecal bacteria was less effective beneath the cultivated land than beneath the nearby wastewater stabilisation lagoons. Agronomic concerns are associated essentially with the effects of the major inorganic constituents of sewage effluent on crop yields and soil structure, as affected by the accumulation of salts, and on the danger of accumulation of trace organic and inorganic constituents in soil, crops and the underlying groundwater.

9.4.9 Mining activities

A range of groundwater pollution problems can be associated with mining activities. The nature of the pollution depends on the materials being extracted and the post-extraction processing. Coal, salt, potash, phosphate and uranium mines are major polluters (Todd, 1980). Metalliferous mineral extraction is also important, but stone, sand and gravel quarries, although more numerous and widespread, are much less important chemically. Both surface and underground mines usually extend below the water table and often major dewatering facilities are required to allow mining to proceed. The water pumped, either directly from the mine or from specially constructed boreholes, may be highly mineralised and its usual characteristics include low pH (down to pH 3) and high levels of iron, aluminium and sulphate. Disposal of this mine drainage effluent to surface water or groundwater can cause serious impacts on water quality for all uses (Table 9.7).

Pollution of groundwater can also result from the leaching of mine tailings and from settling ponds and can, therefore, be associated with both present and past mining activity. The important features of liquid and solid mine waste are similar to those described in sections 9.4.2 and 9.4.3, except that trace elements are likely to be an important constituent (Table 9.7). Examples of groundwater pollution from mine waste are described by Morrin *et al.* (1988) and from the leaching of salt from the tailings of the potash mines in Alsace, France by Meybeck *et al.* (1989). Mining activities can also have an indirect negative impact on groundwater quality where continuous large scale groundwater abstraction lowers the water table sufficiently to permit saline intrusion.

Disposal of the highly saline water produced with the oil from production wells has long been a problem (Meybeck *et al.*, 1989). In the USA, until the 1960s, the brine was usually placed in evaporation ponds. Lined pits functioned as intended, but saline water often reached shallow aquifers from unlined pits. In established oil fields, plumes of polluted groundwater remained for a long time after the pits themselves had been abandoned. Unlined pits were banned by the USA regulatory agencies in favour of re-injection into the oil-bearing formations, and there are estimated to be more than 70,000 brine disposal wells in the USA (Everett, 1980). Economic considerations have led to the use of abandoned oil production wells for brine disposal but, because they were not designed for injection, there have been many instances of faulty wells allowing brines to leak into important freshwater aquifers above the intended disposal zone. This type of groundwater pollution is obviously restricted to oil producing regions, but in such areas of the USA, for example, brines represent one of the major causes of groundwater pollution. An example of groundwater pollution by brine disposal in Arkansas is described by Everett (1980).

9.4.10 Groundwater resource management

The most important water quality change resulting from the management of groundwater resources is saline intrusion, which occurs where saline water displaces or mixes with freshwater in an aquifer. The problem is encountered in three possible circumstances:

- where there is upward advance (upconing) of saline waters of geological origin,
- where there is lateral movement from bodies of saline surface water, and
- where there is invasion of sea water into coastal or estuarine aquifers.

Under natural conditions, fresh groundwater flows towards, and discharges into, the sea. The position of the saline water–freshwater interface in a coastal or island aquifer is governed by the hydrostatic equilibrium between the two fluids of different densities. Sea water intrusion results from the development of these aquifers. Increasing groundwater withdrawals lower water levels and reduce flow towards the sea. Lowering of groundwater levels changes the hydrostatic conditions and causes local upconing of saline water below abstraction wells. Where groundwater withdrawal from a well or group of wells is sufficient to cause regional lowering of water levels or reversal of hydraulic gradients, then lateral movement of the interface will occur. The resulting pollution of the aquifer can be difficult to reverse.

There are important coastal aquifers throughout the world, and many oceanic islands are completely dependent on groundwater for their drinking water supplies. The thin lenses of freshwater on small islands are highly vulnerable to saline intrusion (and other forms of pollution), and development of these resources is often possible only by carefully controlled pumping from shallow skimming wells, accompanied by a well-developed operational surveillance programme. Avoiding saline intrusion is essential since there may be no alternative source of water supply except for the very expensive option of desalination.

Quality problems can also be caused by declining groundwater abstraction. In many of the major cities of northern Europe, groundwater usage increased for over a century with industrial growth to a peak in the 1940s and 1950s, producing falls in groundwater levels of tens of metres. Since then, decline of traditional industries and a switch to municipal water supplies has produced a dramatic decrease in private industrial groundwater abstractions, and water levels have begun to recover. The water table beneath a city may then rise, taking into the groundwater body pollutants of all types left in the dewatered part of the aquifer during the decades of industrial activity.

9.5   Assessment strategies

Groundwater bodies are always less accessible than surface water bodies. Consequently, obtaining the essential information on groundwater quality is technically difficult and costly. Significant limitations in groundwater quality assessment usually have to be accepted and need to be recognised in the interpretation and use of the monitoring results. This is often not appreciated by those responsible for establishing water quality goals or groundwater resource management strategies. Consequently, the information expectations placed on water quality assessments may be far beyond any ability to supply the information (Sanders *et al.*, 1983). It is essential, therefore, for the designer of a groundwater quality assessment programme to understand and define the information objectives, and to appreciate the several types of monitoring that can exist.

9.5.1 Types of groundwater assessment

The fundamental requirement of a groundwater assessment to define the spatial distribution of water quality, applies almost invariably, regardless of the specific objective of the assessment. Three major categories of water quality assessment i.e. monitoring, survey and surveillance are defined in Chapter 2, and the various sub-categories listed in Table 2.1. Surveillance is generally related to the acceptability of water for a given use and/or the control of associated treatment processes (Foster and Gomes, 1989). Sampling for surveillance normally comprises frequent or continuous measurements on pumped water, and there is no need for samples to be representative of conditions in the aquifer. In all other cases, the requirement is to obtain analytical results from samples which are uncontaminated by the processes of sample collection and analysis and are representative of *in situ* conditions at specific points in the groundwater system.

The types of water quality assessment activity listed in Table 2.1 are not mutually exclusive. Considerable overlap exists, and there are also other approaches to defining categories of assessment (Everett, 1980). Some of the types of water quality assessment listed in Table 2.1 have inherent implications of scale. The most common types of national or regional assessment cover large areas, with sampling points tens or hundreds of kilometres apart. In contrast, emergency or impact monitoring associated with a single point source may be limited to the immediate vicinity of the actual, or potential, contaminant plume. The sampling points, whether few or many, may be confined within distances as low as a few tens or hundreds of metres. In some very detailed monitoring activities carried out for research programmes, the intensity of sampling points may be very great with, for example, sampling through an aquifer at very close depth intervals (0.1–0.5 m) to determine vertical variations in groundwater quality and mechanisms of solute transport (Foster *et al.*, 1986; Geake and Foster, 1989).

These differences in scale of operations are important and should be appreciated in establishing a monitoring strategy. They should not, however, be viewed too rigidly. Water quality assessment activities may have to be expanded in response to the results obtained. This could require any combination of extending the area covered, increasing station density, increasing sampling frequency and increasing the number of variables. For example, a basic survey, background or trend monitoring of a large groundwater basin may identify groundwater pollution which requires further investigation, and a programme of monitoring to be developed around a pollution source in a small part of the area. In some circumstances, this might lead on to assessment for enforcement or case preparation because legal action is being taken. Research may be associated with any type of assessment activity.

9.5.2 Establishment of an assessment strategy

Between the policy decision to establish a groundwater assessment programme and effective operation are a large number of decisions and steps. Simplifying assumptions must inevitably be made to reduce the complexity of groundwater quality to a level at which practical solutions are possible, allowing for the requirements for rigour implied by the information objectives. The number and type of simplifying assumptions depends on the purpose of the programme, and also on the financial resources available. Difficult decisions, implying the allocation of funds and staff, have to be made about the selection of sampling points, frequency of sampling and choice of variables. It follows that, depending on the assumptions and choices made, there are many levels of design that could be applied. The aim of the following sections is to present a general strategy for the establishment of a groundwater quality assessment programme. The purpose of such a strategy is to provide a framework for the planning and implementation of a programme that focuses on the most important water quality issues in the area covered, allows for properly designed but cost-effective monitoring of the issues, and interprets and presents the results in a way that allows pollution control measures and/or groundwater management decisions to be made.

The general strategy described here follows that shown in Figure 2.2. The wide range, world-wide, of hydrogeological conditions, groundwater uses, water quality issues and sources and scales of groundwater pollution described in the earlier sections of this chapter, means that there is an almost infinite variety of possible assessment programmes. Nevertheless, within the framework of the general strategy described here, it should be possible to establish a programme to meet all situations. In addition, the strategy outlined here can be applied in the review of existing programmes to ensure that the objectives and design of the programme can be modified as new types of pollutants become important and as improved knowledge and understanding of the hydrogeology becomes available. This is particularly important if a responsible authority wishes to move up through the stages of assessment defined below.

Proper establishment of an assessment programme can be considered as requiring two main stages. Firstly, a period of initial assessment (preliminary surveys) which may be very short (in the case of emergency surveys) or as long as a year, including surveys of pollutant sources, investigation of hydrogeological conditions (either by desk study of existing data or by field studies) and frequent sampling of available groundwater sources for a wide range of variables. This stage identifies the principal features of the groundwater quality and defines any seasonal fluctuations which may need to be taken into account. The second part of the programme entails longer-term groundwater monitoring in which the location and number of stations, sampling frequency and number of variables is established, to minimise costs but still meet the objectives of the programme. The components of each of these stages (see Figure 2.2) as they apply to groundwater are described in the following sections.

9.5.3 Defining objectives

The vital importance of groundwater resources in use and re-use for a range of different purposes requires aquifer management strategies in relation to both quantity and quality. It is unrealistic and unnecessarily expensive to measure all possible variables continuously and throughout the area. Establishing groundwater assessment implies difficult choices in respect of these options. Many authors have emphasised the need to define clearly the objectives of an assessment programme before beginning the design (Nacht, 1983; Canter *et al.*, 1987; Foster and Gomes, 1989). A number of different general reasons for groundwater assessment can be recognised in accordance with the categories of assessment given in Table 2.1:

- To develop an understanding of regional groundwater quality as an aid to better knowledge of the groundwater regime for optimal management of groundwater resources.
- To determine long-term trends in groundwater quality and to relate observed trends to human activities as a basis for informed decision making.
- To identify and monitor the locations of major pollutant sources (e.g. landfill site, mining operation or sewage disposal facility) and the movement of the pollutant in the aquifer, in relation to the design of aquifer restoration, treatment works or legal proceedings.
- To determine compliance with regulations and standards.
- To assess the effectiveness of pollution control measures, such as groundwater protection zones.
- To determine regional groundwater quality variations and background levels for studying natural processes and as a reference point for large scale and long-term anthropogenic impacts.
- To study groundwater recharge, for example chloride and tritium profiles.
- To determine flow paths and rates of groundwater movement using injected tracers.
- To determine the quality of groundwater, particularly with respect to its possible use as a source of drinking water or other non-potable uses. In the case of a major, extensively used aquifer in a highly developed region (possibly subject to both natural and artificial recharge) this could include a wide range of possible influences on groundwater quality, some of which could be specific for particular uses.
- To determine the extent and nature of an accidental pollution event (e.g. chemical leakage).
- To determine groundwater quality in the vicinity of public supply sources, threatened by point source pollution or saline intrusion, to protect the integrity of the supply and maintain its use.
- To calibrate and validate groundwater quality models which may have been developed for pollution control or resource management, for example saline intrusion, contaminant migration, prediction of nitrate trends.

The above reasons for groundwater quality assessment are only a part of the objectives. Those which are appropriate to the situation should be incorporated in a written statement of the objectives, which should also include reference to the techniques that will be employed, the data that will be obtained, the approach to data interpretation for provision of information, and the use to which the information will be put.

9.5.4 Selecting and defining area

Selection of the area in which the assessment is to be established will be governed firstly by the objectives of the programme and then by a combination of administrative and hydrogeological considerations. In many cases the monitoring extends to the area of jurisdiction of the agency responsible for carrying out the monitoring, and may coincide with national, regional, state or district boundaries, or with the boundaries of major catchments. Often, however, political or administrative boundaries are very different from hydrological boundaries, and in any case groundwater divides may not coincide with surface water catchments. Pollutants from an adjoining area may enter from sources which cannot be directly studied and which are beyond the jurisdiction of any pollution control measures that may be indicated as a result of the assessment. In the case of very large catchments or major aquifers, this may even include national boundaries.

Of the assessment categories defined in Chapter 2, trend and operational monitoring are likely to be extended over the whole basin or administrative area. The extent of monitoring activities for the other categories will be related to the water quality issue or issues, and water uses, embraced by the objectives of the programme. Operational surveillance related to diffuse pollution from agriculture is likely to be extensive, concentrating on potable supplies in crop growing areas and any adjacent areas down the hydraulic gradient. Operational surveillance of saline intrusion is generally restricted to coastal areas. Operational surveillance or emergency monitoring related to point source pollution from a landfill or an industrial spill is much more restricted to the area close to the source. In many such cases it may not be possible to define the area to be studied with any degree of certainty, and preliminary surveys (see section 9.5.5) are required. In the meantime, monitoring is concentrated on potable supply sources immediately down gradient of the source, making some general estimates of the possible extent of the contaminant plume based on the aquifer type and regional hydrogeology.

9.5.5 Preliminary surveys

In all cases of the establishment of an assessment programme, an initial survey is required, the extent of which depends on the objectives of the programme, the complexity of the hydrogeology and the number and nature of the water quality issues to be addressed. For the purposes of trend monitoring, the preliminary survey may be restricted to a desk study review of

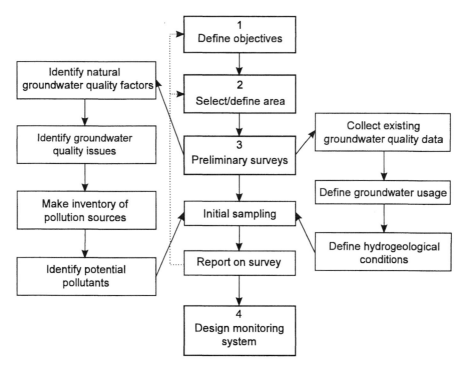

Figure 9.21 Components of preliminary surveys for groundwater quality assessment

existing data. In contrast, emergency surveys at a major spill will almost certainly need detailed and expensive field investigations which may have to be done very rapidly. The principal components of the preliminary survey are shown in Figure 9.21 and described in the following paragraphs.

Collecting existing groundwater quality data
The first task is to collect all existing groundwater quality data. Some data may be held by the organisation which is proposing to carry out the assessments, especially if that is the regional or national authority responsible for water. Other sources include universities or research institutions, consultants' reports, published scientific literature, and other government departments. A certain amount of ingenuity and persistence may be required to locate fragmented and dispersed data, and judgement will often be needed to distinguish reliable from unreliable data. This information is used to identify the principal features of the natural groundwater quality in the region or around the defined area, and to provide an indication of possible anthropogenic influences which may be related to the objectives of the assessment, thus giving a first indication of possible sources of pollution.

Table 9.9   Hydrogeological data requirements for groundwater monitoring system design

Data	Possible source material
Aquifer locations, depths and aerial extent	Geological and hydrogeological maps and reports
Physical properties of aquifers, especially transmissivity	Test pumping, geological and hydrogeological maps
Areal distribution of groundwater levels	Borehole archives and observation wells
Areal distribution of depth to groundwater	Borehole archives and observation wells
Areas and magnitudes of natural groundwater recharge	Climatic and hydrological data, soils and land-use
Areas and magnitudes of artificial recharge	Irrigation records, municipal recharge schemes
Areas and magnitudes of natural groundwater discharge	Streamflow and water level data
Locations and magnitudes of groundwater abstractions	Borehole archives, municipality and irrigation department records, well owners
Directions and velocities of groundwater flow	Geological and hydrogeological maps, water level and transmissivity data

Source: Modified from Everett, 1980

Collecting hydrogeological data
Concurrently, and often from many of the same sources, existing hydro-geological information must be collected. Depending on the location and size of the defined area, this may comprise an enormous body of data, some of which may already have been interpreted and produced in report form or published works. The data requirements are summarised in Table 9.9. Again, considerable persistence may be needed to track down this information in the various national or regional organisations responsible for collecting and holding it. The objective of this is to obtain as full as possible an understanding of the hydrogeological conditions in the area designated for assessment. The principal requirement, especially where point sources of pollution are concerned, is a knowledge of the groundwater flow regime, without which assessment could be misleading and/or financially wasteful because samples are taken from the wrong locations or depths.

Defining groundwater use
Evaluation of the impact of actual or potential pollution requires knowledge of the use of the resource. In this context, it is important to define the quantities of groundwater being, or projected to be, abstracted, the locations of the major centres of pumping and the type of use to which the water is put. The impact that different pollutants have on the range of water uses is given in

Table 1.3. In the case of rural water supplies, small groundwater abstractions may be spread widely over the region or area of concern. In the case of large abstractions for municipal or industrial use, major wells or well fields may draw down groundwater levels sufficiently to modify locally the ground-water flow regime; this must be accounted for in designing the programme. Information on groundwater usage may come from national or regional agencies, municipalities, public water utilities, irrigation or agriculture departments, and individual industrial or agricultural users.

Identifying quality issues and potential pollutants
A further major component of the preliminary survey is the identification of the most important water quality issues in the area. In the case of a major, extensively used aquifer, several water quality issues may be important. Groundwater quality may be affected by agricultural, industrial and urban influences. The principal activities that could potentially have an impact on groundwater quality have been described in section 9.4 and are listed in Table 9.7. The identification of waste disposal practices and the distinction between point, line and diffuse sources of pollution is particularly important at this stage of the programme.

For many assessment objectives it may be necessary to follow up a general appraisal of the main water quality issues in the area with a detailed inventory of pollution sources, including identification of potential pollutants and char-acterisation of the possible sub-surface contaminant load. This then becomes a groundwater pollution risk assessment, as described by Foster and Hirata (1988). Such a risk assessment combines the evaluation of likely sub-surface contaminant load (from whichever potential groundwater pollution sources have been identified) with appraisal of the vulnerability of the aquifer to pol-lution. This appraisal is based on existing hydrogeological knowledge, principally concerning the occurrence of groundwater, depth to water and overall lithology (Foster and Hirata, 1988). This and other empirical methods of assessment of the potential for groundwater pollution typically involve the development of numerical indices, with larger numbers denoting greater risk. Canter *et al.* (1987) review nine such methods, relating to surface impound-ments, landfills, hazardous waste sites, brine disposal and pesticides. One of the most comprehensive numerical methods for assessing vulnerability of groundwater to pollution based on local hydrogeological conditions is the DRASTIC rating scheme devised by the United States Environmental Protection Agency (US EPA) (Aller *et al.*, 1987). However, the DRASTIC scheme has several major weaknesses and, therefore, must be used only by very experienced hydrogeologists. A comprehensive approach to risk assessment related to possible microbiological pollution of potable ground-water supplies is given by Lloyd and Helmer (1991).

To meet the objectives of assessing potable water quality from a heavily used aquifer in a highly developed region, the preliminary survey may need to include an extensive and detailed survey of agricultural, urban and rural waste disposal and sanitation practices and industrial activities. Information will need to be collected from a range of national and/or regional organisa-tions, and field surveys carried out to establish such factors as fertiliser and pesticide use. Figure 9.22 is an example of a survey form used in Barbados to establish which pesticide compounds were commonly in use and their application rates. Having identified the most widely used compounds, an appraisal of their mobility and persistence from existing literature enables an initial selection to be made of the compounds most likely to be leached to groundwater and which should, therefore, be included as variables for monitoring in the programme.

Identification of potential pollutants from a landfill may be particularly difficult. A wide range of trace elements and organic pollutants may be involved and comprehensive records may be lacking for materials disposed at the site, especially for old or disused sites. When studying pollution from a spill or leak at an industrial site, only a single compound may be involved. However, a local survey to identify other sources of the same compound may be required, especially if the monitoring is related to possible legal proceedings.

Most of the discussion in the preceding paragraphs has been directed towards basic surveys. For background and trend monitoring on a regional scale a preliminary survey is necessary to locate stations not affected by immediate local pollution sources. This may present few problems in devel-oping countries, but may be especially difficult in highly industrialised ones.

Initial sampling

The preliminary survey may require a short and intensive programme (see Table 2.2) of groundwater sampling, and this is recognised as a distinctive category of groundwater monitoring. This initial sampling programme may be essential:

- to identify the principal features of natural groundwater quality if there is no, or very little, existing groundwater quality data,
- to act as an invaluable supplement to land-use, sanitation or industrial surveys as a means of identifying potential pollutants, and
- to identify or confirm seasonal, lateral or vertical variations in ground-water quality that must be taken into account in the programme.

ENVIRONMENTAL ENGINEERING DIVISION
MINISTRY OF HEALTH

Name of Farm/Plantation: _____

Address: _____

Name of Manager: _____

Telephone number: _____

Rainfall area: High _____ Intermediate _____ Low _____

Area under vegetable production: _____

Area under sugarcane production: _____

Do you rotate between areas of cane and vegetable production? _____
If yes, how often? _____

Is there a water well on your property? _____

Is it used for irrigation purposes? _____

How do you dispose of old/unwanted pesticides? _____

How do you dispose of cleansing water from spraying equipment? _____

Do you think it is necessary to monitor water, soil and food for
pesticide residues? _____

Any other comments? _____
_____
_____
_____

(please fill in table opposite)

Insecticides used	Active ingredient	Concentr. g l⁻¹ (kg)	Amount used/year	Frequency	Years used
1.					
2.					
3.					
4.					
5.					
6.					
7.					
8.					

Fungicides used	Active ingredient	Concentr. g l⁻¹ (kg)	Amount used/year	Frequency	Years used
1.					
2.					
3.					
4.					
5.					
6.					
7.					
8.					

Herbicides used	Active ingredient	Concentr. g l⁻¹ (kg)	Amount used/year	Frequency	Years used
1.					
2.					
3.					
4.					
5.					
6.					
7.					
8.					

Name other pesticides that have been used in the past:

Figure 9.22 Example of agricultural survey form used in Barbados to establish pesticide use and the associated risk of groundwater pollution

To meet these clearly defined objectives, the preliminary survey is likely to comprise frequent sampling, over a short period of time, of a large number of points for a moderate to high range of variables, the exact combination of which will depend on the circumstances and the resources available. Sampling in preliminary surveys is almost invariably restricted to existing wells, boreholes and springs (the locations of which will have been established earlier, at the time of collecting existing hydrogeological information). In most assessment programmes, this initial sampling can be completed within a year or often much less. In the case of emergency surveys of a spill, particularly if hydrogeological conditions were such that major public supply sources were threatened, the initial sampling might be completed in a matter of days. Where the main potential pollutant is faecal pathogens from unsewered sanitation, which is possibly the most common groundwater pollutant in many developing countries, the preliminary survey will always include bacteriological analysis of groundwater sources (Lloyd and Helmer, 1991).

All of the information collected in the preliminary survey must be analysed and summarised in report form. This may include, as recommendations, the design of the monitoring network, or information to be used as the basis for the design (Figure 9.21).

9.5.6 Design of groundwater quality assessment

Design of a programme to assess groundwater quality includes much more than the station location. Design consists essentially of the choice of sampling stations, sampling frequency, range of variables, methods of data production and interpretative approaches required to produce the information needed for decision making. Design, therefore, involves choices concerning all of the steps in the overall strategy outlined in Figure 2.2. The designer and operator must see the system as a whole, rather than as several separate activities. This has not always been the case in the past and, as a result, monitoring network design has probably received more attention in the literature than the overall assessment system design. Nevertheless, network design and the decisions to be made in the process remain a critically important component of the design of assessment programmes. General considerations of network design are discussed in Chapter 2, and specific problems related to groundwater bodies are further discussed below. The factors which determine network design for groundwater bodies are summarised in Table 9.10.

Number and location of sampling stations
As indicated in Table 9.10, the number and location of sampling stations is a function of the objectives and scale of assessment, the hydrogeological

Table 9.10 Factors which determine sampling network design for groundwaters

Sampling point		Sampling frequency	Choice of variables
Type	Density		
Assessment objectives	Assessment objectives	Assessment objectives	Assessment objectives
Hydrogeology (complexity)	Hydrogeology (complexity)	Hydrogeology (residence time)	Water uses
	Geology (aquifer distribution)	Hydrology (seasonal influences)	Water quality issues
	Land use		Statutory requirements
	Statistical considerations	Statistical considerations	
Costs	Costs	Costs	Costs

complexity, land-use distribution and economic considerations. The latter will inevitably constrain the comprehensiveness of the proposed network. Relating the general principles of Table 2.2 to groundwater, the possibilities range from a small number of stations dispersed over a region or aquifer to provide background and trend monitoring, or a small number of stations for emergency surveys around a spill, to a medium to high number of stations for multi-purpose monitoring or operational surveillance of potable water quality.

In the Netherlands, for example, a national groundwater quality monitoring programme has been established, principally directed towards diffuse sources of pollution (van Duijvenbooden, 1993), and following the general strategy outlined here. The country is densely populated, with intensive agricultural and industrial development, and is dependent on relatively shallow and vulnerable aquifers for its water supplies. Consequently, the relatively large number of 380 monitoring points provides an overall station density of 1 per 100 km^2, with emphasis on areas of importance for drinking water supplies. The sites were chosen to give adequate coverage of the range of soil types, land use and hydrogeological conditions in the country, and avoided the influence of any localised sources of pollution (van Duijvenbooden, 1993). The national groundwater monitoring network of the former Czechoslovakia described in section 9.6.5 had a similar average station density of 1 per 250 km^2. Much wider spacing of sampling stations can be anticipated in larger and/or less developed countries.

In all cases, the designer of the sampling network must consider the importance of lateral and vertical variations in hydraulic conductivity in relation to the network scale. The required number of sampling points per unit area and unit depth of aquifer must be regarded as a function of the hydraulic heterogeneity of the groundwater system (Ward, 1979). Designing the network needs hydrogeological expertise to ensure that adequate account is taken of the complexity of the groundwater flow system.

It is highly unlikely that existing wells will be in suitable locations in relation to most types of point source monitoring, even though analyses from existing supply wells may have been the first indication of the problem. Hydrogeological conditions, especially groundwater flow rates and directions, will be the main consideration in locating observation wells in relation to identified point sources of pollution. Great care should be exercised, however, in the interpretation of limited hydrogeological data in relation to siting observation wells, especially where fissure flow predominates. Extreme heterogeneity and anisotropy of hydraulic conductivity is possible, and the plume of pollutant may take a form and direction very different to that which might be expected from regional groundwater flow based on water level contours. Monitoring wells may completely miss the plume unless this possibility is appreciated, and a wider initial network installed accordingly.

A monitoring scheme will always need to strike a balance between costs and comprehensiveness. Under the USA Resource Conservation and Recovery Act (RCRA), regulations for a single point source require a minimum of three monitoring wells down-gradient and one up-gradient. It is recommended that the three down-gradient wells should be placed in a triangular arrangement, skewed down-gradient, with two inside the existing plume and one beyond it. This will provide data on the spatial variation of groundwater levels and allow for observation of the plume migration rate. Although well-intended, such specific regulations may be of limited value (Canter et al., 1987), or even dangerously misleading. It may not be possible to determine the hydraulic gradient without drilling some observation boreholes, the hydraulic gradient may be transient and reverse seasonally, or the source itself may produce a local recharge mound which masks the regional hydraulic gradient. In the case of extreme heterogeneity as described above, and in karst aquifers, these simple guidelines cannot apply. In all situations of point source monitoring, phasing of the network installation should be envisaged and budgeted for, so that the results from the initial locations can help to optimise the siting and type of additional sampling stations.

Type of groundwater sampling stations
A major consideration in network design for groundwater bodies is the type of sample station in relation to the vertical distribution of water quality.

Existing groundwater quality assessment programmes, for a range of the objectives listed, depend entirely on samples taken from the pump discharge as it comes to ground level. This sample may not be representative of hydro-chemical conditions in the aquifer, because of transformations resulting from the temperature and pressure changes caused by lifting it, perhaps from great depth, to ground level. These processes include the entry of atmospheric oxygen, precipitation of pH controlled variables and loss of volatile compounds. The borehole or well may penetrate through, and draw water from, a considerable thickness of an aquifer or even more than one aquifer, and the resulting pumped sample may be a mixed one. The public supply or irrigation boreholes from which pumped samples are usually taken were designed and constructed for producing water rather than for monitoring it. This is not a problem for many assessment objectives since the mixed or aggregate pumped sample measures the quality of the water that goes to the user and is, therefore, appropriate in relation to objectives related to water use and acceptability. Samples from pumping boreholes and wells are also the basis of background and trend monitoring programmes on a regional, national or continental scale, such as the GEMS (Global Environment Monitoring System) global network (WHO, 1992). Existing boreholes selected for such programmes should be those with the most reliable information, so that a short screen length over the most appropriate depth interval of aquifer and most comprehensive geological log, facilitate subsequent interpretation of the results.

To meet other objectives, a knowledge of the vertical distribution of groundwater quality is important to provide, for example, a three dimensional picture of the development of a contaminant plume or a saline intrusion problem. In detailed monitoring of a pollution incident, adequate definition of sample depth requires a knowledge of the design of the well and the hydraulic conditions within the well and in the surrounding aquifer. However, construction of the borehole itself may have disturbed the local groundwater flow regime. This is especially likely for boreholes open over much of their length and situated in recharge or discharge areas where there are significant vertical components of groundwater flow (Figure 9.23). The same might occur close to major abstraction boreholes. A borehole may penetrate two or more aquifers separated by impermeable strata. Different hydraulic heads may produce upward or downward components of flow, and the borehole itself may induce cross-contamination by permitting flow from one aquifer to the other. Such boreholes are generally very misleading for monitoring purposes, and representative groundwater samples are unlikely to be obtained, whatever sampling method is used. Similar problems can occur

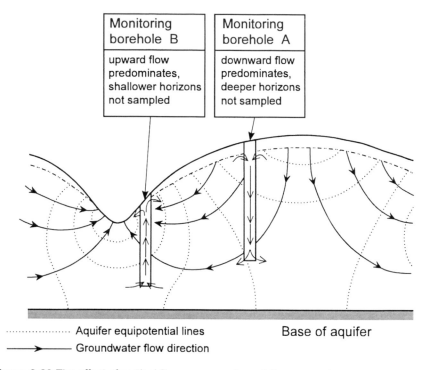

Figure 9.23 The effect of vertical flow components on fully-screened or open monitoring boreholes (After Foster and Gomes, 1989)

in fissured aquifers in which groundwater flow is restricted to a few major horizons which may have very different groundwater quality.

The installation of purpose-drilled observation boreholes to specified depths, with known screened intervals, offers the best chance of obtaining samples that are reasonably representative of conditions in the aquifer. If preferred drilling methods (Scalf *et al.*, 1981; Barcelona *et al.*, 1985; Neilsen, 1991) are employed with suitable casing materials (Barcelona *et al.*, 1983), then the sample bias will be minimised and restricted largely to the effects of sampling itself. Construction procedures for monitoring boreholes follow the same general sequence as for supply boreholes (Keely and Boateng, 1987). This comprises drilling, installation of well screen, filter pack and plain casing and the construction of a sanitary seal to prevent contamination by surface water. The drilling process may introduce fluid or water of different chemistry that contaminates the aquifer close to the borehole. Therefore, a period of cleaning and development pumping may be required after completion, to restore natural hydrochemical conditions. Observation boreholes are normally of smaller diameter (50–100 mm) and open to the aquifer over a much more limited depth interval (1–5 m). They also allow the measurement of water levels and the performance of hydraulic tests on the aquifer.

Design of monitoring boreholes must take into consideration (Foster and Gomes, 1989):

- the objectives of, and financial provision for, the programme,
- the anticipated variables of concern and their probable concentrations,
- the available, or proposed, sampling equipment, and
- the nature of the groundwater flow regime.

Single-level observation boreholes of the type described above are the simplest and most commonly used, and provide adequately representative samples to meet most monitoring objectives at modest cost. Inert materials (pvc, PTFE or stainless steel) can be used where the variables to be monitored include industrial organic compounds or pesticides.

Where the assessment objective requires a knowledge of the vertical distribution of water quality, multi-level sampling installations are required. The simplest solution, technically, is to construct clusters of single observation boreholes completed at different depths in the aquifer (Figure 9.24) or, where necessary, in different aquifers. This is very expensive (Table 9.11), especially for moderate to deep installations. Alternatively, a nest of small-diameter observation points, each sealed above and below, can be installed in a single borehole (Figure 9.24). This approach produces economies in drilling, but placing and sealing each observation point correctly and developing them requires great care. These difficulties of execution, combined with drilling diameter limitations, normally restrict such installations to 2–5 sampling points in a vertical profile (Table 9.11). However, this type of installation has provided reliable monitoring data in many instances. The 380 groundwater sampling stations in the national network in the Netherlands (van Duijvenbooden, 1993) are purpose-built installations of this type (Figure 9.24C), with three separate, short, screened intervals at about 10, 15 and 25 m depth.

More recent developments in sampling technology have seen the increasing use of multi-port installations (Figure 9.24). The borehole is completed with a single plastic (pvc or PTFE) lining tube with ports or small screens at many depth intervals. Sampling through small-diameter, permanently attached inert plastic tubes is achieved by suction at shallow water level depths (Pickens *et al.*, 1978) or syringe type positive-displacement pumps where water levels are deeper (Cherry, 1983). Further variations of multi-level sampling employ gas-driven pumps permanently attached to the small ports or screens, in which case hydraulic measurements are not possible, or installations in which the special ports are opened and sampled by a gas or

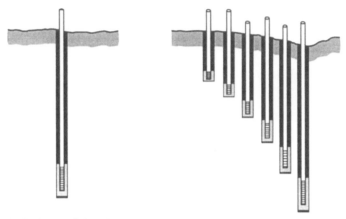

A. A single standpipe piezometer
   installed in a single drillhole

B. A nest of single standpipe piezometers,
   each installed in a single drillhole

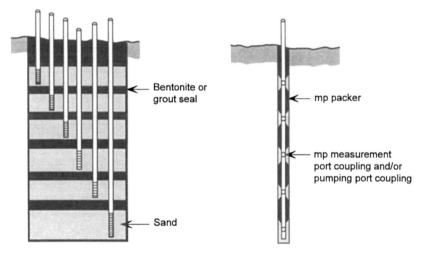

C. A nest or group of piezometers of
   various types, all installed in a
   single drillhole

D. A series of multiple port piezometers
   installed in a single drillhole

Figure 9.24 Types of groundwater monitoring installations

electric sampler introduced into the borehole. Multi-level installations are
required to achieve representative samples in the difficult groundwater
systems that have vertical flow and fissure flow.

Groundwater sampling
A detailed discussion of groundwater sampling techniques is outside the
scope of this book. Comprehensive reviews are given by Scalf *et al.* (1981),

Table 9.11 Comparison of groundwater monitoring installations

Type of installation	Vertical sampling points	Water level measurements	Hydraulic testing	Inert materials	Drilling costs	Material costs	Overall costs
Supply borehole	Integrates over screen interval	Disturbed by pumping	Data may exist	No	None (already exists)	None	Very low
Single piezometer	1	Yes	Yes	Yes	Medium	Low	Low
Cluster of single piezometers	Several	Yes	Yes	Yes	Very high	High	High
Nest of piezometers in single borehole	2–5	Yes	Yes	Yes	High	High	Medium
Multi-port sampling systems	Many	Some types	Some types	Yes	Medium	High	High

Foster and Gomes (1989) and Neilsen (1991). Driven by the increasing interest in groundwater pollution and aquifer restoration and related legal and quality assurance considerations, groundwater sampling, particularly in relation to unstable variables at low concentrations, is the focus of considerable current research. New developments and case studies are continuously reported, with new equipment always being advertised by specialist companies. Any detailed review quickly becomes outdated, but some general comments on traditional and novel methods of sampling can be made.

Choice of sampling methods is an important component of monitoring system design, and is closely tied to the selection of types of sampling points. The choice is governed by the same criteria, i.e. selection of variables (which is in turn determined by the objectives of the assessment programme), the water quality issues to be addressed, the hydrogeological conditions and financial considerations. The most commonly used methods are sampling the discharge of existing production boreholes, grab sampling from non-pumping boreholes and, much less frequently, sampling during borehole drilling.

Discharge samples are normally collected in bottles from a well-head tap or directly from the pump outflow. Where this is not possible, samples are often taken from the nearest tap in the distribution system, which may be after the water has passed through a well-head storage tank. Samples collected in this way have serious limitations for water quality assessment (Table 9.12). Nevertheless, such samples are suitable for surveillance of potable water quality, provided these limitations are appreciated. Pump discharge samples may also be of value for monitoring where there are no vertical variations in water quality, or if an average sample integrated over the whole screened

Table 9.12 Summary of characteristics of groundwater sampling methods in boreholes

Method	Hydrogeological representativity	Sample modification			Relative cost
		Contamination	Degassing	Atmospheric contact	
Production borehole discharge	Poor. No control over sample depth. Mixing and dilution	Moderate from well materials	Severe	Moderate to severe	Very low
			<—————Loss of unstable variables—————>		
Grab sample from non-pumping borehole	Unreliable. No control over sample depth. Vertical flows	Moderate to severe, cross-contamination. Moderate from well materials	Moderate	Moderate to high	Low
During borehole drilling	Moderate to good control over sample depth, with temporary casing	Moderate to severe from drilling fluid. Some cross-contamination	Severe	Severe	High
			<—————Loss of unstable variables—————>		

Source: Modified from Foster and Gomes, 1989

section of the aquifer is appropriate. This is likely to be the case in background and trend monitoring on a regional scale (Clark and Baxter, 1989), provided the construction details of the borehole are well known, and the pumped sample is taken after an adequate period of pumping to allow the borehole outflow to reach hydrochemical equilibrium with inflow from the aquifer.

Grab sampling by bailers or depth samplers (Foster and Gomes, 1989) involves lowering of the device to a known depth in the borehole water column, closing it and raising it to the ground surface. Due to their cheapness and ease of use, such samplers have long been the mainstay of monitoring programmes but, like discharge samples, they have serious limitations (Table 9.12).

In some circumstances, groundwater samples collected during drilling may provide information on vertical variations in quality. The feasibility of doing this depends on the drilling method. The percussion and air-flush rotary methods permit the collection of water samples as drilling progresses (by bailer and air-lift pumping respectively), whereas rotary methods using mud- or water-flush do not. Samples collected during drilling are, however, prone to contamination downward from higher levels in the borehole and from the drilling fluid or compressed air (Table 9.12).

None of these traditional methods is likely to achieve the necessary precision and reliability for the assessment of groundwater pollution. Improved methods should be introduced where the need for better reliability is economically justified or where unstable variables of public health significance are involved. Table 9.13 summarises the features and suitability for groups of variables of some of the improved sampling techniques.

Sampling frequency
Sampling frequency is a function of the type and objectives of the assessment programme, the water quality issues to be addressed, the nature of the groundwater body and the logistical and financial resources for sample collection and analysis (Table 9.10). Optimum frequencies for groundwater sampling have often been set either by regulation or from statistical approaches based on analogies with surface water bodies (Nelson and Ward, 1981; Casey *et al.*, 1983). Because of the generally long residence times and relatively slow rate of evolution of groundwater quality, less frequent sampling is required than for surface water bodies. Basic surveys, background and trend monitoring could be based on annual samples for large groundwater bodies and quarterly samples for smaller aquifers. The national

Table 9.13   Summary of characteristics of borehole sampling pumps

Pumping mechanism	Principle of operation	Advantages	Limitations	Relative cost
Suction lift • peristaltic • manual vacuum • centrifugal	Sample withdrawn by suction applied directly to water or via collection bottle	Highly portable Suitable for all variables given adaptation and accessories Appropriate for small diameter piezometers Gentle delivery with low pumping rates Adaptable for purging prior to sampling Pump-sample contact limited in most cases	Sampling depth limited to 8 m Degassing and aeration difficult to control	Low but increases if inert materials used
Gas-driven • double tube • continuous delivery	Positive gas pressure drives water from borehole backflow prevented by check valves	Unlimited depth Can be made of inert materials Efficient for purging prior to sampling Flow rates can be controlled Can be combined with *in-situ* sorption or depth-control packers Suitable for permanent installation	High purity inert gas needed to avoid contamination Entrance of gas in discharge line may cause degassing/ volatilisation Portability reduces for deep boreholes	Low to moderate, but increases if inert materials used
Positive displacement submersible (A) • electric centrifugal • piston pump • bladder pump	Water driven continuously from borehole by: gears or rotor assembly; gas operated plunger; gas operated diaphragm	Unlimited depth All variables - given adaptation and use of inert materials Appropriate for small diameter piezometers Can be combined with *in-situ* sorption or depth control Flow rates can be controlled Efficient for purging prior to sampling Low purity gas acceptable because no contact with sample	Moderate portability Electric centrifugals not reliable for unstable variables Piston pumps of continuous delivery difficult to clean/maintain	Moderate to high, especially if inert materials used

Continued

Table 9.13 Continued

Pumping mechanism	Principle of operation	Advantages	Limitations	Relative cost
Positive displacement submersible (B) • manual inertial • mechanical inertial	Sample-riser tube and foot-valve assembly moved vertically, filling with water on downstroke and raising water on upstroke, with inertia maintaining downstroke	Portable, but can also be dedicated Durable Good for borehole development, purging, and hydraulic testing Flow rates can be controlled Can be made of inert materials Well suited for unstable variables	Sampling depth limited (manual to 40 m) Deeper boreholes require motor drive, reducing portability and increasing cost Time consuming Not yet well proven and documented Plastic foot valves wear with heavy use	Low for home-made manual version Increases considerably with motor drive and inert materials, which may be difficult to obtain

Source: Foster and Gomes, 1989

groundwater assessment system in the former Czechoslovakia, described in section 9.6.5, took account of aquifer size and depth in determining sampling frequency.

More frequent sampling is required for pollution source studies, which should be related to the type of pollution source and the local hydrogeological conditions. Very frequent sampling may be required, for example, in the case of a point source of pollution close to a potable water supply in a highly vulnerable aquifer with rapid fissure flow. The nature of the source (Figure 9.8) has a direct influence on the type of pollutant plume that may be formed. The size, shape and rate of plume movement depends on the source characteristics, groundwater flow system and the chemical and microbiological attenuation of the pollutants in the plume. The size and shape of a plume from a continuous pollutant source can be estimated (Barcelona *et al.*, 1985) using a relationship developed by Todd (1980) for defining capture zones to pumping boreholes. From this, the number of sampling points in the flow path required to assess the movement of the plume can be decided. Sampling frequency can then be obtained from the hydrogeological characteristics at the site by calculating groundwater flow velocity from the Darcy equation and the effective porosity of the aquifer material. A nomogram for translating the hydraulic data into sampling frequency at various flow path lengths is given in Figure 9.25.

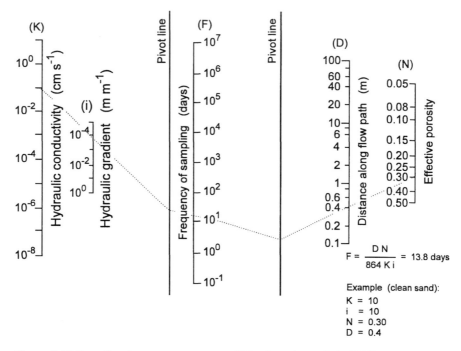

Figure 9.25 Sampling frequency nomogram (After Barcelona *et al.*, 1985)

Choice of variables

Where the assessment objective is related primarily to groundwater pollution, the choice of variables depends on the quality constraints imposed by the principal water use, together with the likelihood of the variables thus defined, being present in troublesome concentrations from natural or anthropogenic origins. Some guidance as to the likely pollutants from a range of activities is given in Table 9.7, and a full discussion of the selection of variables for water quality assessment is given in Chapter 3.

Monitoring in the unsaturated zone

Traditional monitoring systems based on pumped samples suffer from serious inadequacies with respect to the early detection of diffuse pollution (Parker and Foster, 1986). As a result of the considerable residence times of groundwater carrying pollutants through the unsaturated zone (section 9.2), and the possibility of marked water quality stratification below the water table, a large part of an aquifer system may be seriously polluted before contaminants of diffuse origin are detected in a monitoring network based on pumped samples.

In unconfined aquifers, the unsaturated zone occupies a key position between the land surface, close to which most contaminants are discharged, and the groundwater table, from below which water supplies are drawn (Foster and Gomes, 1989). Monitoring of water quality in the unsaturated zone can give early warning of pollution and, especially if this zone is thick, could allow time for control measures to be taken to protect water supplies. Unsaturated zone monitoring is becoming more widely used to detect potential groundwater pollution from diffuse agricultural sources and from waste disposal facilities. A comprehensive discussion of unsaturated zone sampling is given by Wilson (1983).

The most common method of unsaturated zone sampling is by suction or tension. This is an *in situ*, non-destructive method in which water is drawn into a sample chamber through porous cups or plates by applying a vacuum. Once water has filled the chamber, it is drawn to the ground surface either by suction lift (limited to 8 m) or by a gas-driven pump. The suction method has the advantages of being cheap, being easy to install, being made from inert materials and allowing repeat sampling. It has, however, significant operational difficulties caused by clogging, and limitations in respect of sample representativeness due to adsorption of some variables, by-pass of infiltrating water in structured or fissured materials, and uncertainty about what volume of material is being sampled.

Much less frequently used are free-draining pan lysimeters, which consist of a collecting tray placed in the unsaturated zone from a trench or tunnel. The impermeable tray of inert material intercepts downward flow, which is drawn off into a container and sampled regularly. Representativeness is improved, especially in fissured materials by the larger area sampled, and generally by the absence of suction through porous cups. Installation, however, causes more disturbance of natural conditions and is limited to shallow depths by the access requirements.

Destructive sampling, involving the centrifuge extraction of pore water from undisturbed drill cores, is effective in the unsaturated zone. This is an established field method for the study of water quality in both unsaturated and saturated zones (Edmunds and Bath, 1976; Kinniburgh and Miles, 1983) and gives much better control over sample depth and greater confidence about sample origin than other unsaturated zone methods. Repeat coring through the unsaturated zone has been employed as a tool in research studies to observe the downward migration of, for example, nitrate and tritium peaks in the unsaturated zone (Geake and Foster, 1989) in studies of both diffuse pollution and recharge. It is, however, expensive and is not suitable for frequent, routine monitoring. The method also presents additional problems in the sampling of some classes of pollutant. In the case of organic pollutants, pore-water extraction by centrifugation is unlikely to be effective because of the loss of sample by volatilisation and because, particularly in the case of pesticides, insufficient sample volume is produced to permit laboratory analysis to the detection limits required. Methods based on solvent extraction have been used on soils and are being developed, but as yet are far from being applicable to routine monitoring.

Indirect methods of groundwater quality monitoring
In some circumstances, for specific objectives and particular variables, indirect methods of monitoring groundwater quality may be employed. The use of fluid conductivity logging in observation wells to monitor the three dimensional development of saline intrusion is an example. The use of geo-physical methods is limited to situations where groundwater quality differences are sufficiently strong to cause physical contrasts. Thus, the measurement of ground resistivity by surface geophysics may be used in some hydrogeological situations to assess the lateral spread of salinity through an aquifer. For point source pollution incidents involving volatile hydrocarbons, the use of soil gas detection methods may provide a cost-effective means of studying the development of a contaminant plume. Both of these indirect methods depend, as all such methods do, on adequate control being provided from some direct sampling by investigation drilling and construction of permanent monitoring points.

9.5.7 Field operations
Although the questions of where, when and what to sample for are determined in the design of an assessment programme, and the question of how to sample is largely covered by consideration of the design of observation boreholes and sampling methods in relation to the assessment objectives, the field operations of the monitoring activities require careful planning, execution, supervision and evaluation.

Where purpose-built observation wells are required for the programme, as in the case of emergency or impact surveys of a point source, careful super-vision of construction based on a clearly defined drilling contract is required. The contract must have detailed specifications for well design, drilling method, well completion materials and the methods for, and cleaning of, the well. It is particularly essential that no drilling fluid is used that will continue to affect the groundwater quality after the well has been cleaned out in preparation for sampling. Very clear and detailed records must be kept by contractor and supervisor to ensure that there are no subsequent uncertainties about the hydraulic conditions in the well, particularly the positions of open or slotted sections, that might compromise the representativeness of samples.

In both preliminary surveys and subsequent establishment of a permanent monitoring system, the construction of observation wells may provide data that are vital to the understanding of hydrogeological conditions in the area of interest. These data range from geological descriptions (logs) of the strata penetrated to detailed pumping tests to determine the hydraulic charac- teristics of the aquifer. Every effort should be made to collect and record these data in the most cost-effective way. In some cases of detailed local monitoring of point source pollution, especially where aquifer restoration measures are involved or contemplated, the construction of dedicated moni- toring wells may be a major expense running to many thousands of US dollars. The proper preparation and supervision of drilling contracts is essen- tial to ensure that the money is well spent and, where appropriate, to satisfy legal or case preparation requirements. Strict requirements in relation to data collection during well construction may also apply in many cases of research investigations. A comprehensive general description of well construction and hydrogeological data collection is given by Driscoll (1986), and useful dis- cussions of well construction and data collection related specifically to monitoring are included in Everett (1980), Scalf *et al.* (1981), Barcelona *et al.* (1985) and Neilsen (1991).

Field operations also include the collection of hydrogeological data during the lifetime of the monitoring system to assist in the interpretation of the ana- lytical results. This comprises principally the routine measurement of groundwater levels in observation wells (Table 9.14), either continuously by means of autographic recorders or more often by regular spot measurements. These data are required to provide an indication of the persistence of ground- water flow patterns. For large-scale regional monitoring, the flow system (Figure 9.7) is unlikely to change seasonally or in response to effects such as groundwater abstraction sufficiently to affect the interpretation of water quality data. A rare instance where this may not be the case is the recovery of groundwater levels resulting from declining abstraction (see section 9.4.10). On a local scale, however, changes in water table or piezometric level could be very important in the interpretation of monitoring results. This applies particularly to saline intrusion, and salinity resulting from irrigation, where the cause of the problem and its control are inextricably linked to water level changes. In emergency and impact surveys at a very local scale, modified abstraction, including the closure of polluted or threatened high- discharge public supply sources could have a significant effect on local

Table 9.14   Hydrogeological data requirements during monitoring

Data	Frequency	Source	Application
Water levels	Continuous, weekly, monthly	Observation wells	Monitor changes in groundwater flow regime
Pump discharge	Rates, aggregates	Production wells (Public supply, irrigation or industrial)	Mass balance in point source pollution Monitoring saline intrusion
Rainfall, evaporation	Daily, weekly, monthly	Climate stations, meteorological department	Calculate infiltration for diffuse pollution inputs Estimate leachate volume from landfills

groundwater flow patterns. Failure to determine properly these changes by water level measurements could lead to erroneous interpretation of the water quality data. In extreme cases, the results of recording changes in groundwater flow patterns could lead to a requirement for additional sampling stations.

Within the assessment of some types of point source pollution, collection of data on the abstraction rate of polluted wells may be required (Table 9.14). Aggregate discharge data and regular pollutant concentrations may permit a simple mass balance estimation of the amount of pollutant removed by pumping.

Hydrogeological data requirements are somewhat different in relation to the assessment of diffuse pollution. Interpretation of nitrate concentrations in the unsaturated zone requires a knowledge of infiltration rates so that inputs of nitrogen by leaching from the soil can be estimated (Foster and Hirata, 1988) and related to nitrogen application rates in cultivation practices. Climatic and hydrological data are also essential for the estimation of leachate volumes from solid waste disposal sites (Everett, 1980; Foster and Hirata, 1988).

When taking samples from pump discharge, a further consideration is to ensure that the sample is representative of hydrochemical conditions in the aquifer, rather than those in the well. In the case of public supply or irrigation wells which are operating continuously (or at least for long periods) at high discharge rates, it is relatively easy to collect a sample after a sufficiently long period of pumping to be certain that the water has been drawn from the aquifer rather than from within the well itself. If, however, specialist sampling pumps with low discharge rates are used to sample non-pumping observation wells, then "purging" the well of standing water is required before sampling. Some workers have recommended "rule-of-thumb" purging guidelines based on removal of three, five or ten well-volumes. Others (Gibb *et al.*, 1981; Barcelona *et al.*, 1985) have recommended calculation of the purging requirement from the geometry and hydraulic performance of the well. In either case, the well purging operation should be verified by the in-line field measurement of parameters such as Eh, pH, temperature and electrical conductivity until reasonable stability indicates that the required high proportion of aquifer water is being drawn.

Choice of field or laboratory analysis

The choice between carrying out analyses in the field or in the laboratory is part of the assessment design process. This choice is based on the stability of the selected variables, the difficulties of transporting samples, their ease of analysis and several logistical and economic considerations. Some variables (Eh, pH, dissolved oxygen) can only be properly analysed in the field, either in-line or on-site. Other variables lend themselves to easy measurement in the field as general indicators of particular aspects of water quality, such as electrical conductivity used as a guide to overall salinity. The choice between field and laboratory also involves practical considerations. The area to be studied may be remote from a suitable laboratory, in which case field analyses become essential. This applies particularly to bacteriological monitoring. The level of skill and training of the staff who will carry out the field sampling, the financial resources and transport arrangements also affect how much of the analytical work can be done in the field.

9.5.8 Data treatment

Groundwater assessment systems, for whatever objective, will generate an enormous quantity of data. Often, in the past, much less thought and research effort have gone into the vital steps in the system beyond laboratory analysis (see Figure 2.2). The result is poorly archived and inaccessible data, a major and common weakness in assessment programmes (Wilkinson and Edworthy, 1981). In the data treatment phase of groundwater quality assessment, water quality and hydrogcological data need to be converted into information that can be interpreted and used. The conversion of data into useful information (involving storage, retrieval and analysis) is described in Chapter 10.

9.5.9 Interpretation and reporting

Interpretation and reporting are parts of the process of converting data into information, and are closely linked with data treatment, often overlapping. In the case of emergency surveys and operational and early warning surveillance in relation to water used for public supplies, the analytical results must be interpreted and passed on to managers almost immediately in order to protect consumers, if necessary by shutting down the supply. In many, but not all, cases this is possible because the monitoring agency and the managing agency are the same organisation. In these circumstances, the amount of statistical treatment and interpretation of the data is minimal. In contrast, for regional background and trend monitoring and research studies, interpretation is often dependent on the accumulation of a large body of data and, therefore, occurs some time after data collection (see Table 2.2).

For most types of monitoring, the product is information presented in a form that permits an evaluation of groundwater quality with respect to the original objectives, allowing management decisions to be made (see Figure 2.2). These decisions can often be subjective, following an evaluation of the monitoring information and taking account of financial, operational and even political factors. Decisions relate both to water use management and to pollution control (see Figure 2.2). The former could include ceasing or modifying abstraction from polluted wells, blending water from more than one source before putting it into supply, installing water treatment plants, permanently abandoning wells and seeking alternative groundwater or surface water sources, or changing the use to which the abstracted water is put. The latter could include, for example, establishment of protection zones of modified land-use practice around public supply wells, closure and/or removal of point sources of pollution, improved sanitary protection measures at the well itself, and remedial measures to remove pollutants from an aquifer.

The reporting stage of assessment also includes evaluation of the programme itself in relation to the originally stated objectives. The evaluation applies to the whole programme and changes to all steps must be considered, including whether the area needs to be re-defined, whether additional hydrogeological data are required and how they should be collected, whether the monitoring system needs to be redesigned in any way, and whether improvements are required in any of the subsequent steps. In time, the objectives of the assessment need to be reviewed to ensure that they remain relevant to changing economic circumstances and development, and the shift in emphasis in water quality issues that this will bring.

9.6   Examples of groundwater assessment

The outline of the assessment strategy given above is, of necessity, generalised to encompass all types and stages of assessment. The interested reader, charged with the task of establishing an assessment programme or reviewing an existing system, may still find the possible permutations involved somewhat bewildering. The best way to overcome this is to summarise briefly a number of illustrative examples of assessment programmes, covering some of the most important types of assessment and water quality issues. Unfortunately, however, examples of good design, execution, interpretation and presentation of groundwater assessment which follow the general strategy outlined above are difficult to find. Most assessment of groundwater quality has been developed on a much more *ad hoc* basis. Nevertheless, the following examples illustrate how a general overall strategy has been applied to the establishment of groundwater assessments to meet specific objectives related to some of the principal water quality issues described in section 9.4.

9.6.1 Diffuse source pollution from agriculture

South Dade County in Florida is one of the few localities in the continental USA in which fruit and vegetables are grown during the winter (Waller and Howie, 1988). The soils of the area are thin and not naturally fertile. They are irrigated and require large inputs of fertilisers and pesticides to maintain profitable crop production. One method that is used to increase both nutrient and moisture retention capacity is the application to the soil of sludge from treated domestic wastewater. This intensively cultivated agricultural area overlies permeable limestone of the unconfined Biscayne aquifer, which is the sole source of potable and irrigation water in South Dade County. The local and State regulatory agencies became concerned about the possible impact of the agricultural activities on groundwater quality, and in 1985 a study was initiated.

The objective of the assessment was thus well defined from the outset. The area to be assessed comprised South Dade County, with particular emphasis on the 30,000 ha of cultivated land. About 85 per cent of the area is underlain by Rockdale soil composed of crushed limestone and organic debris and the remaining 15 per cent is underlain by the freshwater deposits of the Perrine marl. The hydrogeology of the area is relatively well known. Water levels are shallow and the unsaturated zone varies from < 1 m to 4.5 m thick. The highly fissured Biscayne limestone has measured hydraulic conductivities of

Table 9.15 Variables and sampling methods selected, South Dade County, Florida, USA

Variable group	Variables	Sampling method
Physical properties	Conductivity, alkalinity, pH	Tygon® tube and centrifugal pump
Nutrients	Nitrate, phosphate, dissolved organic carbon	Tygon tube and centrifugal pump, Teflon® bailer
Micronutrients	Copper, iron, magnesium, manganese, potassium, zinc	Tygon tube and centrifugal pump
Sludge-derived contaminants	Arsenic, cadmium, chloride, chromium, lead, mercury, nickel, sodium	Tygon tube and centrifugal pump
Organic constituents	Acid extractable and base-neutral compounds, volatile organic compounds, chlorophenoxy herbicides, organophosphorus insecticides, organochlorine compounds	Teflon tube, peristaltic pump and vacuum chamber Teflon bailer Teflon tube, peristaltic pump and vacuum chamber

Source: Waller and Howie, 1988

> 3,000 m day^{-1}, and the vertical permeability greatly exceeds the horizontal permeability. The aquifer in the area under investigation is 18–37 m thick.

The thickness of the unsaturated zone, variations in lithology and water levels, the hydraulic gradient and the flow directions determined the locations and depths of the wells. The selected network of stations comprised two types. Baseline stations consisted of single monitoring wells to determine the groundwater quality in the upper part of the saturated zone over the whole area. Two of the baseline sites were selected to provide the regional background quality as they were hydraulically up gradient of the agricultural development area. At five test fields, representative of the two soil types and a range of crops, multiple-depth clusters of wells were installed for intensive monitoring of water quality. Details of the construction of the wells are given in Waller and Howie (1988).

The choice of variables was based on historical water quality information, theoretical geochemistry, site inspections, county and state agricultural records and discussion with the local agricultural community. Appropriate sampling methods were then selected (Table 9.15), and a quality assurance programme was established. Agricultural practices are dictated by the climatic regime, therefore, a schedule of eight samples per year was chosen to cover the onset of the summer rains, rising water table, high water table and low water table, and to co-ordinate with agrochemical applications.

Preliminary results indicated that the long history of agricultural activity had little effect on groundwater quality. The few instances of high concentrations of variables were related more to the storage and disposal of agricultural chemicals rather than their application to the fields.

9.6.2 Waste disposal to landfill

At Borden, approximately 80 km north west of Toronto, Canada, an abandoned landfill on an unconfined aquifer of fluvio-glacial sand has been the focus of intensive monitoring and related groundwater studies (Cherry, 1983). Landfilling operations took place from 1940 to 1976, with little record of the quantities or types of waste brought to the site. By the end of this period, the landfill covered 5.4 ha, with a thickness of 5 to 10 m. Investigation drilling in the fill material established that it mostly consisted of ash, wood and construction debris, with lesser amounts of domestic and commercial food wastes.

A range of groundwater monitoring devices has been used at the Borden site (Cherry, 1983). Hydrogeological studies commenced in 1974 with the installation of piezometers, open to the aquifer over a single, relatively short interval, and water table standpipes, open over a single but somewhat longer section. Large numbers of these were installed, and provided information on groundwater levels, flow directions and the lateral extent of the leachate plume. A broad, fan-shaped plume covered an area of about 39 ha and extended about 700 m north of the landfill in the direction of groundwater flow (Figure 9.26). The landfill had caused deterioration of groundwater quality in the shallow, fluvio-glacial sand aquifer, which was not used for water supply. The zone of contamination was separated by a clay and silt layer from a deeper aquifer which was used for water supplies. It appeared unlikely that the deeper aquifer would be affected, but the long-term impact on groundwater quality could not be predicted from this reconnaissance study, and it was decided to close the landfill. A long-term programme of assessment and research was initiated at the abandoned landfill site.

Within this extended programme, additional piezometers and standpipes were installed, supplemented by multi-level groundwater sampling points of three types (Cherry, 1983) to enable regular water samples to be taken from many depths within the aquifer. The natural groundwater quality at the site is characterised by very low overall mineralisation with, for example, chloride concentrations < 10 mg l^{-1}, and this stable, relatively non-reactive inorganic constituent was used as a convenient indicator of the extent of contamination. Figure 9.27 shows the location of sampling points and chloride distribution along a north–south section through the centre of the plume. The

A.   Water table contours

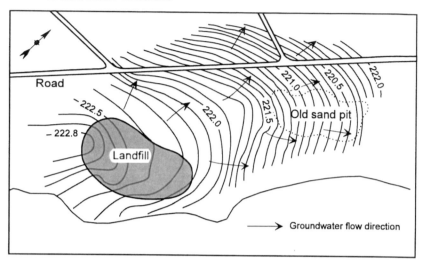

B.   Chloride contours

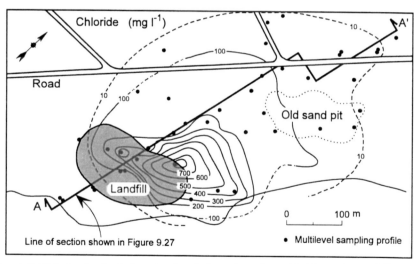

Figure 9.26 Lateral extent of contaminant plume based on chloride contours at the Borden landfill site, Canada (After Cherry, 1983)

greatest chloride concentrations occurred in the middle of the aquifer, 50 to 150 m down gradient of the landfill, and the plume extended to the base of the shallow aquifer. Water samples taken from within the underlying clay had chloride concentrations at the background value of 10 mg l⁻¹, confirming that the clay acts as an impermeable barrier protecting the underlying deeper aquifer.

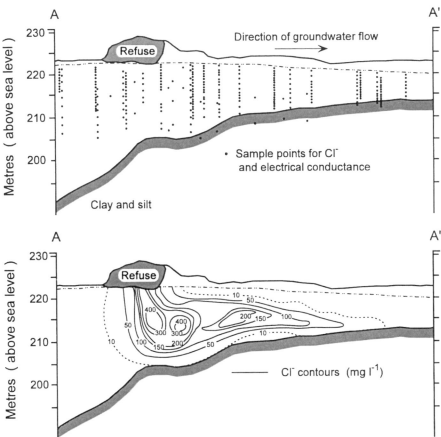

Figure 9.27 Multi-level sampling points and vertical variation of chloride in the plume at the Borden landfill site, Canada. Line of section A – A' shown in Figure 9.26. (After Cherry, 1983)

Provision of such a dense network of sampling points permitted detailed investigation of the three-dimensional distribution of permeability and hydraulic head in the aquifer, of the seasonal modifications to groundwater flow patterns, and of dispersion, together with their effects on contaminant transport and plume characteristics (Cherry, 1983). As a result of the wide range of research activities there, the Borden site is probably the most intensively monitored of all landfills, even though the impact on water supplies is minimal, and has been the model for many such monitoring exercises.

9.6.3 Bacteriological assessment of rural water supplies

The World Health Organization's (WHO) *Guidelines for Drinking-Water Quality. Volume 3* (WHO, 1985) has emphasised the importance of microbiological assessment of the risk of pollution of drinking water as an important, world-wide aspect of water quality control. Within the United Nations Water Decade much effort was directed towards the construction of new supplies, and only recently has more attention been given to the investigation and protection of the installations which supply drinking water. A risk assessment methodology has been developed by Lloyd and Helmer (1991) which aims: (i) to test and evaluate the approach described in the WHO guidelines (WHO, 1985), (ii) to provide a scientific basis for strategies of remedial action, and (iii) to develop a monitoring infrastructure that will ensure drinking water supplies are kept under continuous public health assessment. This risk assessment has been developed as a series of steps in which the causes and sources of pollution are identified and remedial strategies proposed. The principal steps are shown in Figure 9.28, and similarities to Figure 2.2 can be noted. The overall strategy has been developed from pilot studies in Peru, Indonesia and Zambia. In Indonesia, the Gunung Kidul district was chosen for a pilot project (Lloyd and Suyati, 1989). The population is 702,000, about 80 per cent of which is described as living in rural areas. An inventory of water supplies in the district (Figure 9.28: step 1) recorded over 21,000 public installations.

The next step was the planning of inspection visits, and priority was given to the facilities serving the largest populations. A general guide to the suggested frequency of such visits is given in Table 9.16. Sanitary inspection forms were designed for each type of installation listed in the inventory. The forms were based on models included in WHO (1976), and provided for:

- identification of all potential sources of contamination,
- quantification of the level of risk to each facility,
- a graphical means of explaining the risks to the users, and
- guidance to the user on remedial action required.

The check list on the form is completed by the sanitary worker, with the help of the operator or community representative, to give a sanitary inspection risk score, and identify the potential sources of pollution (Figure 9.28: step 4). Details of the forms and risk assessment are given in Lloyd and Helmer (1991). At the same time, a water sample was collected for bacteriological analysis (step 3/4) to confirm whether pollution was occurring (step 5).

The analysis serves to detect actual pollution of the groundwater and the sanitary survey identifies possible sources and establishes the risk of pollution. The two activities are thus complementary, and can be combined (step 7) in a graphical presentation (Figure 9.29) which identifies the installations that most urgently require remedial action (step 8). The graphical method also provides for supervision of field staff and evaluation of the method. Where gross faecal contamination does not correlate with a high risk

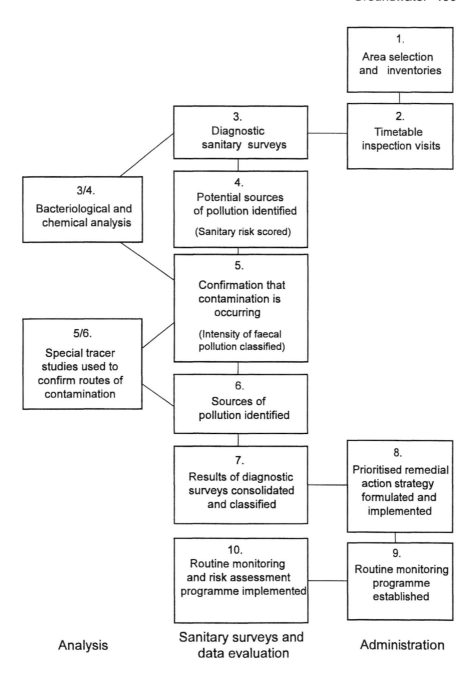

Figure 9.28 Steps used in establishing a microbiological monitoring programme for drinking water supply wells (After Lloyd and Helmer, 1989)

Table 9.16   Suggested frequency of groundwater source surveillance for bacteriological pollution

Population served by source	Maximum interval between sanitary inspections	Maximum interval between bacteriological samples
> 100,000	1 year	1 day
50,000 to 100,000	1 year	4 days
20,000 to 50,000	3 years	2 weeks
5,000 to 20,000	3–5 years	1 month
< 5,000 Community dug wells	Initial, then as situation demands	As situation demands
Deep and shallow tubewells	Initial, then as situation demands	As situation demands
Springs and small piped supplies from tubewells	Initial, and every 5 years, or as situation demands	As situation demands

Source: Lloyd and Helmer, 1991

determined from the sanitary inspection, then a need for urgent re-sampling is indicated (Figure 9.29). A broad spread of points from lower left to upper right of the graph for all types of water supply facilities indicates general agreement between sanitary inspections and bacteriological analyses in identifying polluted sources. This is confirmation of the robustness of the assessment system, but should not be seen as a reason to dispense with bacteriological analysis. A principal objective of the original pilot project was to establish an effective system that could be extended to other provinces over a period of years. This is now being done (steps 9 and 10), and it is anticipated that additional field experience will result in improvements to the basic methodology (Lloyd and Suyati, 1989).

9.6.4   Trend monitoring

A state groundwater quality assessment programme was initially developed in California in 1974. Some 24 priority groundwater basins from a total of about 500 were designated "priority one" on the basis of population, water use, existing water quality problems, available groundwater resources and alternative sources of supply. Of the remaining basins, 180 were classed as "priority two" and 300 as "priority three" (Canter *et al.*, 1987).

Development of the assessment system began with a pilot programme in four basins. An inventory of existing wells was carried out and annual sampling was initiated by the State Department of Water Resources.

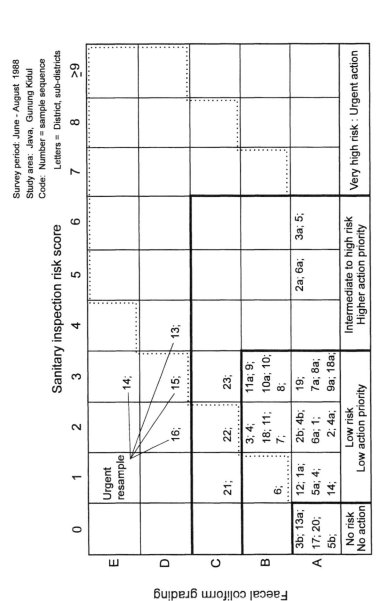

Figure 9.29 Risk analysis from sanitary inspection and bacteriological examination of a handpumped deep tubewell used for drinking water supplies (After Lloyd and Helmer, 1991)

Concentrations of major ions were found to be comparable to historical values. Trace elements and nutrients were added to the selected variables. There was little historical data for these, and some exceeded the state drinking water standards. A structured programme for the establishment of monitoring in the remaining priority basins is currently under way. This comprises an inventory of existing water quality assessment programmes, including details of each well, and design of an improved network, using existing wells where possible, and selecting variables and monitoring frequency appropriate to each basin.

9.6.5 National groundwater quality assessment programme in former Czechoslovakia

The purpose of the former Czechoslovak national groundwater quality assessment programme was to collect background data as a baseline for evaluating the current state and forecasting the changes and trends in groundwater quality due to natural processes and human impacts. The national assessment system was in operation from 1985, and comprised 180 boreholes in shallow aquifers, 98 boreholes in deeper aquifers and 44 springs. The monitoring stations in the deeper aquifers consisted of two or three boreholes, each completed so as to monitor separately individual aquifers in complex sedimentary sequences. The network provided an average monitoring area per station of about 250 km^2. Twenty-two variables for which national drinking water standards were given were included in the basic set and, occasionally, selected organic pollutants were sampled and analysed. Sampling frequency was twice yearly for the shallow aquifers and springs, and annually for the deeper ones. As a result of political changes within former Czechoslovakia, this programme may have been modified recently.

When the programme was established, three categories of monitoring station were designated according to the GEMS classification (Meybeck, 1985). Regional monitoring programmes were established in the middle Elbe region of Bohemia and the Danube lowlands of Slovakia. The former was in operation from 1980 over an area of 3,000 km^2 of intensively cultivated farm land. The regional network in Bohemia was aimed at determining the impact of agriculture on groundwater quality. The areal and vertical distribution of nitrate and other variables were being studied. At selected stations, the sampling frequency was 6–12 times per year; at the remainder it was 2–4 times. The impact on groundwater quality from increasing use of inorganic nitrogen fertiliser, organic fertiliser, farm slurries and from septic tanks was well demonstrated by this programme (Pekney *et al.*, 1989).

Intensive monitoring at a few selected pilot stations was subsequently established to study the evolution of groundwater quality and support the regional programme. Nests of boreholes (Figure 9.24b) with short screens at different depths, were installed in shallow alluvial deposits overlying impermeable bedrock. The unsaturated zone was also being sampled.

9.7 Conclusions and recommendations

The discussion in the preceding sections provides an indication of the broadness of scope of groundwater assessment, encompassing a range of quality issues, purposes, types, scales and levels of assessment. The need to define the objectives at the very beginning has been consistently emphasised, since this will determine the variables to be analysed, the scope of the monitoring network and the financial resources required. Water quality assessment should be seen in the wider context of the management of water resources, comprising both quality and quantity aspects. Unless there is a legal and administrative framework, together with an institutional and financial commitment to appropriate follow-up action, the usefulness of the information obtained from assessment is severely limited. Wilkinson and Edworthy (1981) have identified four main reasons why groundwater assessment systems yield inadequate information:

- The objectives of the assessment were not properly defined.
- The system was installed with insufficient hydrogeological knowledge.
- There was inadequate planning of sample collection, handling, storage and analysis.
- Data were poorly archived.

Despite these inadequacies, data are often used for long-term predictions of water quality and as a basis for decisions on capital expenditure. The same is often true for surface water assessment programmes, and it is hoped that the discussion in this and the preceding chapters will assist in putting assessment on a firmer technical basis, so that sound quality management decisions can be made from relevant and reliable information.

The particular problems of monitoring groundwater bodies relate to the general complexity of flow and contaminant transport in both saturated and unsaturated zones. This is a product of the physical, chemical and biological interactions between soil, rock and water, and the generally slow movement of groundwater compared to other water bodies. The outcome is complex lateral and vertical variations in water quality which are difficult to sample properly and which present major difficulties for interpretation. The need for adequate hydrogeological expertise throughout the assessment and water quality management process cannot be over-emphasised.

To meet assessment objectives related primarily to water use, sampling of pump discharges is usually the simplest and most economical method and remains generally adequate, although there is often scope for improvement in collection and handling of samples and in the choice of variables. However, sampling pump discharges has severe limitations where the objectives are related either to the provision of early warning of a pollution threat to supply wells from diffuse or point sources, or to defining the precise distribution of pollution in an aquifer. Production well samples may be completely inadequate for this purpose because there is insufficient control over the sample depth combined with possible loss of unstable variables. Purpose-built observation wells, open over a selected, short interval and sampled by a device appropriate to the chosen variables are required.

As a result of the generally slow movement of groundwater compared to surface water, once a groundwater body has become polluted it may remain affected for a very long time. Serious groundwater pollution can result in the use of an important aquifer being discontinued, at least locally, for decades. The cost of restoration measures may run into tens of millions of US dollars. Considerable research effort is now devoted to groundwater quality problems, including assessment, and the results have been used in the preparation of this guidebook.

Subjects requiring further assessment effort include sampling for volatile organic compounds in groundwater. Sampling for pesticides occurring at very low concentrations in the unsaturated zone presents particular difficulties. However, such samples are likely to become of increasing importance in first the study, and then the monitoring, of possible diffuse sources of agricultural pesticides. Other future groundwater quality assessment requirements can be anticipated in relation to increasing wastewater re-use for irrigation, problems such as methane generation and other pollutants from reclaimed urban industrial sites, and from rising groundwater levels.

9.8 References

Aller, L., Bennett, T., Lehr, J.H., Petty, R.J. and Hackett, G. 1987 *DRASTIC: A Standardised System for Evaluating Groundwater Pollution Potential Using Hydrogeologic Settings.* EPA/600/2-85/018, US Environmental Protection Agency, Ada, Oklahoma, 455 pp.

Back, W. 1966 *Hydrochemical Facies and Groundwater Flow Patterns in the Northern Part of the Atlantic Coastal Plain.* Professional Paper 498-A, United States Geological Survey, Washington D.C., 42 pp.

Barcelona, M.J., Gibb, J.P., Helfrich, J.A. and Garske, E.E. 1985 *Practical Guide for Groundwater Sampling.* ISWS Contract Report 374, Illinois State

Water Survey, Champaign, Illinois, 94 pp.

Barcelona, M.J., Gibb, J.P. and Miller, R.A. 1983 *A Guide to the Selection of Materials for Monitoring Well Construction and Groundwater Sampling*. ISWS Contract Report 327, Illinois State Water Survey, Champaign, Illinois, 78 pp.

Beard, M.J. and Giles, D.M. 1990 Effects of discharging sewage effluents to the Chalk aquifer in Hampshire. In: *Proceedings of the International Chalk Symposium*, 4-7 September 1989, Thomas Telford, London, 597-604.

Bouwer, H. 1991 Groundwater recharge with sewage effluent. *Water Science and Technology*, **23**, 2099-2108.

Canter, L.W., Knox, R.C. and Fairchild, D.M. 1987 *Groundwater Quality Protection*. Lewis Publishers, Chelsea, Michigan, 562 pp.

Casey, D., Nemetz, P.N. and Uneyo, D.H. 1983 Sampling frequency for water quality monitoring: Measures of effectiveness. *Water Resources Res.*, **19**(5), 1107-1110.

Chapelle, F.H. 1993 *Groundwater microbiology and geochemistry*. J. Wiley and Sons, New York, 424 pp.

Cherry, J.A. [Ed] 1983 Migration of contaminants in groundwater at a landfill: A case study. *J. Hydrology*, **13**, No 1/2, Elsevier, Amsterdam, 1-198.

Chilton, P.J., Lawrence, A.R. and Barker, J.A. 1994 Pesticides in groundwater: some preliminary observations on behaviour and transport in tropical environments. In: N.E. Peters, R.J. Allan and V.V. Tsirkunov [Eds] *Hydrological, Chemical and Biological Processes of Transformation and Transport of Contaminants in Aquatic Environments*, IAHS Publication No. 219, International Association of Hydrological Sciences, Wallingford, UK, 51-66.

Chilton, P.J., Lawrence, A.R. and Stuart, M.E. 1995 The impact of tropical agriculture on groundwater quality. In: H. Nash and G.J.H. McCall [Eds] *Groundwater Quality*, Chapman & Hall, London, 113-122.

Chilton, P.J. and West, J.M. 1992 Aquifers as environments for microbial activity. In: *Proceedings of the International Symposium on Environmental Aspects of Pesticide Microbiology*, Sigtuna, Sweden, 293-304.

Clark, L. and Baxter, K.M. 1989 Groundwater sampling techniques for organic micropollutants: UK experience. *Quart. J. Eng. Geol.*, **22**(2), 159-168.

Conway, G.R. and Pretty, J.N. 1991 *Unwelcome Harvest*. Earthscan Publications, London, 645 pp.

Davis, S.N. and De Wiest, R.J.M. 1966 *Hydrogeology*. John Wiley, New York, 463 pp.

Downing, R.A. and Williams, B.P.J. 1969 *The Groundwater Hydrology of the Lincolnshire Limestone*. Water Resources Board, Reading, UK, 160 pp.

Driscoll, F.G. 1986 *Groundwater and Wells*. 2nd edition, Johnson Division, St Paul, Minnesota, 1089 pp.

Edmunds, W.M. 1973 Trace element variations across an oxidation-reduction barrier in a limestone aquifer. In: E. Ingerson [Ed.] *Proceedings of the Symposium on Hydrogeochemistry and Biogeochemistry*, Tokyo, 1970, Clarke Co., Washington D.C., 500-527.

Edmunds, W.M. and Bath, A.H. 1976 Centrifuge extraction and chemical analysis of interstitial waters. *Environ. Sci. Technol.*, **10**, 467-472.

Edmunds, W.M., Cook, J.M., Darling, W.G., Kinniburgh, D.G., Miles, D.L., Bath, A.H., Morgan-Jones, M. and Andrews, J.N. 1987 Baseline geochemical conditions in the Chalk aquifer, Berkshire, UK: A basis for groundwater quality management. *Applied Geochemistry*, **2**(3), 251-274.

Ehrlich, H.L. 1990 *Geomicrobiology*. Second edition. Marcel Dekker, New York, 646 pp.

Everett, L.G. 1980 *Groundwater Monitoring*. General Electric Company, Schenectady, New York, 440 pp.

Foster, S.S.D. 1985 Groundwater pollution protection in developing countries. In: G. Matthess, S.S.D. Foster and A.C. Skinner [Eds] *Theoretical Background, Hydrogeology and Practice of Groundwater Protection Zones*. IAH International Contributions to Hydrogeology Volume 6, Heinz Heise, Hannover, 167-200.

Foster, S.S.D., Bridge, L.R., Geake, A.K., Lawrence A.R. and Parker, J.M. 1986 *The Groundwater Nitrate Problem*. Hydrogeological Report 86/2, British Geological Survey, Keyworth, UK, 95 pp.

Foster, S.S.D., Gale, I.N. and Hespanhol, I. 1994 *Impacts of Wastewater Use and Disposal on Groundwater*. Technical Report WD/94/55, British Geological Survey, Keyworth, UK, 32 pp.

Foster, S.S.D. and Gomes, D.C. 1989 *Groundwater Quality Monitoring: An Appraisal of Practices and Costs*. Pan American Centre for Sanitary Engineering and Environmental Science, Lima, 103 pp.

Foster, S.S.D. and Hirata, R. 1988 *Groundwater Pollution Risk Assessment*. Pan American Centre for Sanitary Engineering and Environmental Sciences, Lima, 73 pp.

Foster, S.S.D., Ventura, M. and Hirata, R. 1987 *Groundwater Pollution: An Executive Overview of the Latin America-Caribbean Situation in Relation to Potable Water Supply*. Pan American Centre for Sanitary Engineering and Environmental Sciences, Lima, 38 pp.

Freeze, R.A. and Cherry, J.A. 1979 *Groundwater*. Prentice-Hall, Englewood Cliffs, New Jersey, 604 pp.

Geake, A.K., Foster, S.S.D., Nakamatsu, N., Valenzuela, C.F. and Valverde, M.L. 1986 *Groundwater Recharge and Pollution Mechanisms in Urban*

Aquifers of Arid Regions. Hydrogeological Report 86/11, British Geological Survey, Wallingford, UK, 55 pp.

Geake, A.K. and Foster, S.S.D. 1989 Sequential isotope and solute profiling in the unsaturated zone of the British Chalk. *Hydrolog. Sci. J.*, **34**(1/2), 79-95.

Gibb, J.P., Schuller, R.M. and Griffin, R.A. 1981 *Procedures for the Collection of Representative Water Quality Data from Monitoring Wells.* Cooperative Groundwater Report No. 7, Illinois State Water Survey and Illinois State Geological Survey, Champaign, Illinois.

Gowler, A. 1983 Underground purification capacity. In: *Groundwater in Water Resources Planning*, Proceedings of the Koblenz symposium. IAHS Publication 142, Volume 2, International Association of Hydrological Sciences, Wallingford, UK, 1063-1072.

Hem, J.D. 1989 *Study and Interpretation of the Chemical Characteristics of Natural Water*. Water Supply Paper 2254, 3rd edition, US Geological Survey, Washington, D.C., 263 pp.

Idelovitch, E. and Michail, M. 1984 Soil-aquifer treatment - a new approach to an old method of wastewater reuse. *J. Water Poll. Control Fed.*, **56**(9), 36-943.

Keely, J.F. and Boateng, K. 1987 Monitoring well installation, purging and sampling techniques - 1: Conceptualisation, 2: Case histories. *Groundwater*, **25**, 300-313, 427-439.

Kinniburgh, D.G. and Edmunds, W.M. 1986 *The Susceptibility of UK Groundwaters to Acid Deposition*. Hydrogeological Report 86/3, British Geological Survey, Wallingford, UK, 208 pp.

Kinniburgh, D.G. and Miles, D.L. 1983 Extraction and chemical analysis of interstitial water from soils and rocks. *Environ. Sci. Technol.*, **17**, 362-368.

Kovda, V.A. 1983 Loss of productive land due to salinization. *Ambio*, **12**(2), 91-93.

Lawrence, A.R. and Foster, S.S.D. 1987 *The Pollution Threat from Agricultural Pesticides and Industrial Solvents*. Hydrogeological Report 87/2, British Geological Survey, Keyworth, UK, 30 pp.

Lawrence, A.R. and Foster, S.S.D. 1991 The legacy of aquifer pollution by industrial chemicals: technical appraisal and policy implications. *Quarterly Journal of Engineering Geology*, **24**, 231-239.

Lewis, W.J., Farr, J.L. and Foster, S.S.D. 1980 The pollution hazard to village water supplies in eastern Botswana. *Proc. Instn. Civ. Engrs.*, **69**, Part 2, Thomas Telford, London, 281-293.

Lewis, W.J., Foster, S.S.D. and Drasar, B.S. 1982 *The Risk of Groundwater Pollution by On-site Sanitation in Developing Countries*. IRCWD Report 01/82, IRCWD, Duebendorf, 79 pp.

Lloyd, B. and Helmer, R. 1991 *Surveillance of Drinking Water Quality in Rural*

508 Water Quality Assessments

Areas. Published for WHO/UNEP by Longmans Scientific and Technical, Harlow, 171 pp.

Lloyd, B. and Suyati, S. 1989 A pilot rural water surveillance project in Indonesia. *Waterlines*, **7**(3), 10-13.

Lvovitch, M.I. 1972 World water balance: general report. In: IASH/UNESCO/WMO *Proceedings of Symposium on World Water Balance*, Reading 1970, IASH Proceedings No. 2, International Association of Hydrological Sciences, Wallingford, UK, 401-415.

Madison, R.J. and Brunett, J.O. 1985 Overview of the occurrence of nitrate in groundwater of the United States. In: *National Water Summary 1984.* Water Supply Paper 2275, US Geological Survey, Washington DC, 93-105.

Matthess, G. 1982 *The Properties of Groundwater.* J. Wiley, New York, 406 pp.

Matthess, G., Pekdeger, A. and Schroter, J. 1985 Behaviour of contaminants in groundwater. In: G. Matthess, S.S.D. Foster and A.C. Skinner [Eds] *Theoretical Background, Hydrogeology and Practice of Groundwater Protection Zones.* IAH International Contributions to Hydrogeology Volume 6, Heinz Heise, Hannover, 1-86.

Meybeck, M. 1985 The GEMS/Water programme 1978-83. *Wat. Qual. Bull.*, **10**(4), 167-173.

Meybeck, M., Chapman, D. and Helmer, R. [Eds] 1989 *Global Freshwater Quality: A First Assessment.* Blackwell Reference, Oxford, 306 pp.

Morrin, K.A., Cherry, J.A., Dave, N.K., Lim, T.P. and Vivyurka, A.J. 1988 Migration of acidic groundwater seepage from uranium-tailings impoundments: 1. Field study and conceptual hydrogeochemical model. *J. Contaminant Hydrology*, **2**, 271-303.

Nacht, S.J. 1983 Groundwater monitoring system consideration. *Groundwater Monitoring Review*, **3**(2), 33-39.

NRC (National Research Council) 1993 *In Situ Bioremediation: When does it work?* National Academy Press, Washington, DC.

Neilsen, D.M. [Ed.] 1991 *Practical Handbook of Groundwater Monitoring.* Lewis Publishers, Chelsea, Michigan, 717 pp.

Nelson, J.D. and Ward, R.G. 1981 Statistical considerations and sampling techniques for groundwater quality monitoring. *Groundwater*, **19**(3), 617-625.

OECD 1986 *Water Pollution by Fertilizers and Pesticides.* Organisation for Economic Co-operation and Development, Paris, 144 pp.

Parker, J.M. and Foster, S.S.D. 1986 Groundwater monitoring for early warning of diffuse pollution. In: D. Lerner [Ed.] *Monitoring to Detect Changes in Water Quality Series*, Proceedings of the Budapest Symposium, IAHS

Publication No. 157, International Association of Hydrological Sciences, Wallingford, UK, 37-46.

Pekney, V., Skorepa, J. and Vrba, J. 1989 Impact of nitrogen fertilisers on groundwater quality - some examples from Czechoslovakia. *J. Contaminant Hydrology*, **4**(1), 51-67.

Perlmutter, N.M. and Lieber, M. 1970 *Disposal of Plating Wastes and Sewage Contaminants in Groundwater and Surface Water, South Farmingdale, Nassau County, Long Island, New York.* Water Supply Paper 1879G, US Geological Survey, Washington, DC.

Pescod, M.B. 1992 *Wastewater Treatment and Use in Agriculture.* FAO Irrigation and Drainage Paper No 47, Food and Agriculture Organization of the United Nations, Rome, 125 pp.

Pescod, M.B. and Alka, U. 1984 Urban effluent re-use for agriculture in arid and semi-arid zones. In: *Re-use of Sewage Effluent*, Thomas Telford, London, 71-84.

Pickens, J.F., Cherry, J.A., Grisak, G.E., Merritt, W.F. and Risto, B.A. 1978 A multi-level device for groundwater sampling and piezometric monitoring. *Groundwater*, **16**, 322-327.

Price, M. 1985 *Introducing Groundwater.* George Allen and Unwin, London, 195 pp.

Sanders, T.G., Ward, R.C., Loftis, J.C., Steele, T.D., Adrian, D.D. and Yevjevich, V. 1983 *Design of Networks for Monitoring Water Quality.* Water Resources Publications, Littleton, Colorado, 323 pp.

Scalf, M.R., McNabb, J.F., Dunlap, W.F., Cosby, R.L. and Fryberger, J. 1981 *Manual of Groundwater Sampling Procedures.* National Water Well Association, Worthington, Ohio, 93 pp.

Schwille, F. 1981 Groundwater pollution in porous media by fluids immiscible with water. In: W. van Duijvenbooden, P. Glasbergen and H. van Lelyveld [Eds] *Quality of Groundwater*, Proceedings of an International Symposium, Noordwijkerhout, Studies in Environmental Science No. 17, Elsevier, Amsterdam, 451-463.

Schwille, F. 1988 *Dense Chlorinated Solvents in Porous and Fractured Media.* Lewis Publishers, Chelsea, Michigan, 146 pp.

Thomson, J.A.M. and Foster, S.S.D. 1986 Effect of urbanisation on groundwater of limestone islands: an analysis of the Bermuda case. *J. Inst. Water Eng. Sci.*, **40**(6), 527-540.

Todd, D.K. 1980 *Groundwater Hydrology.* 2nd edition, John Wiley, New York, 535 pp.

UNEP (United Nations Environment Programme) 1989 *Environmental Data Report 1989/90*, Blackwell Reference, Oxford, 547 pp.

UNESCO/WHO 1978 *Water Quality Surveys. A Guide for the Collection and*

Interpretation of Water Quality Data. Studies and Reports in Hydrology, 23, United Nations Educational Scientific and Cultural Organization, Paris, 350 pp.

van Duijvenbooden, W. 1993 Groundwater quality monitoring in the Netherlands. In: W.M. Alley [Ed.] *Regional Groundwater Quality*, Van Nostrand Reinhold, New York, 515-536.

Waller, B.G. and Howie, B. 1988 Determining nonpoint source contamination by agricultural chemicals in an unconfined aquifer, Dade County, Florida; Procedures and Preliminary Results. In: A.G. Collins and A.I. Johnson [Eds] *Groundwater Contamination: Field Methods*. ASTM Special Technical Publication 963, American Society for Testing and Materials, Philadelphia, 459-467.

Ward, R.C. 1979 Statistical evaluation of sampling frequencies in monitoring networks. *J. Water Pollut. Control*, **51**, 2291-2300.

WHO 1976 *Surveillance of Drinking-Water Quality*. World Health Organization, Geneva, 135 pp.

WHO 1985 *Guidelines for Drinking Water Quality. Volume 3*. World Health Organization, Geneva, 121 pp.

WHO 1992 *GEMS/Water Operational Guide*. Third edition. World Health Organization, Geneva.

Wilkinson, W.B. and Edworthy, K.J. 1981 Groundwater quality systems - money wasted? In: W. van Duijvenbooden, P. Glasbergen and H. van Lelyveld [Eds] *Quality of Groundwater*. Proceedings of an International Symposium, Noordwijkerhout. Studies in Environmental Science No. 17, Elsevier, Amsterdam, 629-642.

Williams, G.M., Ross, C.A.M., Stuart, A., Hitchman, S.P. and Alexander, L.S. 1984 Controls on contaminant migration at the Villa Farm Lagoons. *Quart. J. Eng. Geol.*, **17**(1), 39-54.

Williams, W.D. 1987 Salinization of rivers and streams: an important environmental hazard. *Ambio*, **16**(4), 180-185.

Wilson, L.G. 1983 Monitoring in the vadose zone: Part 3. *Groundwater Monitoring Review*, **3**(1), 155-166.

WRI (World Resources Institute) 1987 *World Resources 1986: An Assessment of the Resource Base that Supports the Global Economy*. Basic Books Inc., New York.

Chapter 10*

DATA HANDLING AND PRESENTATION

10.1 Introduction

Data analysis and presentation, together with interpretation of the results and report writing, form the last step in the water quality assessment process (see Figure 2.2). It is this phase that shows how successful the monitoring activities have been in attaining the objectives of the assessment. It is also the step that provides the information needed for decision making, such as choosing the most appropriate solution to a water quality problem, assessing the state of the environment or refining the water quality assessment process itself.

Although computers now help the process of data analysis and presentation considerably, these activities are still very labour intensive. In addition, they require a working knowledge of all the preceding steps of the water quality assessment (see Figure 2.2), as well as a good understanding of statistics as it applies to the science of water quality assessment. This is perhaps one of the reasons why data analysis and interpretation do not always receive proper attention when water quality studies are planned and implemented. Although the need to integrate this activity with all the other activities of the assessment process seems quite obvious, achievement of this is often difficult. The "data rich, information poor" syndrome is common in many agencies, both in developed and developing countries.

This chapter gives some guidelines and techniques for water quality data analysis and presentation. Emphasis is placed on the simpler methods, although the more complex procedures are mentioned to serve as a starting point for those who may want to use them, or to help in understanding published material which has used these techniques.

For those individuals with limited knowledge of statistical procedures, caution is recommended before applying some of the techniques described in this chapter. With the advent of computers and the associated statistical software, it is often too easy to invoke techniques which are inappropriate to the data, without considering whether they are actually suitable, simply because the statistical tests are readily available on the computer and involve no computational effort. If in doubt, consult a statistician, preferably before proceeding too far with the data collection, i.e. at the planning and design phase of the assessment programme. The collection of appropriate numbers of samples from representative locations is particularly important for the

This chapter was prepared by A. Demayo and A. Steel

final stages of data analysis and interpretation of results. The subject of statistical sampling and programme design is complex and cannot be discussed in detail here. General sampling design for specific water bodies is discussed in the relevant chapters and the fundamental issues relating to statistical approaches to sampling are described briefly in Appendix 10.1.

10.2 Handling, storage and retrieval of water quality data

Designing a water quality data storage system needs careful consideration to ensure that all the relevant information is stored such that it maintains data accuracy and allows easy access, retrieval, and manipulation of the data. Although it is difficult to recommend one single system that will serve all agencies carrying out water quality studies, some general principles may serve as a framework in designing and implementing effective water quality data storage and retrieval systems which will serve the particular needs of each agency or country.

Before the advent of computers, information on water quality was stored in manual filing systems, such as laboratory cards or books. Such systems are appropriate when dealing with very small data sets but large data sets, such as those resulting from water quality monitoring at a national level, require more effective and efficient methods of handling. In highly populated, developed countries the density of monitoring stations is frequently as high as one station per 100 km^2. When more than 20 water quality variables are measured each month at each station, between 10^5 and 10^6 water quality data entries have to be stored each year for a country of 100,000 km^2.

Personal computers (PCs) in various forms have entered, and are entering more and more, into everyday life. Their popularity and continuing decrease in price have made them available to many agencies carrying out water data collection, storage, retrieval, analysis and presentation. Computers allow the storage of large numbers of data in a small space (i.e. magnetic tapes, discs or diskettes). More importantly, they provide flexibility and speed in retrieving the data for statistical analysis, tabulation, preparation of graphs or running models, etc. In fact, many of these operations would be impossible to perform manually.

10.2.1 Cost-benefit analysis of a computerised data handling system

Although the advantages of a computerised data storage and retrieval system seem to be clear, a careful cost-benefit analysis should be performed before proceeding to implement such a system. Even if, in many cases, the outcome of the analysis is predictable (i.e. a computerised system will be more beneficial) performing a cost-benefit study allows an agency to predict more effectively the impact such a system would have on the organisation as a whole.

Some of the benefits of a computerised system over a manual system for storing and retrieving water quality data are:

- Capability of handling very large volumes of data, e.g. millions of data entries, effectively.
- Enhanced capability of ensuring the quality of the data stored.
- Capability of merging water quality data with other related data, such as water levels and discharges, water and land use, and socio-economic information.
- Enhanced data retrieval speed and flexibility. For example, with a computerised system it is relatively easy to retrieve the water quality data in a variety of ways, e.g. by basin or geo-political boundaries, in a chronological order or by season, by themselves or associated with other related data.
- Much greater variety in data presentation. There is a large variety of ways the data can be presented if they are systematically stored in a computer system. This capability becomes very important when the data are used for a variety of purposes. Each application usually has its own most appropriate way of presenting data.
- Much greater access to statistical, graphical and modelling methods for data analysis and interpretation. Although, theoretically, most of these methods can be performed manually, from a practical point of view the "pencil, paper and calculator" approach is so time consuming that it is not a realistic alternative, especially with large data sets and complex data treatment methods.

In addition to the above, a computerised water quality data storage and retrieval system allows the use of geographic information systems (GIS) for data presentation, analysis and interpretation (see section 10.7.2). The use of GIS in the field of water quality is relatively recent and still undergoing development. These systems are proving to be very powerful tools, not only for graphically presenting water quality data, but also for relating these data to other information, e.g. demography and land use, thus contributing to water quality interpretation, and highlighting facets previously not available.

Amongst the financial costs associated with the implementation of a computerised data and retrieval system are:

- Cost of the hardware. For a computerised system to perform data analysis and presentation effectively and efficiently it must have, not only a computer with an adequate memory, but also the supporting peripherals, such as a printer (e.g. a high quality dot matrix, inkjet or laser printer) preferably capable of graphics output (a colour plotter can be an advantage). Depending on the size of the data bank and on the number of users, more than one computer and one printer may be necessary. When the data are not accessible in any other way, mechanical failure or intensive use leaving only one computer available, would significantly limit all uses of the data bank.

- Cost of software, including: (i) acquisition of a commercial database package that can be used as a starting point in developing a water quality database; (ii) development of the appropriate data handling routines, within the framework of the commercial database, required by the water quality system; (iii) acquisition of other software, such as word processing, graphical, spreadsheet, statistics and GIS packages.

- Cost of maintenance. It is vital that proper maintenance of the hardware and software be readily available. If such maintenance is not available, the day-to-day operation and use of the system, and ultimately its credibility, can be seriously impaired. As a result of its importance, the cost of maintenance can be significant and thus it is important to consider it in the planning stage. It is not unusual for the annual prices of hardware maintenance contracts to be about 10 per cent of the hardware capital costs.

- Cost of personnel. A computerised system may require personnel with different skills from those typically employed in a water agency. It is, for example, very important to have access to computer programmers familiar with the software packages used by the system, as well as being able to call on appropriate hardware specialists.

- Cost of supplies. A computerised system with its associated peripherals (i.e. disc drives, printers and plotters) requires special supplies, such as diskettes, special paper and ink cartridges. The cost of these can be high when compared with a manual data storage system. The increased output from a computerised system also contributes to a much greater use of paper.

- Cost of training. The introduction of an automated data storage, analysis and presentation system requires major changes in the working practices and, usually, considerable reorganisation in the structure of the water quality monitoring laboratory. These all have substantial cost implications.

10.2.2   Approaches to data storage

When storing water quality data it is very important to ensure that all the information needed for interpretation is also available and can be retrieved in a variety of ways. This implies that substantial amounts of secondary, often repetitive, information also need to be stored. It is usual, therefore, to "compress" input data by coding. This not only reduces the amount of data entry activity, but allows the system to attach the more detailed information by reference to another, descriptive portion of the database. The following represents the minimum set of items which must accompany each water quality result in a substantive database.

1. Sampling location or station inventory
Information relating to the sampling location includes:
- geographical co-ordinates,
- name of the water body,
- basin and sub-basin,
- state, province, municipality, etc., and
- type of water, e.g. river, lake, reservoir, aquifer, industrial effluent, municipal effluent, rain, snow.

An example of a station inventory is given in Figure 10.1. All this information, except geographical co-ordinates and the name of the water body, is usually indicated by appropriately chosen alphanumeric codes and put together in a unique identifier for each sampling location. Additional, non-essential information about the sampling location consists of such items as a narrative description of the location, the agency responsible for operating the station, the name of the contact person for additional information, average depth and elevation.

2. Sample information
Further tabulation describes the type of samples collected at any particular location and provides additional information on the function of the sampling station. For example:
- sampling location,
- date and time of sampling,
- medium sampled, e.g. water, suspended or bottom sediment, biota,
- sample matrix, e.g. grab, depth integrated, composite, replicate (e.g. duplicate or triplicate), split, spiked or blank,
- sampling method and/or sampling apparatus,
- depth of sampling,
- preservation method, if any,
- any other field pretreatment, e.g. filtration, centrifugation, solvent or resin extraction,
- name of collector, and
- identification of project.

```
                                      DATE 90-01-21      II.71


             GLOBAL WATER QUALITY MONITORING - STATION INVENTORY
             ********************************************************


STATION NAME       - JEBEL AULIA RESERVOIR      STATION NUMBER - 078001
COUNTRY            - SUDAN                       OCTANT         - 3
DATE OPENED        - 35-01-01                    LATITUDE       - 15/16/00
REGIONAL CENTRE    - EMRA                        LONGITUDE      -  32/27/00
COLLECTION AGENCY  - 07801                       WATER LEVEL (M)- 376.0
WMO CODE           -                             AVG. DEPTH (M) -   4.5
STATION TYPE       - IMPACT                      WATER TYPE     - LAKE

RETENTION (YRS)              - 0.5              MAX. DEPTH (M) - 77.4
AREA OF WATERSHED (KM**2) - 1500                AREA (KM**2)   - 3000.0
                                               VOLUME (KM**3) - 5.0


NARRATIVE
       AT GAUGING STATION UPSTREAM OF DAM



STATION NAME       - BLUE NILE AT KHARTOUM       STATION NUMBER - 078002
COUNTRY            - SUDAN                       OCTANT         - 3
DATE OPENED        - 06-01-01                    LATITUDE       - 15/30/00
REGIONAL CENTRE    - EMRA                        LONGITUDE      -  32/30/00
COLLECTION AGENCY  - 07801                       WATER LEVEL (M)- 374.0
WMO CODE           -                             AVG. DEPTH (M) -   7.6
STATION TYPE       - IMPACT                      WATER TYPE     - RIVER

UPSTREAM BASIN AREA (KM**2)          - 0085520  RIVER WIDTH (M)    - 450.0
AREA UPSTREAM OF TIDAL LIMIT (KM**2)-           DISCHARGE(M**3/SEC)- 003360


NARRATIVE
       AT GAUGING STATION UPSTREAM BLUE NILE BRIDGE
```

Figure 10.1 Example of sample station inventories, as stored in the GEMS/WATER monitoring programme computer database

Table 10.1 illustrates some coding conventions used by Environment Canada in its computerised water quality data storage and retrieval system when storing information on sampling locations, the type of samples and the function of the sampling station. A further station number is unique to a particular location and consists of a combination of different codes. An example of a station number is:

AL	02	SB	0020
region or province	basin code	sub-basin code	assigned number

Each sample collected is also given an unique sample number, and the following is a typical example:

90	BC	000156
year	region or province	assigned number

A similar, but numerical, system is used in the GEMS (Global Environment Monitoring System) water quality database (WHO, 1992).

Table 10.1   Examples of codes used in a computer database for water quality monitoring results in Canada

Province	Code	Type	Code	Sub-type	Code	Sample	Code	Analysis type	Code
Alberta	AL	Surface water	0	River or stream	0	Discrete	01	Water	00
British Columbia	BC			Lake	1	Integrated	02	Wastewater	20
Manitoba	MA			Estuary	2	Duplicate	03	Rain	30
New Brunswick	NB			Ocean	3	Triplicate	04	Snow	31
Newfoundland	NF			Pond	4	Composite	06	Ice (precipitated)	32
Northwest				Reservoir	5	Split	07	Mixed precipitation	33
Territories	NW					Blank	20	Dry fallout	34
Nova Scotia	NS					Spiked	21	Sediments	50
Ontario	ON	Groundwater	1	Well	0	Spiked	21	Suspended	
Prince Edward				Spring	1			sediments	51
Island	PE			Tile drains	2			Soil	59
Quebec	QU			Bog	3			Biota	99
Saskatchewan	SA								
Yukon Territory	YT								

3. Measurement results

These consist of the information relating to the variable measured, namely:

- variable measured,
- location where the measurement was taken, e.g. *in situ*, field, field laboratory, or regular laboratory,
- analytical method used, including the instrument used to take the measurement, and
- actual result of the measurement, including the units.

To indicate the variables measured, method of measurement used, and where the analysis or the measurement was done, codes are usually used. For example, for physical variables an alphabetic code, for inorganic substances their chemical symbols, and for organic compounds or pesticides their chemical abstract registry numbers can be used.

10.2.3   Information retrieval

The primary purpose of a database is to make data rapidly and conveniently available to users. The data may be available to users interactively in a computer database, through a number of customised routines or by means of standard tables and graphs. The various codes used to identify the data, as described in the previous section, enable fast and easy retrieval of the stored information.

Standard data tables

Standard data tables provide an easy and convenient way of examining the information stored in a database. The request for such tables can be submitted through a computer terminal by a series of interactive queries and "user-friendly" prompts. An example of a standard, summarised set of data is given in Figure 10.2 which gives some basic statistics as well as an indication of the number of data values reported as "less-than" or "greater-than" values (i.e. the L- and G-flagged values; see section 10.3.2 for further explanation). For detailed studies, users may wish to probe deeper into the data and treat the results differently. In this case, customised tables can be produced according to the user's needs (e.g. Figure 10.3). Such tables can often be produced by reprocessing and reformatting the data file retrieved for the standard detailed table of results.

Graphs

In many cases, graphs, including maps, facilitate data presentation, analysis and understanding (see section 10.5). The most common types of graphs may be provided routinely, directly from the database. Also, the user may employ specialised graph plotting packages that are not directly linked to the main database. In this case, the data may be retrieved from the main database and transferred (electronically or on a diskette or tape) to another computer system for further processing. Figure 10.4 is an example of a computer-generated graphic and data processing result.

10.3 Data characteristics

10.3.1 Recognition of data types

The types of data collected in water quality studies are many and varied. Frequently, water quality data possess statistical properties which are characteristic of a particular type of data. Recognising the type of data can often, therefore, save much preliminary, uninformed assessment, or eventual application of inappropriate statistical procedures.

Data sets typically have various, recognisable patterns of distribution of the individual values. Values in the middle of the range of a data set may occur frequently, whereas those values close to the extremes of the range occur only very infrequently. The normal distribution (see section 10.4.4) is an example of such a range. In other distributions, the extreme values may be very asymmetrically distributed about the general mass of data. The distribution is then said to be skewed, and would be termed non-normal. The majority of water quality data sets are likely to be non-normal.

GLOBAL WATER QUALITY MONITORING - STATISTICAL SUMMARY BY STATION STATION NO. 002009 90-01-21 V.44

STATION - RIO CAPIBARIBE
COUNTRY - BRAZIL   DEPTH(M)   1.8
LOCATION - OCT. 5 LAT. 8/01/00 LONG. 35/03/00   ELEVATION(M) 57.0
REGIONAL CENTRE - AMRA   WATER TYPE   RIVER

PERIOD REQUESTED FROM - 85-01-01 TO 87-12-31
PERIOD OF RECORD FROM - 85-02-28 TO 87-12-28

STATISTICS INCLUDE ALL FLAGGED DATA AS REPORTED

	ELEC COND	TEMP	DISS O2	B.O.D.	ALK TOT	PH	SUSP SOL 105 DEG	CL DISS	CU TOTAL	ZN TOTAL	FAEC COL
	02041	02061	08101	08201	10102	10302	10401	17201	29006	30004	36011
	USIE/CM	DEG C	MG/L O2	MG/L O2	MG/L	PH UNITS	MG/L	MG/L CL	MG/L CU	MG/L ZN	NO/100ML
NO. OF L-FLAGGED VALUES	0	0	0	0	0	0	0	0	0	0	1
NO. OF G-FLAGGED VALUES	0	0	0	0	0	0	0	0	0	0	0
NO. OF UNFLAGGED VALUES	32	32	32	29	32	32	31	31	13	22	29
MEAN	712.	27.0	3.8	23.	99.08	7.2	105.	309.	.105	.076	1025.
MINIMUM	111.	23.0	.0	1.	10.50	6.3	16.	99.	.000	.000	1.
DATE OF MINIMUM	85-05-15	86-06-18	87-01-27	85-06-28	87-12-28	86-06-18	87-07-17	87-11-25	87-03-27	87-03-27	86-04-18
10TH PERCENTILE	293.	24.2	.1	2.	72.00	6.8	20.	108.	.002	.003	3.
MEDIAN	708.	27.0	4.0	6.	95.00	7.3	36.	284.	.019	.017	700.
90TH PERCENTILE	1139.	29.0	6.7	63.	132.00	7.6	179.	492.	.194	.278	2400.
MAXIMUM	1350.	31.0	7.0	249.	145.00	8.0	1039.	1065.	1.940	.600	3500.
DATE OF MAXIMUM	86-02-28	86-04-18	86-04-18	87-02-16	85-05-15	86-10-15	85-05-15	87-06-25	85-07-19	87-06-25	85-05-15
STANDARD DEVIATION	308.	1.8	2.3	47.	25.61	.4	215.	203.	.345	.138	967.

Figure 10.2 Example of some standard, summarised water quality data for a single monitoring station from the GEMS/WATER monitoring programme computer database

RESULTADOS DOS PARAMETROS E INDICADORES DE QUALIDADE DAS AGUAS. ANO: 1989

LOCAL : RIO CORUMBATAI, PONTE AO LADO DA USINA TAMANDUPA, EM RECREIO.
CODIGO : 00SP14CR2500 CLASSE: 2 BACIA: PIRACICABA.
NAO ATENDEM AOS LIMITES DA CLASSE (*), DO IT (**).

PARAMETROS	PADROES CONAMA 20/ DEC. 8468♦	FEVEREIRO 16/07:30	ABRIL 13/08:30	JUNHO 15/08:00	AGOSTO 10/07:45	OUTUBRO 19/07:40	DEZEMBRO 11/09:20
TEMP. AMOST. GR.C		24	24	15	17	19	23
pH UNID. pH	6.0 a 9.0	7.0	6.7	8.7	7.1	7.1	8.7
OX. DISSOL. mg/L	5.0	6.4	5.6	8.4	7.0	5.0	* 4.8
DBO(5,20) mg/L	5	2	* 6	2	2	5	* 6
CO.F. NMP/100mL	1.0E+03	* 2.1E+05	* 9.0E+05	* 1.1E+05	* 2.3E+04	* 3.0E+04	2.0E+02
N. TOTAL mg/L		1.16	1.26	1.11	0.81	1.48	2.06
FOSF. TOT. mg/L	0.025	* 0.263	* 0.238	* 0.080	* 0.157	* 0.213	* 0.172
RES. TOTAL mg/L		250	320	108	114	115	97
TURBIDEZ UNT	100	* 120	* 140	15	20	15	* 116
IQA		39	36	52	56	49	51
BARIO mg/L	1.00	< 0.10	< 0.10	< 0.10	< 0.10	< 0.10	< 0.10
CADMIO mg/L	0.001	I < 0.005 II	I < 0.005 II	I * 0.010 **	I < 0.005 II	I < 0.005 II	I * 0.010 **
CHUMBO mg/L	0.03	I < 0.10 II	I < 0.10 II	I < 0.10 II	I < 0.10 II	I < 0.10 II	I < 0.10 II
COBRE mg/L	0.02	< 0.01	* 0.03 **	< 0.01	0.01	0.01	< 0.01
CROMO TOT. mg/L	0.05♦	< 0.05	< 0.05	* 0.06 **	< 0.05	< 0.05	< 0.06
NIQUEL mg/L	0.025	< 0.010	< 0.010	0.020	0.010	* 0.030 **	0.010
MERCURIO mg/L	0.0002	< 0.0001	< 0.0001	...	< 0.0001	< 0.0001	< 0.0001
ZINCO mg/L	0.18	< 0.003	0.08	...	0.01	0.08	0.04
FENOL mg/L	0.001	< 0.001	< 0.001	< 0.001	< 0.001	< 0.001	* 0.002 **
IT		II	0	0	II	0	0
TEMP. AR. GR.C		25	23	18	21	19	28
CO.T. NMP/100mL	5.0E+03	* 1.1E+06	* 1.6E+06	* 5.0E+05	* 1.8E+06	* 1.3E+05	* 1.3E+04
FERRO TOT. mg/L		8.08	13.9	2.28	1.39	1.15	0.47
MANGANES mg/L	0.10	* 0.26	* 0.38	0.08	0.10	0.10	* 0.12
CLORETO mg/L	250	2.6	3.7	4.3	4.1	8.2	4.8
DQO mg/L		50	50	< 14	< 14	< 14	< 14
SURFACT. mg/L	0.50	< 0.04	0.07	0.08	0.09	0.15	0.20
N. NITRATO mg/L	10.0	0.15	0.22	0.28	0.26	0.38	0.30
N. NITRITO mg/L	1.00	0.01	0.04	0.03	0.05	0.12	0.08
N. AMONIAC. mg/L	0.50♦	0.10	0.18	0.30	0.18	0.40	* 0.54
N. KJELDA. mg/L		1.00	1.00	0.80	0.60	1.00	1.89
RES. FIL. mg/L	500	88	74	77	87	103	78
RES. N. FIL. mg/L		162	246	31	27	12	21
ORTFO. SOL. mg/L							
COND. ESP. us/cm		89	85	100	108	124	120
COLORACAO		VERMELHA	MARROM	MARROM	MARROM	AMARELA	MARROM
CHUVAS		SIM	SIM	NAO	NAO	NAO	SIM
VAZAO m3/s MEDIA DIARIA		72.2	12.5	17.0	10.5	7.33	8.33

(I) CONFORMIDADE INDEFINIDA QUANTO AO LIMITE DA CLASSE E (II) QUANTO AO LIMITE DO IT, DEVIDO AO LIMITE DE DETECCAO DO METODO ANALITICO NAO ATINGIR AINDA OS NOVOS LIMITES ESTABELECIDOS PELA RESOLUCAO CONAMA 20.

Figure 10.3 Example of water quality monitoring data formatted for presentation in an annual report. Note that basic data are included together with calculated water quality indices (IQA). (Reproduced by kind permission of CETESB, 1990)

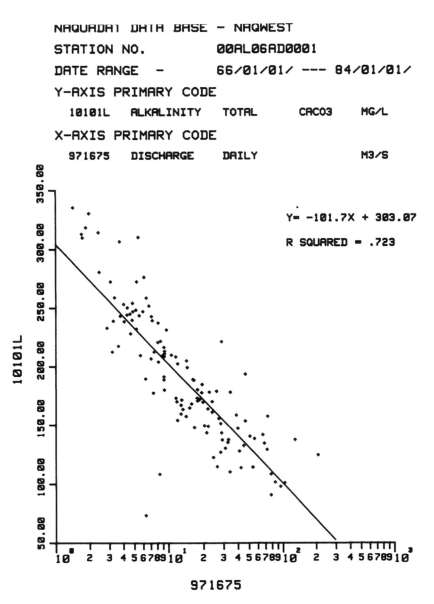

Figure 10.4 Example of statistical analysis (two variables line regression plot) and presentation of discharge and alkalinity data by a computer database of water quality monitoring results

These distinctions are important as many statistical techniques have been devised on the presumption that the data set conforms to one or another of such distributions. If the data are not recognised as conforming to a particular distribution, then application of certain techniques may be invalid, and the statistical output misleading.

Measurement data are of two types:

- Direct: data which result from studies which directly quantify the water quality of interest in a scale of magnitude, e.g. concentration, temperature, species population numbers, time.

- Indirect: these data are not measured directly, but are derived from other appropriately measured data, e.g. rates of many kinds, ratios, percentages and indices.

Both of the above types of measurement data can be sub-divided into two further types:

- Continuous: data in which the measurement, in principle, can assume any value within the scale of measurement. For example, between temperatures of 5.6 °C and 5.7 °C, there are infinitely many intermediates, even though they are beyond the resolution of practical measurement instruments. In water quality studies, continuous data types are predominantly the chemical and physical measurements.

- Discontinuous: data in which the measurements may only, by their very nature, take discrete values. They include counts of various types, and in water quality studies are derived predominantly from biological methods.

Ranked data: Some water quality descriptors may only be specified in more general terms, for example on a scale of first, second, third,...; trophic states, etc. In such scales, it is not the intention that the difference between rank 1 and rank 2 should necessarily be equal to the difference between rank 2 and rank 3.

Data attributes: This data type is generally qualitative rather than quantitative, e.g. small, intermediate, large, clear, dark. For many such data types, it may also be possible to express them as continuous measurement data: small to large, for example, could be specified as a continuous scale of areas or volumes.

Data sets derived from continuous measurements (e.g. concentrations) may show frequency distributions which are either normal or non-normal. Conversely, discontinuous measurements (e.g. counts) will almost always be non-normal. Their non-normality may also depend on such factors as population spatial distributions and sampling techniques, and may be shown in various well-defined manners characterised by particular frequency distributions. Considerable manipulation of non-normal, raw data may be required (see section 10.4.5) before they are amenable to the established array of powerful distribution-based statistical techniques. Otherwise, it is necessary to use so-called non-parametric or distribution-free techniques (see sections 10.4.1 and 10.4.2).

Ranked data have their own branch of statistical techniques. Ratios and proportions, however, can give rise to curious distributions, particularly when derived from discontinuous variables. For example, 10 per cent survivors from batches of test organisms could arise from a minimum of one survivor in ten, whereas 50 per cent could arise from one out of two, two out of four, three out of six, four out of eight or five out of ten; and similarly for other ratios.

10.3.2 Data validation

To ensure that the data contained in the storage and retrieval system can be used for decision making in the management of water resources, each agency must define its data quality needs, i.e. the required accuracy and precision. It must be noted that all phases of the water quality data collection process, i.e. planning, sample collection and transport, laboratory analysis and data storage, contribute to the quality of the data finally stored (see Chapter 2).

Of particular importance are care and checking in the original coding and keyboard entry of data. Only careful design of data codes and entry systems will minimise input errors. Experience also shows that major mistakes can be made in transferring data from laboratories to databases, even when using standardised data forms. It is absolutely essential that there is a high level of confidence in the validity of the data to be analysed and interpreted. Without such confidence, further data manipulation is fruitless. If invalid data are subsequently combined with valid data, the integrity of the latter is also impaired.

Significant figures in recorded data

The significant figures of a record are the total number of digits which comprise the record, regardless of any decimal point. Thus 6.8 and 10 have two significant figures, and 215.73 and 1.2345 have five. The first digit is called the most significant figure, and the last digit is the least significant figure. For the number 1.2345 these are 1 and 5 respectively. By definition, continuous measurement data are usually only an approximation to the true value. Thus, a measurement of 1.5 may represent a true value of 1.500000, but it could also represent 1.49999. Hence, a system is needed to decide how precisely to attempt to represent the true value.

Data are often recorded with too many, and unjustified, significant figures; usually too many decimal places. Some measuring instruments may be capable of producing values far in excess of the precision required (e.g. pH 7.372) or, alternatively, derived data may be recorded to a precision not commensurate with the original measurement (e.g. 47.586 %). A balance must, therefore, be obtained between having too few significant figures, with the data being too coarsely defined, and having too many significant figures, with the data being too precise. As a general guideline, the number of significant figures in continuous data should be such that the range of the data is covered in the order of 50 to 500 unit steps; where the unit step is the smallest change in the least significant figure. This choice suggests a minimum measurement change as being about 2–0.2 per cent of the overall range of the measurements. These proportions need to be considered in the light of both sampling and analytical reproducibility, and the practical management use of the resultant information.

Example
A measure of 0, 1, 2, 3, 4, 5 contains only five unit steps, because each unit step is one, whereas 0.0, 0.1, ..., 5.0 contains 50 unit steps, because the smallest change in the least significant figure is 0.1 and (5.0 – 0.0)/0.1 = 50. Similarly, 0.00 to 5.00 would contain 500 unit steps; and 0.000 to 5.000 would contain 5,000 unit steps. For these data ranges the first consideration would be to decide between one or two decimal places. The first (0–5) and last (0.000–5.000) examples are either too "coarse" or too "fine". For analytical instrument output, these ranges should also be considered in conjunction with the performance characteristics of the analytical method.

For continuous data, the last significant digit implies that the true measurement lies within one half a unit-step below to one half a unit-step above the recorded value. Thus pH 7.8, for example, implies that the true value is between 7.75 and 7.85. Similarly, pH 7.81 would imply that the true value lies between 7.805 and 7.815, and so on. Discontinuous data can be represented exactly. For example, a count of five invertebrates means precisely five, since 4.5 or 5.5 organisms would be meaningless. Even for such data, sampling and sub-sampling should be chosen so as to give between 50 and 500 proportionate changes in the counts as suggested for continuous data.

Data rounding
When more significant figures than are deemed necessary are available (e.g. due to the analytical precision of an instrument) a coherent system of reducing the number of digits actually recorded is also worthwhile. This procedure is usually called "rounding" and is applied to the digits in the least significant figure position. The aim is to manipulate that figure so as to recognise the magnitude of any succeeding figures, without introducing undue biases in the resultant rounded data. A reasonable scheme is as follows:

- Leave the last desired figure unchanged if it is followed by a 4 or less.
- Increase by one (i.e. round up) if the last desired figure is followed by a 5 together with any other numbers except zeros.
- If the last desired figure is followed by a 5 alone or by a 5 followed by zeros, leave it unchanged if it is an even number, or round it up if it is an odd number.

The last rule presumes that in a long run of data, the rounded and unchanged events would balance.

Example
For three significant figures:

2.344x	⇒ 2.34	where x = any figure or none
2.346x	⇒ 2.35	
2.345y	⇒ 2.35	where y = any figure > 0
2.3450	⇒ 2.34	
2.3350	⇒ 2.34	
2.345	⇒ 2.34	
2.335	⇒ 2.34	

Data "outliers"
In water quality studies data values may be encountered which do not obviously belong to the perceived measurement group as a whole. For example, in a set of chemical measurements, many data may be found to cluster near some central value, with fewer and fewer occurring as either much larger or much smaller values. Infrequently, however, a value may arise which is far smaller or larger than the usual values. A decision must be made as to whether or not this outlying datum is an occasional, but appropriate, member of the measurement set or whether it is an outlier which should be amended, or excluded from subsequent statistical analyses because of the distortions it may introduce. An outlier is, therefore, a value which does not conform to the general pattern of a data set (for an example of statistical determination of an outlier see section 10.4.6).

Some statistics, e.g. the median — a non-parametric measure of central tendency (commonest values) (see section 10.4.2), would not be markedly influenced by the occurrence of an outlier. However, the mean — a parametric statistic (see section 10.4.3), would be affected. Many dispersion statistics could also be highly distorted, and so contribute to a non-representative impression of the real situation under study. The problem is, therefore, to discriminate between those outlying data which are properly within the data set, and those which should be excluded. If they are inappropriately excluded the true range of the study conditions may be seriously underestimated.

Inclusion of erroneous data, however, may lead to a much more pessimistic evaluation than is warranted.

Rigorous data validation and automatic input methods (e.g. bar-code readers and direct transfer of analytical instrument output to a computer) should keep transcriptions errors which cause outliers down to a minimum. They also allow the tracing back of any errors which may then be identified and either rectified or removed. If outliers are likely to be present three main approaches may be adopted:

- Retain the outlier, analyse and interpret; repeat without the outlier, and then compare the conclusions. If the conclusions are essentially the same, then the suspect datum may be retained.
- Retain all data, but use only outlier-insensitive statistics. These statistics will generally result from distribution-free or non-parametric methods (see section 10.4.1).
- Resort to exclusion techniques. In principle, these methods presume that the data set range does include some given proportion of very small and very large values, and then probability testing is carried out to estimate whether it is reasonable to presume the datum in question to be part of that range.

It is important not to exclude data solely on the basis of statistical testing. If necessary, the advice and arbitration of an experienced water quality scientist should be sought to review the data, aided by graphs, analogous data sets, etc.

"Limit of detection" problems

Frequently, water samples contain concentrations of chemical variables which are below the limit of detection of the technique being used (particularly for analytical chemistry). These results may be reported as not detected (ND), less-than values (< or LT), half limit-of-detection (0.5 LOD), or zeros. The resulting data sets are thus artificially curtailed at the low-value end of the distribution and are termed "censored". This can produce serious distortion in the output of statistical summaries. There is no one, best approach to this problem. Judgement is necessary in deciding the most appropriate reporting method for the study purposes, or which approach the data set will support. Possible approaches are:

(i)   compute statistics using all measurements, including all the values recorded as < or LT,

(ii) compute statistics using only "complete" measurements, ignoring all < or LT values,

(iii) compute statistics using "complete" measurements, but with all < or LT values replaced by zero,

(iv) compute statistics using "complete" measurements, and all < or LT values replaced by half the LOD value,

(v) use the median which is insensitive to the extreme data,

(vi) "trim" equal amounts of data from both ends of the data distribution, and then operate only on the doubly truncated data set,

(vii) replace the number of ND, or other limited data, by the same number of next greater values, and replace the same number of greatest values by the same number of next smallest data, and

(viii) use maximum likelihood estimator (MLE) techniques.

In order to derive measures of magnitude and dispersion, as a minimum summary of the data set, approach (iv) is simple and acceptable for mean and variance estimates; the biases introduced are usually small in relation to measurement and sampling errors. If < or LT values predominate, then use approach (i) and report the output statistics as < values.

Where trend analysis is applied to censored data sets (section 10.6.1), it is almost invariably the case that non-parametric methods (e.g. Seasonal Kendall test) will be preferable, unless a multivariate analysis (section 10.6.5) is required. In the latter case, data transformation (section 10.4.5) and MLE methods will usually be necessary.

10.3.3 Quality assurance of data

Quality assurance should be applied at all stages of data gathering and subsequent handling (see Chapter 2). For the collection of field data, design of field records must be such that sufficient, necessary information is recorded with as little effort as possible. Pre-printed record sheets requiring minimal, and simple, entries are essential. Field operations often have to take place in adverse conditions and the weather, for example, can directly affect the quality of the recorded data and can influence the care taken when filling-in unnecessarily-complex record sheets.

Analytical results must be verified by the analysts themselves checking, where appropriate, the calculations, data transfers and certain ratios or ionic balances. Laboratory managers must further check the data before they allow them to leave the laboratory. Checks at this level should include a visual screening and, if possible, a comparison with historical values of the same sampling site. The detection of abnormal values should lead to re-checks of the analysis, related computations and data transcriptions.

The quality assurance of data storage procedures ensures that the transfer of field and laboratory data and information to the storage system is done without introducing any errors. It also ensures that all the information needed to identify the sample has been stored, together with the relevant information about sample site, methods used, etc.

An example of quality assurance of a data set is given in Table 10.2 for commonly measured ions, nutrients, water discharge (direct measurement data) and the ionic balance (indirect measurement). Data checking is based on ionic balance, ionic ratio, ion-conductivity relationship, mineralisation-discharge relationship, logical variability of river quality, etc. In some cases, questionable data cannot be rejected without additional information.

Storage of quality assurance results
Data analysis and interpretation should be undertaken in the light of the results of any quality assurance checks. Therefore, a data storage and retrieval system must provide access to the results of these checks. This can be done in various ways, such as:

- accepting data into the system only if they conform to certain pre-established quality standards,
- storing the quality assurance results in an associated system, and
- storing the quality assurance results together with the actual water quality data.

The choice of one of these methods depends on the extent of the quality assurance programme and on the objectives and the magnitude of the data collection programme itself.

10.4   Basic statistics

Statistics is the science that deals with the collection, tabulation and analysis of numerical data. Statistical methods can be used to summarise and assess small or large, simple or complex data sets. Descriptive statistics are used to summarise water quality data sets into simpler and more understandable forms, such as the mean or median.

Questions about the dynamic nature of water quality can also be addressed with the aid of statistics. Examples of such questions are:

- What is the general water quality at a given site?
- Is the water quality improving or getting worse?
- How do certain variables relate to one another at given sites?
- What are the mass loads of materials moving in and out of water systems?
- What are the sources of pollutants and what is their magnitude?
- Can water quality be predicted from past water quality?

When these and other questions are re-stated in the form of hypotheses then inductive statistics, such as detecting significant differences, correlations and regressions, can be used to provide the answers.

Table 10.2 Checking data validity and outliers in a river data set

Sample number	Water discharge (Q)	Elec. Cond. (µS cm⁻¹)	Ca²⁺ (µeq l⁻¹)	Mg²⁺ (µeq l⁻¹)	Na⁺ (µeq l⁻¹)	K⁺ (µeq l⁻¹)	Cl⁻ (µeq l⁻¹)	SO₄²⁻ (µeq l⁻¹)	HCO₃⁻ (µeq l⁻¹)	Σ+¹ (µeq l⁻¹)	Σ-¹ (µeq l⁻¹)	NO₃-N (mg l⁻¹)	PO₄-P (mg l⁻¹)	pH
1	15	420	3,410	420	570	40	620	350	3,650	4,440	4,620	0.85	0.12	7.8
2	18	405	**3,329.4**	370	520	35	590	370	3,520	4,254	4,480	**0.567**	**0.188**	**7.72**
3	35	**280**	2,750	390	**980**	50	**1,050**	260	2,780	4,150	4,090	0.98	0.19	7.5
4	6	515	4,250	**5,200**	620	50	680	510	4,160	**5,440**	5,350	**0.05**	**0.00**	8.1
5	29	395	2,950	420	630	**280**	670	280	2,800	**4,280**	3,770	0.55	0.08	**9.2**
6	170	290	2,340	280	480	65	**930**	250	2,550	3,165	3,730	1.55	0.34	7.9
7	**2.5**	380	3,150	340	530	45	585	**3,240**	**375**	4,065	4,200	0.74	**3.2**	7.6

Questionable data are shown in **bold**

¹ Σ+ and Σ– sum of cations and anions respectively

Sample number

1 Correct analysis: correct ionic balance within 5 per cent, ionic proportions similar to proportions of median values of this data set. Ratio Na⁺/Cl⁻ close to 0.9 eq/eq etc.

2 Excessive significant figures for calcium, nitrate, phosphate and pH, particularly when compared to other analyses.

3 High values of Na⁺ and Cl⁻, although the ratio Na⁺/Cl⁻ is correct — possible contamination of sample? Conductivity is not in the same proportion with regards to the ion sum as other values — most probably an analytical error or a switching of samples during the measurement and reporting.

4 Magnesium is ten times higher than usual — the correct value is probably 520 µeq l⁻¹ which fits the ionic balance well and gives the usual Ca²⁺/Mg²⁺ ratio. Nitrate and phosphate are very low and this may be due to phytoplankton uptake, either in the water body (correct data) or in the sample itself due to lack of proper preservation and storage. A chlorophyll value, if available, could help solve this problem.

5 Potassium is much too high, either due to analytical error or reporting error, which causes a marked ionic imbalance. pH value is too high compared to other values unless high primary production is occurring, which could be checked by a chlorophyll measurement.

6 The chloride value is too high as indicated by the Na⁺/Cl⁻ ratio of 0.51 — this results in an ionic imbalance.

7 Reporting of SO₄²⁻ and HCO₃⁻ has been transposed. The overall water mineralisation does not fit the general variation with water discharge and this should be questioned. Very high phosphate may result from contamination of storage bottle by detergent.

This section describes the basic statistical computations which should be performed routinely on any water quality data set. The descriptions will include methods of calculation, limitations, examples, interpretation, and use in water resources management. Basic statistics include techniques which may be viewed as suitable for a pencil-and-paper approach on small data sets, as well as those for which the involvement of computers is essential, either by virtue of the number of data or the complexity of the calculations involved. If complex computer analysis of the data is contemplated, it is essential that the computer hardware and software is accompanied by relevant statistical literature (e.g. Sokal and Rohlf, 1973; Box *et al.*, 1978; Snedecor and Cochran, 1980; Steel and Torrie, 1980; Gilbert, 1987; Daniel, 1990), including appropriate statistical tables (e.g. Rohlf and Sokal, 1969).

Computer-aided statistical analysis should only be undertaken with some understanding of the techniques being used. The availability of statistical techniques on computers can sometimes invite their use, even when they may be inappropriate to the data! It is also essential that the basic data conforms to the requirements of the analyses applied, otherwise invalid results (and conclusions) will occur. Unfortunately, invalid results are not always obvious, particularly when generated from automated, data analysis procedures.

10.4.1 Parametric and non-parametric statistics

Some examples of parametric and non-parametric basic statistics are worked through in detail in the following sections for those without access to more advanced statistical aids. A choice often has to be made between these statistical approaches and formal methods are available to aid this choice. However, as water quality data are usually asymmetrically distributed, using non-parametric methods as a matter of course is generally a reliable approach, resulting in little or no loss of statistical efficiency. Future developments in the applicability and scope of non-parametric methods will probably further support this view. Nevertheless, some project objectives may still require parametric methods (usually following data transformation), although these methods would usually only be used where sufficient statistical advice and technology are available.

In principle, before a water quality data set is analysed statistically, its frequency distribution should be determined. In reality, some simple analysis can be done without going to this level of detail. It is usually good practice to graph out the data in a suitable manner (see section 10.5), as this helps the analyst get an overall concept of the "shape" of the data sets involved.

Parametric statistics
Just as the water or biota samples taken in water quality studies are only a small fraction of the overall environment, sets of water quality data to be analysed are considered only samples of an underlying population data set, which cannot itself be analysed. The sample statistics are, therefore, estimations of the population parameters. Hence, the sample statistical mean is really only an estimate of the population parametric mean. The value of the sample mean may vary from sample to sample, but the parametric mean is a particular value. Ideally, sample statistics should be un-biased. This implies that repeat sample sets, regardless of size, when averaged will give the parametric value.

By making presumptions about the data frequency distribution (see section 10.4.4) of the population data set, statistics have been devised which have the property of being unbiased. Because a frequency distribution has been presumed, it has also been possible to design procedures which test quantitatively hypotheses about the data set. All such statistics and tests are, therefore, termed parametric, to indicate their basis in a presumed, underlying data frequency distribution. This also places certain requirements on data sets before it is valid to use a particular procedure on them. Parametric statistics are powerful in hypothesis testing (see section 10.4.9) wherever parametric test requirements are met.

Non-parametric statistics
Since most water quality data sets do not meet the requirements mentioned above, and many cannot be made to do so by transformation (see section 10.4.5), alternative techniques which do not make frequency distribution assumptions are usually preferable. These tests are termed non-parametric to indicate their freedom from any restrictive presumptions as to the underlying, theoretical data population. The range (maximum value to minimum value), is an example of a traditional non-parametric statistic. Recent developments have been to provide additional testing procedures to the more traditional descriptive statistics.

Non-parametric methods can have several advantages over the corresponding parametric methods:
- they require no assumptions about the distribution of the population,
- results are resistant to distortion by outliers and missing data,
- they can compute population statistics even in censored data,
- they are easier to compute,

- they are intuitively simpler to understand, and
- some can be effective with small samples.

Non-parametric tests are likely to be more powerful than parametric tests in hypothesis testing when even slight non-normality exists in the data set; which is the usual case in water quality studies.

An illustrative example of specific conductance data from a river is given in Table 10.3, and the techniques used in basic analysis of the data are outlined in the following sections.

10.4.2 Median, range and percentiles

The median M, range R and percentiles P, are non-parametric statistics which may also be used to summarise non-normally distributed water quality data. The median, range and percentiles have similar functions to the mean and the standard deviation for normally distributed data sets.

The median is a measure of central tendency and is defined as that value which has an equal number of values of the data set on either side of it. It is also referred to as the 50th percentile. The main advantage of the median is that it is not sensitive to extreme values and, therefore, is more representative of central tendency than the mean. The range is the difference between the maximum and minimum values and is thus a crude measure of the spread of the data, but is the best statistic available if the data set is very limited. A percentile P is a value below which lies a given percentage of the observations in the data set. For example, the 75th percentile P_{75} is the value for which 75 per cent of the observations are less than it and 25 per cent greater than it. Estimation is based on supposing that when arranged in increasing magnitudes, the n data values will, on average, break the parent distribution into $n + 1$ segments of equal probability. The median and percentiles are often used in water quality assessments to compare the results from measurement stations (see Table 10.10).

Calculations

Number and arrange the data in ascending numerical order. This is often denoted by: $(x_1, x_2, x_3,...,x_n)$ or, $x_1 < x_2 < x_3 <...< x_n$ or, $\{x_a\}$ $a = 1, 2, 3,..., n$ or, $(x_b < x_{b+1})$ $b = 1, 2,..., n - 1$. The arranged values are called the 1st, 2nd,..., nth order statistics, and a is the order number.

Median:
i) If n is odd: $M = \{(n+1)/2\}$th order statistic
    i.e. Median $(x_1, x_2, x_3, x_4, x_5, x_6, x_7) = x_4$
ii) If n is even: $M = $ average$\{(n/2)$th $+ ((n/2)+1)$th$\}$ order statistics
    i.e. Median $(x_1, x_2, x_3, x_4, x_5, x_6) = (x_3 + x_4)/2$

Range: $R = x_{max} - x_{min}$
mid-range $= (x_{max} + x_{min})/2$ is a crude estimate of the sample mean
$range/d_n$ gives an estimate of the standard deviation:

where	d_n	$=$	2:	3; 4;	5
for	n	$\approx$	4;	10; 30;	100

Percentiles:

The ordered statistic whose number is a gives the percentile P as:

$$P = a * 100/(n + 1)$$ See footnote[*]

e.g. the 7th order statistic out of 9 gives: $P = 7 * 100/10 = 70$ percentile
therefore: 70 percentile $P_{70} = x_7$

A particular percentile $P_\%$ is given by the order statistic number:

$$a\% = P\% * (n + 1)/100$$

If $a_\%$ is an integer, the $a_\%$th order statistic is the percentile $P_\%$
If $a_\%$ is fractional, the $P_\%$ is obtained by linear interpolations between the $a_\%$ and $a_\% + 1$ order statistics.

i.e. if $a_\%$ is of the form "j.k" (e.g. 6.60) then:

$$P\% = x_j * (1 - 0.k) + x_{(j + 1)} * 0.k$$

e.g. the 66 percentile P_{66} of 9 order statistics has an order number:

$$a = 66 * 10/100 = 6.6$$

therefore: 66 percentile $P_{66} = x_6 * 0.4 + x_7 * 0.6$

Occasionally, the Mode, or the Modal value is quoted. This is the value which occurs most frequently within the data set. It is a useful description for data which may have more than one peak class, i.e. when the data are multimodal, but is otherwise little used.

Example using Table 10.3
The recommended values for summarising a non-normal data set are the median, some key percentile values, such as 10, 25, 75, 90th, together with the minimum and maximum values.

Step	Procedure	Table 10.3 column	Derived results
i)	Number the data:	(a)	27 values in the data set median will be (27+1)/2 = 14th member of sorted data
ii)	Enter the data:	(b)	
iii)	Sort the data in ascending order:	(c)	
iv)	Derive basic non-parametric statistics		median: 218 maximum value: 430 minimum value: 118

*To avoid confusion with statistical terms throughout this chapter the symbol * has been used to denote a multiplication*

Table 10.3 Basic statistical analysis of a set of specific conductance measurements ($\mu S\ cm^{-1}$) from a river (for a detailed explanation see text)

Column Procedure	(a) No.	(b) Value ($\mu S\ cm^{-1}$)	(c) Sorted values	(d) Class limits L_i	(e) Class values	(f) Class numbers O_i	(g) cfd	(h) $\dfrac{(L_i - \bar{x})}{s}$	(i) A_i	(j) a_i	(k) E_i	(l) E_i	(m) $\dfrac{(O_i - E_i)^2}{E_i}$
				$-\infty \rightarrow 99$		0			0.5000	0.047	1.3	1.3	1.26
	1	396	118				3.6						
	2	360	127	$100 \rightarrow 149$	124.5	2	7.1	−1.68	0.4535	0.087	2.3	2.3	0.05
	3	127	161				10.7						
	4	174	163				14.3						
	5	181	169				17.9						
	6	192	173	$150 \rightarrow 199$	174.5	8	21.4	−1.11	0.3665	0.165	4.4	4.4	2.84
	7	218	174				25.0						
	8	216	181				28.6						
	9	267	183				32.1						
	10	395	192				35.7						
	11	430	208				39.3						
	12	365	216				42.9						
	13	163	218	$200 \rightarrow 249$	224.5	5	46.4	−0.53	0.2019	0.186	5.0	5.0	0.00
	14	173	218				50.0						
	15	169	223				53.6						
	16	183	251				57.1						
	17	218	267				60.7						
	18	208	270	$250 \rightarrow 299$	274.5	5	64.3	0.04	0.0160	0.245	6.6	6.6	0.40
	19	270	272				67.9						
	20	277	277				71.4						
	21	305	305	$300 \rightarrow 349$	324.5	2	75.0	0.61	0.2291	0.152	4.1	4.1	1.08
	22	345	345				78.6						
	23	118	360				82.1						
	24	161	365	$350 \rightarrow 399$	374.5	4	85.7	1.18	0.3810	0.081	2.2	2.2	1.53
	25	223	395				89.3						
	26	272	396				92.9						
	27	251	430	$400 \rightarrow 449$	424.5	1	96.4	1.76	0.4616	0.029	0.8	1.0	0.00
				$450 \rightarrow \infty$		0		2.33	0.4901	0.010	0.3		

(the value 218 at sorted position 14 is the boxed / median value)

$n = 27$

Chi-square: 7.15

The boxed area represents the median value
cfd Cumulative frequency distribution

range:	312
mid-range:	274
range/4:	78 ($n \approx 30$)
25 percentile:	174
50 percentile:	218 ($\equiv$ median)
75 percentile:	305
90 percentile:	395

Hence, with very little calculation, the median (218) and certain percentiles are known, and rough estimates of the mean (274) and the standard deviation (78) are available. In this data set, the median is smaller than the mean estimate and, from the locations of the percentiles within the range, the data seems to be slightly asymmetrically distributed toward the lower values. If this is so, then the data set would not be absolutely normally distributed.

10.4.3 Mean, standard deviation, variance and coefficient of variation

The mean $\bar{x}$, and the corresponding standard deviation s are the most commonly used descriptive statistics. The arithmetic mean is a measure of central tendency and lies near the centre of the data set. It is expressed in the same units as the measurement data. The variance s^2 is the average squared deviation of the data values from the mean. The standard deviation s is the square root of the variance, and is a measure of the spread of the data around the mean, expressed in the same units as the mean. The coefficient of variation cv is the ratio of the standard deviation to the mean and is a measure of the relative variability of the data set, regardless of its absolute magnitudes.

Calculations

Mean:
$$\bar{x} = \frac{1}{n}\sum_{i=1}^{n} x_i$$

where: x_i = values of the data set

  n = number of values in the data set

i.e. $\bar{x}$ = $(x_1 + x_2 + x_3 + ... + x_n)/n$

Variance:
$$s^2 = \frac{1}{(n-1)}\sum_{i=1}^{n}(x_i - \bar{x})^2$$

i.e.
$$s^2 = \frac{(x_1 - \bar{x})^2 + (x_2 - \bar{x})^2 + (x_3 - \bar{x})^2 + ... + (x_n - \bar{x})^2}{n-1}$$

Note, however, that the calculation is more efficiently done by:

$$s^2 = \left\{\sum_{i=1}^{n}(x_i)^2 - (\textstyle\sum x)^2 \big/ n\right\} \Big/ (n-1)$$

which is equivalent, but avoids possibly small differences in deviation from the mean and is suited to simple machine computation.

Standard deviation:

$$s = \sqrt{s^2}$$

Coefficient of variation:

$$cv = \frac{s}{\bar{x}}$$

Uses and limitations

Provided that the data set is reasonably normally distributed, the mean and the standard deviation give a good indication of the range of values in the data set:

68.26 % of the values lie in the range $\bar{x} \pm s$;

95.46 % of the values lie within $\bar{x} \pm 2s$;

99.72 % of the values lie within $\bar{x} \pm 3s$.

The mean and the standard deviation are best used with normally distributed data sets. Because the mean is sensitive to the extreme values often present in water quality data sets, it will tend to give a distorted picture of the data when a few extreme high or low values are present, even if the distribution is normal or approximately normal. For this reason, when used, the mean should always be accompanied by the variance s^2 or the standard deviation s.

Example using Table 10.3

Step	Procedure	Table 10.3 column	Derived results	
v)	Derive basic parametric statistics		mean:	246.6
			variance:	7,627.6
			standard deviation:	87.34
			coefficient of variation:	0.35

For this data, the mean (246.6) is definitely greater than the median (218); thus supporting the initial impression of asymmetry towards the lower values. The relative variability in the data is about 35 per cent of the mean.

10.4.4   Normal distributions

Sometimes it is desirable to be able to describe the frequency distribution of the data set to be analysed, either as a descriptive technique or to test validity of applied techniques or to assess effectiveness of transformations. The frequency distribution is determined by counting the number of times each value, or class of values, occurs in the data set being analysed. This information is then compared with typical statistical distributions to decide which one best fits the data. This information is important because the statistical procedures that can be applied depend on the frequency distribution of the data set. By knowing the frequency distribution, it is also sometimes possible to gain significant information about the conditions affecting the variable of interest. For the purpose of this chapter the data sets are considered in two groups:

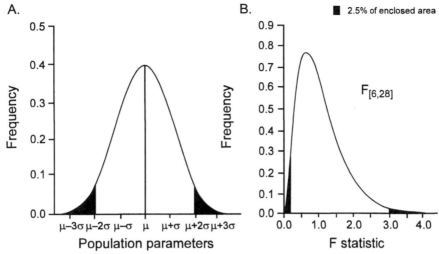

Figure 10.5 Examples of frequency distributions. **A**. Normal; **B**. Non-normal

- Data sets that follow a normal distribution, either in their original form or after a transformation. Parametric methods are used to analyse these data; although appropriate non-parametric tests (where available) would be equally valid.
- All other data sets. Non-parametric methods are recommended to analyse these data sets.

Figure 10.5 gives an example of a normal, and a non-normal frequency distribution.

A wide variety of commonly used statistical procedures, including the mean, standard deviation, and Analysis of Variance (ANOVA), require the data to be normally distributed for the statistics to be fully valid. The normal distribution is important because:

- it fits many naturally occurring data sets,
- many non-normal data sets readily transform to it (see later),
- aspects of the environment which tend to produce normal data sets are understood,
- its properties have been extensively studied, and
- average values, even if estimated from extremely non-normal data, can usually form a normal data set.

Some of the properties of a normal distribution are:

- the values are symmetrical about the mean $\bar{x}$ (the mean, median and mode are all equal),
- the normal distribution curve is bell-shaped, with only a small proportion of extreme values, and

- the dispersion of the values around the mean $\bar{x}$ is measured by the standard deviation s.

Assuming that sampling is performed randomly, the conditions that tend to produce normally distributed data are:

- many "factors" affecting the values: in this context a factor is anything that causes a biological, chemical or physical effect in the aquatic environment,
- the factors occur independently: in other words the presence, absence or magnitude of one factor does not influence the other factors,
- the effects are independent and additive, and
- the contribution of each factor to the variance s^2 is approximately equal.

Checking for normality of a data set

A data set can be tested for normality visually, graphically or with statistical tests.

A. Visual method. Although not rigorous, this is the easiest way to examine the distribution of data. It consists of visually comparing a histogram of a data set with the theoretical, bell-shaped normal distribution curve.

The construction of a histogram involves the following steps:
1. Arrange the data in increasing order.
2. Divide the range of data, i.e. minimum to maximum values, into 5 to 20 equal intervals (classes)*.
3. Count the number of values in each class.
4. Plot this number, on the y axis, against the range of the data, on the x axis.
* As a general guide: within this range, the number of classes can be based on the square root of the number of values (rounded up, if necessary, to give a convenient scale). If too many classes are used for the amount of data available, many classes have no members; if too few classes are used, the overall shape of the distribution is lost (see the example below).

If the histogram resembles the standard, bell-shaped curve of a typical normal distribution (Figure 10.5), it is then assumed that the data set has a normal distribution. If the histogram is asymmetrical, or otherwise deviates substantially from the bell-shaped curve, then it must be assumed that the data set does not have a normal distribution.

Example using Table 10.3

Step	Procedure	Table 10.3 column	Derived results
vi)	Choose number of classes and class intervals:	(d)	$\sqrt{n}$ = 5.2; therefore, number of classes = 6 range/6 = 52; therefore, number of values in each class = 50 class intervals: 100–149 150–199 200–249 250–299 300–349 350–399 400–449
vii)	Enter the class values:	(e)	124.5, 174.5, ... ,424.5
viii)	Count the number of values in each class:	(f)	2,8,5,5,2,4,1

The numbers of values in each class show the tendency for values to be located towards the minimum end of the range. The resultant histogram is shown in Figure 10.6A. For illustration, subsidiary histograms have been included which have either too many (Figure 10.6B), or too few classes (Figure 10.6C). In this particular example the visual method is not very conclusive and, therefore, other methods should be used to determine if the data set has a sufficiently normal distribution.

B. Graphical method. This is a slightly more rigorous method of determining if the data set has a normal distribution. If the data are normally distributed, then the cumulative frequency distribution *cfd* will plot as a straight line on normal probability graph paper.

To apply this method the following steps are necessary:
1. Arrange the data in ascending order.
2. Give each value a rank (position) number (symbol *i*). This number will run from 1 to *n*, *n* being the number of points in the data set (column a).
3. Calculate cfd_i for each point from:

$$cfd_i(\%) = \frac{i}{(n+1)} * 100$$

(Note the similarity to the percentile calculations)
4. Plot the value of each data point against its *cfd* on normal probability graph paper.

If the plot is a straight line (particularly between 20 and 80 per cent), then the data set has a normal distribution. Figure 10.7 illustrates some generalised frequency distributions, and the form that the resultant probability plot takes.

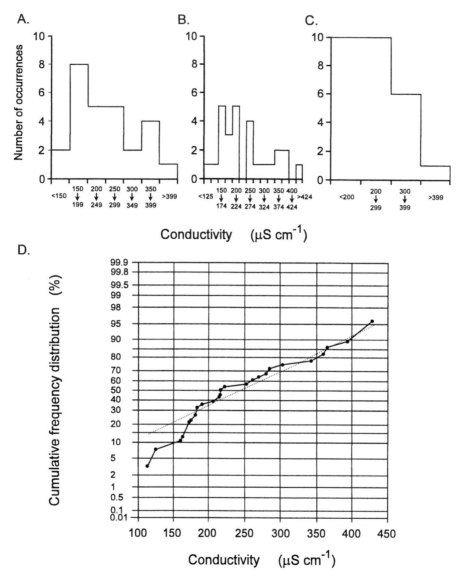

Figure 10.6 Graphical tests for normality in a data set. **A – C**. Histograms of the specific conductance data in Table 10.3, illustrating the different frequency distribution shapes produced by choosing different class intervals. **D**. The cumulative frequency distribution of the data from Table 10.3

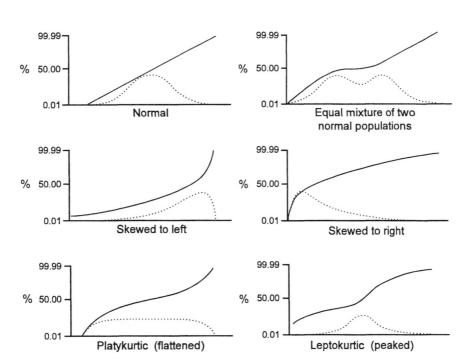

Figure 10.7 Graphical presentations of generalised frequency distributions and resulting probability plots

Example using Table 10.3

Step	Procedure	Table 10.3 column	Derived results
ix)	Calculate the *cfd*	(g)	3.6, 7.1, ..., 96.4

The results are also shown plotted in Figure 10.6D. Although there is some evidence of the data skewness (compare with Figure 10.7), the majority of the data fit sufficiently to a straight line for normality to be accepted, and parametric testing may be used without further manipulation.

C. Statistical tests. The previous two methods of testing for normality contained an element of intuition in deciding whether or not the histogram has a bell-shaped curve (visual method) or how well the *cfd* points fitted to a straight line. However, more rigorous tests of the normality of the data can be used such as the Chi-Square Goodness of Fit Test, *G*-Test, Kolmogorov-Smirnov Test for Normality, and the Shapiro-Wilk or *W*-Test for Goodness of Fit. The latter test is recommended when the sample size is less than or equal to 50, while the Kolmogorov-Smirnov test is good for samples greater than 50; although some modification is necessary to the Kolmogorov-Smirnov test if the parameters of the comparison frequency distribution have been estimated from the data itself. An alternative for larger numbers (≥ 50) is the D'Agostino D-test, which complements the W-test (Gilbert, 1987).

The Chi-square and G-tests should be used when the sample size is equal to or greater than 20 (although the critical value tables usually also accommodate smaller sample sizes). The tests are often not very effective with very small data sets, even if the data are extremely asymmetrical. In such instances, tests specifically designed for small numbers (e.g. the W-test) are a better approach and are now more routinely used. These tests are usually available with commercial statistical packages for computers. The Chi-square and W-test are described below.

An alternative, or supplementary approach to testing for normality, is to consider the extent to which the data set appears as a distorted normal curve. Two main distortions may be considered:

- skewness, which is a measure of how asymmetrically the data are distributed about the mean, (and so the median and mean would generally not coincide), and
- kurtosis, which measures the extent to which data are relatively more "peaked" or more "flat-topped" than in the normal curve.

This approach involves third and fourth power calculations and, because the numbers to be handled are large (if coding is not used); it is only recommended with the aid of a computer.

Intuition and the experience of the investigator play a key role in the selection and application of these statistical tests. It should also be noted that these normality tests determine whether the data deviate significantly from normality. This is not exactly the same as checking that the data are actually distributed normally.

Chi-Square Goodness of Fit Test

As the chi-square χ^2 test has traditionally been used to test for normality, and is simple, it is presented here as an illustration. The chi-square test compares an observed distribution against what would be expected from a known or hypothetical distribution; in this case the normal distribution.

The procedure for the test is as follows:
1.   Divide the data set into m intervals of hypothetical, probable distribution (see below).
2.   Calculate the expected numbers E_i in each interval. (Traditionally for χ^2, $E_i \geq 5$ has been required, but this is probably more rigorous than strictly necessary. A reasonable approach is to allow $E_i \geq 1$ in the extreme intervals, as long as in the remainder $E_i \geq 5$).
3.   Count the number of observed values O_i in each interval.

4. Calculate the test statistic χ^2 from:

$$\chi^2 = \sum_{i=1}^{m} \frac{(O_i - E_i)^2}{E_i}$$

5. Compare with the critical values of the chi-square distribution given in statistical tables. If: $\chi^2_{\alpha[v]} \rangle \chi^2$ then the data set is consistent with a normal distribution. Alpha, α, is the level of significance of the test. In water quality assessments α is usually taken at 0.05–0.10; v is the number of degrees of freedom $d.f.$

For frequency distribution goodness-of-fit testing:

$v = m -$ (number of parameters estimated from the data) $- 1$

where m is the total number of intervals into which the data was divided. For the normal curve, two parameters (μ, σ) (see below) are estimated (by the mean $\bar{x}$ and the standard deviation s respectively). The $d.f.$ are, therefore, $v = m - 3$.

The expected frequency E_i is calculated from the hypothetical frequency distribution which specifies the probable proportion of the data which would fall within any given band, if the data followed that distribution. For example, from section 10.4.3, for normally distributed data, 68.26 per cent of the data would lie within ± 1 standard deviation of the mean. All statistical tables contain the areas of the normal curve ($\equiv$ proportions of the data set), tabulated with respect to "standard deviation units" sdu. These are equivalent to: $(x - \mu)/\sigma$, where x is the datum, μ is the parametric mean, and σ is the parametric standard deviation. In practice the estimators are used, i.e. the data mean $\bar{x}$ and standard deviation s.

Example using Table 10.3

Step	Procedure	Table 10.3 column	Derived results
x)	Decide number, and limits of intervals L_j:	(d)	9 intervals (from previous workings) i.e. 7 original, plus 2 extremes $-\infty$, 100, 150, ..., 400, 450, ∞
xi)	Express limits, L_j, as sdu's:	(h)	$(L_j - \bar{x})/s$: $(0 - 246.6)/87.34$ $(100 - 246.6)/87.34$ etc. $-2.82, -1.68, ..., 2.33, \infty$
xii)	Look-up normal curve areas A_i proportional to interval limits in sdu's:	(i)	$A_1, A_2, ..., A_m$ $0.4976, 0.4535, ..., 0.4901, 0.5$
xiii)	Calculate interval areas a_i (taking care when the interval contains the mean!):	(j)	$a_1, a_2, ..., a_m$ $0.5 - A_9, A_8 - A_7, A_7 - A_6,$ $A_6 + A_5,$ $A_1 - A_2, A_2 - A_3, A_3 - A_4, A_4 - A_5.$ $0.001, 0.029, ..., 0.090, 0.045$

xiv)	Calculate	(k)	$n * a_1, n * a_2, ..., n * a_m$
	expected numbers		$E_1 = 27 * 0.045 = 1.2$,
	E_i in each interval:		$E_2 = 27 * 0.090 = 2.4$, etc.
xv)	Aggregate any	(l)	
	adjacent classes so		
	that no $E_i < 1$:		
xvi)	Calculate χ^2 by	(m)	$\chi^2 = 7.57$
	the formula given:		

After the aggregation step, eight intervals remain. There are, therefore, five degrees of free-dom for the critical value. For $\alpha = 0.05$, the critical value of $\chi^2_{0.05[5]} = 11.07 \rangle \chi^2_{test} = 7.57$. It can be concluded that the data distribution is not significantly different from normal.

Note: if $\chi^2_{test} > \chi^2_{crit}$, it would have positively shown that at the chosen level of significance α the data were not drawn from a normal distribution. In the present case, it is only possible to conclude that the data set is consistent with a normal distribution. This allows the possibility that a re-test with more data could indicate non-normality.

For the purpose of this example, the class limits used previously were adhered to, but any desired limits may be used with similar procedures. The classes need not be of equal size. The classes may also be chosen by having desired proportions of the data within them. For this, the tables of the normal curve are used to read off the relevant *sdu* [i.e. $(x_L - \mu)/\sigma$] with the desired proportion ($\equiv$ area). The limit is then back-calculated as $x_L = x + sdu * s$.

The chi-square test is non-specific, as it assesses no specific departure from the hypothetical, test distribution. This property weakens the power of the test to identify non-normality in certain data distributions. Distributions which are markedly skewed, for example, but otherwise can be approximated by a normal curve (even if this theoretically implies "impossible" negative values), may not generate critical values of χ^2 and so give a misleading result for this particular analysis.

Table 10.4 gives an example of a series of counts of the invertebrate *Gammarus pulex* in samples from the bottom of a stony stream. Similar data analysis procedures to the previous examples have been applied to the counts.

Example using Table 10.4

Step i)	32 values in the data set
	median will be the average of (32/2)th and
	(32/2 + 1)th = 16th and 17th members of the data set.

Steps ii)–iv)	median:	4.5	
	maximum value:	16	
	minimum value:	0	
	range:	16	
	mid-range:	8	
	range/4:	4.0	$(n \approx 30)$
	mode:	3	

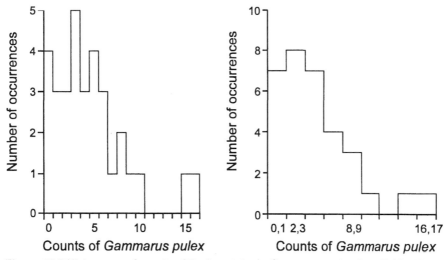

Figure 10.8 Histograms of counts of the invertebrate *Gammarus pulex* from Table 10.4 showing the different frequency distributions obtained by choosing different class intervals

Although the data type is discontinuous, the following can be calculated for illustrative purposes only:

25 percentile:	2.0	
50 percentile:	4.0	(= median)
75 percentile:	6.0	
90 percentile:	9.7	
Step v)	mean:	4.6
	variance:	15.5
	standard deviation:	3.93
	coefficient of variation:	0.86

Although the mean is close to the median, their position with respect to the mid-range and the percentiles location within the range, suggest that the distribution of the data is skewed. Note also how well the non-parametric statistics estimate their parametric equivalents. As the variance is much greater than the mean, a clumped or aggregated population is suggested and a negative binomial distribution is probably the most suitable model (Elliott, 1977) (see section 10.4.5).

Steps vi) – ix)   17 class intervals were chosen: one for each possible count value. From the data numbers, about six classes would probably have been sufficient. An example with nine is also shown.

Figure 10.8 shows the frequency histograms for the *Gammarus pulex* data. It can be seen that the data set appears markedly non-normal, regardless of the number of class intervals. As a probability plot is a technique for continuous data types, it is not illustrated since the *G. pulex* counts are a discontinuous data type.

Table 10.4 also contains the results of a χ^2 analysis of the hypothesis that the data set is normally distributed. Considerable aggregation of the histogram classes is necessary in order to meet the restrictions on E_i (although new class intervals could have been designed):

Steps x) – xvi) χ^2:   1.15

ν: 3

$$\chi^2_{0.05[3]} = 7.82 \rangle \chi^2 = 1.15$$

Table 10.4 Basic statistical analysis of counts of *Gammarus pulex* from a stony stream (for a detailed explanation see text)

(a) Numbers and values	(b)	(c) Sorted values	(d) Class limits L_i	(e) Class values	(f₁) Class numbers O_i	(h) $\dfrac{(L_i - \bar{x})}{s}$	(f₂) Class numbers O_i	(l) E_i	(m) $\dfrac{(O_i - E_i)^2}{E_i}$	(n) x_i	(o) Shapiro-Wilk W-test x_i^2	(p) a_i	(q) W'
1	3	0	$-\infty \rightarrow -0.5$							0	0	0.4188	6.701
2	1	0	$-0.5 \rightarrow 0.5$	0	4	-1.29	4	4.8	0.13	0	0	0.2898	4.347
3	3	0								0	0	0.2462	2.462
4	1	0								0	0	0.2141	1.927
5	4	1	$0.5 \rightarrow 1.5$	1	3	-1.04				1	1	0.1878	1.315
6	2	1								1	1	0.1651	1.156
7	10	1								1	1	0.1448	0.869
8	0	2	$1.5 \rightarrow 2.5$	2	3	-0.79	6	4.8	0.32	2	4	0.1265	0.506
9	6	2								2	4	0.1093	0.437
10	2	2								2	4	0.0931	0.372
11	7	3								3	9	0.0777	0.155
12	6	3	$2.5 \rightarrow 3.5$	3	5	-0.53				3	9	0.0629	0.126
13	9	3								3	9	0.0485	0.097
14	5	3								3	9	0.0344	0.069
15	15	3								3	9	0.0206	0.021
16	1	[4]								4	16	0.0068	0.000
17	0	4	$3.5 \rightarrow 4.5$	4	3	-0.28	8	6.2	0.52	4	16		
18	5	4								4	16		
19	8	5								5	25		
20	5	5	$4.5 \rightarrow 5.5$	5	4	-0.02				5	25		
21	2	5								5	25		
22	0	5								5	25		
23	0	6								6	36		
24	4	6	$5.5 \rightarrow 6.5$	6	3	0.23	6	6.2	0.00	6	36		
25	5	6								6	36		
26	3	7	$6.5 \rightarrow 7.5$	7	1	0.48				7	49		
27	8	8	$7.5 \rightarrow 8.5$	8	2	0.74	4	4.9	0.18	8	64		

Continued

Table 10.4 Continued

Column Procedure	(a)	(b) Numbers and values	(c) Sorted values	(d) Class limits L_i	(e) Class values	(f₁) Class numbers O_i	(h) $\dfrac{(L_i-\bar{x})}{s}$	(f₂) Class numbers O_i	(l) E_i	(m) $\dfrac{(O_i-E_i)^2}{\bar{E_i}}$	(n) x_i	(o) Shapiro-Wilk W-test x_i^2	(p) a_i	(q) W
	28	16	8	8.5 → 9.5	9	1	0.99				8	64		
	29	3	9	9.5 → 10.5	10	1	1.25				9	81		
	30	6	10	10.5 → 11.5	11	0	1.50				10	100		
	31	3	15	11.5 → 12.5	12	0	1.76				15	225		
	32	4	16	12.5 → 13.5	13	0	2.01	4	5.2	0.26	16	256		
				13.5 → 14.5	14	0	2.26							
				14.5 → 15.5	15	1	2.52							
				15.5 → 16.5	16	1	2.77				147	1155		20.559
				16.5 → ∞			3.03							

Chi-square   1.15

$W_{crit, 32}$:

n: 32
d: 479.7
k: 16.0
W: 0.881

$W_{crit, 32}$: 0.93

The boxed area represents the median value

[1] Column letters indicate the same procedures as used in Table 10.3

As the critical value of χ^2 exceeds the test value, the χ^2 test indicates that the data set is consistent with a normal distribution! Caution in this respect has been noted above, and in view of the distribution shown by the histogram, it would be appropriate to undertake further testing in this case.

The G-test is a more critical test than χ^2, but is essentially similar. A better approach is to adopt the Shapiro and Wilk W-test, as it is a powerful test specifically designed for small sample numbers (i.e. $n \le 50$).

W-test of Goodness of Fit

An aspect of the power of this test is its suitability for various data types and its freedom from restrictive requirements. Its main disadvantage is its limitation to sample sets of 50 or less, and that it is slightly more complex to compute.

The procedure for the test is as follows:
1.   Compute the sum of the squared data value deviations from the mean:

$$d = \sum_{i=1}^{n}(x_i - \bar{x})^2 = \sum_{i=1}^{n}x_i^2 - \frac{(\sum x_i)^2}{n}$$

2.   Arrange the data in ascending order.
3.   Compute k where:
      $k = n/2$ if n is even; $k = (n-1)/2$ if n is odd
4.   Look-up W-test coefficients for n: $a_1, a_2, \ldots a_k$; in statistical tables.
5.   Compute the W statistic by applying these coefficients to the ranges between the two ends of the order statistics:

$$W = \left\{\sum_{i=1}^{k}a_i\left(x_{(n-i+1)} - x_i\right)\right\}^2 \Big/ d$$

6.   Compare this value with the critical percentile $W_{\alpha[n]}$. If $W < W_{\alpha[n]}$, then the data set does not have a normal distribution.

Example using Table 10.4

Step	Procedure	Table 10.4 column	Derived results
xvii)	Calculate d, k:	(n,o)	$d = 479.7$ $k = 32/2 = 16$
xviii)	Look-up W coeff's:	(p)	$a_1, a_2, \ldots, a_k$ 0.4188, 0.2898, ..., 0.0068
xix)	Calculate part W:	(q)	$a_1 * (x_{32} - x_1) + a_2 * (x_{31} - x_2) +$ $a_3 * (x_{30} - x_3) + \ldots + a_{16} * (x_{17} - x_{16})$ $0.4188 * (16 - 0) + 0.2898 * (15 - 0) +$ $\ldots + 0.0068 * (4 - 4) = 20.559$
xx)	Calculate W by the formula given:		$W = (20.559)^2/d$ $= 0.881$

The critical value for W for significance level $\alpha = 0.05$ and $n = 32$, is $W_{0.05[32]} = 0.930$. As $W_{test} = 0.881 < W_{crit} = 0.930$, the hypothesis that the data could have come from a normal population is rejected. The raw data would, therefore, be better approximated by an alternative, non-normal distribution and are not suitable for parametric statistical analysis.

If a computer is available the data in Table 10.4 would, therefore, be suitable for considering tests of skewness and kurtosis. These could also provide further evidence for deciding for, or against, a normal distribution.

10.4.5 Non-normal distribution

If tests reveal that a data set has a non-normal distribution, then either apply distribution-free (non-parametric) statistics or first process the data so as to render them normal if possible (in which case parametric testing is also valid). This procedure is known as transformation.

Transformation is a process converting raw data into different values; usually by quite simple mathematics. A common transformation is to use the logarithms, or square roots, of data values instead of the data values themselves. These transformed values are then re-assessed for consistency with a normal distribution. Various transformations may be used, but there will usually be one which most efficiently converts the data to normality. The particular choice is often decided by reference to the frequency distribution of the parent data population, and its parameter values (e.g. μ, σ). This allows some general guidance to be given for common distributions:

	Frequency distribution or data type	Conditions	Transformation
1.	Logarithmic	$x_i > 0$	$x_i \Rightarrow \ln(x_i)$
	e.g. measurement data	$x_i \geq 0$	$x_i \Rightarrow \ln(x_i + 1)$
2.	Positive binomial	$s^2 < \bar{x}$	
	$P = (p+q)^n$	$p + q = 1$	
	e.g. percentages, proportions		$p_i \Rightarrow \sin^{-1}\sqrt{p_i}$
	e.g. some count data	see below	

	$p < 0.1$	$0.1 \leq p < 0.4$	$0.4 \leq p \leq 0.6$	$0.6 < p < 0.9$	$p > 0.9$
$n < 10$	No approximations available in this region				
$10 \leq n \leq 30$		use $\sin^{-1}\sqrt{p}$		use $\sin^{-1}\sqrt{p}$	
$n \geq 30$	use Poisson approx.	distribution is approximately normal			

3.	Poisson	see above:	
	e.g. count data	$s^2 = \bar{x}$	
		$n > 10$	$x_i \Rightarrow \sqrt{x_i}$
		$n \geq 10$	$x_i \Rightarrow \sqrt{(x_i + 0.5)}$
4.	Negative binomial	$s^2 > \bar{x}$	
	$P = (q - p) - k$	$p + q = 1$	
	e.g. count data	$k > 5$	$x_i \Rightarrow \sinh^{-1} \sqrt{\left\{ \dfrac{x_i + 0.375}{k - 2*(0.375)} \right\}}$
		$2 \leq k \leq 5$	$x_i \Rightarrow \ln(x_i + k/2)$
		$x_i > 0$	$x_i \Rightarrow \ln(x_i)$
		$x_i \geq 0$	$x_i \Rightarrow \ln(x_i + 1)$

If a distribution such as the negative binomial is suspected as being the most appropriate for the data under analysis, the initial goodness of fit comparisons use defined procedures for estimating the parameters of the distribution being tested. Once these have been satisfactorily determined, the appropriate transformation for further analysis can be carried out.

In some studies, a series of investigations into a particular water quality variable allow systematic analysis of the means and variances of the resulting data sets. For normal data, the mean and the variance are independent, and data plots of one against the other should show no obvious trend. If the mean and variance are in some way proportional, then the data are generally amenable to treatment under the Taylor power law. This states that the variance of a population is proportional to a fractional power of the mean:

$$\sigma^2 = a\,\mu^b$$

For count data, a is related to the sampling unit size; and b is some index of dispersion, varying from 0 for regular distribution to ∞ for a highly contagious distribution. The appropriate transformation is:

$$x_i \Rightarrow x_i^p \qquad \text{where } p = 1 - b/2$$

This includes some of the transformations already given. For a Poisson distribution, for example $\sigma^2 = \mu$, so $a = b = 1$. The appropriate transformation exponent p is, therefore, $p = 1 - 1/2 = 0.5$. The transformation is then:

$$x_i \Rightarrow \sqrt{x_i} \qquad \text{(as } x^{0.5} \equiv \sqrt{x}).$$

Once an appropriate transformation has been identified and confirmed as suitable (by methods such as those outlined above), the data are transformed, and the transformed data are then analysed. For example, if the logarithmic transformation is to be used:

$$(x_1, x_2, \dots, x_n) \frac{\text{Transform}}{\text{Take logs}} \Rightarrow (\ln(x_1), \ln(x_2), \dots, \ln(x_n))$$

The mean is then calculated as:

$$\overline{\ln(x)} = \frac{\ln(x_1) + \ln(x_2) + \dots \ln(x_n)}{n}$$

and used as required in further analysis. All subsequent analysis is done on the transformed data, and their transform statistics. Once all these procedures are complete, the descriptive statistics are un-transformed to give results in the original scale of measurement. For the present example:

$$\overline{\ln(x)} \xrightarrow[\text{Anti - log}]{\text{Untransform}} \Rightarrow \bar{x}$$

In general, this mean will not be the same as the arithmetic mean calculated from the raw data. For the present example:

$$\bar{x}_a = \frac{(x_1 + x_2 + \dots + x_n)}{n} \qquad (\bar{x}_a: \text{arithmetic mean})$$

$$\bar{x}_1 = \exp\left\{ \frac{\ln(x_1) + \ln(x_2) + \dots + \ln(x_n)}{n} \right\} \qquad (\bar{x}_1: \text{logarithmic mean})$$

$$\equiv (x_1 * x_2 * \dots x_n)^{\frac{1}{n}}$$

Thus the mean from the transformation is derived from the product of the raw data, whereas the arithmetic mean is derived from their sum. For logarithmic transformations, the mean from the transformation is the geometric mean. The geometric mean of water quality data is always somewhat less than its arithmetic mean. In logarithmic transformations, for example, the arithmetic mean is close to:

$$\bar{x}_a \approx \exp\left\{ \overline{\ln(x)} + \left(1.15 * s_1^2 \right) \right\}$$

where s_1^2 is the variance of the transformed data (which is usually small).

Transformation is an essential technique in water quality data analysis. It may, however, be difficult to accept the relevance of a test which is valid on transformed data, when it is apparently not valid on the raw data! In fact, many transformations are simply a recognition that although water quality measurements are most often recorded on a linear basis, the fundamental variable characteristic is non-linear; pH is a typical example.

Since most transformations used in water quality assessment are non-linear, statistics associated with the reliability of the estimates (e.g. confidence limits (see section 10.4.8)) will be asymmetric about the statistic in question. In a logarithmic transformation, for example, the confidence limits represent percentages in the un-transformed data. Thus, for un-transformed data, the mean may be quoted as: $\bar{x} \pm C$, an equi-sided band, centred on the mean. In the transformed example, a similar specification would be: $\overline{\ln(x)} \pm \ln(C)$. Once un-transformed back to the original scale of measurement, this would become: $\bar{x}_1 * C$ and $\bar{x}_1/C$, an asymmetric band about the derived mean. This distinction is important when interpreting summary statistics.

10.4.6 Probability distributions

In statistical terms, any water quality variable which is: (i) the concentration of a substance in water, sediment or biological tissues or (ii) any physical characteristic of water sediments or life forms present in the water body, and the abundance or the physiological status of those life forms, is taken to be due to a stochastic process. In such a process the results are not known with certainty and the process itself is considered to be random, and the output a random variable. For normally distributed water quality data sets (with or without transformation) the probability distribution addresses this attribute of randomness, and allows predictions to be made and hypotheses to be tested. This type of test can be used, for example, to estimate the number of times a water quality guideline may be exceeded, provided that the variable follows a normal distribution (with or without transformation).

The probability of an event which may or may not happen is an indication of how likely it is that it will actually occur. The probability is usually given a number between 0 and 1; where 0 represents the event being impossible, and 1 represents the event being certain. The particular value assigned is derived either by inference from knowledge of all possible outcomes and their likelihoods, or by experience from a great many experimental events.

The normal curve is an example of a probability density function *pdf*. It indicates by its value, $\phi(x)$ the relative frequency with which any given x will occur. The normal curve *pdf* is fully specified by three parameters: the mean μ, the standard deviation σ and a scaling coefficient c:

$$\phi(x) = c * e^{-0.5*((x-\mu)/\sigma)^2}$$

where e is the base of naperian or natural logarithms. At any value x_i, this curve provides a point estimate, $P(x_i)$ of the probability of the value x_i arising out of the infinite possibilities for a continuous, measurement variable. The curve must, therefore, allow for $P(x_1$ or x_2 or ... or $x_\infty) = P(x_1) + P(x_2) + ...+ P(x_\infty)$, from before. This is equivalent to the total area under the curve, i.e. the cumulative probability function $\Phi(x)$ which is:

$$\Phi(x) = \int_{-\infty}^{+\infty} \phi(x)dx$$

As this represents all possibilities, it must equal one. This allows the scaling factor to be precisely defined. The integral value is: $\Phi(x) = c\sigma\sqrt{(2\pi)}$. Thus by choosing $c = 1/(\sigma\sqrt{(2\pi)})$, $\Phi(x) = 1$ as required, the normal curve is:

$$\phi(x) = \frac{1}{\sigma\sqrt{(2\pi)}}e^{-0.5*((x-\mu)/\sigma)^2}$$

and is, as a result, fully defined by the mean μ and the standard deviation σ. In practice the population parameters are usually unknown, and the best estimates of them: the sample mean $\bar{x}$ and standard deviation s, would be substituted for them.

These relationships emphasise the links between the normal curve and probabilities, and that it is the areas under the curve which directly represent the probabilities relating to x. Thus the area enclosed by x_a, $\phi(x_a)$, x_b, $\phi(x_b)$ represents the probability that any x chosen at random would be in the interval x_a, x_b. This may be extended to the concept of cumulative probabilities. For a particular value $x = i$, the area enclosed by the axes and the curve $\phi(x)$ from $x = -\infty$ to $x = i$ represents the probability that an event $< i$ would be drawn at random. Similarly, the area from $x = i$ to $x = +\infty$ represents the probability of a value $> i$ occurring. If generalised, this concept has clear relevance to pollution control standards.

Individual normal curves differ one from another only in their location, and their spread about that location; but they are fundamentally the same. If x is transformed to $z = (x - \bar{x})/s$, another normal curve results which has a mean of 0, and a standard deviation of 1.

$$\phi(z) = \frac{1}{\sqrt{(2\pi)}}e^{-0.5*z^2}$$

This is the standard normal curve and is tabulated in all statistical tables, along with the associated areas:

$$\Phi(z) = \int_{-\infty}^{z_i}\phi(z)dz; \qquad -\infty \leq z_i \leq +\infty$$

The variable z is the "standard deviation unit" previously used and is known as the z-score. The standard normal curve is applicable to all normally distributed data, by un-transforming z back to $x = \bar{x} + sz$, using the data statistics. The standard normal curve is, therefore, the required generalised function, with $\phi(z)$ and $\Phi(z)$ the point and cumulative probability functions respectively.

Example
Using the specific conductance data of Table 10.3, what is the probability that the value of 375 μS cm^{-1} will be exceeded?
As the mean is 246.6 and the standard deviation is 87.3:

$$z = (x - \bar{x})/s = (375 - 246.6)/87.3 = 1.47$$

The value 375 is 1.47 standard deviations above the mean. The cumulative probability function is:

$$\Phi(z) = \int_{1.47}^{\infty} \phi(z)dz$$

and from previous results, this is equal to:

$$1 - \Phi(z) = \int_{-\infty}^{1.47} \phi(z)dz$$

since $\Phi(z)$ is usually given in this form in statistical tables. From such a table, $\Phi(1.47) = 0.9292$, hence $P(x \geq 375) = 0.071$, or 7.1 per cent. At this particular location, therefore, a conductance value greater than 375 μS cm^{-1} can be expected about 7 times out of every 100 samples taken.

If a specific conductance of 527 μS cm^{-1} had been encountered in the example of Table 10.3, it is possible to determine how likely that event was. For the value of 527 the z-score = 3.21. Thus $P(z \geq 3.21) = 0.0007$, or 7 in 10,000 samples. It is, therefore, highly improbable that a value of 527 μS cm^{-1} would have been encountered naturally in a set of 27 samples, and it is very likely that this value should be regarded as an outlier.

Note: as $\phi(z)$ is symmetrical about zero, sometimes the function $\Phi(z)$ is given in statistical tables as:

$$\Phi(z) = \int_0^{z_i} \phi(z)dz; \qquad 0 \leq z_i \leq \infty$$

and its value will lie between 0.0000 and 0.5000. Due to the symmetry, this is exactly equivalent to the form given above if 0.5000 is added to it.

10.4.7 Error statistics

In the conclusions of a water quality assessment, part of the summarising information usually includes, for example, some of the descriptive statistics mentioned above. If conclusions and possible management actions are to be based on this output, it is important that an indication is given of how reliable the estimates are. In the examples used immediately above a mean value has been estimated and quoted. However, if another 27 specific conductance values had been taken at the same time, it is unlikely that the mean calculated from them would have been exactly the same as the one used. A possible approach is to average the available means. This process would produce the "mean of the means", i.e. $x = (\bar{x}_1 + \bar{x}_2)/2$, and an estimate of their spread about that mean, $s_{\bar{x}} = (\bar{x}_1 - \bar{x}_2)/1.13$ which is the standard deviation estimated from the range of two samples. The standard deviation of the mean, $s_{\bar{x}}$ is known as the standard error of the mean. As with other standard deviations, probability allows estimates of the reliability of the true data mean.

If the environment being assessed is stable, the means of a series of similar samples of a water quality variable show less variation than the data sets themselves. The extent to which this reduction of variation occurs is, however, dependent also on the variability of the data sets. If data are highly variable, then it is likely that means estimated from it will show a much greater relative spread than the same number of means estimated from another, much less variable water quality data set. By experiment and theory the standard error of the mean is: $s_{\bar{x}} = s/\sqrt{n}$.

This important result applies to a single estimate of the mean from a normal data set as well as to a series of estimates of the mean.

Example using Table 10.3

	mean:	$\bar{x}$ = 246.6
	standard deviation:	s = 87.3
	sample no.:	n = 27
therefore:	standard error	
	of the mean:	$s_{\bar{x}}$ = 16.8

An important aspect of the normal distribution is that estimates of means are either normally distributed, or tend to be so, regardless of the underlying frequency distribution of the raw data set. Means from normal data are always normally distributed regardless of the sample size whereas means from extremely non-normal data tend to normality with sufficient sample numbers (it is not possible to give a precise estimate of sufficient sample numbers; as few as 5, or at least 50, may be required). For large sample numbers, standard normal probabilities can be applied.

The above results suggest that in 95 out of 100 similar investigations the mean would lie in the approximate range:
$$\bar{x} \pm 2 * s_{\bar{x}} \equiv 246.6 \pm 2 * 16.8 = 213 \text{ to } 280 \text{ } \mu S \text{ cm}^{-1}$$
If this range is too great for the water quality assessment objectives, the precision may only be increased at the cost of extra sampling (see Appendix 10.1). If the mean was to be determined 95 times out of 100 as lying within a range of $\pm$ 10, then $s_{\bar{x}}$ = 5. If the characteristics of the variable are assumed to remain constant, the number of samples required would be: $n = (s/s_{\bar{x}})^2 = 305$! The estimated mean and standard deviation of a sample are, in principle, unchanged with different sample sizes, whereas the precision of the mean is inversely proportional to the number of samples taken.

Similar standard errors exist for the other main descriptive statistics, although they are usually not distributed so normally:

Statistic	Standard error	Restrictions
Mean:	$s_{\bar{x}} = s/\sqrt{n}$	Any normal population
		Non-normal with large n
Standard deviation:	$s_s = 0.707 * s_{\bar{x}}$	Normal populations, $n > 15$
Median:	$s_m = 1.253 * s_{\bar{x}}$	Normal populations, $n > 50$

10.4.8 Confidence limits

Since the error statistics discussed above give some indication of the reliability of the data estimates, they can be used to define a range about the statistic of interest, with some confidence that this range covers the "true" mean μ. The bounds of this confidence interval are the confidence limits of the statistic. Thus, for a data set such as that in Table 10.3, it is actually the confidence interval: $x - 1.96\sigma \leq \mu \leq x + 1.96\sigma$ which was required; although for illustrative purposes the sample standard deviation s was used instead of the population (or data set) standard deviation σ. However, the population standard deviation can never be known in water quality studies.

An approach to correct this simplification, is to use a transformation of the z-score type. The mean is transformed to a new statistic t by:
$$t = (\bar{x} - \mu)/s_{\bar{x}} = (\bar{x} - \mu)/(s/\sqrt{n}).$$
As this still contains μ it cannot be determined exactly, but its relative distribution (i.e. its *pdf*) can be worked out. This now forms part of all statistical tables as the Student's t. The t distribution is practically a standard normal curve for $n > 30$, but is distinctly different when n is small. Similarly, the areas beneath the curve can be equated to probabilities. This then allows a confidence limit to be defined, even though it is only the sample standard deviation which is known. For example, it is possible to define a confidence interval such that there is 95 per cent confidence (i.e. confidence that on 95 surveys out of every 100) that the interval will cover the true mean μ. This range is $\bar{x} \pm C_L$, where C_L is the confidence limit. As before:
$$\bar{x} - t_\alpha s_{\bar{x}} \leq \mu \leq \bar{x} + t_\alpha s_{\bar{x}} \equiv \bar{x} - t_\alpha(s/\sqrt{n}) \leq \mu \leq \bar{x} + t_\alpha(s/\sqrt{n})$$
where $\alpha = $ the significance level which is equivalent to:
$$1 - (\% \text{ confidence})/100.$$
The appropriate t value is looked-up in statistical tables for the required significance and degrees of freedom in the sample standard deviation s. For all error statistics the degrees of freedom $= (n - 1)$.

Example using Table 10.3

mean:	$\bar{x} = 246.6$
standard deviation:	$s = 87.3$
sample no:	$n = 27$

therefore:	d.f.:	$v = 26$
by look-up:		$t_{0.05[26]} = 2.056$
therefore:	confidence limit	$C_L = 34.5$

Therefore, there is 95 per cent confidence that 246.6 ± 34.5 (i.e. from 212.1 to 281.1) covers the true mean μ. This range is slightly wider than the first approximation, since allowance has now been made for only being able to estimate σ.

Unlike the mean, some estimates are never normal unless drawn from a normal population. This makes their confidence intervals more complicated to calculate. For example, for variances (and hence, standard deviations) the quantity $v * s^2/\sigma^2$ is distributed as chi-square χ^2. The tables of χ^2 are given as probabilities of exceeding the respective values of χ^2. Interval estimates, therefore, require a difference between appropriate values of χ^2. Thus $\chi^2_{0.025} - \chi^2_{0.975}$ is the 0.95 (i.e. 95 per cent) probability that a randomly drawn χ^2 lies between those two values. For a 95 per cent confidence interval of the variance:

$$\chi^2_{0.975[v]} \leq v * s^2 / \sigma^2 \leq \chi^2_{0.02[v]}$$

$$\equiv \left(v * s^2\right) / \chi^2_{0.025[v]} \leq \sigma^2 \leq \left(v * s^2\right) / \chi^2_{0.975[v]}$$

and
$$\sqrt{\left(v * s^2\right) / \chi^2_{0.025[v]}} \leq \sigma \leq \sqrt{\left(v * s^2\right) / \chi^2_{0.975[v]}}$$

For the example in Table 10.3, this gives: $68.8 \leq \sigma \leq 119.6$. The sample standard deviation, $s = 87.3$, is asymmetrically placed within this interval. These formulae indicate the need for large samples for σ to be estimated precisely.

This method of estimating the confidence interval does not necessarily yield the smallest possible size of confidence interval, because the χ^2 involved were placed symmetrically, whereas the distribution of variances is skewed. Statistical tables of the shortest, unbiased, confidence interval for the variance are available. For $\alpha = 0.05$, and $v = 26$, these tables provide two coefficients: $f_1 = 0.6057$ and $f_2 = 1.825$. The lower confidence limit, $C_1 = f_1 * s^2$ and the upper, $C_u = f_2 * s^2$. Therefore, by taking square roots $67.9 \leq \sigma \leq 117.9$. For the median, the 95 per cent confidence limits for continuous data and $n \geq 25$, are the $\{(n+1)/2 \pm (z * \sqrt{n})/2\}$th order statistics, where z is the z-score equivalent to the desired confidence probability.

For 95 per cent confidence, $z = 1.96$, hence for the example in Table 10.3:

	median	$m = 246.6$
	sample no:	$n = 27$
by look-up:		$z_{0.95} = 1.96$
therefore:	lower confidence limit	$C_1 = \{14 - 0.98 * 5.2\} =$
		$\{8.9\}$th order statistic
by interpolation, as before:		$C_1 = 182.8$
similarly:		$C_u = 272.5$

By being able to define probability areas around various statistics, even though the population parameters are unknown, opportunities also arise to test (quantitatively) assumptions about those statistics.

10.4.9 Hypothesis testing

One of the most important requirements for a successful water quality assessment programme is to have clear, quantifiable objectives. In order to use statistics to analyse the data collected, these objectives must be stated as questions that can be statistically answered, i.e. as hypotheses. For example, questions such as:

- Has there been any change or trend in variable x from time t_1 to t_2?
- Is there a difference between water quality at point M and point N on a given river?

These questions can be reduced to the following hypotheses:

- Null hypothesis (H_0): there is no change (or trend) or
- Alternate hypothesis (H_A): there is a change (or trend).

The questions can then be answered using statistical methods such as the t-test, F-test and the Analysis of Variance (ANOVA).

Example
The objective of a water quality study is to: "Describe water quality conditions on the Flat River and identify concerns related to the potential impact of the Canada Tungsten Mines on this waterway". This statement can be reduced to one or more testable hypotheses:
H_0: There is a measurable effect as a result of the discharge from the mining operation.
H_A: There is no measurable ecological effect as a result of the discharge from the mining operation.

In any hypothesis testing there are two possible types of errors which can occur:
Type I error: H_0 is rejected when in reality it is true.
Type II error: H_0 is accepted when in reality it is false.

	TRUE	FALSE
ACCEPT	OK	TYPE II ERROR
REJECT	TYPE I ERROR	OK

In water quality work the acceptable risk, alpha α, that a type I error will occur is usually set at 5 or 10 per cent. The value $1 - \alpha$ is equivalent to the acceptance level of the test. Working at a confidence of 0.9 (i.e. 90 per cent) means that there is a 10 per cent chance that a type I error will occur. The working value for ß, the risk of making a type II error, is usually also 10 per cent. The quantity $1 - ß$ is a measure of the power of a test. For test purposes, ß should ideally be as small as possible. Unfortunately as ß is reduced, α increases! Therefore, an appropriate compromise is necessary.

Hypothesis testing can be used, for example, to make inferences (e.g. to find out if there is a significant difference between the mean calcium concentrations in two water bodies) or to test assumptions upon which parametric statistics are based, e.g. normality or the homogeneity of variances. Statistical significance is not the same as biological or chemical significance. Data sets may be shown to be statistically significant without having practical consequences. Also significant biological/chemical differences may not give significant statistical differences under certain circumstances. Informed judgement by trained water quality scientists is still necessary because statistical hypothesis testing is only an aid to judgement, not a replacement for it.

Homogeneity of variances

If statistics are used to compare two sets of data, e.g. the mean values, then the variances s^2 of the two data sets must be approximately equal for the test to be valid. This "homogeneity of variances" can be tested using the F-test which computes the F_s value and compares it to tabulated values for the test conditions.

Example
1. Let $H_0 : s_1^2 = s_2^2$ and $H_A : s_1^2 \neq s_2^2$. Thus the null hypothesis, H_0, is that the two variances are not significantly different. If this is rejected by the test, it must be concluded that the two variances are different at the level of significance chosen.
Take the values $s_1^2 = 4.6$, $s_2^2 = 1.9$, and $n_1 = 25$ and $n_2 = 10$
2. F_s is calculated as the ratio of the larger variance over the smaller one:
$$F_s = s_1^2/s_2^2 = 4.6/1.9 = 2.42$$
3. This number is then compared to $F_{\alpha/2[v1,v2]}$, where $v_1 = n_1 - 1$ and $v_2 = n_2 - 1$, obtained from a table of critical values of the F distribution (available in statistical tables).
$$F_{0.025[24,9]} = 3.61 \text{ and } F_{0.05[24,9]} = 2.90$$
Since this is a two-tailed test (s_1^2 can be greater or lesser than s_2^2) the F value of 3.61 represents a probability of $\alpha = 0.05$, and that of 2.90 is $\alpha = 0.10$. Since the calculated F_s value of 2.42 is less than both critical values of 3.61 and 2.90, the conclusion is that the two variances are not significantly different at the 90 or 95 per cent confidence level.

Detecting significant differences

Many water quality studies attempt to determine pollutant impacts on a given water body. Impacts can be assessed, for example, from comparisons of the "disturbed" site with a site presumably unaffected by the disturbance, i.e. a control. The objective of the study can then be addressed by testing for significant differences between the mean values of the water quality variables at the two sites. For "normal" data sets the Student's t-test is commonly used. The equivalent non-parametric procedure is the Mann-Whitney U Test.

Student's t-test

This test assumes that the variances of the two data sets being compared are approximately the same (see above), and that the variables are independently distributed. This application of the *t*-test, called the 2-sample *t*-test, can be used in water quality management to test for compliance with water quality objectives (or standards or guidelines) or for assessing the effectiveness of water pollution control measures. The test may be used for a wide variety of purposes as it relies on the universal property that a deviation divided by a standard deviation is distributed as *t* with $n - 1$ degrees of freedom; and that the variance of a difference is the sum of the individual variances:

$$\sigma^2_{\bar{x}_1 - \bar{x}_2} = \sigma^2_{\bar{x}_1} + \sigma^2_{\bar{x}_2}$$

For different sample sizes, the variance of the difference will be:

$$\sigma^2_{\bar{x}_1 - \bar{x}_2} = \sigma^2_{\bar{x}_1}/n_1 + \sigma^2_{\bar{x}_2}/n_2 = \sigma^2\left(\frac{1}{n_1} + \frac{1}{n_2}\right) \quad \text{for equal variance}$$

The test is based on:

$$t = \frac{(\bar{x} - \bar{y}) - (\mu_x - \mu_y)}{s_{x-y}} = \frac{(\bar{x} - \bar{y}) - (\mu_x - \mu_y)}{s\sqrt{(1/n_x + 1/n_y)}}$$

The null hypothesis is that $\mu_x = \mu_y$, so $(\mu_x - \mu_y) = 0$; therefore:

$$t_{test} = \frac{(\bar{x} - \bar{y})}{s_w\sqrt{(1/n_x + 1/n_y)}} \qquad \text{with:} \, \nu = n_1 + n_2 - 2$$

where: $\bar{x}$ = the mean concentration of the water quality variable for the first data set

$\bar{y}$ = the mean concentration of the water quality variable for the second data set

n_x and n_y are the number of samples in the two data sets

s_w = the pooled standard deviation = √(pooled variance)

t ≡ the Student's *t* statistic

and the pooled variance is:

$$s_w^2 = \frac{(n_x - 1)s_x^2 + (n_y - 1)s_y^2}{n_x + n_y - 2}$$

The two data sets have means which are not statistically different when:

$$t_{test} \langle \, t_{(1-\alpha)[\nu]}]$$

Example

Given two different years of water quality data collected at different but equal time intervals, use the two-sample t-test to determine if the means are significantly different at 95 per cent and 90 per cent confidence levels.

Data: $n_y = 19$ $n_x = 22$
        $\bar{y} = 8.2$ mg l^{-1} $\bar{x} = 9.4$ mg l^{-1}
        $s_y^2 = 5.4$ (mg l^{-1})^2 $s_x^2 = 6.2$ (mg l^{-1})^2

$$s_w^2 = \frac{(19-1)5.4 + (22-1)6.2}{19+22-2} = 5.83$$

$$t = \frac{\bar{x}-\bar{y}}{s_w\sqrt{(1/n_x + 1/n_y)}} = \frac{9.4-8.2}{2.41*\sqrt{\left(\frac{1}{22} + \frac{1}{19}\right)}} = 1.59$$

From the Student's t tables:
   $t_{0.05[39]} = 2.02$
   $t_{0.10[39]} = 1.68$

The t_{test} value of 1.59 is lower than both critical values and, therefore, there is no significant difference between the two means, at either the 90 per cent or 95 per cent confidence level (i.e. the null hypothesis is accepted).

Mann-Whitney U-Test (also called the Wilcoxon two-sample test)

This is the equivalent non-parametric procedure for the testing of significant differences between two data sets. The method does not require that the data represent precise measurements. As long as the data can be systematically ordered the test is applicable. For example, the times of a series of events could be measured, or the order in which they occurred could be recorded. As with many non-parametric methods, the data have to be arranged in rank order sequence. The lowest value of the smaller of the two sample sizes is allocated to the first rank. The data are then systematically allocated to the subsequent ranks, maintaining the distinction between the two samples.

Example

The two samples {25;31;37;42} and {28;29;37;37;38;39}, would be allocated as:

Rank:	1	2	3	4	5	6	7	8	9	10
Sample 1:	25			31	37					42
Sample 2:		28	29			37	37	38	39	

In the event of tied values, either within or between the samples, the average of the ranks is assigned. In the above example, there are three values 37, i.e. in R_5 in sample 1, and R_6, R_7 in sample 2. Each sample is, therefore, assigned $(5+6+7)/3 = 6$. Therefore, sample 1 has Ranks, R_1: 1;4;6;10; sample 2 has R_2: 2;3;6;6;8;9.

When $n \leq 20$ the U statistic is computed as follows:

$$U = n_1 n_2 + \frac{n_1(n_1+1)}{2} - \sum_{i=1}^{n_1} R_{1,i} \quad \text{or} \quad U = n_1 n_2 + \frac{n_2(n_2+1)}{2} - \sum_{i=1}^{n_2} R_{2,i}$$

where: n_1 is the size of the smaller sample;
         n_2 is the size of the larger sample.

In this example, with $n_1 = 4$ and $n_2 = 6$; $\Sigma R_1 = 21$, $\Sigma R_2 = 34$: $U_{largest} = 13$, and $U_{smallest} = 11$

Then calculate: $U_1 = n_1 n_2 - U_{smallest}$

therefore: $U_1 = 24 - 11 = 13$

The maximum of $\{U_{largest}$ and $U_1\} = U_{max}$ is chosen and compared with critical values of U tabulated in statistical tables (the level of significance $\alpha/2$ is used because both tails of the U distribution are being used).

If: $U_{0.5\alpha[n_1,n_2]} \rangle U_{max}$ the probability is that differences in the values in the two data sets are a result of random variations, i.e. the null hypothesis H_0 is true, at a given level of confidence $(1 - \alpha)$.

When $n > 20$, a Student's t is calculated instead, and compared with tabulated values. If there are no tied ranks:

$$t_{test} = \frac{U_{max} - \frac{n_1 n_2}{2}}{\sqrt{\frac{n_1 n_2 (n_1 + n_2 + 1)}{12}}}$$

If there are tied values in the ranks, this formula is modified to:

$$t_{test} = \frac{U_{max} - \frac{n_1 n_2}{2}}{\sqrt{S}}$$

where:

$$S = \frac{n_1 n_2 \left\{ (n_1 + n_2)^3 - (n_1 + n_2) - \sum_{j=1}^{m} T_j \right\}}{12(n_1 + n_2)(n_1 + n_2 - 1)}$$

and T is a function of the number of values in the jth group of ties t_j, such that $T_j = t_j^3 - t = (t_j - 1)t_j(t_j + 1)$. For example, if there were three values in a group of ties, $T = 24$. All such groups are computed and summed. For the simple example above, there is only the one group of three tied values (i.e. the value 37). The value t_{test} is then compared to $t_{0.5\alpha[\infty]}$ in the statistical tables. If: $t_{0.5\alpha[\infty]} > t_{test}$, then H_0 is again accepted.

10.5   Basic graphical methods

Water quality data can be presented in either tabular or graphical form. Although graphs are not as good as tables in conveying details of quantitative information, such as the exact values of data points, they are better for showing the general behaviour of a data set. Graphical presentation of data has several other advantages, including:

- Large data sets can be shown effectively within a small area.
- The qualitative, e.g. patterns, correlations and trends, as well as some quantitative aspects, e.g. outliers, are highlighted.
- Graphs are more effective in capturing the attention of the reader, especially for non-technical audiences. Many people are much more receptive to visual presentation rather than written information.

The use of graphs and illustrations to the maximum possible extent is strongly encouraged for water quality data presentation and interpretation. The use of computers makes the production of graphs available in a way which was previously impossible. This section discusses the general principles for graphical presentation of water quality data and presents some of the more common graphs used in this field. There are three basic levels of graphical presentation:

- Total reproduction of all the data, usually for technical purposes.
- Selected graphs and maps interpreting water quality for specialised uses.
- Summary graphical presentations for non-specialists, educational purposes, etc.

Summary presentations are usually designed in relation to their specific purpose and are, therefore, not discussed in detail here.

10.5.1 Basic rules for graphical presentation of water quality data

When using graphs, it is essential that the data are presented clearly and accurately. The key to producing good graphs is to examine the data and to decide what features and relationships should be shown. The types of information which can be shown graphically vary widely, but the most common in the water quality field are:

- Time series: the values of water quality variables are plotted against the date the respective sample was collected, in chronological order (temporal variation).
- Seasonality of the data (temporal variation).
- Statistical summary of water quality characteristics.
- Correlations between two or more variables.
- Spatial and temporal comparisons of water quality variables.
- Ancillary information.

Choosing the most suitable graphical presentation is an iterative process. The first step is to determine exactly what the illustration should show. Then the type of graph most appropriate to the purpose and the data set can be selected. For example, logarithmic scales are recommended when it is important to understand per cent change, exponential factors or when reported values cover more than one order of magnitude. The types of illustrations that are used in presenting water quality data can be grouped into: line, pie and bar graphs, box plots, maps and combinations thereof.

After choosing the appropriate type of graph, it is important to select the size, weight of lines, size of lettering, symbols and colour (if any) so that the graph is clear, easily read and understood. By choosing appropriate visual presentation tools, graphs can be designed to attract and hold the attention of the reader. However, over-elaborate or complex graphs will rapidly lose the attention of all but the most dedicated readers.

Planning a good graph

The most important elements to consider in preparing a graph are:

- Content: The graph should contain the minimum elements consistent with imparting the desired information. Over-complexity will obscure the important features. If necessary, it is usually better to use linked, but separate, graphs for complicated subjects.
- Size of graph: The size should be selected with careful consideration for ease of construction and of reading. It is preferable that a graph fits into a page or less. Graphs that extend over two pages become difficult to read and lose accuracy. Graphs which use an unusual size of paper may be difficult to reproduce and to handle.
- Title/legend: A graph should always have a title or legend which is explanatory but not too long. The legend should contain sufficient information to interpret the graph without having to read all the surrounding text.
- Axes labels: The axes should always have labels (including the units of measurement) and the labels should be parallel to the respective axis. It is also acceptable to use common computer graphics software which labels axes horizontally only.
- Axes numbers: These should be large and easy to read.
- Scale: The scale used for the axes should be chosen such that the graph is an appropriate size and interpolation is convenient. It is recommended that the same scale be maintained when multiple charts are used to compare trends.
- Scale breaks: These should be used only when necessary.
- Labels: The lines of line graphs, bars of bar graphs, and sectors of pie graphs should also be labelled. Whenever possible, the labels should be placed in a separate area. Any notes or markings in the data region of a graph should be kept to a minimum.
- Bar or column widths: These should be greater than the spaces between them.
- Grid lines: Grid lines are helpful when conveying quantitative information. These should always be thinner than the lines of the actual graph.
- Shading and patterns: Shading on divided bars or columns should be ordered from darkest at the bottom to lightest at the top, unless clarity is lost because of insufficient differences between adjacent levels of shading. In such situations different types of patterns might also be necessary.

10.5.2 Time series plots (variable versus time)

The time series plot is the most basic graph of water quality data. It consists of plotting the values of one or more water quality variables (y axis) against the date of sampling (x axis). This type of graph assists both the data inter-preter and the reader to visualise changes in the water quality variables over a period of time. In addition to showing the general distribution of the data, it can point out possible trends, cyclical variations and outliers. Connecting the points, as in Figure 7.7, helps this visualisation. Without connecting the points, the time course is much less evident; particularly if the data are relatively sparse.

The time, on the x axis, can be represented in any time units ranging from minutes to years (e.g. Figures 4.4, 4.12, 4.14, 6.12, 6.20, 7.18, 8.6, 9.14) depending on the available data and the purpose of the graph. For example, an x axis divided in weeks or months will be natural for plotting the data col-lected from a water quality monitoring programme with weekly or monthly sampling schedules. The units of the y axis are chosen in relation to the water quality variable being represented, e.g. mg l^{-1}, μg l^{-1} or ng l^{-1} when plotting concentrations, kg hr^{-1} or t day^{-1} when plotting loads, or no. ml^{-1} or no. per 100 ml when plotting bacterial counts.

In Figure 4.4 two water quality variables, discharge and suspended sediments, have been plotted on the same graph, using the same time scale for the x axis, but different scales for the y axis. This graph shows not only the change with time of the two variables, but also allows some interpretation of the interaction between them, particularly in relation to the exceptional storm event causing increases in discharge. When time series information is averaged (e.g. monthly averages for a complete year) the average values can be represented as bars spanning the appropriate width on the time scale as in Figure 6.5.

Changes in water quality variables over time at a series of sampling loca-tions can be represented in a three-dimensional plot. These plots are particularly useful where horizontal and vertical variations have manage-ment implications (such as with respect to water withdrawal points along a river or at different depths in a reservoir). An example of a three-dimensional representation of light penetration (and hence an indication of suspended material, particularly planktonic algae) in a reservoir is given in Figure 10.9.

10.5.3 Seasonal diagrams

The seasonal variation of water quality variables is a common, natural phenomenon. Probably the most important factor governing the seasonality of water quality is the cyclical nature of climatic changes, but that may only be obvious in water quality studies covering two or more years.

Figure 10.10 shows different ways of plotting several years of water quality data such that the seasonality of the data is emphasised. In Figures

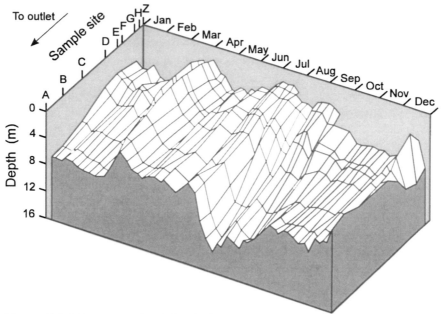

Figure 10.9 A three-dimensional plot of light penetration in the Wahnbach reservoir, Germany, showing horizontal and vertical variations throughout 1988 (After LAWA, 1990)

10.10A, 10.10B and 10.10C the data from 20 to 30 year periods are grouped by month to show the range of variation encountered in any one month; using percentiles (Figure 10.10A), range-lines (Figure 10.10B) and confidence intervals (Figure 10.10C). Figure 10.10D combines water quality data and flow information to explain the seasonal variations in major ion concentrations. The arrows are used to suggest seasonal progressions. The K^+, and HCO_3^- reach their highest annual concentrations in the December to March period, i.e. the time of the lowest discharge. During this period, the groundwater flow is expected to be the major ion source. Discharges increase dramatically in April and ion concentrations consequently decline, except for potassium. Concentrations continue to decline, through dilution and flushing of accumulated salts with increases in discharge. Declining discharges in the autumn are accompanied by increasing levels of ions which return to base-flow concentrations in December.

Temporal variations in lakes are commonly illustrated as time-depth contours (isopleths) (Figures 7.19 and 8.7) showing seasonal variations throughout the depth of the lake. The distance between isopleths represents the rate of change of the plotted variable with depth or time.

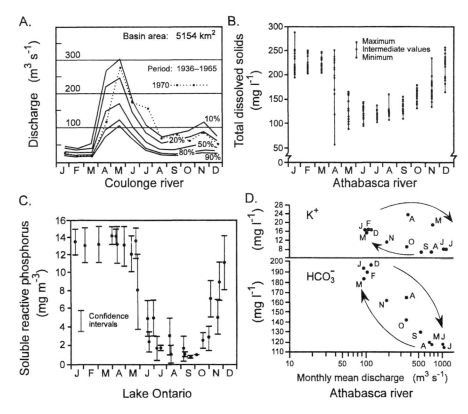

Figure 10.10 Different graphical presentations of water quality data illustrating seasonal variations. **A**. The use of percentiles; **B**. The use of range-lines; **C**. The use of confidence intervals; **D**. Seasonal progression illustrated by plotting concentration against discharge. (After Dobson, 1984; Blachford *et al.*, 1985; Annuaire Hydrologique, 1987, 1988, 1989)

10.5.4 Spatial plots (variable versus position)

In some assessments it is the relative positions of the sampled sites and their results which is of primary interest.

Figure 7.15 shows the vertical profiles of average silica concentrations in Lake Geneva for each month. The plots use the x axis for the variable of interest, with the y axis for the depth. This maintains the natural y axis orientation to the sampling position. Figure 7.17A uses a similar depth orientation for temperature data from Lake Ontario. Figure 7.12 shows the results of Zn analyses on 1 cm depth sections of a vertical sediment core. The centimetre intervals are represented as bars. Similar bars may be used instead of points on the graph when the precise depth of the sample is only accurate to a particular depth interval. Points are more suitable when the sample depth is only a small proportion of the depth interval between samples as in Figure 4.20. Figure 7.12 is also designed to show more detail of the relative speciation of the variable studied. Figure 4.20 shows how four different sites may be compared by over-lay plotting.

Although vertical profiles are used mostly for lake and reservoir waters and sediment cores they can also be useful for groundwater assessments (see Figure 9.18).

Figures 4.16, 5.3, 5.4, 5.8 and 6.21 illustrate longitudinal profiles of water quality variables in rivers. By plotting the data this way, the downstream development of any changes may be shown. This is particularly clearly illustrated if the changes are shown with respect to the positions of potentially major influences on the water quality, such as major cities, tributaries or industrial effluents (Figures 5.4 and 5.8).

Maps and cross-sections

Isopleths can also be used on maps to illustrate the spatial distribution of variables, particularly in groundwaters. Figures 9.13 and 9.26 show horizontal distribution and Figures 9.16 and 9.27 show vertical distribution in cross-sections. Isopleths may also be used to show the horizontal distribution of variables in lake water and sediments, provided sufficient sediment samples have been taken (e.g. Figures 4.11, 4.18 and 7.20). Such plots are also particularly useful for illustrating the extent of a plume of a contaminant from a point source.

Regional variations in water quality can be illustrated on maps by the use of different colours representing different indices of quality for river stretches, as in Figures 6.34 and 6.35.

10.5.5 Summarising water quality characteristics

Box plots are a good way to illustrate the distribution and summary statistics of a data set and are also useful in detecting outlying values. The convention most commonly used in producing a box plot is to show the lowest and highest values of a data set, together with the first quartile (lower 25th), the second quartile (median 50th), and the third quartile (upper 75th) values. Different versions of the box plot are illustrated in Figure 10.11.

In the simple, original form (Figure 10.11A) the box is drawn using lower and upper quartiles as its limits. The minimum and maximum values are connected to the box by using two straight lines called "whiskers". In Figure 10.11B, the whiskers are drawn up to (only) 1.5 times the length of the box. Observations that fall beyond the whiskers, up to 3 times box width in either direction, are shown using a different symbol. Values beyond these limits are

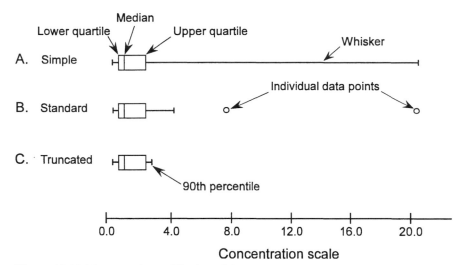

Figure 10.11 Three versions of the box plot

shown by still another symbol. Sometimes used, but not recommended, is the truncated box plot (Figure 10.11C). In this graph, instead of showing the minimum and maximum values, the whiskers are drawn only up to the 10th and 90th percentiles. Therefore, the lower and upper 10 per cent of the data are not illustrated. Figure 10.12 shows a series of box plot graphs presenting the variation in the groundwater quality of various aquifers, thus allowing a quantitative comparison of data from the different locations.

Rosette diagrams are used to illustrate the ionic composition of water, and examples are contained in Figure 10.13 for the relative major ion concentrations for selected sites in the Peace River sub-basin, Canada. Theoretically, a rosette diagram can be used to represent the complete ionic composition of water, i.e. major ions, trace elements, phosphorus and nitrogen species, but in practice the diagram has been used to illustrate major ions, only. The major ion concentrations at a specific location or over a wide area, such as a river basin, can be illustrated by using these diagrams. The values can be individual measurements or the means of observations taken over a relatively long period of time, e.g. years. The conditions that apply in using the mean values to summarise a data set (see section 10.4.3) also apply here.

In rosette diagrams one half of a circle is used to indicate per cent equivalent concentrations of the four major cations (Ca^{2+}, Na^+, K^+ and Mg^{2+}), and the other half is used to show the per cent equivalent concentrations of the four major anions (HCO_3^-, CO_3^{2-}, Cl^- and SO_4^{2-}). Each half circle is divided into four equal sectors and the area of each sector corresponds to 25 per cent relative concentration. The length of the bisecting radius in each

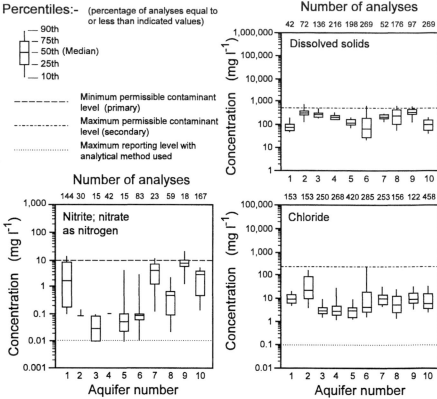

Figure 10.12 Use of box plots to summarise groundwater quality data for ten different aquifers in Maryland, USA (After Wheeler and Maclin, 1987)

of the eight sectors represents the equivalent concentration of the respective ion. The concentration in meq l⁻¹ (milliequivalents per litre) of the individual ions is calculated as a percentage of the total anion or cation concentration. Values of relative concentration up to 25 per cent lie within the circle, whereas values greater than 25 per cent extend beyond the circle.

Trilinear diagrams (e.g. Figure 9.11) are similar to rosette diagrams and can be used to represent the major ion chemistry at a specific location or over a wide area, such as a river basin. Average values over long periods of time, e.g. years, are also best to construct these diagrams. In Figure 9.11 the left triangle has been used to represent cations, the right triangle for anions, and the central diamond-shaped field for the total, or combinations, of anion and cation concentrations.

To obtain a trilinear diagram, the relative concentrations of anions and cations are calculated and plotted as points in the right and the left lower triangles, respectively. The points are then extended into the central,

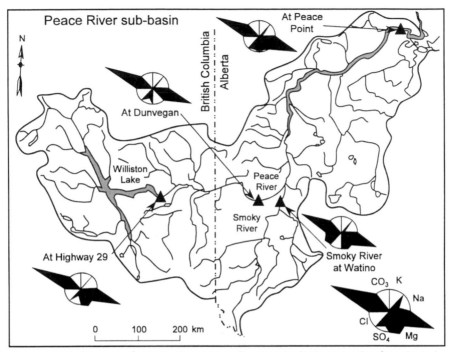

Figure 10.13 The use of rosette diagrams to illustrate the ionic proportion for selected sites in the Peace River sub-basin, Canada (After Blachford *et al.*, 1985)

diamond-shaped field by projecting them along lines parallel to the edges of the central field. The area of the circles in the central field is proportional to the dissolved solids concentrations of the particular type of water. Increasing chloride concentrations such as those due to urban/industrial activities move the points towards the right. An increase in sulphate concentration due, for example, to acid rain or acid mine drainage moves points upwards and to the right. In the diamond-shaped area, the calcium-bicarbonate freshwaters plot towards the left and saline waters towards the right. Trilinear diagrams have also been used to illustrate changes of ionic concentrations in water as stream flow volume changes at several sampling points, and similarities and differences in the composition of water from certain geologic and hydrologic units. Other forms of diagram used to represent ionic composition can be found in Hem (1989).

Bar graphs can be used effectively to portray summary statistics, e.g. mean values. Examples are shown in Figures 4.6, 4.17, 5.8, 7.9, 7.21 and 9.20. By using multiple bars, Figure 4.17 shows concentration and its time variation at seven different sites, all illustrated in a single figure. Bar in-fills help to distinguish the various periods to which the mean values apply.

A simple method of showing proportions of various water quality variables is to use a "pie" diagram, which is a circle segmented to represent the various proportions. Figure 4.9 contains an example of this approach. Although good at giving an appropriate impression of the data, quantitative information is less well conveyed. One approach is to scale the circle so that some indication of the actual magnitudes involved may also be given. This method is also used in Figure 4.9.

Figures 6.10 and 6.14 show probability distributions for a variety of water quality variables. These illustrate the proportions of the samples which exceed particular concentrations. They also give insight into the form of the underlying frequency distributions (see section 10.4.4), and hence the types of statistical techniques likely to be appropriate to the data.

10.5.6   Data correlations

The correlation between discharge and concentration should always be tested for quantitative measurements in rivers since it gives important information on the origin and/or behaviour of the variables measured (see Figure 6.11 and section 6.3.3). In Figure 10.4, the discharge data are shown plotted on a logarithmic scale. Although a straight regression line is produced, the inter-relation between the discharge and alkalinity is actually non-linear. Figure 6.13B also shows similar data plotted with linear scales. This illustrates more graphically the non-linearity of the data. In other cases such plots show a large scatter of points, which indicates that no correlation exists between the variables plotted.

In addition to the temporal correlation between discharge and sediment transport illustrated in Figure 4.4, correlations between water quality variables can be more directly shown by plotting one variable against the other, as in Figures 6.13C, 7.13, 10.4 and 10.14. The linear scales, and virtually straight lines of Figure 10.14, indicate linear correlations between the specific conductance and the three quality variables considered.

10.5.7   Spatial and temporal comparison of water quality variables

Apart from the three-dimensional plots as illustrated in Figure 10.9, spatial and temporal comparisons of water quality variables often rely on the combination of several of the presentation techniques discussed above.

Bar graphs are effective in illustrating how the concentrations, or other measurements (e.g. loads), of various water quality variables compare with each other, and also in showing spatial and temporal variations. Figure 4.17 shows clearly the changes that occurred in the copper content of the

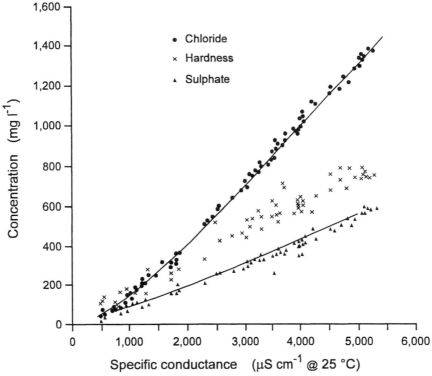

Figure 10.14 Regression plot to show the relationship of specific conductance to chloride, hardness and sulphate in the Gila River, USA (After Hem, 1989)

suspended matter over four time periods for each of the stations studied whilst still allowing comparison between the different river stations.

Figure 7.21 shows the combination of a map and a bar graph indicating the range of dioxin residues in bird's eggs at various sites around the Great Lakes. Figure 10.13 is similar except that instead of using bars, rosettes have been used to show the mean values of major ions at several locations in a river basin. Figure 6.25 shows PCBs in the deposited sediments of major inflows to Lake Geneva, by pointing to the inflow position and indicating the concentration by the diameter of the circles. Figure 7.22 illustrates both the spatial and temporal trends of sediment mercury in Lake Ekoln by combining the contour map with the temporal sediment history at selected sites.

The increased availability of computer based geographic information systems (GIS) has led to the development of a large number of applications in the field of water quality. Over the coming years the range of applications is expected to grow significantly. Two examples of GIS output are shown in

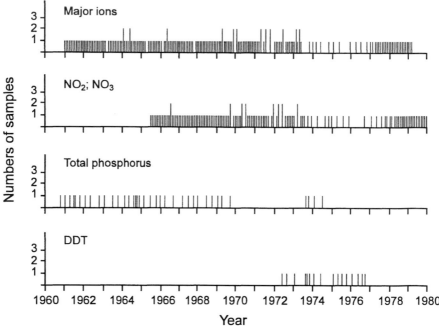

Figure 10.15 Graphical illustration of the frequency of sampling of selected variables in the Athabasca River, Canada (After Blachford *et al.*, 1985)

Figures 7.10 and 7.11. The first of these two figures is a map of soil sensitivity to acid deposition in Eastern Canada. The second figure, shows the wet sulphate deposition for 1982–1986 over the same regions. The use of GIS allows the overlaying of such mapped data in order to identify areas of co-incident high soil sensitivity and sulphate deposition, i.e. the southern part of the mapped areas. Geographic information systems are described in more detail in section 10.7.2.

10.5.8   Ancillary information

This group includes all information that may be relevant to understanding the results of a water quality monitoring programme, such as sampling information.

Figures 4.10, 9.27 and 10.15 give details of the sampling ranges for some water quality studies. Figures 4.10 and 9.27 give position information, in plan and profile respectively. Figure 10.15 shows the frequency and duration of sampling for a few selected water quality variables. The details of sampling patterns are necessary to understand and interpret the data.

Figure 10.16 is an example of a time-discharge curve and shows the percentage of water quality samples that were taken in 10 per cent discharge intervals. This illustrates whether or not the data set being examined is

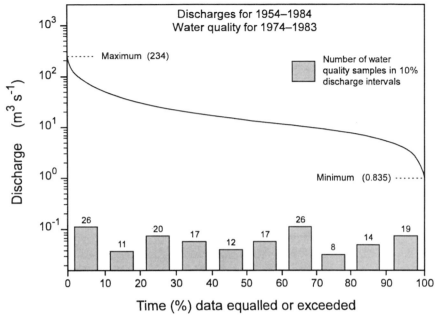

Figure 10.16 An example of a time-discharge curve indicating the percentage of water quality samples taken in 10 per cent discharge intervals (After Demayo and Whitlow, 1986)

representative of all discharge regimes. This is particularly important when describing the state of a river system, calculating summary statistics and making long-term projections. When considering river flux the sampling effort should ideally be proportional to the discharge distribution, i.e. more biased towards the rare events of high water discharge.

10.6   Data analysis and interpretation methods

The techniques described in this section have more stringent data requirements, e.g. frequency and length of sampling period, when compared with the methods described in sections 10.4 and 10.5.

10.6.1 Trend analysis

One of the most widely used approaches when interpreting water quality information is the analysis of trends. Trends require regular measurements over a long time span, such as monthly measurements over periods of years or annual measurements over decades or more. In trend analysis the regressor variable x, whether it be sampling location or time, is controlled by the investigator to ensure it is measured with negligible error (i.e. it is a non-random variable). If both the x and y variables are random, correlation analysis (see section 10.6.4) should be used instead.

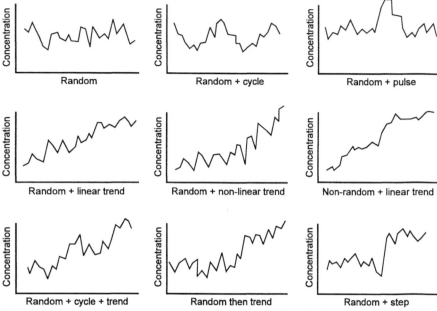

Figure 10.17 Graphical representation of different forms of trends (Based on Gilbert, 1987)

Before proceeding with any trend calculation it is recommended that a scatter plot of the data set is produced. Figure 10.17 shows some of the possible forms that a scatter plot can take. In some cases this plot may indicate the presence, and even the type of trend (e.g. linear, curvilinear, step). Two main types of trends are usually identified before applying statistical analysis to the data: monotonic or step trends. The former is a change in one direction only and the latter assumes data collected before a specific time are from a distinctly different data population than the data collected after that time. The trends analysis procedures applied also depend on whether the data set is censored or not (see section 10.3.2). The recommended procedures are given in Table 10.5 and discussed further in Hirsch *et al.* (1991) and Loftis *et al.* (1989). For annual sampling the Mann-Kendall test is recommended and for normally distributed and independent data sets, the trend line can be calculated using the least squares method. For seasonal sampling either the Seasonal Kendall method or analysis of covariance (ANOCOV) on ranks are recommended. It is beyond the scope of this book to show the details of these methods as they are described in many standard statistical texts (e.g. Sokal and Rohlf, 1973; Snedecor and Cochran, 1980). A variety of computer software packages such as SAS (SAS Institute Inc., 1988) and SYSTAT (Wilkinson, 1988) can also perform trend analysis on data sets.

Table 10.5   Options for testing for monotonic and step trends in uncensored and censored water quality data

	Not flow adjusted	Flow adjusted
Uncensored data: Monotonic trends		
Fully parametric	Regression of *C* on time and season[1]	Regression of *C* on time, season[1] and *Q*
Mixed	Regression of deseasonalised[3] *C* on time	Seasonal Kendall[2] on residuals from regression of *C* on *Q*
Non-parametric	Seasonal Kendall[2]	Seasonal Kendall[2] on residuals from LOWESS of *C* on *Q*
Uncensored data: Step trends		
Fully parametric variance	Analysis of co-variance *C* on season and group (before and after)	Analysis of co-variance *C* on season, *Q* and group
Mixed	Two-sample *t*-test on deseasonalised[3] *C*	Seasonal Rank-sum[4] on residuals from regression of *C* on *Q*
Non-parametric	Seasonal Rank-sum[4]	Seasonal Rank-sum[4] on residuals from LOWESS of *C* on *Q*
Censored data: Monotonic trends		
Fully parametric	TOBIT regression of *C* on time and season[5]	TOBIT regression of *C* on time, season[5] and *Q*
Non-parametric	Seasonal Kendall[2]	No test available
Censored data: Step trends		
Fully parametric	TOBIT analysis of covariance of *C* on season[5] and group	TOBIT analysis of covariance of *C* on season, *Q* and group
Non-parametric	Seasonal Rank-sum[4]	No test available

C Concentration
Q Discharge (streamflow) or a transformation of discharge
LOWESS Locally weighted scatter plot smoothing
[1] Regression on season using a periodic function of time of year
[2] Seasonal Kendall test is the Mann-Kendall test for trend done for each season (the Seasonal Kendall test statistic is the sum of the several test statistics)

[3] Deseasonalising can be done by subtracting seasonal medians
[4] Seasonal Rank-sum test is the Rank-sum test done for each season (the Seasonal Rank-sum test statistic is the sum of the several test statistics)
[5] TOBIT Regression on season is using a periodic function of time of year

Source: After Hirsch *et al.*, 1991

10.6.2   Time series analysis

Time series analysis encompasses a wide range of mathematical techniques designed for preliminary data exploration, detection of trends and forecasting. It is beyond the scope of this section to discuss the numerous methods available. Instead the reader is referred to Box and Jenkins (1970), Young (1984) or Hipel (1985) for a comprehensive treatment of time series analysis. This section discusses, briefly, the most prominent feature of time series data: autocorrelation. It is recommended that seasonality, which is another common feature of time series data, is analysed using the Seasonal Kendall test, as this does not require normality, and allows the inclusion of missing, equal and ND (not detected) results; although data collected over a few years are usually required.

Autocorrelation

The presence of autocorrelation in a series means that a data point is significantly affected by the adjacent points. The basis for the series may be time or space. Testing for autocorrelation is important because, if it is present, conventional statistical analyses such as t or F-tests lose their validity. Autocorrelation is difficult to detect if there are any missing values or if the data set is small ($n < 50$).

There are a wide variety of methods for calculating the presence of autocorrelation. These can be found in Box and Jenkins (1970), Montgomery and Reckhow (1984) or Gilbert (1987). One of the simpler methods is to plot the difference between the observed and calculated values (the residuals) from a regression analysis. Regular patterns indicate the presence of autocorrelation and a lack of any discernible pattern would indicate the absence of autocorrelation.

10.6.3 Regression analysis

Section 10.4 was concerned with the description and testing of hypotheses for one variable only or univariate statistics. This section discusses bivariate and multivariate statistical methods which use statistics to make inferences about two or more variables, respectively. The particularly important methods are correlation and regression analysis. Although the two techniques are closely linked, they are intended for different purposes, and are often confused. Correlation is basically the study of the association between two or more functionally independent variables, whereas regression analyses the relationships between independent variables and their dependent variables as indicated below:

Purpose	y random, x fixed or decided	y random, x random
Dependence of one variable on another	Model I Regression	Model II Regression[1]
Association of variables	(Correlation only to establish r^2 proportion of variance explained by regression)	Correlation

Note[1]: Model II regression is most easily undertaken by the Bartlett three-group method.

In standard regression analysis it is generally assumed that the determining (independent) variable is measured without error. All errors of measurement are then assigned to the determined (dependent) variable. This standard case is known as a Model I regression. If, however, it is clear that there is a dependent relationship between two variables, but both are subject to measurement error, then standard (Model I) techniques are no longer applicable. This type of analysis is known as a Model II regression.

Although all realistic measurement is subject to error, it is fortunate that if choice can be exercised in the values of the independent variable used in the regression, then this has the effect of reducing to zero the mean associated errors. Model I regression analysis may then, validly, be applied (Mandel, 1964).

The relationship between two normally distributed variables x and y can be expressed mathematically in a variety of ways such as:

Linear:	$y = a + bx$
Non-linear:	$y = a + bx + cx^2 + \ldots$
Inverse:	$y = a + b/x + c/x^2 + \ldots$
Exponential:	$y = a.\exp(bx), y = a + b.\exp(x), \ldots$
Logarithmic:	$y = a + b.\ln(x)$
Trigonometric:	$y = a + b.\cos(x), y = a + b.\sin(x), \ldots$
Combinations of above, etc.	

By the use of such transformations, the regression of x and y is linearised. How well x and y fit one of these expressions can be assessed by calculating the correlation coefficient r or the relative mean square deviation $rmsd$. The rmsd value, which is calculated from the differences between the actual data points and the respective calculated values according to the mathematical expression(s) chosen, is a regression analysis procedure.

In the simplest case, the linear regression statistic represents the value y_0 when $x = 0$ and is also the incremental increase in y with incremental increase in x, i.e. $\delta y/\delta x$ the slope or regression coefficient of the regression line. The regression line always passes through the point $(\bar{x}, \bar{y})$, so $\bar{y} = \bar{y}_0 + b\bar{x}$. Each of the regression statistics have associated error statistics which allow them to be tested for significance, and have confidence limits assigned to them:

Statistic	Standard error	d.f.
Regression coefficient:	$s_b = \sqrt{\left\{ \dfrac{s_{y.x}^2}{\Sigma x^2} \right\}}$	$v = n - 2$

Sample mean:
$$S_{\bar{y}} = \sqrt{\left\{\frac{s_{y.x}^2}{n}\right\}} \qquad v = n-2$$

where: $s_{y.x}^2$ is the variance of y once regression is taken into account.

Prediction of the dependent variable is usually the aim of regression and it is essential to have some estimate of the reliability of such predictions. It is, therefore, necessary to be able to set confidence limits on predictions, so the standard errors of predicted values are also required:

Statistic	Standard error	d.f.
Estimated y_i:	$S_{y_i} = \sqrt{s_{y.x}^2 \left[\frac{1}{n} + \frac{(x_i - \bar{x})}{\sum x^2}\right]}$	$v = n-2$
Predicted y_i:	$S_{y_i} = \sqrt{s_{y.x}^2 \left\{1 + \frac{1}{n} + \frac{(x_i - \bar{x})}{\sum x^2}\right\}}$	$v = n-2$

where: estimated y_i is the local, average y_i for the particular x_i
the point $\bar{y}_i, \bar{x}_i$ is on the regression line, and
predicted y_i is a particular y_i for x_i.

The particular feature of these formulations is the appearance of the value $(x_i - \bar{x})$ in the standard errors. The standard error, therefore, increases the further x_i is from $\bar{x}$. As a result the confidence interval boundaries are not parallel to the regression line, but diverge further and further from it. The further the independent variable x_i is from its mean, the worse the reliability of the regression becomes.

Example
A pilot study was conducted to determine whether chloride concentrations in water were related to effluents from a pulp and paper mill using bleaching processes. It was hypothesised that if chloride concentrations are a function of the effluent discharge then downstream concentrations should decrease with distance from the mill. The hypotheses are:
   H_0: $p = 0$ There is no trend in the Cl^- concentrations as a function of distance.
   H_A: $p < 0$ There is a trend in the Cl^- concentrations as a function of distance.
   To test this hypothesis water samples were collected for chloride analysis. The river reach was divided into 10 m segments. Random sampling without replication was then conducted. Least squares regression analysis was used on the data set to define the functional relationship between distance and Cl^- concentrations.
   The chloride concentrations and distances downstream are listed in Table 10.6. A plot of chloride concentration versus distance from the pulp mill approximates a straight line, and so the computer package SYSTAT was used to perform a linear, least squares regression on the data set (Table 10.6). The equation obtained (significant at the 99.9 per cent confidence level) was: C = 995 – 0.624 * X

Table 10.6   Testing for a longitudinal trend: least squares regression analysis of chloride concentrations in surface waters with distance downstream of a pulp mill using bleaching processes (for further details see text)

Sample No.	Distance from mill (m)	Cl⁻ concentration (mg l⁻¹)
1	10	1,131
2	160	980
3	780	567
4	630	604
5	220	740
6	350	780
7	240	806
8	510	697
9	130	865
10	300	708

Dependent variable: Chloride   *No. of Samples:* 10   *R:* 0.879

Variable	Coefficient	Std. error	T	P(2-tail)
Constant	995.6	48.2	20.67	0.000
Distance	−0.624	0.119	−5.224	0.001

where C is the chloride concentration in mg l⁻¹ and X is the distance downstream in metres. The regression also shows that there was a highly significant relationship between Cl⁻ and distance downstream.

10.6.4 Correlation analysis

In water quality studies correlation analysis is used to measure the strength and statistical significance of the association between two or more random water quality variables. Random in this case means that the variables are not under the control of the investigator and are, therefore, measured with an associated error. An example of a random variable is the concentration of calcium in the water column. A non-random variable would be, for example, sampling time or location. Correlation analysis is also used as a "Goodness of Fit" test by determining how closely a regression equation fits the sample points.

Calculation of the correlation coefficient r is available on most commercially available statistical packages. However, the capabilities of these packages to calculate r and perform other statistical analyses can be very different. Some of the factors which should be considered when choosing a statistical package are:

- Variety of mathematical expressions allowed: the more variety, including the option of setting one's own expressions, the better the package is.

- Maximum number of data points allowed: the more points the better the package is.
- The speed of calculation. Since some of the calculations can be quite lengthy, the package which performs them the fastest is preferable.
- Type (e.g. table, graph, etc.) and quality of the output. This is a very important consideration. The output should be easy to understand, of good quality and flexible so that the computer package can be used for various applications.
- User manual. This consideration is also very important because the manual can help the user select the correct statistical procedure, and aid in the interpretation of the output. User manuals that include statistical theory, descriptions of the algorithms and step-by-step examples are preferable.

Bivariate correlations — tests of association

The strength of the association between two random variables can be determined through calculation of a correlation coefficient r. The value of this coefficient ranges from -1 to 1. A value close to -1 indicates a strong negative correlation, i.e. the value of y decreases as x increases. When r is close to 1 there is a strong positive correlation between x and y, both variables increase or decrease together. The closer the value of r is to zero the poorer the correlation.

When the data sets meet the assumptions of normality and independence, Pearson's Product Moment Correlation Coefficient r is normally used. A non-parametric equivalent to r is Spearman's Rank-order Correlation Coefficient r_s.

A. Pearson's Product Moment Correlation Coefficient, r

Calculation:

$$r = \frac{n\sum_{i=1}^{n} x_i y_i - \sum_{i=1}^{n} x_i \sum_{i=1}^{n} y_i}{\sqrt{\left(n\sum_{i=1}^{n} x_i^2 - \left(\sum_{i=1}^{n} x_i\right)^2\right)\left(n\sum_{i=1}^{n} y_i^2 - \left(\sum_{i=1}^{n} y_i\right)^2\right)}}$$

In order to calculate r, Σx_i, Σy_i, Σx_i^2, Σy_i^2, $\Sigma x_i y_i$ and n, the total number of samples, are needed. The correlation coefficient r can be tested for the level of significance by the Student's t-test. Because r can be either positive or negative, the sign of the resulting test values is ignored.

The t test form is: $\qquad\qquad t_{test} = r\dfrac{n-2}{1-r^2}$ $\qquad\qquad$ with $v = n - 2$

Example
Table 10.7A shows the calculation of Pearson's Product Moment Correlation Coefficient to determine whether a statistically significant correlation exists between the dissolved oxygen (DO) and water temperature data given in the upper part of Table 10.7A. As the test value of t is much greater than even $t_{0.001[10]}$ the conclusion is that there is a negative correlation between dissolved oxygen and surface water temperature which is highly significant ($P < 0.001$).

B. Spearman's Rank Correlation Coefficient

Calculation:
1.   The x and y values are arranged in increasing order by x.
2.   The difference in rank, $x_i - y_i$, between the x_i and y_i values of each original pair is computed for the n pairs of the data set.
3.   The correlation coefficient r is calculated from:

$$r_s = 1 - \frac{6\sum_{i=1}^{n}(x_i - y_i)^2}{n(n^2 - 1)}$$

Example
Table 10.7B shows the calculation of Spearman's Rank Order Correlation Coefficient to test the strength of the relationship between the dissolved oxygen and temperature data in Table 10.7B. The test result is in excess of the tabulated critical value, and the conclusion is that a highly significant ($P < 0.01$) negative correlation exists between the two variables. Note: In this instance, the $r_{s,crit}$ values are only tabulated for $\alpha = 0.05$ and 0.01.

Multiple correlations

The multiple correlation coefficient $R(y,x_1,x_2,...)$ measures the strength of the relationship between a dependent variable and a set of independent ones. The coefficient of multiple determination R^2 is an estimate of the proportion of the variance of y that can be attributed to its linear regression on the observed independent variables $x_1, x_2,...,x_n$. Due to the complexity of the calculations, multiple correlations are now usually only performed on a computer. The output of these calculations normally consists of the coefficients for each x (i.e. the slopes), the y-intercept (i.e. the constant), R^2, and the results of an analysis of variance (ANOVA). It may also contain the results of t-tests on the standard regression coefficients generated for each independent variable.

Regression coefficients are used as a measure of the relative importance of each x variable in determining y. However, these coefficients should be interpreted carefully. If the independent variables are correlated with one another, which is often the case with water quality data, the results may instead merely reflect the order in which the variables were processed by the software program. In this case the first variable processed usually has the highest regression coefficient because, up to that point, only the constant

Table 10.7   Correlation test between dissolved oxygen (DO) and temperature in a river
A. Pearson's Product Moment Correlation Coefficient, r

No.(n)	DO (x)	x^2	Temp (y)	y^2	$x * y$
1	13.8	190.44	0.8	0.64	11.04
2	12.6	158.76	6.0	36.00	75.60
3	11.9	141.61	7.2	51.84	85.68
4	10.2	104.04	14.2	201.64	144.84
5	8.9	79.21	21.3	453.69	189.57
6	10.7	114.49	13.9	193.21	148.73
7	11.6	134.56	9.9	98.01	114.84
8	12.3	151.29	6.2	38.44	76.26
9	14.7	216.09	0.3	0.09	4.41
10	13.4	179.56	2.2	4.84	29.48
11	14.2	201.64	−0.2	0.04	−2.84
12	13.8	190.44	1.0	1.00	13.8

$$\sum_{i=1}^{n} x_i = 148.1 \quad \sum_{i=1}^{n} x_i^2 = 1{,}862.13 \quad \sum_{i=1}^{n} y_i = 82.8 \quad \sum_{i=1}^{n} y_i^2 = 1{,}079.44 \quad \sum_{i=1}^{n} x_i * y_i = 891.41$$

$$r = \frac{12(891.41) - (148.1)(82.8)}{\sqrt{\left(12(1{,}862.13) - (148.1)^2\right) * \left(12(1{,}079.44) - (82.8)^2\right)}} = -0.99$$

Significance test:

$$t_{test} = r\sqrt{(n-2)/(1-r^2)}$$

$$= -0.99\sqrt{(12-2)/(1-0.99^2)} = -22.14$$

$$|t_{test}| = 22.14 > 4.587 = t_{0.001[10]}$$

B. Spearman's Rank Order Correlation Coefficient, r_s

No. (n)	DO (x)	R_1	Temp (y)	R_2	$R_1 - R_2$	$(R_1 - R_2)^2$
1	14.7	1	0.3	11	−10.0	100.0
2	14.2	2	−0.2	12	−10.0	100.0
3	13.8	3.5	0.8	10	−6.5	42.25
4	13.8	3.5	1.0	9	−5.5	30.25
5	13.4	5	2.2	8	−3.0	9.0
6	12.6	6	6.0	7	−1.0	1.0
7	12.3	7	6.2	6	1.0	1.0
8	11.9	8	7.2	5	3.0	9.0
9	11.6	9	9.9	4	5.0	25.0
10	10.7	10	13.9	3	7.0	49.0
11	10.2	11	14.2	2	9.0	81.0
12	8.9	12	21.3	1	11.0	121.0

$$r_s = 1 - \frac{6*568.5}{12*143} = -0.99 \qquad\qquad \sum_{i=1}^{n}(R_1 - R_2)_i^2 = 568.5$$

Significance test: $|r_s| = 0.99 > 0.712 = r_{s0.01[12]}$

Table 10.8   Concentrations of calcium, chloride and sulphate ions in samples of sub-surface (1 m) drainage water

Specific conductance ($\mu S\ cm^{-1}$)	Calcium (mg l^{-1})	Chloride (mg l^{-1})	Sulphate (mg l^{-1})
156	18	12	5.6
260	36	10.1	4.6
361	47	22	8.2
388	51	25	9.6
275	37	13.3	5.0
286	35.7	13.4	8.7
321	43	14.5	5.2
317	43	14.3	4.8
309	40.6	16.9	8.5
313	37.9	19.5	8.8

term had been used to explain any of the variance in y. As more variables are included, there is less and less unaccounted-for variance, and the coefficients are subsequently smaller. For a more detailed discussion on this topic see Snedecor and Cochran (1980).

Example
Concentrations of calcium, sulphate and chloride ions are all associated with specific conductance in water. Using the data in Table 10.8, a multiple regression was used to describe the functional relationship between specific conductance (the dependent variable, y) and calcium, chloride and sulphate (the regressor or independent variables, x_1, x_2, and x_3). Following this, correlation analysis was used to test how closely the regression equation fitted the observed values of y, and whether or not a significant proportion of the total variance had been explained by the equation. The results of the multiple regression and multiple correlations analyses are as follows:

Dependent Variable: Specific Conductance   No. of Samples: 10   R: 0.996
Adjusted Squared Multiple R: 0.989   Standard Error of Estimate: 6.636

Variable	Coefficient
Constant	11.349
Chloride	1.027
Sulphate	3.926
Calcium	6.260

A pair-wise correlation analysis revealed significant correlations among the regressor variables. Consequently, the standard regression coefficients are not included in the output.
The equation generated by the multiple regression procedure is therefore:
    Specific Conductance = 11.35 + 6.260 ∗ Ca + 1.027 ∗ Cl + 3.926 ∗ SO$_4$
The multiple correlation coefficient R of 0.996 indicates that there is a very close association between the observed specific conductance values and the values calculated by the above equation. The statistical significance of R^2 can be evaluated with an F test:

$$F = \frac{(n-k-1)R^2}{k(1-R^2)}$$ where n = number of samples and k = number of variables.

In this case, $F = (10-3-1)0.993/(3(1-0.993)) = 283.8$ (p = 0.000). Therefore, R is highly significant.

10.6.5 Principal components analysis

Investigations of water quality often require that numerous variables be examined simultaneously. It is not unusual for data sets to consist of a large number of samples each containing, for example, the concentrations of major ions, organic compounds, and/or nutrients. Simultaneous analysis of the data with conventional methods would be, at best, very difficult. Plotting the data on one graph, if it were possible, would be extremely laborious and the resulting plot of doubtful value.

Principal Components Analysis (PCA) was developed to help summarise and facilitate the interpretation of multi-variable data sets (Gauch, 1982). The term "principal component" is based on the concept that of the n descriptors, x_1, x_2, ..., x_n describing the attributes of each sample, e.g. water quality variables describing the characteristics of the water column, there exists a fundamental group of independent descriptors which determine the values of all x points. These fundamental descriptors are called "components", with the most important of these termed "principal components". The components must meet two conditions (although departures are tolerated if PCA is used for descriptive purposes only):
- the descriptors are normally distributed, and
- they are uncorrelated.

Principal Component Analysis reduces the multi-dimensionality of a complex data set to two or three dimensions by computing principal components or factors. This computation is achieved by transforming the observations from each sample (e.g. concentrations of contaminants) into a "linear combination" of contaminant concentrations. Principal Component Analysis produces several important outputs:
- A correlation or covariance matrix. This is a summary table of the correlations for each pair of variables in the data set.
- *Eigenvalues*: the variances accounted for by the component.
- *Eigenvectors*: that specify the directions of the PCA axes.

Further details of principal components analysis can be found in Morrison (1965) and Guertin and Bailey (1970).

Other multivariate analysis techniques

Factor analysis is often preferred over PCA because it can perform a variety of multivariate analyses, including PCA. A "factor" is different from a "component". While PCA is a linear combination of observable water quality variables, factors can include unobservable, hypothetical variables. Factor analysis is available with some computer software such as the SAS package (SAS Institute Inc., 1988).

Table 10.9 The application of principal component analysis to the distribution of organic contaminants in Lake Ontario

	Factor 1	Factor 2	Factor 3	Factor 4
eigenvalue	6.4	3.4	2.1	1.3
% of total variability	37.7	20.3	12.1	7.4
Hexachlorobenzene (HCB)	−0.01	0.20	0.72*	0.21
Alpha-BHC (BHC)	−0.10	0.90*	0.03	0.03
Lindane (ALIN)	0.14	0.23	0.26	0.82*
Heptachlorepoxide (HEX)	−0.10	0.91*	−0.01	−0.02
cis-Chlordane (CHA)	0.05	0.68*	−0.01	0.11
Dieldrin (DEO)	0.06	0.77*	0.21	0.41
Polychlorinated Biphenyl (PCB)	0.06	−0.17	0.77*	−0.25
m-Dichlorobenzene (mDCB)	0.99*	−0.07	−0.01	0.07
p-Dichlorobenzene (pDCB)	0.83*	−0.13	0.33	0.07
o-Dichlorobenzene (oDCB)	0.97*	−0.14	0.05	−0.04
1,2,4-Trichlorobenzene (T4CB)	0.96*	0.08	−0.01	0.02
1,2,3-Trichlorobenzene (T3CB)	0.97*	−0.01	0.01	−0.02
1,2,3,4-Tetrachlorobenzene (TeCB)	0.96*	0.02	0.003	0.003
Pentachlorobenzene (QCB)	0.76*	0.07	−0.03	−0.15
1,3,5-Pentachlorobenzene (T5CB)	0.53*	−0.04	0.22	0.32
p,p'-DDE (DDE)	0.12	0.05	0.81	0.31
Endrin (END)	−0.13	0.11	−0.02	0.88*

* Significant associations
Source: R. Stevens and M. Neilson, unpublished data

Example
The distribution of contaminants in the Great Lakes was investigated by collecting water samples during four lake cruises. The samples were analysed for organochlorine compounds, polychlorinated biphenyls and chlorobenzenes. A correlation analysis performed on the data set showed a large number of inter-correlations among the 17 different substances. Due to the complexity of the data set, the PCA option of the Factor Analysis (using the package SAS) was used to find general patterns in the distribution of contaminants in each of the Great Lakes, and to reduce the number of variables required for subsequent analyses. The results of the PCA revealed that four factors accounted for 77.5 per cent of the total variance (Table 10.9). The eigenvalues also showed, for example, that 50 per cent of the explained variance was due to factor 1, and so on. Chlorobenzenes were all moderately to strongly correlated with the factor 1. Alpha-BHC, dieldrin, heptachlorepoxide and cis-chlordane were associated with the factor 2; p,p'-DDE, total PCBs, and hexachlorobenzene with factor 3; and endrin and lindane with factor 4. As a result future patterns of variation could be deduced by measuring only the principal components of factor 1.

In studies using so many simultaneously analysed variables, data analysis is virtually always complicated by multiple co-variance; and informative graphic presentation is extremely difficult. Factor (and PCA) analysis not only copes with the co-variance but, by associating groups of variables with the various factors, it becomes possible to provide a quantitative graphical summary of the study. Figure 10.18 shows how such a presentation provides

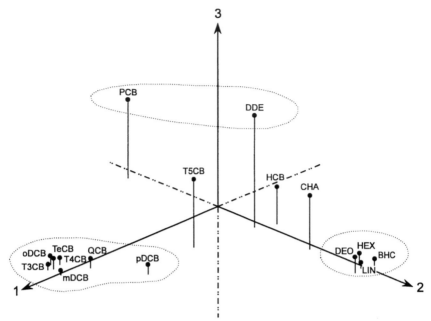

Figure 10.18 Three-dimensional plot of three rotated factors for organic micropollutants in Lake Ontario showing the association between the different chemical compounds. For further explanation see text and Table 10.9 (After Stevens and Neilson, 1989)

an immediate impression of the group associations. Such data treatment is most suitable for specialised and technical audiences and is too complex for general presentation.

10.6.6 Water quality indices

A water quality index is an indicator of the quality of water obtained by aggregating several water quality measurements into one number. This simplified expression of a complex set of variables can be used for a variety of purposes, including:

- Communicating water quality information to the public and decision makers.
- As a planning tool for managing water uses.
- Assessing changes in the quality of the water resource.
- Identifying water quality problems for which special studies are needed.
- Evaluating the performance of pollution control programmes.

Different water quality indices are currently in use or under development. They can be grouped as follows:

- General water quality indices: Various physical, chemical and microbiological variables are aggregated to produce an overall index of water quality (see Chapter 6).

- Specific use-quality indices: Variables are aggregated on the basis of their water quality importance to specific water uses to produce, for example, an Irrigation Water Quality Index.
- Planning indices: These indices are similar to use-quality indices, but they also include social and economic factors.
- Chemical indices: These are calculated by grouping water quality variables according to their chemical nature, e.g. toxic metals or chlorinated phenols indices.
- Biological indices: Selected biological characteristics including community diversity and the presence of indicator organisms (e.g. faecal coliforms) are used to produce indices such as the diversity or Saprobic indices (see Chapter 5).
- Trophic state indices: These are calculated from variables related to the trophic status of a lake or reservoir (e.g. transparency, dissolved oxygen, macrophyte abundance, chlorophyll a, P-total and N-total).

An example of a complex water quality index has been developed by Environment Canada. The index is based on a series of micro-organism, biochemical and toxicity screening tests, collectively known as the "Battery of tests", for environmental hazard assessment and priority setting (Dutka *et al.*, 1988). The results from each test are scored using a pre-determined point awarding system, and then totalled. Water and sediment samples with the most points are deemed to contain the greatest potential hazard to humans or aquatic organisms. The point system conveys a general impression of water quality to non-experts, who need have no detailed knowledge of aquatic ecosystems.

A simpler water quality index, based only on some easily measured physical, chemical and biological variables, is in use in the highly industrialised state of São Paulo, Brazil (see IQA values in Table 10.3). Further details of the method of calculation are available in CETESB (1990). The index values are assigned water quality levels from excellent to inadequate and are also used to construct water quality maps for the state, which highlight areas of significant change from year to year (a similar approach to Figures 6.34 and 6.35).

10.7 Advanced data analysis and management techniques

This section describes methods which have practical application in water quality management. The methods represent a progression from data collection and assessment to the sustainable development of a water body or water basin. In general these methods require more advanced use of computers and the associated hardware.

10.7.1 Models

Water quality models are designed to simulate the responses of aquatic ecosystems under varying conditions. They have been applied to help explain and predict the effects of human activities on water resources, such as lake eutrophication, dissolved oxygen concentrations in rivers, the impacts of acid rain on natural water bodies, and the fate, pathways, impacts and effects of toxic substances in freshwater systems (Fedra, 1981; Straskrabá and Gnauck, 1985). Mathematical models are very useful tools for water quality management because they allow:

- the identification of important variables in a particular aquatic system, and help interpretation of the system's processes,
- forecasting of the impacts of developments on water bodies, and
- policy testing and analysis.

The high degree of complexity, spatial and functional heterogeneity, non-linearity, complex behavioural features (such as adaptation and self-organisation) and the considerable stochastic element of natural systems, make model development a difficult and highly skilled task. Data requirements for model calibration and for model use pose additional constraints on their widespread use. This complexity, together with the limited knowledge of the processes taking place in water bodies, require that a high degree of simplification and a number of assumptions must be built into any model. The model user must be aware of the model's limitations, and its assumptions, in order to draw appropriate conclusions. At present, highly predictive models are not general, and general models are not highly predictive.

Mathematical models belong to one of two basic classes, theoretical (or deterministic) and empirical (e.g. Smith, 1982).

Theoretical models

If the physical, chemical and/or biological mechanisms underlying a process are well understood, a steady-state or dynamic model can be developed. Steady-state models cannot be used for predicting system responses over time, and therefore, have limited water management value. Time-variable models, on the other hand, can handle variable input loads, and can be useful for establishing cause–effect relationships. When compared to empirical models, theoretical models are generally more complex. They require a longer period of observation for calibration and the number of variables and parameters to be measured are greater. They also require a significant amount of time for validation. Due to their complexity, and because our understanding of aquatic systems is usually incomplete, these types of models are used less frequently than empirical models.

Empirical models

Empirical or statistically-based models are generated from the data analysis of surveys at specific sites. The relationships identified are then described in one or more mathematical equations. These models can be built relatively quickly when compared to theoretical models, and they are easier to use because they have fewer data requirements. Sometimes empirical models have to be generated from incomplete or scattered information about the aquatic system. For example, the model may be supported by observations made over a limited range of conditions or a relatively short time period. In such cases the model output should be interpreted with caution. It is also important to remember that such models are not directly transferable to other geographic areas or to different time scales.

Examples of water quality models

Hundreds of water quality models have been developed (e.g. Park *et al.*, 1975; Collins *et al.*, 1979; DiToro *et al.*, 1983; Lang and Chapra, 1982; Booty and Lam, 1989). Some of them are site or problem specific while others are more general, such as multimedia models. There is no single model that can be applied to all situations. Some examples of models are described below.

Water Analysis Simulation Programme (WASP) is a theoretical model which is applicable to a wide variety of water quality problems and can be adapted for site specific uses (DiToro *et al.*, 1983). It is a time-variable model that can be applied to one, two or three dimensions. The input data consist of loads, boundary conditions, mass transfer rate, kinetic rates and concentrations of organic compounds, trace elements and phytoplankton. The output lists variable concentrations.

AQUAMOD-3, SALMO and MSCLEANER are multi-layer, multi-compartment, kinetic models of lake and reservoir processes (Collins *et al.*, 1979; Straskrabá *et al.*, 1980; Benndorf and Recknagel, 1982). Input data include climatic, hydraulic and biological variables. Main outputs are the time variation of the biological communities (mainly phytoplankton and zooplankton) and nutrients such as phosphorus and nitrogen in thermally stratifying water bodies.

10.7.2 Geographic Information Systems

Water quality is affected by a variety of natural and anthropogenic factors including hydrology, climate, geology, pesticide use, fertiliser application, deforestation, chemical spills, and urbanisation. As the data are often measured in different units, at different temporal and spatial scales, and because the data sources are very diverse (e.g. conventional maps, Landsat imagery, and tabular data obtained in the field), the spatial analysis of these and other factors is a time-consuming and complicated task when performed manually. The computer-based geographic information system is a tool that allows analysis of all these various types of data (Marble, 1987; Walsh, 1987). It encodes, analyses and displays multiple layers of geographically referenced information.

Since GIS are capable of combining large volumes of data from a variety of sources, they are a useful tool for many aspects of water quality investigations. They can be used to identify and to determine the spatial extent and causes of water quality problems, such as the effects of land-use practices on adjacent water bodies. They can also:

- help to determine location, spatial distribution and area affected by point-source and non-point source pollution,
- be used to correlate land cover and topographical data with a variety of environmental variables including surface run-off, drainage, and drainage basin size,
- be used for assessing the combined effects of various anthropogenic (e.g. land-use) and natural (e.g. bedrock, precipitation and drainage) factors on water quality, and
- be incorporated into water quality models.

The two basic types of geographic data structures used in most GIS are raster and vector. The raster data structure is analogous to a grid placed over an image. This structure allows for the efficient manipulation and analysis of data, and is preferred for overlay operations. Geographic features such as rivers, roads and boundaries are represented as lines or vectors. A typical GIS consists of:

1.   A data input sub-system which collects and processes spatial data from a variety of sources, including:
     - digitised map information: this data can be either imported from pre-existing databases or obtained directly from a map with a digitiser,
     - coded aerial photographs, satellite and remote sensing images, and
     - geographically referenced data such as tables of concentrations of water quality variables (attributes) at given locations or watershed boundaries.

2.   A data storage and retrieval sub-system which:
     - organises the data for quick retrieval, and
     - permits rapid and accurate updates and corrections.

3.   A data manipulation and analysis sub-system which transforms the various data into a common form that permits subsequent spatial analysis. Tabular data, stored on computer systems for example, are

converted first to a map scale common with other data, and then all data are analysed together. The analysis usually consists of overlaying various data sets by intersection, mutual exclusion, or union, and then performing some statistical calculations on the overlaid data sets.

4.   A data reporting sub-system which displays the database, the results of statistical computations, and produces hybrid maps. Outputs include tabular data, colour images, graphs and single or composite maps.

Two of the better known GIS packages in use in connection with water resources are SPANS (TYDAC Technologies Inc., 1989) and ARCINFO. They require investments in the order of tens of thousands of US dollars for hardware and software (i.e. at least microcomputers based on the latest microprocessor technology), a high level of computer expertise, and the appropriate databases. Examples of the output from a GIS package are shown in Figures 7.10 and 7.11.

10.7.3   Expert systems

An expert system attempts to capture the decision-making abilities of a specialist or "expert" in a particular field of endeavour into a computer program in such a way that the user can be confidently guided through the necessary steps required to solve a complex problem (Weiss and Kulikowski, 1984; Fenly, 1988). This human experience which is encoded into an expert system includes knowledge, experience and problem-solving ability. Development of such systems requires highly trained personnel.

Expert systems are tools which can be used in a variety of water quality activities including planning, diagnosis, interpretation, and analysis. For example, following a chemical spill an appropriate expert system could quickly provide the information necessary for its containment, clean-up and possible environmental effects. It would also provide a list of the data which must be collected about the spill to satisfy existing regulations. The advantage of the system is that even a non-specialist could be guided to the correct actions, which would otherwise have required considerable expertise and experience.

Currently, few expert systems have been developed for water quality. One such system is RAISON (Regional Analysis by Intelligent Systems on a Microcomputer). This system was developed at the National Water Research Institute, Burlington, Ontario, Canada to predict the effects of SO_2 and NO_x emission reductions on water quality in eastern Canada. The results of these predictions are eventually to be used by decision makers to help set adequate emission standards. RAISON offers users the capacity to examine and to model regional statistics using a tightly integrated spreadsheet, database, GIS, programming language and expert system shell. The system was designed to run on commonly available microcomputers. The system is now being adapted for a variety of other water quality applications including the analysis of data from the GEMS/WATER programme (WHO, 1991).

Expert system structure

Expert systems contain four basic components:

1.   The knowledge base: this consists of facts and heuristics. The facts constitute a body of information that is widely shared, publicly available, and generally agreed upon by experts in a field. Heuristics are "rules of thumb" based on experience, or compiled knowledge. Rules are used by the expert system for selecting a limited number of alternative solution paths the system will follow. One of the most commonly used structures for knowledge representation is production rules. They consist of IF–THEN conditional statements. Essentially, when input data used for a particular problem match the data in the IF part of the rule, statements in the THEN part of the rule are actuated. For example: IF PCBs are detected in groundwater THEN the groundwater is contaminated.

2.   The inference system: interprets the rules contained within the knowledge base to reduce the time an expert system must spend searching for a specific conclusion. Without such rules an expert system would consider all possible paths during the solution process. For large scale problems this could lead to unacceptably large numbers of possibilities to be evaluated, the so called combinatorial explosion.

3.   The working memory: is the database. It also keeps track of data inputs and new facts which were inferred during the problem solving process.

4.   A user interface: typically a video work-station that permits easy interaction between the user and the expert system. An important component of the user interface is the explanation facility, which explains the line of reasoning used to arrive at a solution, explains reasons for system queries and elaborates on "rules".

Advantages and disadvantages of expert systems

Expert systems can help non-experts achieve expert-like results by making the knowledge of a few experts available to many. Other advantages are:

■ Expert systems are always available for use. They never sleep, can be run 24 hours a day, seven days a week, and never get tired or bored.
■ They may lead the expert to discover new knowledge.
■ Expert systems can manage very complex rules from a variety of sources, such as legislation, manuals, rules, regulations, and guidelines.

- Expert systems can handle uncertainty and make their knowledge explicit.
- They can work with incomplete knowledge and incomplete, and possibly inconsistent, data.
- Expert systems are capable of explaining how, and why, a particular answer was obtained.

The development of expert systems for water quality applications has only recently begun. At present, there are no hard and fast rules as to whether an application is suitable for an expert system. Therefore, caution should be exercised in implementing such systems because:

- They are expensive and time-consuming to build. However, their potential usefulness assures that expert systems will be a growth area in the water resources field.
- In order to make efficient use of an expert system some basic understanding of the system is required and this may involve quite advanced levels of training.
- The acquisition of the required knowledge is difficult because the number of experts is small, they may not be available, or because the expert may never have conceptualised the process by which a particular conclusion was reached.

10.8   Examples of the application of data analysis and presentation
The range of possible applications of data analysis and presentation is enormous. Therefore, only a few examples are given below but many more specialised topics are covered with examples in Chapters 4 to 9.

10.8.1 Description of water quality
One of the most common objectives for a water quality study is to characterise the water quality of a water body or a river basin. This is usually accomplished by determining the temporal and spatial distribution of the physical, chemical, and biological variables as well as their inter-relationships.

The Mackenzie River Basin with a drainage area of 2×10^6 km^2 is the largest in Canada and the river, approximately 4,000 km long, is the longest in Canada. Several agencies collect water quality data in this basin and data were available for 448 river and lake sampling stations for the period 1960–1979. The objectives of the data analysis were to:

- Determine the inter-variable relationships within each sub-basin, including discharge.
- Characterise and compare the physical and chemical composition of river and lake water in each sub-basin.
- Determine spatial and seasonal variation.

Table 10.10 Summary statistics of physical variables measured at selected sites in the Athabasca River sub-basin of the Mackenzie River basin

Variable	n	Mean	Minimum	Maximum	s.d.	50th	75th	90th
						Percentiles		
Turbidity (NTU)								
Upstream from Sunwapta	20	24	1.7	88	26	8.0	40	63
At Athabasca	201	33	< 0.1	1,100	98	7.0	34	74
At Fort McMurray	79	14	4.0	30	15	15		
At Bitumount	5	40	1.0	650	77	22	43	90
Fond du Lac River	31	1.3	0.2	5.6	1.2	1.0	1.7	2.4
Total dissolved solids (mg l^{-1})								
Upstream from Sunwapta	20	85	52	113	13	86	93	101
At Athabasca	191	169	57	288	48	160	213	234
At Fort McMurray	78	143	117	168	25	143		
At Bitumount	3	164	96	280	48	146	205	241
Fond du Lac River	31	18	12[1]	28	3	17	19	21
Colour (units)								
Upstream from Sunwapta	20	9	5	40	10	5	8	25
At Athabasca	200	29	5	200	26	20	30	60
At Fort McMurray	77	41	< 5	200	28	30	60	80
At Bitumount	0							
Fond du Lac River	30	7	< 5	20	3	5	10	10
Specific conductivity ($\mu S\ cm^{-1}$)								
Upstream from Sunwapta	20	160	124	217	22	157	172	192
At Athabasca	203	306	117	501	83	292	383	421
At Fort McMurray	79	236	190	270	27	243	248	
At Bitumount	6	295	181	476	81	256	373	427
Fond du Lac River	31	40	26	94	13	37	42	51

n   Number of observations
s.d. Standard deviation

[1]   Qualified result, i.e. values less than the detection limit were used in the calculation
Source: After Blachford *et al.*, 1985

- Determine trends.
- Evaluate the compliance with water quality guidelines for the protection of aquatic life.

The data (range, median and important percentiles) for physical, inorganic and organic variables were summarised in tables. The tables were complemented by many graphs. Table 10.10 and Figure 10.19 are examples of such tables and graphs.

10.8.2   Prediction of water quality

One of the most important aspects of water quality management is the ability to predict the changes in water quality resulting from natural phenomena,

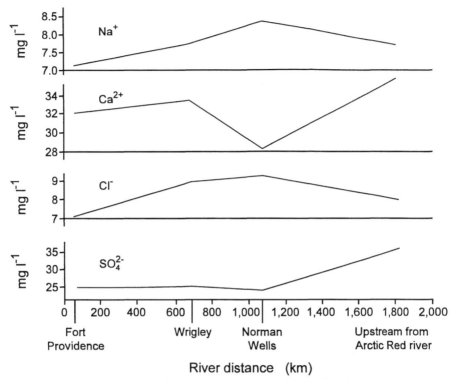

Figure 10.19 Longitudinal profiles of average concentrations of major ions in the Mackenzie River, Canada (After Blachford *et al.*, 1985)

changes in land and water use, and from human activities. Such forecasting is an essential tool in making decisions about economic and social development in a given region of the country. The forecasting of water quality can be done by determining trends and extrapolating them into the future or by mathematical models (see section 10.7.1). Extrapolating trends into the future can be used successfully only if no major changes in water and land use, and in environmental conditions, are anticipated in and around the water body. When there are significant changes expected or contemplated for land uses, water uses and environmental factors, mathematical models should be used to make predictions about the associated changes in water quality.

The freshwater River Thames is an important recreational river and wastewater disposal facility. It is also used as a major water supply source, particularly in its downstream reaches. To satisfy varied, and occasionally conflicting, interests, active quantity and quality management is required. An important aspect of that management is mathematical modelling of the river discharges and various important water quality variables. Figure 10.20 is an

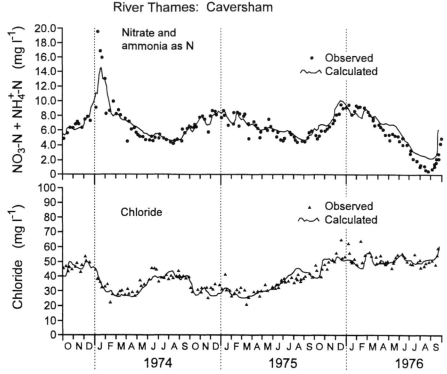

Figure 10.20 Comparison of simulated and observed nitrate and chloride concentrations in the River Thames at Caversham, UK, 1974–1976 (Based on Thomson, 1979)

example of the output of a river quality model which takes account of discharges, loadings and within-river processes. The resulting chloride and nitrate concentrations are illustrated. The data presented cover years which include significant periods of flood, drought and normal discharge. Such models can assist in providing short- and long-term assessment of the possible implications for downstream abstraction, of upstream discharge variation or catchment loading. In this instance the predicted values were a good representation of the observed results over an extended period of time.

10.8.3 Compliance with water quality guidelines

The quality of surface waters in the eastern coastal basin of New Brunswick, Canada was assessed for its ability to support a variety of uses. This assessment was done by comparing the concentrations of ten chemical variables to water quality guidelines for various uses of water. Table 10.11 shows the results of this analysis for the protection of aquatic life.

Table 10.11 Non-compliance analysis of surface water data in relation to maximum acceptable concentration guidelines for the protection of freshwater life in the Eastern Coastal Basin, New Brunswick, Canada

Variable	Guideline value	Number of		Median of	
		Non-compliant values	Included values	Excluded[1] values	Non-compliant values
Chromium as Cr	≤ 0.04 mg l⁻¹	0	17	0	
Copper as Cu	≤ 0.005 mg l⁻¹	130	379	57	0.028
Dissolved Iron as Fe	≤ 0.300 mg l⁻¹	2	219	0	0.36
Iron as Fe	≤ 0.300 mg l⁻¹	101	376	0	0.41
Lead as Pb	≤ 0.03 mg l⁻¹	10	320	32	0.068
Nickel as Ni	≤ 0.025 mg l⁻¹	2	22	0	0.030
Dissolved oxygen as O₂	≥ 4.0 mg l⁻¹	1	498	0	2.0
pH	6.5 – 9.0	35	163	0	6.1
Total phosphate as P	< 0.100 mg l⁻¹	12	177	0	0.295
Zinc as Zn	≤ 0.030 mg l⁻¹	74	420	0	0.050
Total		367	2,591	89	

[1] Excluded values refer to two types of data: i) values of < x, where x is > guideline, and ii) values of > y, where y < guideline. In both instances, the actual value may be at, above or below the guideline and, therefore, compliance or non-compliance could not be determined
Source: After Belliveau, 1989

Variability of groundwater quality has been monitored in ten principal aquifers in Maryland, USA since 1938. Figure 10.12 shows a graphical summary of the results for dissolved solids, nitrate plus nitrite (as nitrogen) and chloride. Also shown in each of the graphs are the US Environmental Protection Agency drinking water standards (Wheeler and Maclin, 1987).

10.8.4   Water pollution control measures
Figure 10.21 shows the dissolved oxygen (DO) results from the monitoring programme of the St. Croix River. This river forms the border between the province of New Brunswick, Canada and the state of Maine, USA. The results show that DO concentrations have significantly improved following the introduction of a secondary treatment facility in 1977–1978.

Similar improvements in biological water quality over 20 years in Nordrhein-Westfalen, Germany, following the installation of sewage treatment facilities, have been demonstrated using a water quality classification scheme in Figures 6.34 and 6.35.

10.8.5   State of the Environment (SOE) reporting
A State of the Environment (SOE) report is an environmental summary assessment used to inform decision makers, environmental organisations,

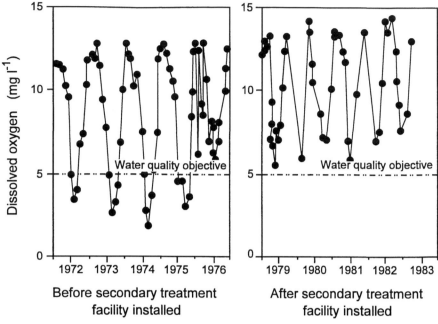

Figure 10.21 Dissolved oxygen concentration in the St Croix River, Canada before and after installation of secondary wastewater treatment facilities (After Kwiatowski, 1986)

scientists and the general public about the quality of the natural environment. State of the Environment reports usually include the stresses being imposed by human activities and the sustainability of these activities in view of their impacts on both human health and ecosystems. A full SOE report, therefore, normally consists of concise and focused articles on:

- the current state of the environment,
- changes and trends in the environment,
- links between human and environmental health and human activities, including the economy, and
- the actions taken by societies to protect and restore environmental quality.

These four themes can be applied at local, regional, national or global scales. For water quality topics, SOE reporting units include individual lakes or rivers, watersheds, or drainage basins.

State of the Environment reports can also address specific issues. They can deal with specific variables (e.g. PCBs in the environment), with specific environmental problems (e.g. eutrophication or effects of agriculture on water quality), or with broader topics, such as compliance with water or environmental quality objectives. Due to their usually broad scope, the preparation of an SOE report calls for the use of many of the data presentation and analysis techniques described in this chapter. Their preparation also requires good data and information bases as well as a multi-disciplinary group of experts. Examples of SOE reports are Environment Canada (1977), Bird and Rapport (1986), Great Lakes Water Quality Board (1989), Meybeck *et al.* (1989), CETESB (1990) and National Rivers Authority (1994).

10.9   References

Annuaire Hydrologique 1987, 1988, 1989 Ministère de l'Environnement, Québec, Directions du Milieu Hydrique, 162 pp.

Belliveau, P.E. [Ed.] 1989 *Historical Overview of Water Quality in New Brunswick, 1961 – 1979.* Environment Canada, Inland Waters Directorate, Atlantic Region Water Quality Branch and Water Planning and Management, Moncton, New Brunswick/Dartmouth, Nova Scotia, 230 pp.

Benndorf, J. and Recknagel, F. 1982 Problems of application of the ecological model SALMO to lakes and reservoirs having various trophic states. *Ecol. Modelling*, **17**, 129-145.

Bird, P.M. and Rapport, D.J. 1986 *State of the Environment Report for Canada.* Minister of the Environment, Ottawa, Canada, 263 pp.

Blachford, D.P., Demayo, A. and Gummer, Wm. 1985 *Water Quality.* Mackenzie River Basin Study Report, Supplement 9, Minister of Supply and Services, Ottawa, 201 pp.

Booty, W.G. and Lam, D.C.L. 1989 *Freshwater Ecosystem Water Quality Modelling.* RRB-89-12, NWRI Contribution No. 89-63, National Water Research Institute, Canada Centre for Inland Waters, Burlington, Ontario, 44 pp.

Box, G.E.P., Hunter, W.G. and Hunter, J.S. 1978 *Statistics for Experimenters. An Introduction to Design, Data Analysis, and Model Building.* John Wiley and Sons, Toronto, 653 pp.

Box, G.E.P. and Jenkins, G.M. 1970 *Time Series Analysis Forecasting and Control.* Holden-Day Series in Time Series Analysis. Holden-Day, San Francisco, 553 pp.

CETESB 1990 *Relatório de Qualidade das Aguas Interiores do Estado de Sao Paulo 1989.* Companhia de Tecnologia de Saneamento Ambiental, Sao Paulo, 164 pp.

Collins, C.D., Leung, D.K., Boylen, C.W., Albanese, G., de Caprariis, P. and Forstner, M. 1979 The aquatic ecosystem model MS CLEANER. In: S.E. Jorgensen [Ed.] *State-of-the-art of Ecological Modelling.* International Society for Ecological Modelling, Copenhagen, 579-602.

Daniel, W.W. 1990 *Applied Non-Parametric Statistics.* 2nd edition. PWS-Kent Publishing Company, Boston, 635 pp.

Demayo, A. and Whitlow, S. 1986 Graphical presentation of water quality data.

In: *Monitoring to Detect Changes in Water Quality Series*, Proceedings of the Budapest Symposium, July 1986, IAHS Publ. No. **157**, 13-27.

DiToro, D.M., Fitzpatrick, J.J. and Thomann, R.V. 1983 *Documentation for Water Quality Analysis Simulation Program (WASP) and Model Verification Program (MVP)*. U.S. Environmental Protection Agency, Large Lakes Research Station, Grosse Ile, MI, EPA-600/3-81-044, 158 pp.

Dobson, H.F.H. 1984 *Lake Ontario Water Chemistry Atlas*. Scientific Series No. 139, National Water Research Institute, Inland Waters Directorate, Canada Centre for Inland Waters, Burlington, Ontario, 59 pp.

Dutka, B.J., Munro, D., Seidl, P., Lauten, C. and Kwan, K.K. 1988 *Use of a Battery of Biological Tests to Assess Aquatic Quality in Canadian Prairie and Northern Watersheds*. National Water Research Institute Contribution #88-107, Canada Centre for Inland Waters, Burlington, Ontario, 23 pp.

Elliott, J.M. 1977 *Some Methods for the Statistical Analysis of Samples of Benthic Invertebrates*. Second edition. Scientific Publication No. 25, Freshwater Biological Association, Windermere, 160 pp.

Environment Canada 1977 *Surface Water Quality in Canada — An Overview*. Water Quality Branch, Inland Waters Directorate, Environment Canada, Ottawa, Canada, 45 pp.

Fedra, K. 1981 *Mathematical Modelling — A Management Tool For Aquatic Ecosystems?* RR-81-2. International Institute for Applied Systems Analysis, Laxenburg, Austria, 14 pp.

Fenly, C. 1988 *Expert Systems. Concepts and Applications*. Advances in Library Information Technology, Issue Number 1. Cataloguing Distribution Service, Library of Congress, Washington, D.C., 37 pp.

Gauch Jr., H.G. 1982 *Multivariate Analysis in Community Ecology*. Cambridge University Press, Cambridge, 298 pp.

Gilbert, R.O. 1987 *Statistical Methods for Environmental Pollution Monitoring*. Van Nostrand Reinhold Co., New York, 320 pp.

Great Lakes Water Quality Board, 1989 *Report on Great Lakes Water Quality*. Report to the International Joint Commission. International Joint Commission, Windsor, Ontario, 128 pp.

Guertin, W.H. and Bailey, J.P.J. 1970 *Introduction to Modern Factor Analysis*. Edwards Brothers, Inc., Ann Arbor.

Hem, J.D. 1989 *Study and Interpretation of the Chemical Characteristics of Natural Waters*. Water Supply Paper 2254, 3rd edition, U.S. Geological Survey, Washington D.C., 263 pp.

Hipel, K.W. [Ed.] 1985 *Time Series Analysis in Water Resources*. Monograph Series No. 4. American Water Resources Association, Bethesda, MD, 609-832 pp.

Hirsch, R.M., Alexander, R.B. and Smith, R.A. 1991 Selection of methods for the detection and estimation of trends in water quality. *Water Resources Res.,* **27**(5), 803.

Kwiatowski, R.E. [Ed.] 1986 St. Croix River. In: *Water Quality in Selected Canadian River Basins - St. Croix, St. Lawrence, Niagara, Souris and the Fraser Estuary.* Scientific Series No. 150. Inland Waters Directorate, Water Quality Branch, Ottawa, 46 pp.

Lang, G.A. and Chapra, S.C. 1982 *Documentation of SED-A Sediment-Water Column Contaminant Model.* NOAA Technical Memorandum ERL GLERL-41m Ann Arbor, Michigan, 49 pp.

LAWA 1990 *Limnologie und Bedeutung ausgewählter Talsperren in der Bundesrepublik Deutschland.* Länderarbeitsgemeinschaft Wasser, Weisbaden, 280 pp.

Loftis, J.C., Ward, R.C., Phillips, R.D. and Taylor, C.H. 1989 *An Evaluation of Trend Detection Techniques for use in Water Quality Monitoring Programs.* EPA/600/3-89/037, U.S. Environmental Protection Agency, Washington D.C., 139 pp.

Mandel, J. 1964 *The Statistical Analysis of Experimental Data.* John Wiley, New York, 410 pp.

Marble, D.F. 1987 Geographic Information Systems: An overview. In: W.J. Ripple [Ed.] *Geographic Information Systems for Resource Management: A Compendium.* American Society for Photogrammetry and Remote Sensing and American Congress on Surveying and Mapping, Falls Church, Virginia, 288 pp.

Meybeck, M., Chapman, D. and Helmer, R. [Eds] 1989 *Global Freshwater Quality: A First Assessment.* Blackwell Reference, Oxford, 306 pp.

Montgomery, R.H. and Reckhow, K.H. 1984 Techniques for detecting trends in lake water quality. *Water Resources Bull.,* **20**(1), 43-52 pp.

Morrison, D.A. 1965 *Multivariate Statistical Methods.* Holt, Rinehart and Winston, Inc., New York.

National Rivers Authority 1994 *The Quality of Rivers and Canals in England and Wales (1990 to 1992).* Water Quality Series No. 19, HMSO, London, 38 pp + map.

Park, R.A., Scavia, D. and Clesceri, L.S. 1975 CLEANER, the Lake George Model. In: C.S. Russel [Ed.] *Ecological Modelling in a Management Context.* Resources for the Future, Washington, 49-81.

Rohlf, F.J. and Sokal, R.R. 1969 *Statistical Tables.* W.H. Freeman & Co., San Francisco, 253 pp.

SAS Institute Inc. 1988 *SAS/STATtm User's Guide,* Release 6.03 Edition, Cary, N.C., 1028 pp.

Smith, V.H. 1982 The nitrogen and phosphorus dependence of algal biomass in lakes: An empirical and theoretical analysis, *Limnol. Oceanogr.*, **27**(6), 1101-1112.

Snedecor, G.W. and Cochran, W.G. 1980 *Statistical Methods*. 7th edition. The Iowa State University Press, Iowa, 507 pp.

Sokal, R.R. and Rohlf, F.J. 1973 *Introduction to Biostatistics*. W.H. Freeman and Company, San Francisco, 368 pp.

Steel, R.G.D. and Torrie, J.H. 1980 *Principles and Procedures of Statistics. A Biometrical Approach*, 2nd edition. McGraw-Hill Book Company, New York, 633 pp.

Stevens, R.J.J. and Neilson, M.A. 1989 Inter- and intra-lake distributions of trace organic contaminants in surface waters of the Great Lakes. *J. Great Lakes Res.*, **15**(3), 377-393.

Straskrabá, M. and Gnauck, A.H. 1985 *Freshwater Ecosystems. Modelling and Simulation*. Developments in Environmental Modelling, 8, Elsevier Science Publishers, Amsterdam, 309 pp.

Straskrabá, M., Dvoráková, M., Kolárová, N., Potucek, J. and Schwabik, S. 1980 Periodicity, sensitivity of the periodic solutions and parameter optimization of a theoretical ecosystem model. *ISEM J.*, **2**, 42-62.

Thomson, G.D. 1979 A model for nitrate-nitrogen transport and denitrification in the River Thames. *Wat. Res.*, **13**, 855-863.

TYDAC Technologies Inc. 1989 *SPANStm Spatial Analysis System, Version 4.0*. TYDAC Technologies Inc., Ottawa, Ontario, 3 volumes.

Walsh, S.J. 1987 Geographic Information Systems for natural resource management. In: W.J. Ripple [Ed.] *Geographic Information Systems for Resource Management: A Compendium*. American Society for Photogrammetry and Remote Sensing and American Congress on Surveying and Mapping, Falls Church, Virginia, 288 pp.

Weiss, S.M. and Kulikowski, C.A. 1984 *A Practical Guide to Designing Expert Systems*. Chapman & Hall, London, 174 pp.

Wheeler, J.C. and Maclin, L.B. 1987 *Maryland and the District of Columbia Ground-Water Quality*. U.S. Geological Survey Open File Report 87-0730. U.S. Geological Survey, Denver, Colorado.

WHO 1991 *GEMS/WATER 1991-2000 The Challenge Ahead*. WHO/PEP/91.2, World Health Organization, Geneva.

WHO 1992 *GEMS/Water Operational Guide*. Third edition. World Health Organization, Geneva.

Wilkinson, L. 1988 *SYSTAT: The System for Statistics*. SYSTAT, Inc., Evanston, Il.

Young P. 1984 *Recursive Estimation and Time-Series Analysis. An Introduction*. Springer-Verlag, Berlin, 300 pp.

Appendix 10.1

BASIC DESIGN FOR SAMPLING PROGRAMMES

General guidance on sampling design (location, frequency, sampling equipment, number of samples) has been given in the relevant chapters for different levels of assessment in rivers, lakes, reservoirs and groundwaters. However, in many water bodies (e.g., complex reservoirs) the issues of where to sample and how many samples to take in order to give representative results for the whole water body, need more than an intuitive choice of sampling design. Nevertheless, the task of deciding the optimum number of samples to take and the most suitable locations in a water body in order to characterise its water quality in a meaningful way, and with the most economic use of resources, can be quite daunting. Statistically based methods of sampling design can help this task and also ensure that the data collected are appropriate for later statistical analysis and interpretation. A full coverage of this subject is beyond the scope of this guidebook but some basic principles are described briefly in this appendix. A full description of the sampling process is given in Keith (1988). Gilbert (1987) also gives a comprehensive account of sampling theory, with many excellent examples based on environmental studies. Much of the source material is based on standard texts such as Cochran (1963) and Snedecor and Cochran (1980).

Basic sampling design naturally falls into seven aspects: (i) reasons to sample, (ii) what to sample, (iii) how to sample, (iv) when to sample, (v) where to sample, (vi) how many samples to take, and (vii) sampling evaluation. The issues of what, where, when and how to sample are defined by the assessment programme objectives and are discussed in detail in relation to rivers, lakes, reservoirs and groundwaters in Chapters 6, 7, 8 and 9. Some guidance for obtaining statistically valid measurements is given below.

Where to sample
Although random sampling is often suggested as the basis of a sampling design, it is rarely the most appropriate approach, particularly when there is already some knowledge of the nature of the water body and the characteristics of the variable being monitored. The choice of sample site is, therefore, usually based on:
- *Judgement sampling:* Sample site location is based on experience of the type of water body and the variables, together with their anticipated distribution pattern. Although judgement sampling can provide good estimates

of many variable types, it is difficult to provide a quantified assessment of the applicability of the estimates so derived to the whole water body.

- *Probability sampling:* Sample sites and numbers are based on statistical probability theory, and so are ostensibly based on an objective methodology. However, as will be seen later, in practice many subjective aspects have to be involved. Pure, sampling theory may also lead to logistically impractical schemes. The main options are:
  - *Simple random*: Sites are chosen at random across the surface of the water body, and within the water mass.
  - *Stratified random*: Sample sites are selected randomly within areas chosen relative to the more homogeneous components of an otherwise heterogeneous variable. For example, randomly sampling within the epi-, meta- and hypolimnion of a stratified reservoir or within multiple basins of a dendritic water body.
  - *Systematic sampling*: This is usually the method of choice for practical purposes. Sample sites are chosen at appropriate locations. These locations are chosen on the basis of experience and/or judgement. Samples are usually drawn at prescribed, not necessarily regular, points along horizontal and vertical transects.
  - *Cluster sampling*: Specialised sampling, e.g. for fish or water weeds. Usually involves taking all specimens within an isolated sampling area.
  - *Double sampling*: Used where a more simple variable, which bears a known relationship to a more complex one, may be sampled more intensively to provide an estimated distribution of the lesser sampled, more complex variable, e.g. measuring conductivity instead of comprehensive anion/cation analysis.

How many samples to take

A. Random sampling

In schemes which have to use discrete sample sites or volumes, the numbers of samples required to attain various monitoring objectives may be basically derived from the relevant probability distributions and error statistics (see sections 10.4.6, 10.4.7, 10.4.8). These generally derive: $(\bar{x} - \mu)/(s/\sqrt{n}) = t$, where t is the Student's t probability distribution function, $\bar{x}$ is the sample mean, μ is the "true" mean, s is the sample standard deviation, and n is the number of samples.

Bearing in mind that such formulations apply to at least near-normally distributed and independent samples (see section 10.4.4), by simple rearrangement: $n = (t*s/(\bar{x} - \mu))^2$.   If $d = \bar{x} - \mu$, then $n = (t*s/d)^2$.

In order to use such a formulation, it is necessary (because of the terms it contains) to cast the sampling objective in the form of: How many samples n are necessary from a population with variance s^2 to provide an estimate of the population mean within $\pm d$ units of the true mean, with a probability of p per cent? (90 or 95 per cent confidence levels are commonly used). Even this specification involves two difficulties. The main difficulty lies in having already available an acceptable estimate of the variance of the population about to be sampled! The secondary problem is that t is also dependent on the number of samples involved n, so it is only possible to solve the equation by successive trials.

Example 1
Suppose the sampling and measurement variability of a mixed reservoir population of algae produced a standard deviation of 3 mg m^{-3} chlorophyll *a*. Assuming this value remains true, how many samples would be necessary to ensure that the measured mean was within 2 mg m^{-3} chlorophyll *a* of the actual population mean, with 95% confidence?

Thus, $s = 3$, $d = 2$ and $p = 95$. The probability of obtaining a difference greater than d is $\alpha = (100 - p)/100 = 0.05$ (this change is convenient for the manner in which t is usually tabulated). As there is no initial estimate of sample numbers, the formula containing t can only be used with an estimated value for the degrees of freedom df, which along with α determines t. For present purposes, $df = n - 1$. Due to the form of the t distribution (it is nearly normal, and relatively constant for values of df greater than about 30), it is convenient to start our trial solutions with $df = 30$.

Thus:
$t_{0.05[30]}$ = 2.042. Then: $n_1 = (2.042*3/2)^2 = 9.4 \Rightarrow 9$
$t_{0.05[8]}$ = 2.306. Then: $n_2 = (2.306*3/2)^2 = 12.0 \Rightarrow 12$
$t_{0.05[11]}$ = 2.201. Then: $n_3 = (2.201*3/2)^2 = 10.9 \Rightarrow 11$
$t_{0.05[10]}$ = 2.228. Then: $n_4 = (2.228*3/2)^2 = 11.2 \Rightarrow 11$

As $n_4 = n_3$, the solution has been found: 11 samples.

If the solutions oscillate between two figures, then the higher figure should be chosen for the basis of the sample collection as these formulations give theoretical, minimum sample numbers. Any spatial correlation amongst samples, for example, will usually require greater numbers of samples for similar statistical accuracy (see below).

The ease of the example computation above rested almost entirely on the pre-knowledge of the variable standard deviation. If that value is not known or cannot be reasonably estimated, then the computation cannot be made. As having a reasonable estimate of the sample variance is so essential, efforts are usually made to provide a reasoned value. The most likely approaches include: preliminary survey, previous monitoring data, expert judgement, and reasoned assessment.

Even the first two of these approaches also require a subjective judgement that the previous survey results are still applicable, particularly with qualities which are potentially highly variable in horizontal and vertical extent, and time. It is also usually true that many of the variables of interest in some water bodies do not have similar relative variability, or that their relative variabilities change differently through the year. At any sampling instant, therefore, the relatively most variable of the qualities of interest must be used to set the sample number calculation. If these differ widely, then a sub-sampling approach may be used without loss of efficiency.

An example of a reasoned assessment when there is no direct information available on quality variability can be based on a diatom bloom in a lake or reservoir. In such cases the variable of interest (i.e., chlorophyll a) often has a more readily apparent maximum and minimum value at any particular sampling time. For example, a Spring dominance by a diatom in a temperate reservoir might produce possible biomass measurements up to about 100 mg m^{-3} chlorophyll a. If the reservoir is reasonably mixed at this time, it might be assumed that it is unlikely that the minimum value sampled will be much less than 50 mg m^{-3}. The sample range R is, therefore, $100 - 50 = 50$ mg m^{-3} chlorophyll a. Assuming distribution of the results to be somewhat non-normal, then $s = 0.3*R$ (see section 10.4.2). Thus the estimate of s is 15 mg m^{-3} chlorophyll a which can be used in the sample number calculation.

The assessment of sample numbers may also be simplified by considering the problem in more relative terms. For example: How many samples are necessary for a variable with a coefficient of variation (see section 10.4.3) of 40 per cent so that, with 90 per cent confidence, the mean is estimated to within 10 per cent of its true value? Casting the problem in this way will always help for those water quality measurements in which sample variability tends to be related to sample concentration (i.e., high variability is associated with high values and low variability with low values). It is then easier to assume that previous results, for example, will still hold for a subsequent sampling occasion.

Example 2
Using the values from the previous paragraph: cv = $s/\bar{x}$ = 0.4; δ = d/μ = 0.1, and α = 0.1; then $n = (t*cv/\delta)^2$. Starting once again with df = 30:

$t_{0.1[30]}$ = 1.697. Then: $n_1 = (1.697*0.4/0.1)^2$ = 46.1 = 47
$t_{0.1[46]}$ = 1.681. Then: $n_2 = (1.681*0.4/0.1)^2$ = 45.2 = 46
$t_{0.1[45]}$ = 1.680. Then: $n_3 = (1.680*0.4/0.1)^2$ = 45.2 = 46

Thus, at least 46 samples would be required to meet the specified sampling objective.

The examples show how the numbers of samples required rapidly mount up if overly stringent accuracy and associated confidence in the results are applied to inherently variable sites or circumstances. Once again, considerable judgement may be necessary in balancing the assessment needs and the sampling resources available.

One usual response to the need for large sample numbers is to spread the sampling over a number of sample sites located within the water body. If the sites are vertically mixed, and samples from them are not spatially correlated, then spreading the sample numbers over a series of sampling sites is a natural development. However, such a simple approach is not appropriate if the sample values are not independent but are correlated, as will frequently be the case if many sample sites are used. For example, if there is a tendency for a high result for a given variable at a particular site to be accompanied by another high result for the same variable at a neighbouring site (or a low value with a low value) then the true variability will be greater than that simply measured. If the lesser variance is then used in the formulae given above it will result in an under-estimate of the numbers of samples required to meet the sampling objective.

If experience or previous results allow some estimate of a uniform spatial correlation coefficient r amongst sampling sites, then:

$$n*n_s = (ts/d)^2*(1+r(n_s-1)),\qquad \text{where } n_s \text{ is the number of sampling sites.}$$

With positive correlation $r > 0$, so a factor $(1 + r(n_s - 1))$ as many samples will be required. If, for example, $r = 0.1$ and $n_s = 10$ virtually twice as many samples relative to the basic number would be required. Clearly, with only a single sampling site, or no correlation, the formula becomes the same as that used before. Moreover, serial correlation may also exist, for example amongst sample sites spaced along a transect or within samples taken relatively close together in time, and further allowance may have to be made for that situation. Furthermore, if the means at each sample site differ, but are presumed to be measurements of the same global mean, then: $s = \sqrt{(s^2 + n*s_s^2)}$ where s_s^2 is the inter-site means variance. Where such complexity occurs, it is highly desirable that expert assistance is involved in the sampling design.

Example 3
Nitrate samples taken at monthly intervals from three, neighbouring groundwater sampling wells, show a standard deviation of 60 mg l^{-1} and a correlation coefficient of 0.45, but independence from month to month. Could those samplings give an estimate of the annual mean nitrate value to within 20 mg l^{-1}, with 90% confidence?

From the formula above, $n*3 = (t_{0.1[3*n-1]}*60/20)^2*(1+0.45(3-1))$. Iterative solution of this equation leads to $n = 16$ samples at each site. Therefore, the monthly sampling is insufficient. Assuming that adding a few more sampling stations would reduce the site-to-site correlation coefficient to about 0.4, but all else remained the same; how many stations are necessary to achieve the original sampling aim?

Then, $12*n_s = (t_{0.1[12*ns-1]}*60/20)^2*(1+0.4(n_s-1))$.

Re-arranging: $n_s = (1 - 0.4)/(12/(t_{0.1[12*ns-1]}*3)^2 - 0.4)$ and solving by successive approximation gives $n_s = 7$; so a further four sites would be necessary. If, more realistically, there was also correlation between sample estimates at succeeding time intervals, then further modification is necessary. A reasonable approximation is:

$n*n_s = (ts/d)^2*(1+r(n_s - 1))*(1 + 2\Sigma r_k)$, where r_k are the correlations between samples $i,i+1$; $i,i+2$; $i,i+3$; etc.; and the sum is taken to $n-1$.

If, for this example, it is assumed that $r_1 = 0.6$; $r_2 = 0.3$; $r_3 = 0.1$, and all other $r_k = 0$; then $(1+2*(0.6+0.3+0.1) = 3$ times as many samples would be required. Correlations clearly reduce the amount of information gained from each individual sample.

B. Stratified random sampling

In reservoirs and lakes the phenomenon of summer thermal stratification is a further source of complexity in sampling design because of its potential support of vertical water quality differences (see Chapters 7 and 8). The most usual basis for design in such circumstances is stratified random sampling. In this method, a series of non-overlapping vertical strata are pre-chosen, so that the quality to be sampled may be presumed to be very much less variable within any stratum than between the strata. The strata are often chosen on the basis of criteria such as vertical temperature or oxygen gradients. The epilimnion, metalimnion and hypolimnion may, therefore, form appropriate strata. However, for some organisms such as buoyant blue-green algae or zooplankton, further stratification may be necessary. Significant bottom depressions or general bathymetric complexity may also form the basis for stratum choice. The weight, w_i to be given to any stratum is the proportion of the total sampling units which may be sampled that is contained within the stratum. For example, for the epilimnion w_{epi} could represent the volume of the epilimnion as a proportion of the total reservoir volume, i.e., $w_{epi} = Vol_{epi}/Vol_{total}$.

Samples are drawn from random depths within these pre-defined strata (note: random depths are not haphazard depths, but are chosen according to an appropriate randomising scheme). Within any stratum, the numbers of samples to be taken relative to the total are usually set by one of two procedures: proportional allocation, or optimised allocation.

In proportional allocation, the number of samples in the ith layer is $n_i = n*w_i$. The numbers of samples are simply allocated on the basis of the relative weights w_i of the various strata. The proportional allocation scheme then leads to $n = t^2(\Sigma w_i s_i^2)/d^2$ as an estimate of the total number of samples to be taken at the site, where s_i^2 is the quality variance in the ith layer.

In optimised allocation, the numbers of samples taken from any layer reflects both the variability of the quality within that layer, and the difficulty (cost) of sampling from that layer. Then the optimum number of stratum samples would be $n_i = n*(w_i s_i /\sqrt{c_i})/\Sigma(w_i s_i/\sqrt{c_i})$, where c_i is the cost of sampling in the ith layer. This formula suggests taking more samples where the stratum is larger, the stratum quality is more variable and the sampling cost is least. If the costs are similar for each layer, then: $n_i = n*w_i s_i / \Sigma(w_i s_i)$. Optimised allocation then leads to $n = t^2 (\Sigma w_i s_i)^2 / d^2$ as an estimate of the total numbers of samples to be taken. It is usually the case that optimised allocation requires fewer samples for a given precision and confidence of estimate.

Example 4
Consider a lake, with soluble reactive phosphorus (SRP) concentrations and distribution similar to those illustrated in Figure 7.19 for Heilingensee in late August, 1987. How many samples are required to provide an estimate of the mean amount of phosphorus in the lake to within 10 mg P m^{-3}, with 90 per cent confidence, if other sample variances relative to the concentration ranges hold?

i	Depth range (m)	w_i	Concn. range (mg P m^{-3})	Prev. data s_i	$w_i s_i$	Optimal allocation n_i	$w_i s_i^2$	Propn. allocation n_i
1	0–5	0.7	30–50	6.4	4.5	8	28.7	241
2	5–7	0.2	50–1,000	171.6	34.3	61	5,889.4	69
3	7–10	0.1	1,000–2,000	260.7	26.1	47	6,796.4	34
Σ					64.9	116	12,714.5	344

Optimal: $n = t^2_{0.1[115]} (\Sigma w_i s_i)^2 / d^2 = 116$ samples
Proportional: $n = t^2_{0.1[\infty]} (\Sigma w_i s_i^2) / d^2 = 344$ samples

Samples would then be drawn from random sites on the lakes, and at random depths within the strata. The overall mean would then be $\bar{x}_t = \Sigma w_i \bar{x}_i$, where $\bar{x}_i$ is the ith stratum mean. The total SRP in the lake would also be estimated as $SRP_{total} = Vol_{total} * \bar{x}_t$. It will be noticed how substantially the optimal allocation strategy has reduced the sampling load necessary for the specified accuracy and confidence.

If no previous variability statistics exist, then a possible approach to provide rough estimations of the stratum standard deviations and sample numbers is as follows. Calculate s_i for each stratum by $(SRP_{max} - SRP_{min})_i /4$; the divisor being chosen relative to about 30 samples in the absence of any other data (see section 10.4.2). This leads to: $s_i = 5$; 238; 250 for the three strata. Then $n_i = 7$; 99; 52; (total =158). Therefore, even this simple procedure gives a sample programme quite close to that of the more specific optimal allocation strategy previously considered. Since for stratum 1 only 7 samples are suggested, a more appropriate divisor of, for example, 3 ($\cong 7 - 10$ samples) might be used instead. Similarly for stratum 2, a divisor of 5 ($\cong 50 - 100$ samples) could be used. Recalculation on these factors leads to $n = 126$, with $n_i = 9$; 71; 46 respectively. These approximations show that very useful sampling guidance can be derived from reasoned estimation, even when accurate, critical data are not available.

Example 4 might also be considered on the basis of whether random sampling is the best approach. An alternative procedure would be to use systematic sampling, where sites are chosen to give appropriate coverage of the water-mass, and then sampled at regular depths so as to allow reasonable depth profiles to be drawn. These could, for example, then be planimetrically integrated. A possible problem with this approach is the difficulty of ascribing appropriate confidence levels to the result. Nevertheless, it illustrates the need always to consider alternative sampling approaches as part of the initial sample planning stage.

Advanced statistical analysis may also be used to identify spatial and temporal patterns amongst sampling results drawn from special surveys. These patterns may then be used as the basis for more general sampling site allocation. This is a particularly important aspect for dealing with the variability associated with reservoirs (see section 8.2.1). A good example of such an approach is given in Thornton *et al.* (1982) (see section 8.5.2) Essentially similar approaches to sampling number estimation may be made in rivers (Montgomery and Hart, 1974).

Sampling evaluation

A very important part of any sampling exercise is to review the extent to which the programme has achieved the desired objectives. Once the samples have been taken, contemporary information is available as to distributions and variability. If the required precision has not been achieved, these new data may be used in many of the foregoing equations to establish how many extra samples are required, and how the sampling strategy may be further optimised. If necessary, any extra sampling may then be immediately undertaken.

If the water quality variable under investigation is not likely to be normally distributed (e.g., count data), or is found on the basis of first sampling to be distinctly non-normal, then it would usually be necessary to apply an appropriate transformation (see section 10.4.5) so that these normal distribution based formulae may be applied.

References

Cochran, W.G. 1963 *Sampling Techniques*. Second edition. Wiley and Sons, New York, 413 pp.

Gilbert, R.O. 1987 *Statistical Methods for Environmental Pollution Monitoring*. Van Nostrand Reinhold Company, New York, 320 pp.

Keith, L.H. [Ed.] 1988 *Principles of Environmental Sampling*. American Chemical Society. 458 pp.

Montgomery, H.A.C. and Hart, I.C. 1974 The design of sampling programmes for rivers and effluents. *Wat. Pollut. Control*, **73**, 77-101.

Snedecor, G.W. and Cochran, W.G. 1980 *Statistical Methods*. 7th edition. Iowa State University Press, Ames, 507 pp.

Thornton, K.W., Kennedy, R.H., Magoun, A.D. and Saul, G.E. 1982 Reservoir water quality sampling design. *Water Resources Bull.*, **18**, 471-480.

Index